Fluid Flow and Solute Movement in Sandstones: The Onshore UK Permo-Triassic Red Bed Sequence

Geological Society books refereeing procedures

The Society makes every effort to ensure that the scientific and production quality of its books matches that of its journals. Since 1997, all book proposals have been refereed by specialist reviewers as well as by the Society's Books Editorial Committee. If the referees identify weaknesses in the proposal, these must be addressed before the proposal is accepted.

Once the book is accepted, the Society Book Editors ensure that the volume editors follow strict guidelines on refereeing and quality control. We insist that individual papers can only be accepted after satisfactory review by two independent referees. The questions on the review forms are similar to those for *Journal of the Geological Society*. The referees' forms and comments must be available to the Society's Book Editors on request.

Although many of the books result from meetings, the editors are expected to commission papers that were not presented at the meeting to ensure that the book provides a balanced coverage of the subject. Being accepted for presentation at the meeting does not guarantee inclusion in the book.

More information about submitting a proposal and producing a book for the Society can be found on its web site: www.geolsoc.org.uk.

It is recommended that reference to all or part of this book should be made in one of the following ways:

BARKER, R. D. & TELLAM, J. H. (eds) 2006. *Fluid Flow and Solute Movement in Sandstones: The Onshore UK Permo-Triassic Red Bed Sequence*, Geological Society, London, Special Publications, **263**.

BLOOMFIELD, J. P., MOREAU, M. F. & NEWELL, A. J. Characterization of permeability distributions in six lithofacies from the Helsby and Wilmslow sandstone formations of the Cheshire Basin, UK. *In*: BARKER, R. D. & TELLAM, J. H. (eds) *Fluid Flow and Solute Movement in Sandstones: The Onshore UK Permo-Triassic Red Bed Sequence*, Geological Society, London, Special Publications, **263**, 83–102.

GEOLOGICAL SOCIETY SPECIAL PUBLICATION NO. 263

Fluid Flow and Solute Movement in Sandstones: The Onshore UK Permo-Triassic Red Bed Sequence

EDITED BY

R. D. BARKER and J. H. TELLAM
The University of Birmingham, UK

2006
Published by
The Geological Society
London

THE GEOLOGICAL SOCIETY

The Geological Society of London (GSL) was founded in 1807. It is the oldest national geological society in the world and the largest in Europe. It was incorporated under Royal Charter in 1825 and is Registered Charity 210161.

The Society is the UK national learned and professional society for geology with a worldwide Fellowship (FGS) of 9000. The Society has the power to confer Chartered status on suitably qualified Fellows, and about 2000 of the Fellowship carry the title (CGeol). Chartered Geologists may also obtain the equivalent European title, European Geologist (EurGeol). One fifth of the Society's fellowship resides outside the UK. To find out more about the Society, log on to www.geolsoc.org.uk.

The Geological Society Publishing House (Bath, UK) produces the Society's international journals and books, and acts as European distributor for selected publications of the American Association of Petroleum Geologists (AAPG), the American Geological Institute (AGI), the Indonesian Petroleum Association (IPA), the Geological Society of America (GSA), the Society for Sedimentary Geology (SEPM) and the Geologists' Association (GA). Joint marketing agreements ensure that GSL Fellows may purchase these societies' publications at a discount. The Society's online bookshop (accessible from www.geolsoc.org.uk) offers secure book purchasing with your credit or debit card.

To find out about joining the Society and benefiting from substantial discounts on publications of GSL and other societies worldwide, consult www.geolsoc.org.uk, or contact the Fellowship Department at: The Geological Society, Burlington House, Piccadilly, London W1J 0BG: Tel. +44 (0)20 7434 9944; Fax +44 (0)20 7439 8975; E-mail: enquiries@geolsoc.org.uk.

For information about the Society's meetings, consult *Events* on www.geolsoc.org.uk. To find out more about the Society's Corporate Affiliates Scheme, write to enquiries@geolsoc.org.uk.

Published by The Geological Society from:
The Geological Society Publishing House, Unit 7, Brassmill Enterprise Centre, Brassmill Lane, Bath BA1 3JN, UK

(*Orders*: Tel. +44 (0)1225 445046, Fax +44 (0)1225 442836)
Online bookshop: www.geolsoc.org.uk/bookshop

The publishers make no representation, express or implied, with regard to the accuracy of the information contained in this book and cannot accept any legal responsibility for any errors or omissions that may be made.

British Library Cataloguing in Publication Data

A catalogue record for this book is available from the British Library.

ISBN-10 1-86239-204-8
ISBN-13 978-1-86239-205-5

Typeset by Type Study, Scarborough, UK

Printed by The Cromwell Press, Wiltshire, UK

Distributors

North America
For trade and institutional orders:
The Geological Society, c/o AIDC, 82 Winter Sport Lane, Williston, VT 05495, USA
Orders: Tel. +1 800 972-9892
           Fax +1 802 864-7626
           E-mail gsl.orders@aidcvt.com

For individual and corporate orders:
AAPG Bookstore, PO Box 979, Tulsa, OK 74101-0979, USA
Orders: Tel. +1 918 584-2555
           Fax +1 918 560-2652
           E-mail bookstore@aapg.org

India
Affiliated East-West Press PVT Ltd, G-1/16 Ansari Road, Darya Ganj, New Delhi 110 002, India
Orders: Tel. +91 11 2327-9113/2326-4180
           Fax +91 11 2326-0538
           E-mail affiliat@vsnl.com

Contents

Preface

The Permo-Triassic sandstones are the second most-used aquifer type in the UK. However, as a fractured fluvial and aeolian red-bed sequence, they also represent an aquifer type that is common internationally. Much research has also been undertaken on such sequences in the context of hydrocarbon exploitation, but, although reference is made to this research, the emphasis of the papers included in this volume is on aquifer properties and water migration in shallow terrestrial settings.

We would like to thank the following colleagues for undertaking reviews: J. Aldrick, D. Allen, S. Banwart, J. Bloomfield, R. Brassington, W. Burgess, M. Carey, N. Cassidy, R. Davison, P. Degnan, J. Dottridge, M. Ford, K. Hiscock, J. Ingram, R. Ixer, B. Klink, M. Lovell, R. Mackay, W. Owens, K. Privett, M. Shepley, J. Smith, G. Stuart, S. Taylor, S. Thornton, P. Turner, A. Weller, J. West, G. Williams, M. Worthington, P. Younger, and two others who preferred to remain anonymous.

Material for this volume draws partly on a meeting of the Environmental and Industrial Geophysics Group and the Hydrogeological Group of the Geological Society of London, held in London in 2003, and partly on a number of additional papers solicited by the editors or submitted independently as contributions to this thematic volume.

J. H. Tellam & R. D. Barker

Towards prediction of saturated-zone pollutant movement in groundwaters in fractured permeable-matrix aquifers: the case of the UK Permo-Triassic sandstones

JOHN H. TELLAM & RONALD D. BARKER

Hydrogeology Research Group, Earth Sciences, School of Geography, Earth and Environmental Sciences, Birmingham University, Birmingham B15 2TT, UK (e-mail: J.H.Tellam@bham.ac.uk; R.D.Barker@bham.ac.uk)

Abstract: The UK on-shore Permo-Triassic sandstones are fluvial and aeolian red beds showing a nested cyclic architecture on scales from millimetres to 100s of metres. They are typical of many continental sandstone sequences throughout the world. Groundwater flows through both matrix and fractures, with natural flow rates generally of less than 200 m year^{-1}. At less than 30 m horizontal distances, below important minimum representative volumes for both matrix and fracture network permeability, breakthroughs are likely to be multimodal, especially close to wells, with proportionately large apparent dispersivities. 'Antifractures' – discontinuities with permeability much less than that of the host rock – may have a dominating effect. Where present, low-permeability matrix (e.g. mudstones) will significantly affect vertical flow, but will rarely prevent eventual breakthrough. Quantitative prediction of breakthrough is associated with large uncertainty. At scales of 30 to a few 100s of metres, multimodal breakthroughs from a single source become less common, although very rapid fracture flow has been recorded. At distances of hundreds of metres to a few kilometres, there is evidence that breakthroughs are unimodal, and may be more immediately amenable to quantitative prediction, even in some cases for reacting solutes. At this and greater scales, regional fault structures (both slip surfaces and granulation seams) can have major effects on sub-horizontal solute movement, and mudstones and cemented units will discourage vertical penetration. The aquifer has limited oxidizing capacity despite the almost ubiquitous presence of oxides, limited reductive capacity and limited organic sorption capacity. It has a moderate cation-exchange capacity, and frequently contains carbonate. Mn oxides are important for sorption and oxidation, but are present in limited quantity. Relationships between hydraulic and chemical properties are largely unknown. 'Hard' evidence for the solute transport conceptual model presented above is relatively limited. To be able to predict to a reasonably estimated degree of uncertainty requires knowledge of: the geological, and thence the hydraulic and geo-chemical, structure of the complex sandstone architecture (including the correlations between these properties); the development of suitable investigation techniques (especially geophysical) for mapping the structures; and the development of modelling tools incorporating matrix, fractures, 'antimatrix' and antifracture elements, each with associated hydraulic and possibly geochemical properties. In common with solute movement studies in most aquifer types, much more geological characterization needs to be undertaken. Although new investigation and modelling tools are being developed specifically for (shallow) hydrogeological applications with some considerable success, much greater advantage could be taken of importing techniques from other disciplines, and in particular from oil exploration and development.

The development of a quantitative understanding of groundwater solute movement is an important goal for all aquifer types. This paper is concerned with evaluating the progress towards this goal in an example fractured permeable matrix aquifer – the UK Permo-Triassic fluvial–aeolian sandstone sequence. Permian and Triassic continental sandstones were widely distributed prior to the break-up of Pangaea. They now occur in NW Europe, East Greenland and in many locations across the continental USA. They are also well known in Gondwanaland, especially South America, Africa, Antarctica and Australia. Although the character of the sandstones varies with location, in broad terms the UK sandstones are typical fractured continental red-bed sequences with well-marked sedimentary structures. After the (Cretaceous) chalk, they constitute the most heavily used set of aquifers in the UK; they are also extensively used elsewhere for water supply, and form hydrocarbon reservoirs

From: BARKER, R. D. & TELLAM, J. H. (eds) 2006. *Fluid Flow and Solute Movement in Sandstones: The Onshore UK Permo-Triassic Red Bed Sequence*. Geological Society, London, Special Publications, **263**, 1–48.
0305–8719/06/$15 © The Geological Society of London 2006.

particularly in the shelf areas surrounding the UK.

The development of the attributes that affect reacting solute movement in present-day groundwater systems is determined by the interaction of three main geological factors: depositional environments; palaeo-groundwater–rock interactions; and stress (Fig. 1). In most groundwater systems, solute movement, as observed, depends strongly on the scale(s) of heterogeneity of the sequence, the scale of measurement and the scale of interest (e.g. Dagan 1989), and it is well recognized that this needs to be considered when evaluating evidence. Accordingly, the structure of this paper (and this Special Publication) reflects these issues: first, the geological setting is briefly described, then flow and, finally, solute movement. In the latter two sections, the general approach is to consider evidence in order of increasing scale of investigation, and in the final discussion scale is a major consideration. Aquifer geometry and presence of overlying deposits will not be directly considered: although these have a major effect on flow patterns and water chemistry, they are not intrinsic to the phenomenon of solute movement in the sandstone. In addition, the unsaturated zone and non-aqueous phase liquid movement will not be covered, but papers on these topics are included within the volume (Binley *et al.* 2006; Gooddy & Bloomfield 2006; Privett 2006; Rees 2006; Taylor & Barker 2006). Evidence is drawn from the very considerable literature, 'white', 'grey' and unpublished: because of the difficulty of obtaining much of the latter two, the selection is unavoidably biased. Wherever relevant, an attempt has been made to include reference to the very significant petroleum geology literature, a resource that often is not fully exploited in hydrogeology: it is, however, important to note that observations from deep systems with a different geological history are not necessarily directly transferable.

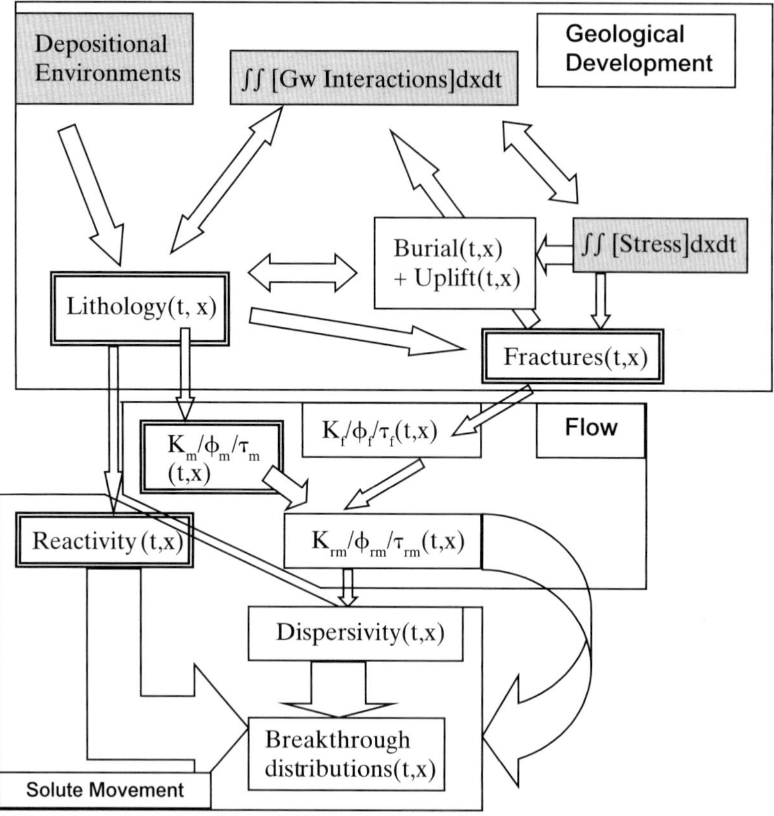

Fig. 1. Geological factors and their relationships with solute movement ($\iint[\,]dxdt$ = integrated effect over space and time; K is hydraulic conductivity; ϕ is porosity; τ is tortuosity; subscripts f, m, and rm are fracture, matrix and rock mass; double line boxes indicate properties that may be amenable to geophysical measurement).

Geological considerations

Introduction

An outline stratigraphy is given in Table 1, and the distribution of outcrop is indicated in Figure 2 (for comprehensive details see, for example, Warrington *et al.* 1980 and Benton *et al.* 2002). In general, the sequence, up to over 1 km in thickness at its maximum development, is underlain by deposits of low permeability, sometimes Permian in age but often Carboniferous or older. Usually it is immediately overlain by either the Triassic Mercia Mudstone Group (mudstones and evaporates with occasional thin sandstones) or by Quaternary deposits (tills to outwash gravels). The Permian and Triassic sandstone sequences are broadly similar in origin and lithology, and, for the purposes of this review, will be considered together.

Depositional environments

During the Permo-Triassic, the area that is now the UK migrated from approximately 10° to 30° N of the equator. The climate was semi-arid to arid, with flash flood episodes bringing debris from the eroding Variscan mountains in the south (present continental Europe and SW England). Northward-flowing braided river systems deposited increasingly finer-grained sediments in tectonically active, often half-graben, subsiding basins. Locally this northwards trend of decreasing grain size is reversed, and conglomerates are deposited as a result of steep slopes (e.g. in southern Scotland: Akhurst *et al.* 2006). The main depositional environments are summarized in Table 2.

The wide range of fluvial and aeolian environments produced a variety of sedimentary structures, including plane lamination, cross-lamination, trough and planar tabular cross-stratification, water escape structures, imbricate gravels, debris flows, palaeosols and desiccation cracks (e.g. Thompson 1970a; Steel & Thompson 1983) (Fig. 3). Bed size varies from less than a few centimetres to at least several metres.

From the 1960s, vertical depositional 'cycles', or sequences, have been recognized at various scales in the fluvial-dominated parts of the succession, with cycle thicknesses from less than 1 to more than 100 m (Fitch *et al.* 1966; Thompson 1970a; Wills 1970, 1976): such conceptualizations are now embedded within sequence stratigraphy (e.g. Vail *et al.* 1991; Jensen *et al.* 1996; Howell & Mountney 1997; Mountney & Thompson 2002). Thus, for example, Wills (1970) recognized 'microcycles' at the scale of a few beds, 'miocycles' comprising groups of microcycles up to approximately 100 m thick at most and 'magnacycles' comprising groups of miocycles (usually equivalent to formations). Each microcycle represents a wet to dry transition. Thus, an *ideal* microcycle sequence according to Thompson (1970a) and Wills (1970) might be:

top   Aeolian sandstones
          Mudstone, with desiccation cracks
          Finer-grained plane-laminated sandstone
          Medium-grained cross-laminated sandstone
          Coarse (pebbly) channel deposit
base Erosion surface.

Miocycles were defined by Wills (1970), rather subjectively, using indicators of maximum water velocity with high-energy conditions passing upwards into lower energy conditions. At the largest scale (magnacycles), the main formations in any one locality may be paired into a cycle: for example, in the Triassic Sherwood Sandstone Group of the Midlands (Table 1) the first magnacycle would comprise the Kidderminster Formation (coarser, pebbly) and the Wildmoor Sandstone Formation (finer, non-pebbly): the overlying Bromsgrove Sandstone Formation (coarser, pebbly) would form the first part of the upper magnacycle.

In recent years the significance of the cycles described by Wills (1970) has become evident and they can be related to orbital forcing mechanisms (Clemmensen *et al.* 1994). Descriptions have become rather more sophisticated, and similar conceptualizations are now widely applied to Triassic stratigraphy and sedimentology on a global basis (Szurlies *et al.* 2003). In the UK Triassic the distinctive wetting–drying cycles (Mountney & Thompson 2002) are most conveniently explained using orbital forcing mechanisms.

In strong contrast to the practice in petroleum geology, such geologically oriented conceptualizations have seldom been applied in water resources and non-nuclear pollution studies in the UK Permo-Triassic sandstones (but compare Bloomfield *et al.* 2006 and Bouch *et al.* 2006).

Lithologically, the sandstones range from lithic arkoses to quartz arenites. Detrital clays and mica are commonly present, although a few sequences are effectively 'clean' (e.g. the Penrith Sandstone, Cumbria: Lovell *et al.* 2006). Organic carbon contents are typically less than 0.1% (Steventon-Barnes 2001; Shepherd *et al.* 2006). The generally less than 1 m-thick

Table 1. *Outline stratigraphy of the UK Permo-Triassic sandstone sequence (after Howard unpublished 2002). Triple line at formation boundary indicates non-deposition: no thicknesses of formations implied. The Sherwood Sandstone Group roughly corresponds to the named Triassic formations in the table. MMG = Mercia Mudstone Group*

	Northwest			East Midlands	West Midlands					Southwest
	Cheshire Basin	East Irish Sea Basin	Carlisle and Eden basins	East Midlands	Worcester Basin	Stafford Basin	Knowle Basin	Needwood Basin	Hinkley Basin	Wessex Basin
Mid Triassic / Early Triassic	MMG Helsby Sst Fm	MMG Ormskirk Sst Fm	MMG Kirklinton Sst Fm	MMG	MMG Bromsgrove Sst Fm	MMG Bromsgrove Sst Fm	MMG Bromsgrove Sst Fm	MMG Bromsgrove Sst Fm	MMG Bromsgrove Sst Fm	MMG Otter Sst Fm
	Wilmslow Sst Fm	Calder Sst Fm	St Bees Sst Fm	Nottingham Castle Sst Fm	Wildmoor Sst Fm	Wildmoor Sst Fm	Wildmoor Sst Fm	Hawksmoor Fm	Polesworth Fm	Budleigh Salterton Pebble Beds Fm
	Chester Pebble Beds Fm	St Bees Sst Fm			Kidderminster Fm	Kidderminster Fm	Kidderminster Fm			
Late Permian	Kinnerton Sst Fm	St Bees Shale and Evaporite	St Bees and Eden Shale	Lenton Sst Fm	Bridgnorth Sst Fm	Bridgnorth Sst Fm	Beccias	Beccias	Beccias	Aylesbeare Mdst Gp Exeter Gp
	Manchester Marl/Bold Fm			Zechstein GP						
Early Permian	Collyhurst Sst Fm	Collyhurst Sst Fm	Penrith Sst & Brockram	Beccias						
					Beccias	Beccias				Beccias

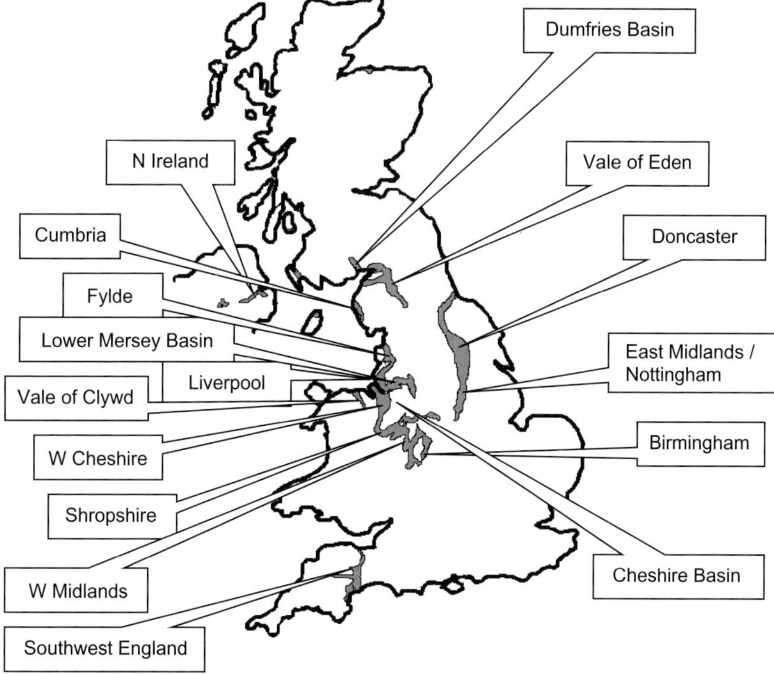

Fig. 2. The main outcrops of Permo-Triassic sandstone in the UK, and locations of aquifers mentioned in the text.

mudstones within the sequence have a similar mineral composition (clays, fine quartz, feldspar, mica and haematite): illite, kaolinite, chlorite and smectite have all been recognized (Burley 1984; Hough *et al.* 2001).

Stress (burial and uplift)

The basins in which the Permo-Triassic sequence was deposited were often active tectonically during deposition (Poole & Whiteman 1955, 1966; Audley-Charles 1970; Plant *et al.* 1999). Typically they were asymmetric grabens, controlled by reactivation of basement faults: in the Midlands and NW England this results in the longer axes of the basins being aligned roughly N–S. Maximum burial depths vary from basin to basin, but are usually no more than a few kilometres, and sometimes much less (e.g. Burley 1984; Plant *et al.* 1999). Components of uplift/inversion occurred at various times from the Jurassic onwards. Generally, in all the UK on-shore Permo-Triassic basins, bedding dips are typically less than 10°.

As with many porous sandstones, faults display a range of morphologies, including single deformation bands, zones of deformation bands, and slip surfaces (e.g. Fisher & Knipe 1998; Manzocchi *et al.* 1998). Deformation bands (or granulation seams: Aydan 1978; Aydin & Johnson 1978; Fowles & Burley 1994) are common, often as swarms (Fig. 4a). Slip surfaces now exist as open, clay-filled or cataclastite-filled features (Fig. 4b).

Jointing occurs within the sandstone sequence to varying degrees. Only a few sets of quantitative data are available (Barnes *et al.* 1998; Gutmanis *et al.* 1998; Wealthall *et al.* 2001; Jeffcoat 2002), and some of these are discussed below. Fracturing becomes less frequent with depth (see below).

Dewatering structures are not uncommon, but large-scale sedimentary dykes appear to be rarely noted.

Palaeo-groundwater interactions/diagenesis

Although locally very variable, in outline the diagenesis of the sandstones has included the following phases (e.g. Burley 1984; Metcalf *et al.* 1994; Strong *et al.* 1994; Milodowski & Gillespie 1997; Milodowski *et al.* 1999):

Table 2. *Fluvial and aeolian depositional environments seen in the UK Permo-Triassic sandstone sequence (after Benton* et al. *2002)*

Environment		Lithologies	Structures

Fluvial	Fan	Poorly sorted gravels	*Sheet floods*
			  Poorly stratified or imbricate
			  Upwards fining
			Debris flows
			  Matrix supported
			  Chaotic internal organization

	High-energy braided river	Texturally mature gravels with interbedded sands	*Conglomerates*
			  Crude stratification
			  Large foresets
			Sandstones
			  Cross-stratified

	Lower energy braided river	Sands with subordinate silts and muds	*Sandstones*
			  Upper-phase plain stratification
			  Planar tabular stratification
			  Trough cross-stratification
			  Upwards-fining cycles
			Mudstones
			  Usually <1 m thick, and often much less

	Meandering river channel and overbank	Sands with 30–60% silts with palaeosols	Upwards-fining cycles *Sandstones* Lateral accretion cross-stratification
	floodplain		Variety of smaller scale cross-stratified, rippled and plane-bedded sandstones Thin overbank sands
Aeolian	Dunes	Texturally mature sands	Large-scale cross-stratification, asymptotic foresets up to 30° inclination, sets separated by sharp, generally planar bounding surfaces
	Sheets	Sand	Bi-modally laminated on mm-scale scours, granule lags, thin cross-laminated units
	Damp interdune	Silty sandstones	Irregular wavy or lenticular lamination, slump structures, with adhesion ripples dominant

- Early diagenesis. It would appear that in many areas, early diagenesis included:
  - dissolution of unstable silicates; and
  - precipitation of
    - clays (e.g. smectite/illite),
    - K-feldspar,
    - non-ferroan carbonate,
    - haematite and
    - gypsum.
- Burial diagenesis. Subsequent burial, typically to less than a few kilometres, compacted the sandstone particularly where cements were not pervasive and converted at least some of the smectite to illite. As many of the sandstones have present porosities of approximately 0.2–0.3, cementation probably frequently occurred prior to burial. During this phase, some precipitation of ferroan dolomite and ankerite occurred.
- Post-inversion diagenesis. Following inversion, dissolution of carbonate and sulphate occurred, with weathering of some of the feldspars to clay. This process continues, but at rates that are unlikely to be hydrogeologically significant except in very rare cases (e.g. Bath *et al.* 1987).

In some cases, the mineralogy allows the past presence of saline groundwaters to be inferred (Milodowski *et al.* 1998). Saline groundwaters are common in the deeper parts of many of the sandstone basins, the majority being derived from evaporite dissolution (Edmunds 1986; Barker 1990; Tellam 1995*a*; Bottrell *et al.* 2006). Especially in central England, mineralization is sometimes present (e.g. baryte with Cu, and, less abundantly, Pb, Co, Zn, Mn, V, Ni and As in the Cheshire Basin: Ixer & Vaughan 1982; Naylor *et al.* 1989; Plant *et al.* 1999. This may be the

Fig. 3. Example outcrop sections of the UK Permo-Triassic sandstone sequence. (**a**) Dawlish, SW England; (**b**) Torbay, SW England; and (**c**) Runcorn, NW England.

result of flow of reducing brines from the underlying Carboniferous up fault planes into oxidized (evaporite-derived) brines in the sandstone during uplift in the Tertiary (Naylor *et al.* 1989; Tellam 1995*a*; Milodowski *et al.* 1998). Although localized, these mineral deposits may play a wider role as sources for dissolved metals (Kinniburgh *et al.* 2006). The distribution of the mineralization and the often-associated bleaching (removal of the ferric oxides) strongly suggest it to be associated with the faulting, and to be affected by the presence of low-permeability beds (Rowe & Burley 1997). In a few places, in otherwise unaffected aquifers, natural hydrocarbons have been observed to issue from the sandstones (Rowe & Burley 1997; Plant *et al.* 1999).

The main pattern in cementation appears to be: local dissolution of carbonates and feldspars (especially from the upper 50–100 m of the sandstone sequence) (Travis & Greenwood 1911; Burley 1984; Kinniburgh *et al.* 2006); the presence of remnant gypsum cements in some low-flow systems (e.g. Jackson & Lloyd 1983); and the absence of carbonates in some formations (e.g. Edmunds & Morgan-Jones 1976; Walton 1981).

The present-day sandstone sequence

The present-day sequence therefore has the following principal properties:

- at most a moderately well-developed cementation;
- often well-developed sedimentary structures;
- layering, often with a lithological cyclicity at several scales;
- occasional mudstones, usually less than 1 m thick;
- usually shallow bedding dips;
- fracturing, especially at shallower depths;
- frequent occurrence of faults of a range of morphologies:
  - granulation seams, single and swarmed;
  - slip surfaces (well cemented, clay gouge-filled, breccia-filled or empty);
- veining (see below);
- carbonate cementation (although sometimes removed at shallow depths);
- very low organic carbon content;
- clay as a common constituent;
- ferric and often manganese oxides/hydroxides.

(a)

(b)

Fig. 4. Fault outcrops (Wirral Peninsula, NW England). (**a**) A granulation seam 'swarm'. (**b**) A slip surface, with bleached zones (notebook is about 8 cm wide).

Groundwater flow

Matrix flow

Laboratory studies of hydraulic properties. Very many studies have examined the porosity and permeability of laboratory samples of the sandstones, from Roberts (1869) to various papers in this volume (Bloomfield *et al.* 2006; Lovell *et al.* 2006; Pokar *et al.* 2006). Allen *et al.* (1997) summarize much of this work. In general the following observations can be made:

- Porosity
  - distributions appear to be Gaussian, with values in Allen *et al.*'s (1997) database ranging between 0.02 and 0.35, with a median of 0.26 (although it is likely that the data set is slightly biased, with too few results from friable samples with higher than average porosity (and permeability)).
- Permeability
  - values are log-normally distributed;
  - geometric means are often in the range 0.1–10 m day^{-1};
  - there is some variation with formation, at least locally (e.g. Campbell 1982) and between basins;
  - permeability parallel to laminations in the sandstones can be up to at least 50 times that perpendicular to laminations, but the ratio is usually much closer to 1–3 (e.g. Barker & Worthington 1973*a*, *b*; Lovelock 1977; Campbell 1982);
  - except possibly locally, there appears to be no systematic variation in matrix permeability with depth recorded in the literature, despite the fairly frequent dissolution of carbonates from shallow depths;
  - systematic investigation of three-dimensional permeability structures are rare (cf. Prosser & Maskall 1993; Pickup *et al.* 1994; Ringrose & Corbett 1994), as are studies on the relationship between bed form and permeability/ porosity;
  - the few studies available suggest that the permeability of some sandstones is sensitive to change in water chemistry (Braney *et al.* 2001; Mitchener 2003);
  - the logarithm of permeability is often linearly correlated with porosity, albeit with considerable scatter and with only locally validity (e.g. Campbell 1982);
  - correlations between permeability and pore size have been sought and, in some cases, found (Seif el-Dein 1983; Digges la Touche 1998; Bloomfield *et al.* 2001; Lovell *et al.* 2006; Pokar *et al.* 2006), but scaling may be an issue (Worthington 2004);
- mudstone properties have been much less frequently studied. Limited work in the West Midlands and NW England (Curião unpublished data 2002) confirms that intergranular permeabilities are very low (*c.* 10^{-6} m day^{-1}) (Fig. 5), although none of the samples tested contained infilled desiccation cracks, a common sedimentary feature.

Sandstone permeability structure and flow. In general, estimation of a large-scale hydraulic conductivity value for the sandstone has usually involved taking a weighted mean of the hydraulic conductivity data from small-scale tests (e.g. Pokar *et al.* 2006) or finding an effective hydraulic conductivity distribution by calibration of a numerical model against field data. These deterministic approaches are often satisfactory for flow assessment, but usually are not adequate for prediction of the flow components of solute movement. Historical data are often inadequate for estimation of the effect of the local-scale hydraulic conductivity variations on the transport and fate of solutes. For the latter, stochastic approaches are available, although application of these techniques is much rarer in practice in hydrogeology than in petroleum geology. They include the use of geostatistical, process-imitating and structure-imitating models (e.g. Guadagnini & Winter 2004; de Marsily *et al.* 2005).

Publications on geostatistical modelling of the sandstone hydraulic conductivity distributions appear to be lacking in the accessible hydrogeological literature. In petroleum geology, the most significant early application of geostatistics to UK sandstones was the work of Matheron *et al.* (1987) on the coastal Jurassic sandstone outcrop in Yorkshire. This application used a truncated Gaussian method to simulate the distribution of rock types within the sandstone based on the division of a continuous variable into a discrete set of indicators. More recently, the markedly structured nature of many sequences, for example the UK Permo-Triassic sandstone, has lead to suggestions that discontinuous, facies-based approaches are more appropriate (e.g. Aigner *et al.* 1999). Such approaches have been developed and applied in the oil industry. They require detailed geological characterization, making heavy use of core, geophysical and outcrop analogue data, the

J. H. TELLAM & R. D. BARKER

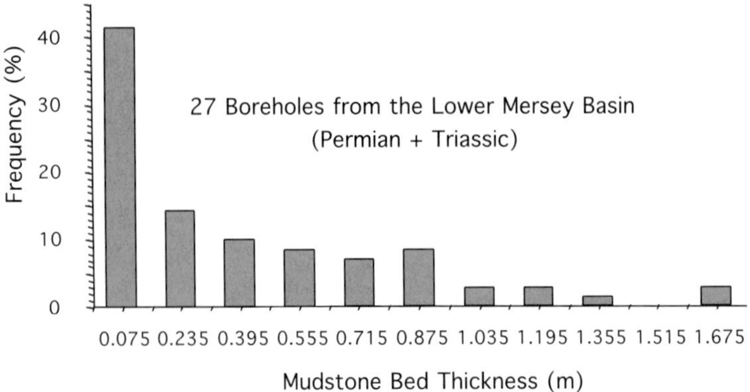

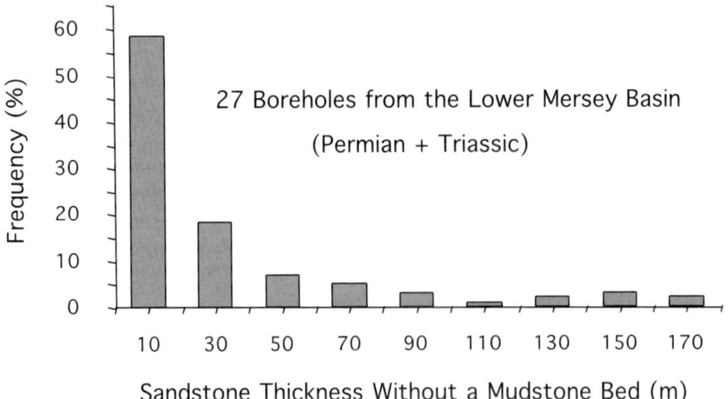

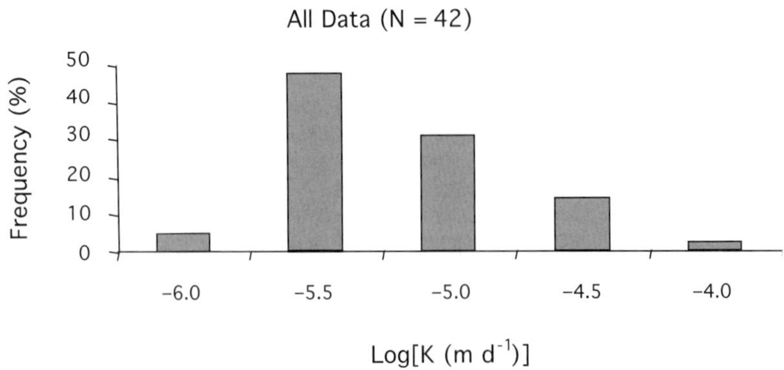

Fig. 5. Mudstone thickness and frequency (Lower Mersey Basin), and permeability (Cheshire Basin and Birmingham) (Curião unpublished data 2002).

latter increasingly collected with geophysical tools (e.g. Aigner *et al.* 1995, 1999; Bryant *et al.* 2000; Pringle *et al.* 2004). There are many practical problems with this approach: for example, quantifying uncertainty, dealing with qualitative data, appropriate definition of facies and calibration of resulting models (e.g. Wong *et al.* 1997; Vaughan *et al.* 1999; Biver *et al.* 2002; Liu & Oliver 2005; Martinius & Naess 2005). Nevertheless, codes that include explicit accounts of bedforms and facies are starting to be used not only in research but also in practical applications (e.g. Smith & Moller 2003; Liwanag 2005).

At the heart of the facies approach is the characterization of the properties and distribution of facies and lithotypes. There are numerous petroleum industry examples of this, but a good hydrogeological example, incorporating both hydraulic and geochemical data is the study of Allen-King *et al.* (1998) and Moysey *et al.* (2003). In the UK, although comparisons of hydraulic properties and sandstone lithologies have long been attempted (e.g. Lovelock 1977), detailed work of this type is uncommon. An exception is the study of Bloomfield *et al.* (2006) who show that various lithofacies of the Wilmslow Standstone and Helsby Sandstone formations (Table 1) have distinctive hydraulic properties. (See also Bloomfield *et al.* 2001.) Bouch *et al.* (2006) similarly present detailed lithofacies data for a predominantly fluvial sequence.

A particular difficulty in practice is the determination of the lateral extents of each facies ('micro'- and 'mio' scales in particular). Generic data are available (e.g. Miall 1990), but they may not be appropriate for a given system. One relatively recent development is the use of sediment deposition models to help constrain sedimentary structure architecture. These models often use process-based physics models and/or empirical rules to develop sediment distributions associated with a particular environment (e.g. Kolterman & Gorelick 1996). Teles *et al.* (2004) have used this approach in modelling an alluvial aquifer in France. Bunch *et al.* (2004) presented a sedimentation model designed to apply to the UK Permo-Triassic sandstone sequence, and current work is concentrating on the SW England outcrop area (Fig. 6) (Bunch 2006). This approach should provide insights into the vertical and horizontal components of facies architecture, but a difficulty for translation of these to yield hydraulic conductivity distributions that is not usually important in studies of more recent sediments is the presence of cements and fractures. The permeability initially present through depositional processes may well affect the development of cementation (and thence also to fracturing – Fig. 1).

A conceptual framework for modelling hydraulic conductivity heterogeneity is supplied by the nested cyclic sedimentation–sequence stratigraphy model described earlier (cf. Jensen *et al.* 1996). However, this model possesses form and noise (randomness) that needs to be placed on a much sounder statistical foundation: this in turn requires much improved geological data collection. At present, the required geological information is largely lacking from the existing water well logs, as Thompson pointed out as long ago as 1970 (Thompson 1970*a*). However, it may be possible to take advantage of the very considerable number of accessible water wells by relogging with geophysical tools (e.g. Turner *et al.* 2001; Houston 2004), including acoustic and optical televiewer logging and resistivity imaging or spectral IP (induced polarization) logging, to gain enough geological detail to quantify the structure and variability. Many possibilities exist for searching for structure using such geophysical approaches (e.g. Yang & Baumfalk 1997). The application of geophysics to hydraulic property estimation is discussed in more detail below.

Although the development of geologically informed stochastic approaches have developed largely to study transport problems, there is potentially substantial benefit to be gained from their application to flow-only studies, notably through a greater awareness of predictive uncertainty.

The effect of mudstone permeability on flow. Little quantitative information is available on the distribution and hydraulic properties of the mudstone beds, but some example data obtained from analysis of cores and geophysical logs are shown in Figure 5: comparison of the geophysical and core data indicate that usually only mudstones thicker than approximately 10 cm can be resolved by geophysical logs run under standard conditions, and that sometimes distinguishing mudstone from mud-pellet-containing beds using natural gamma-ray and resistivity logs alone is difficult. Lateral continuity is generally uncertain. Some mudstones can be traced for at least about 1 km (Thompson 1970*b*), but often it is found that mudstones in boreholes separated by less than a few tens of metres cannot be correlated, although whether this results from limited extent, local erosion or fault displacement is usually uncertain (cf. Stanistreet & Stollhofen 2002). Palaeosols may

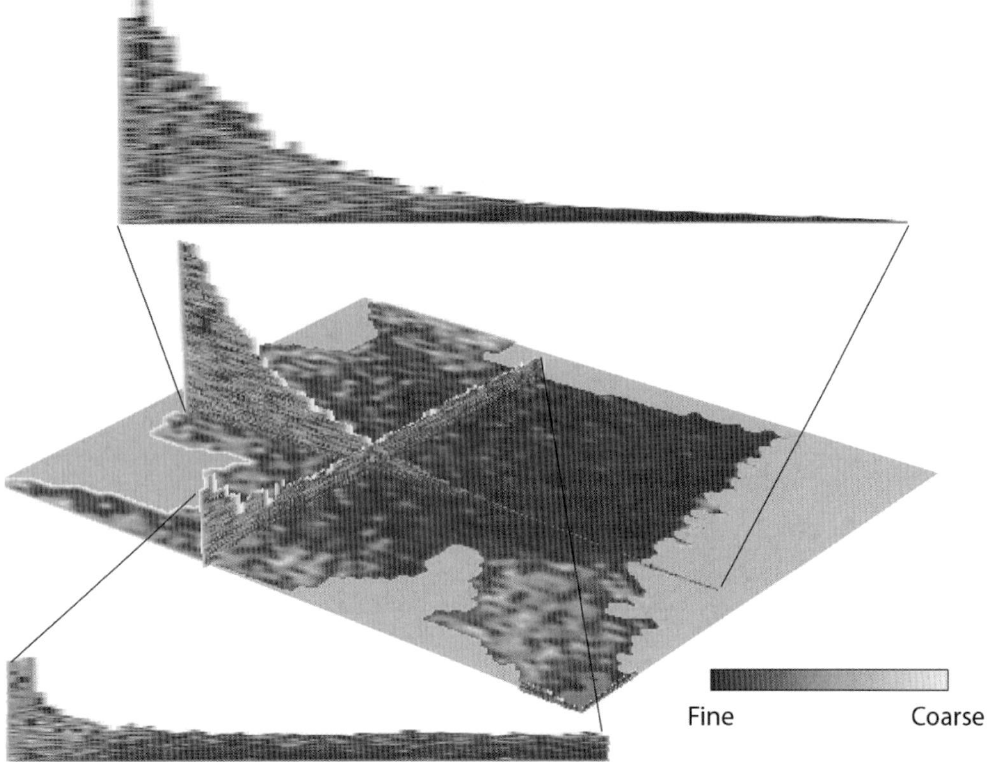

Fig. 6. Modelled alluvial-fan deposition showing grain-size distribution architecture (based on Permian sequence, Torbay, SW England, viewing towards the NW, area approximately 5 × 5 km) (Bunch (2006); see also Bunch *et al.* (2004)).

also have significantly lower permeability than the majority of the sequence (Bouch *et al.* 2006; Newell 2006).

Rushton & Salmon (1993) demonstrated the importance of low-permeability beds, especially mudstones, in controlling vertical head gradients in the sandstones. Vertical head gradients of a few per cent are often seen in boreholes (Brassington 1992; Taylor *et al.* 2003), and usually most head change is seen across mudstone beds (Segar 1993). An extreme example is seen at a location in the southern Cheshire Basin where head differences of approximately 65 m have been recorded in wells of different depths at effectively the same location (Voyce pers. comm. 2002). It is now common practice to model regional flow in the sandstones using layered models. Although the mudstone matrix permeabilities are very low, flow will pass across the mudstone beds to some extent: in general, simple estimates suggest that if the mudstones are of limited lateral extent,

flow will go around them, but if they are hundreds of metres in length, most flow will pass through them, albeit still imparting a directional influence on flow (see also Alexander *et al.* 1987 and the section on Unreacting solute movement later).

Geophysical estimation of hydraulic properties. As mentioned above, geophysics is a potentially powerful tool for mapping geological structures, but it also has potential for more directly mapping the matrix hydraulic property architecture of the sandstones, provided appropriately sensitive methods can be found. Early attempts to estimate hydraulic properties used equations developed for oil reservoir studies. Application of the Archie (1942, 1950) formation factor equations ($\frac{\rho_o}{\rho_w} = F$, where ρ_o is bulk resistivity, ρ_w is pore fluid resistivity and F is the formation factor) to the Permo-Triassic sandstones (e.g. Worthington 1973) quickly showed that typical

UK Permo-Triassic sandstone did not behave as a 'clean' sandstone as it contained electrically conductive material, mainly clay, disseminated throughout the rock (cf. the Penrith Sandstone Formation described by Lovell *et al.* 2006). In this situation, the Archie equations are only valid for high-salinity pore waters.

In order to determine the porosity (ϕ) from the formation factor (F) using Wyllie & Gregory's (1953) equation, $F = \dfrac{a}{\phi^m}$, corrections must be made for the excess conductivity introduced by the clay, and many methods to do this have been suggested (e.g. Patnode & Wyllie 1950; Winsauer *et al.* 1952). In order to employ these equations successfully, values of the excess conductivity together with the constants a and m have to be determined for the local aquifer. These relationships are further complicated when the sandstone is partially saturated (Taylor & Barker 2002, 2006; cf. Dalla *et al.* 2004). So although the relationships between electrical properties and porosity are established, their practical application is not trivial and can be expensive. In addition, the important precise indirect determination of permeability has hitherto proved elusive.

In an attempt to overcome these shortcomings, other geophysical techniques have been used, particularly induced polarization. Some partial success was achieved for the Permo-Triassic sandstone of the Fylde (Fig. 2) when Collar & Griffiths (1976) presented an empirical relation between induced polarization and permeability. Similar results were reported from the Permo-Triassic sandstone of the Lower Mersey Basin by Olorunfemi & Griffiths (1985), who also used induced polarization field surveys to determine groundwater salinity and excess conductivity.

Most recently, it has been shown that there are strong relationships between the spectral resistivity parameters and the pore-size distribution obtained using mercury injection capillary pressure instruments (Scott & Barker 2003, 2005, 2006). This opens up the possibility of estimation of permeability, using both surface-based and borehole-based resistivity measurements.

Cross-borehole surveys may also be used to examine the properties of sandstone over volumes of several tens of cubic metres (Winship *et al.* 2006). Once techniques have been demonstrated on this scale, it is a relatively small step to their application in large-scale field surveys.

Discontinuities

Introduction. The main discontinuities in the UK Permo-Triassic sandstone sequence are joints, faults/granulation seams and veins. Joints and some faults appear to have greater permeabilities than the host sandstone: some faults including granulation seams appear to act as barriers, and presumably this is the case also for veins. The following sections discuss fractures (defined here as discontinuities with permeability apparently greater than the matrix) and antifractures (discontinuities with permeability less than the matrix).

Fractures. Occurrence of fractures. Joints and bedding-plane joints are seen in many outcrops of the sandstones, and are often detected in borehole logs (although apparent frequency can vary significantly with technique used: Jeffcoat 2002) (Fig. 7). There appears to be no systematic study of joint distributions and properties across the UK, but some local data sets are available (Barnes *et al.* 1998; Gutmanis *et al.* 1998; Wealthall *et al.* 2001; Jeffcoat 2002). Hitchmough *et al.* (submitted) recognize six fracture sets for the Cheshire Basin (Table 3).

The distribution of bedding-plane joints may be partly controlled by lithology according to Allen *et al.* (1998), occurring especially in laminated sandstones (cf. suggestions by Gabrielsen *et al.* 1998). Allen *et al.* (1998) cite the example of the well-cemented Helsby Sandstone Formation (Table 1) for which the spacing is approximately 0.1–10 m, and the underlying more friable Wilmslow Sandstone Formation with a spacing of 25–100 m. Bouch *et al.* (2006) present detailed fracture data from the Wildmoor Sandstone Formation of the English Midlands, a similar sandstone to the Wilmslow Sandstone. Allen *et al.* (1998) suggest that bedding-plane fractures are relatively short (less than tens of metres) in channel sandstones, but longer where developed at the junctions with fine-grained sheet deposits (up to a few hundreds of metres) (cf. Table 3).

Borehole geophysical and other data suggest that bedding-plane fractures become less frequent below about 200 m in the Lower Mersey Basin (Fig. 2) (University of Birmingham/NWWA 1981; Campbell 1986) and below approximately 120 m depth in Cumbria (Allen *et al.* 1998): however, there appears to be no frequency–depth correlation above this depth in the Cheshire Basin (Jeffcoat 2002).

Allen *et al.* (1998) suggest that bedding-plane joints have apertures of up to 10 mm, although it is often difficult to estimate apertures in

(a)   (b)

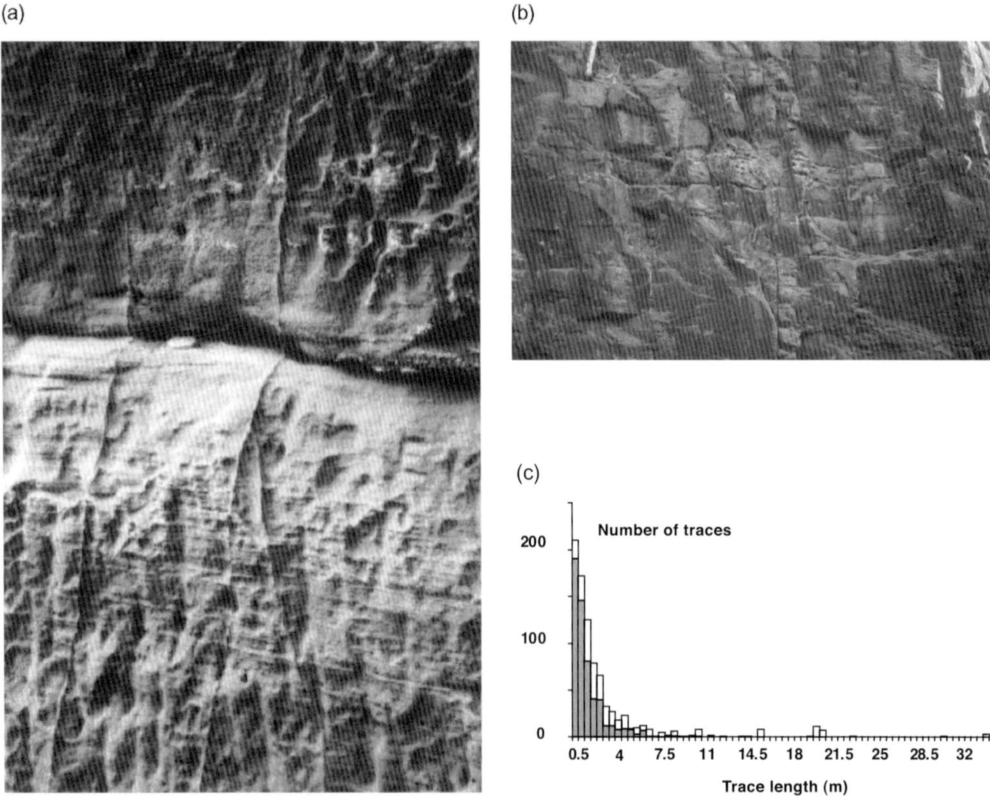

(c)

Fig. 7. (**a**) Example fractures (with granulation seams), southern Cheshire Basin; (**b**) example fractures, northern Cheshire Basin; (**c**) fracture trace length distributions, Cheshire Basin (after Hitchmough *et al.* submitted; unshaded, length censored by edge of outcrop).

Table 3. *Summary discontinuity properties from the Cheshire Basin (Hitchmough* et al. *submitted).* K_H *is horizontal permeability;* K_V *is vertical permeability; MRV is the minimum representative volume*

Set	Dip	Strike	Length	% granulation seams	Description	Potential importance (assuming all 'flowing' fractures have the same, isotropic, transmissivity)
1	Subhorizontal	–	10s of m	0	Bedding planes	Dominates K_H; increases K_V; reduces tortuosity
2	Subvertical	N–S	<10 m	43	Joints, faults, granulation seams	Minor effect on K_H/K_V ratio
3	Subvertical	E–W	<10 m	26	Joints, faults	Little
4*	Various	Various	<10 m	23	Joints, granulation seams, faults	Increases K_V. Increases tortuosity
5	Subhorizontal	–	<10 m	0	Stratification planes	Important in absence of Set 1 for K_H and (slightly less so) for K_V
6	Subvertical	N–S	>10 m	90	Granulation seams; rarely faults, joints	If fractures present, significantly changes K tensor in vertical and N–S directions, increases MRV above 35 $\times$ 35 $\times$ 35 m, marginally increases N–S tortuosity

* All fractures not included in the definitions of the other sets.

eroded and, sometimes, stress-relieved outcrop (cf. Yeo *et al.* 1998). There is little evidence of the shape of the fractures. Allen *et al.* (1998) suggest that joint apertures are much less variable than bedding-plane joint apertures (i.e. in the range of tens of micrometres to millimetres). Fractures with unfilled apertures of millmetre size would have a considerable affect on local flows.

In general, fracture properties would be expected to vary with bed thickness and lithology (Narr & Suppe 1991; Allen *et al.* 1998; Gabrielsen *et al.* 1998) (Fig. 1), and if this is confirmed for the sandstones, then fracture properties as well as matrix properties may be correlated to lithofacies and a cyclic nested distribution may be present. However, fractures will also be considerably influenced by nearby faulting.

In common with many sequences, the UK Permo-Triassic sandstones are often extensively faulted on all scales as outcrop mapping and borehole evidence shows (e.g. Knott 1994; Chadwick 1997) (hence, both fracture and matrix systems show some degree of scale-invariance, at least informally).

In some areas of the sandstones (e.g. southern Lancashire in NW England) longwall mining in the underlying Coal Measures has probably enhanced fracturing, but no detailed quantitative evidence is available. The importance of fractures may be less where weathering is extreme because of both greater intergranular permeabilities and less ability to maintain open apertures, but again little information is available.

Evidence of flow in fractures. Fractures have long been assumed to be important for flow (Tellam 2004), and Table 4 summarizes the evidence now available from geological, geophysical, aquifer testing and modelling studies. Much more detail is provided by the reviews of Allen *et al.* (1997, 1998).

In a few areas, comparisons of matrix and field-scale permeability estimates show that fracture flow must occur to a significant extent (e.g. in the St Bees Sandstone Formation). However, often the evidence is not unequivocal, as indicated in Table 4. Considering only that evidence obtained from borehole studies, there are two main uncertainties: (i) the zone around a borehole may be atypical of the rest of the aquifer because it has been developed or clogged during and after drilling; and (ii) the fractures detected may not be connected on a regional scale (even via

short matrix 'bridges'). These points are considered below.

- It would be difficult to argue that fracture permeability is not increased by well development, but to demonstrate that natural fill material, as well as drilling debris, is removed during development is not straightforward. Outcrops and well faces are prone to erosion, and are hence not ideal sites to search for evidence for fill material. To avoid this problem, Wealthall *et al.* (2001) examined outcrops in hand-dug tunnels in Runcorn, Lower Mersey Basin. They found fracture apertures up to several centimetres wide, filled in every case with sand and/or clay (cf. Hewitt 1898), and speculate that this is the normal situation within the rock. To test this idea, Hough *et al.* (2001) examined quarry and other outcrops of sandstones at approximately 20 sites from southern Scotland to SW England, but found very few such fractures (an example being in the Dumfries Basin in southern Scotland: Kemp *et al.* 2003; see also Akhurst *et al.* 2006). The absence of large infilled fractures does not mean that infill in narrow aperture fractures is not important, and indeed Jeffcoat (2002) reports that around 20% of fractures in her outcrop surveys had obvious fill. The same study found evidence from packer and geophysical log data of flow in less than 10% of fractures (Hitchmough *et al.* submitted). Packer testing of fractures in the Wildmoor Sandstone Formation of the English Midlands in boreholes that were purposely only very gently developed failed to show permeabilities greater than matrix values; yet, less than 100 m distant in the same sequence, normally developed boreholes contained clearly identifiable fractures that were seen to be further developed during production testing (unpublished data; site described by Bouch *et al.* 2006). Sand yields from wells may also provide evidence of fracture development, but even in the rare cases where sand yield is recorded, it is often difficult to determine whether the particles are released from within the formation or from the borehole walls as suggested by Bouch *et al.* (2006). Overall, it would appear that borehole data overestimate the importance of fracture flow, at least in the sandstones with higher matrix permeabilities.
- Using outcrop scan-line survey, borehole geophysics and packer test data from the

Table 4. *Evidence of fracture flow*

Type of evidence	Evidence	Comments	Example references
Direct observation	Flow emerging from fractures at outcrop and in wells	Atypical, eroded fractures?	Brereton & Skinner (1974)
Geological	1. Weathering or deposition at fracture surfaces 2. Bleached zones, formed by the passage of reducing fluids, following fault planes 3. Fracture fill material	1. When was the fracture flow active? 2. Flow occurred in the geological past and may not be occurring now 3. Locally important (Refs 3 and 4), but at most sites the vast majority of fractures appear unfilled or evidence is unclear (Ref. 5)	1. Bouch *et al.* (2006) 2. Burley & Rowe (1997) 3. Wealthall *et al.* (2001) 4. Kemp *et al.* (2003) 5. Hough *et al.* (2001)
Borehole geophysics	Calliper and CCTV indicating fractures, and conductivity, temperature, and flow logs indicating flow.[3] found that <10% of fractures identified by CCTV had associated flow indicators	Positive evidence depends on conductivity, temperature, and head contrasts being present	1. Brereton & Skinner (1974) 2. University of Birmingham/NWWA (1981) 3. Hitchmough *et al.* (submitted) 4. Reeves *et al.* (1975)
Packer tests	High-permeability zones associated with fractures (see Refs 1–4). Ref. 1 interpret two fractures at one site to provide 80% of the total borehole transmissivity	No direct evidence of connectivity of fractures on a larger scale Affected by well development	1. Price *et al.* (1982) 2. Walthall & Ingram (1984) 3. Brassington & Walthall (1985) 4. Walthall & Campbell (1986)
Pumping tests	1. Generally greater average permeability than indicated by laboratory permeability testing[1,2]. E.g. Refs 1 and 6 estimated 90% T from fractures in some cases, but in other cases laboratory K is aproximately pumping test K^2 2. Variable yields at the same site. Ref. 3 reported a site where transmissivity varies by an order of magnitude at boreholes located at the same pumping station 3. Very high transmissivities (Refs 1, 2 and 4) 4. Sand yields (Ref. 2)	1. Laboratory samples probably exclude the highest permeability, most friable sandstones (Ref. 2) Well development may result in overestimation of T (correlations of apparent T and pumping rate have been noted locally (Ref. 5) which could mean development rather than scale-dependence) Pumping test T often greater than T used in regional modelling (Refs 2 and 5) 1,2,3. No direct evidence of connectivity of fractures on a larger scale 4. Sand from fractures or from borehole wall (Ref. 8)	1. Lovelock (1977) 2. Allen *et al.* (1997) 3. Price *et al.* (1982) 4. Bottrell *et al.* (2006) 5. Howard (1988) 6. Heathcote *et al.* (1996) 7. Reeves *et al.* (1975) 8. Bouch *et al.* (2006)
Surface geophysics	1. Lack of success in predicting well yield from surface geophysics based on resistivity/laboratory K correlations	Possibility demonstrated at at least two sites (Refs 1 and 2) where matrix flow was, presumably, dominant.	1. Barker & Worthington (1973*a*, *b*) 2. Worthington & Griffiths (1975)
Modelling	1. Often model permeability greater than laboratory values (Ref. 1) 2. Decrease in permeability with depth (Refs 1–4)	1. Laboratory samples probably exclude the highest permeability, most friable sandstones (Ref. 1). Possible scale-dependency of K? 2. For example, modelling required drop in permeability below 200 m depth, and correlated with reduced fracture frequency as indicated by logging (Ref. 3)	1. Allen *et al.* (1997) 2. Heathcote *et al.* (1996) 3. University of Birmingham/NWWA (1981) 4. Allen *et al.* (1998)

Cheshire Basin, Hitchmough *et al.* (submitted) investigated the properties of the fracture network using the code NAPSAC (Herbert 1992; Hartley 1998) (Table 3). It was estimated that about 9% of the fractures carried significant flow. If all the fracture sets had the same permeability distribution, the fracture network would be very anisotropic, being approximately 23 times more permeable in the horizontal direction than the vertical. With this assumption, the permeability tensor is much more sensitive to the presence and properties of some of the fracture sets than it is to others. One joint set (Set 6, Table 3) contains rare but very long through-going subvertical fractures, which, when present, will radically modify the flow system. The bedding-plane joints (Set 1, Table 3) were found to facilitate vertical flow as well as horizontal flow, as they connect up the shorter vertical fractures (cf. Allen *et al.* 1998). The minimum representative volume for the 'flowing' fracture network permeability, ignoring the rare through-going subvertical fractures, was estimated to have a characteristic length of 35 m. In the present context, the most significant finding was that the system appears to be connected regionally. Some fractures may, of course, be filled, but in other cases matrix flow will allow good connection between close but not intersecting fractures (e.g. Odling & Roden 1997).

Overall, it seems likely that fracture flow does occur in the sandstones, but in most cases where intergranular permeability is relatively high it is not the dominant mode of flow, except locally around pumping wells. However, there are parts of the sequence, especially in northern England and southern Scotland, where matrix permeability is low and fracture flow is dominant. Where it occurs, it will disrupt the effects of the matrix permeability structures, but any correlations between the distributions of permeable fractures and lithology are largely unknown. Application of the 'generalized flow model' pumping test interpretation technique suggested by Barker (1988) would be interesting at a site with detailed fracture and matrix characterization.

Antifractures.

Faults. In hydrocarbon development, faults are often assumed to be barriers in porous sandstone reservoirs until proved otherwise (e.g. Davies & Handschy 2003), and there are many studies of low-permeability fault structures including in the off-shore UK Permo-Triassic sandstone reservoirs (Edwards *et al.* 1993, includes on-shore analogues; Leveille *et al.* 1997).

Experience in the on-shore sequences indicates that many faults here also have restricted permeability perpendicular to the slip plane (e.g. Allen *et al.* 1997), examples including major faults in the Birmingham and Nottingham aquifers (Knipe *et al.* 1993; Yang *et al.* 1999; cf. Trowsdale & Lerner 2003), and the 'bucket' at Nurton, in central England (Fletcher 1989; Hunter-Williams 1995). The most comprehensive demonstration of the importance of faults in regional flows is provided by Seymour *et al.* (2006), who detail a set of case studies across NW England involving flow modelling supported by water-level measurement, geophysics and chemical evidence. In most cases there is no information on the nature of the fault, and the low permeability could be due to clay infill, fault rock or granulation seam 'swarms' (see below).

As yet, predictions of likely fault properties are rarely attempted, despite the availability of oil industry techniques (e.g. Gibson & Bentham 2003) and potentially useful fault property correlations (Huntoon 1986; Knott 1994). For estimating the probability of 'sealing', it would be useful to discover what proportion of pumping tests show barrier–boundary responses.

Although there is convincing evidence of faults acting as barriers, there is also convincing evidence, even from the earliest hydrogeological records (Tellam 2004), of faults acting as high-permeability pathways (Allen *et al.* 1997). Clearly, some faults may have low permeability where mudstone beds have been smeared along their planes or where some form of cataclastite has formed, and other faults may be effectively open fractures. However, it is also possible, as a number of authors have pointed out (e.g. Gutmanis *et al.* 1998), that a fault may have a high permeability in a direction parallel with the fault plane, and a low permeability at right angles to it: faults are not simple single features, and often may comprise a complex of planes of a wide range in permeability (e.g. Edwards *et al.* 1993; Fowles & Burley 1994; Allen *et al.* 1998). As with any fracture, the permeability along the discontinuity may be far from uniform, and flow may be channelled (Yeo *et al.* 1998; Steele & Lerner 2001). Combinations of faults will result in changes in effective large-scale permeability (e.g. Manzocchi *et al.* 1998).

Granulation seams. Granulation seams (or deformation bands: e.g. Main *et al.* 2001; or cataclastic slip bands: e.g. Rowe & Burley 1997) are a type of fault expression common in the UK Permo-Triassic sandstone sequences (Underhill & Woodcock 1987; Fowles & Burley 1994; Main *et al.* 2001), and in many other high-porosity sandstones throughout the world (e.g. Aydin & Johnson 1978, 1983; Pittman 1981; Jamison & Stearns 1982; Antonellini & Aydin 1994; Antonellini *et al.* 1994; Gibson 1998). In the UK Permo-Triassic sandstones, these deformation bands typically comprise approximately 1 mm-wide crushed, almost zero porosity, zones, often lacking haematite. They often occur in swarms, occasionally dominating the host rock and forming a wall-like feature 1 m or so wide often associated with adjacent main slip surfaces (Fig. 4a). Not surprisingly, granulation-seam orientations may parallel main slip-surface orientations (e.g. Bouch *et al.* 2006) (Table 3). Reactivation is common, and complex structures can form (Edwards *et al.* 1993).

Laboratory measurements indicate that the seams typically have permeabilities of 3–6 orders of magnitude lower than the host sandstone (Fowles & Burley 1994; Ballard 2000; Main *et al.* 2001; Bouch *et al.* 2006). Fowles & Burley (1994) also found that the permeability of the sandstone immediately surrounding the seams is increased (by 3–4 times in the case described). Gutmanis *et al.* (1998) found a negative correlation between 'potential flowing zones' and deformation band presence in Cumbria. The effect on flow of the seams, and particularly where they occur in swarms, is likely to be substantial. Fractures appear to cut through the seams, but the latter often restrict the aperture, at least in outcrop (Fig. 7a). It is therefore possible that flow is funnelled towards the fractures leaving zones of limited flow. More information is required on the probability of encountering granulation seams in different lithofacies (not necessarily an easy task as Gabrielsen *et al.* 1998 warn).

Veins. Veins of iron-rich rock up to a few centimetres thick snake through the sandstone in some places (Moore 1896; Walton 1981) (Fig. 8). Walton (1981) suggests their formation is due to pH rise following low pH groundwater flow into carbonate-bearing units (Budleigh Salterton Pebble Beds to Otter Sandstone Formation, in his case; Table 1). Their permeability is unknown, but from visual inspection is likely to be extremely low.

Carbonate veins are also common in some parts of the sequence (Milodowski *et al.* 1998;

Bouch *et al.* 2006). Their permeability is likely to be very low.

Geophysical detection of fractures, antifractures and fracture flow. Often accurate delineation of even major structures is difficult using outcrop and borehole evidence alone, and, locally, where lithological contrasts are present, geophysical surveys have been used in determining faulted structure. For example, Collar (1974) used gravity and resistivity surveys to determine the structure of the Clwyd Basin (Fig. 2), Barker (1974) used resistivity and induced polarization in the Fylde, and resistivity surveys have been widely used in many other areas of the UK and Europe. In recent times, electrical imaging has been increasingly employed (Fig. 9): its greater resolution can improve fault and other structure recognition in the sandstone. Bunch (2000) was unable to detect known sandstone–sandstone faulting at shallow depth using resistivity imaging at one site, but more work on this needs to be done. Reflection seismic data have been decisive in mapping the larger structures in NW England especially (Seymour *et al.* 2006), but often data do not exist, features are too small for conventional interpretation or depths are too shallow.

It is possible that in future, surface geophysics may also be able to provide indirect evidence of fracture flow, although this is likely to be difficult. Griffiths *et al.* (1981) demonstrated that induced polarization could be used in tracing pollution through fractures in the Permo-Triassic sandstone in the East Midlands. However, no direct link to permeability was found. New technology is almost certainly required. It is possible that developing techniques such as spectral resistivity (Scott & Barker 2006) and magnetic resonance sounding (Lubczynski & Roy 2005) may hold the key to future large-scale indirect field characterization of sandstone properties.

Flow through the rock mass

Flow in the rock mass occurs through both matrix and fractures. Some partially filled or filled fractures may have properties intermediate between these extremes. Both fractures and matrix have their low-permeability equivalents – some faults/granulation seams and mudstone or unusually well-cemented beds. Geologically, the development of these features will not have been independent, and there may be correlations in spatial distributions. On the other hand, effects of one permeability type may be moderated by the presence of the second.

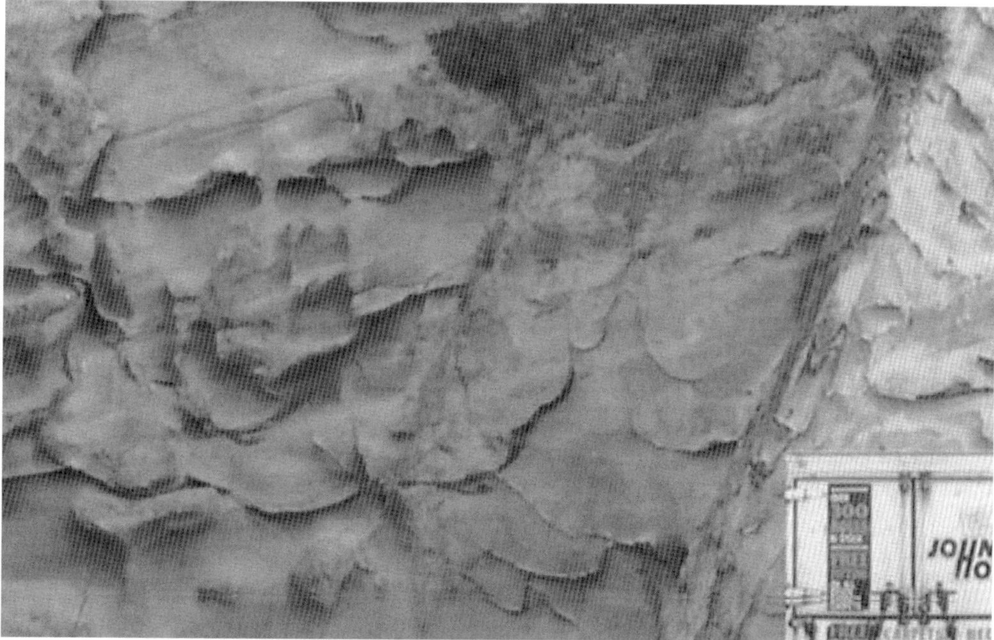

Fig. 8. Veining (looped structures to left of fault), Exeter, SW England.

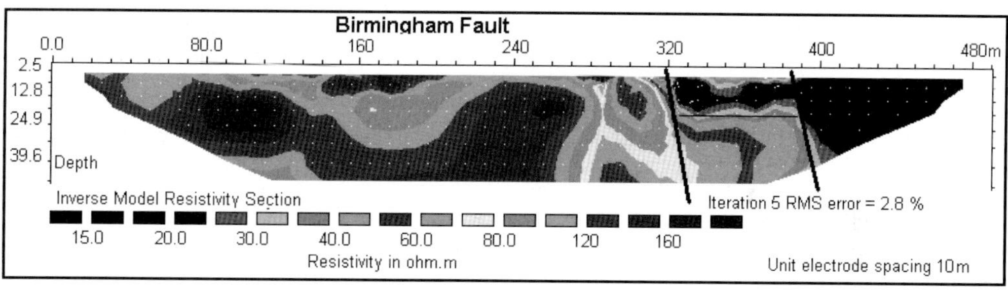

Fig. 9. Resistivity imaging across the Birmingham Fault. Resistivity more than *c.* 60 Ωm (i.e. everything to the left of 320 m, and below the line between 320 and 390 m) indicates Permo-Triassic sandstone; lower resistivities indicate mudstones.

At the regional scale, flow systems have been modelled apparently successfully on many occasions using equivalent porous medium approaches. Many models include some account of specific types of heterogeneity: for example, decrease of fracturing with depth (NW England; University of Birmingham/NWWA 1981), presence of mudstones (Midlands; Rushton & Salmon 1993) and regional faulting (Fylde area, NW England; Seymour *et al.* 2006). Sometimes account is taken of more than one feature (e.g. Furlong 2002), but it is rare that models, even of pumping tests, include all the features described in the previous sections in explicit form (cf. Edwards *et al.* 1993). It thus appears that the system is forgiving when flow calculations are involved, and that flow at large scales at least can be fairly readily represented by an equivalent porous medium. Only major heterogeneities affect the observation network sufficiently to require inclusion, and in some aquifers, where mudstones (Rushton & Salmon 1993) or cementation are not well developed, even layered models are unnecessary – transmissivity models still reproduce field data within their uncertainty limits. Overall, with a few exceptions where the sandstone is much better cemented, the apparent effective permeabilities at the regional scale are often relatively close to laboratory measured values.

At the local (site/well) scale, and possibly at slightly larger scales, rock mass flow, however, often appears to be strongly affected by fractures and matrix heterogeneities (Table 4). It is unclear over what distances local scale becomes regional scale: the method of Streetly *et al.* (2006) offers a possible valuable method for testing at intermediate scales (see also Streetly *et al.* 2000). This question is considered again after solute transport has been discussed.

Unreacting solute transport

Introduction

Much less work has been completed on solute movement than on flow, and the evidence is rather fragmentary. Average linear velocities expected from bulk permeabilities and a range of head gradients are expected to be slow (Fig. 10).

Laboratory scale investigations

Investigations of the movement of non-reacting solutes through laboratory samples have been reported by White (1986), Read *et al.* (1993), Bashar (1997), Braney *et al.* (2001) and Bashar & Tellam (2006). Lovell *et al.* (2006) show how geophysical methods can provide detailed information on sedimentary structure at this scale, and Greswell *et al.* (1998) imaged flow using positron emission projection techniques.

Bashar & Tellam (2006) investigated breakthrough of tracers in relatively low-permeability sandstone cores and slabs from the Wildmoor Sandstone Formation of the West Midlands

(Table 1). For flow parallel to laminations, they found significant tailing of the breakthrough curves to occur as the tracer swept through each lamination at a different velocity (Fig. 11), and positron emission projection images of larger blocks also indicated differential movement through sets of laminae. The breakthrough can be described explicitly by modelling advective movement along laminae with diffusive exchange: dual region first-order transfer analytical models also can reproduce the tailing (Fig. 11d), but the parameters are scale-dependent (Bashar 1997). For solute movement perpendicular to laminations, untailed, Ogata & Banks (1961)-type breakthrough occurs, and this is also the case for unlaminated sandstones. The limited laboratory data available appear to suggest that, for the latter cases, kinematic porosity is very similar to total porosity as measured by water saturation methods.

Measurements of diffusion through the sandstones appear to be rare (Bashar 1997; Braney *et al.* 2001). Bashar (1997) gives measurements for Br and amino-G-acid for 10 samples, with D_i (i.e. free water diffusion coefficient $\times$ porosity/tortuosity) values ranging between 9.9×10^{-11} and 2.4×10^{-10} m^2 s^{-1} for CaBr$_2$ and 3.5×10^{-11}–9.6×10^{-11} m^2 s^{-1} for amino-G-acid, with implied tortuosities (in this case, path length/straight-line path length) of less than 1.5. Unfortunately, there appears to be no diffusion coefficient data for mudstones, and their role as a diffusive sink for solutes is uncertain.

Borehole testing and pore-water sampling

In their review of tracer tests in UK aquifers, Ward *et al.* (1998) discovered only three

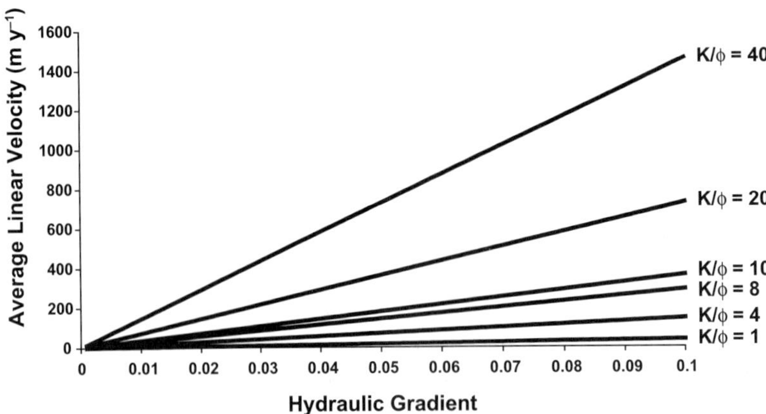

Fig. 10. Average linear velocity ranges for typical ratios of hydraulic conductivity (K) to porosity (ϕ) and typical hydraulic gradients: K values in m day^{-1}.

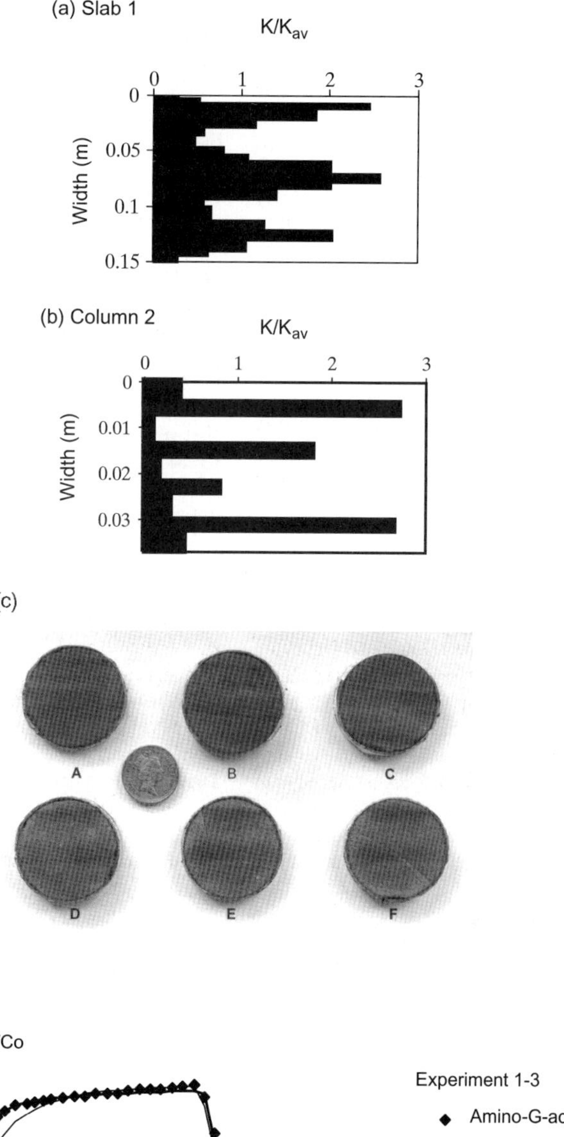

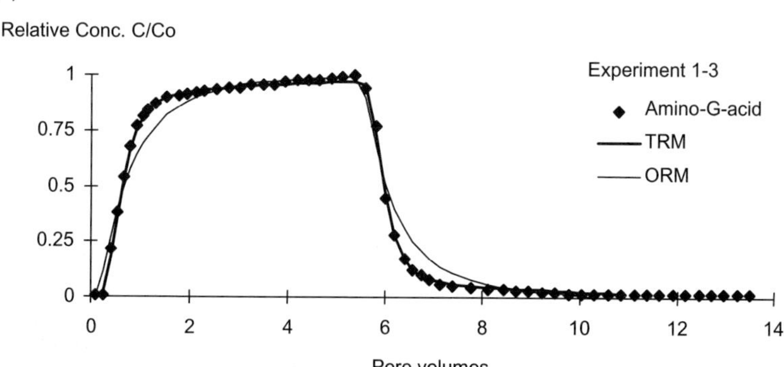

Fig. 11. Laboratory-scale breakthrough of tracers (Bashar 1997; Bashar & Tellam 2006). (**a**) & (**b**) Permeability distributions through a slab and a small column of sandstone showing different lamination-scale permeability structures (K_{av} is the mean K for the sample assuming constant porosity). (**c**) Rhodamine movement through a core, flow parallel to laminations. The experiment was stopped before complete breakthrough and the core sectioned to observe sorbed tracer. (**d**) Non-reacting tracer breakthrough showing tailing, and one-region (ORM) and two-region (TRM) first-order exchange model fits.

locations in the Permo-Triassic sandstones where such tests had been completed, and at this time lack of test data seems to have been typical worldwide for fractured permeable-matrix aquifers (Stafford *et al.* 1998).

One test listed by Ward *et al.* (1998) was at Heath House, near Hodnet, Shropshire, in the NW Midlands [national grid reference SJ 6028 2459] (Coleby 1996). In the test, amino-G-acid was injected into an observation borehole 20 m from an approximately 120 m-deep well that had been abstracting at 3.83 Ml day^{-1} for over a week and was assumed to be close to steady state. Breakthrough commenced at approximately 50 h after injection started, and peak concentrations occurred at 105 h. The breakthrough curve showed a single peak with prominent tailing with 93% recovery of tracer. Although detailed analysis was not attempted, the timing of the peak suggests an effective porosity of 0.14.

The second location was at Haskayne, a research site north of Liverpool in NW England (also called Plex Moss: see also Jones & Lerner 1995) [SJ 357 089]. Here three tests were attempted with injection of a slug of tracer from one 0.4 m-long interval in a piezometer nest approximately 5 m from an open borehole sealed by a packer at the elevation of the base of the injection interval (Streetly *et al.* 2002). Fluorescein and amino-G-acid at the two successful test intervals showed a multimodal breakthrough. As the first peaks appeared much earlier than expected from calculations assuming piston flow through the matrix, they were interpreted as fracture flows, although it is possible that they represent fast intergranular pathways. Unlike the earlier pulses, which had dispersivities of less than 10 cm, the last pulse in both tests was consistent with dispersivities of around 2 m, very large given that the travel distance was only 5 m. An apparent kinematic porosity of approximately 0.2 was determined for these pulses: this is very close to the average porosity for the profile at this site (Segar 1993). The fastest pathways are 10–40 times faster than if the systems were homogeneous. Streetly *et al.* (2002) estimated that there would have been 30–45% recovery had the tests been continued to apparent completion. During the drilling of one of the test boreholes, breakthrough was observed at all 1 m-long intake zones within a piezometer nest 15 m away: calculations assuming homogeneous piston flow suggest that flushing to only 1 m radius from the borehole should have taken place, and again this strongly suggests the presence of transport in (frequent) fractures.

Similar tests have been carried out recently at two sites on the Birmingham University campus, and four examples are shown in Figure 12a–d. Although quantitative interpretations are not yet available, again, over the 7–10 m distances involved, breakthrough curves are multimodal, with the earliest being so fast as to suggest strongly the presence of fracture flow. In one test, no breakthrough was achieved after several pore volumes had been pumped, and this may suggest that there was a barrier (granulation seam or vein) between the injection and recovery boreholes in this interval (see Bouch *et al.* 2006). More typical recoveries of tracer are 60–80% at this site. Vertical tracer tests were also carried out at the first test site, and breakthrough was again recorded despite the presence of intervening mudstone beds (Fig. 12e) (Sauer in prep.).

Qualitatively, these results appear to agree with the conclusions of the two-dimensional (2D) numerical modelling work of Odling & Roden (1997). These workers considered three systems, each with a heterogeneous isotropic matrix permeability, containing: (A) four unconnected en echelon fractures; (B) a poorly connected fracture system based on that observed in a shale; and (C) a connected fracture network based on that observed in a sandstone. The fracture/matrix permeability ratio was set at 10^5, probably rather higher than seen in most UK Permo-Triassic sandstones. It was found that the fracturing increased the flow heterogeneity, even where fractures were unconnected, with maximum–minimum flow rate ratios being 10 in the matrix-only system but more than 10^6 in systems B and C. Enhanced flow occurs in the matrix where there are 'bottlenecks' in the fracture flow systems, and, where the fracture flow predominates, flow in the matrix is reduced resulting in zones of very slow flow. This produces very heterogeneous distributions of contaminants throughout the modelled regions, even for the simple case A, with the downflow ends of fractures acting as solute plume sources. At the downflow constant head boundary, most contaminant was discharged via fractures, despite most mass being held within the matrix at distance from the boundary. The breakthrough curves were multimodal, the matrix breakthrough only being recognized as a distinctly identifiable entity in case A where only four fractures were present. Rubin *et al.* (1996) show that the complex breakthrough at small distances in fractured permeable media becomes similar to that obtained in an ideal non-fractured permeable medium at longer distances, as might be

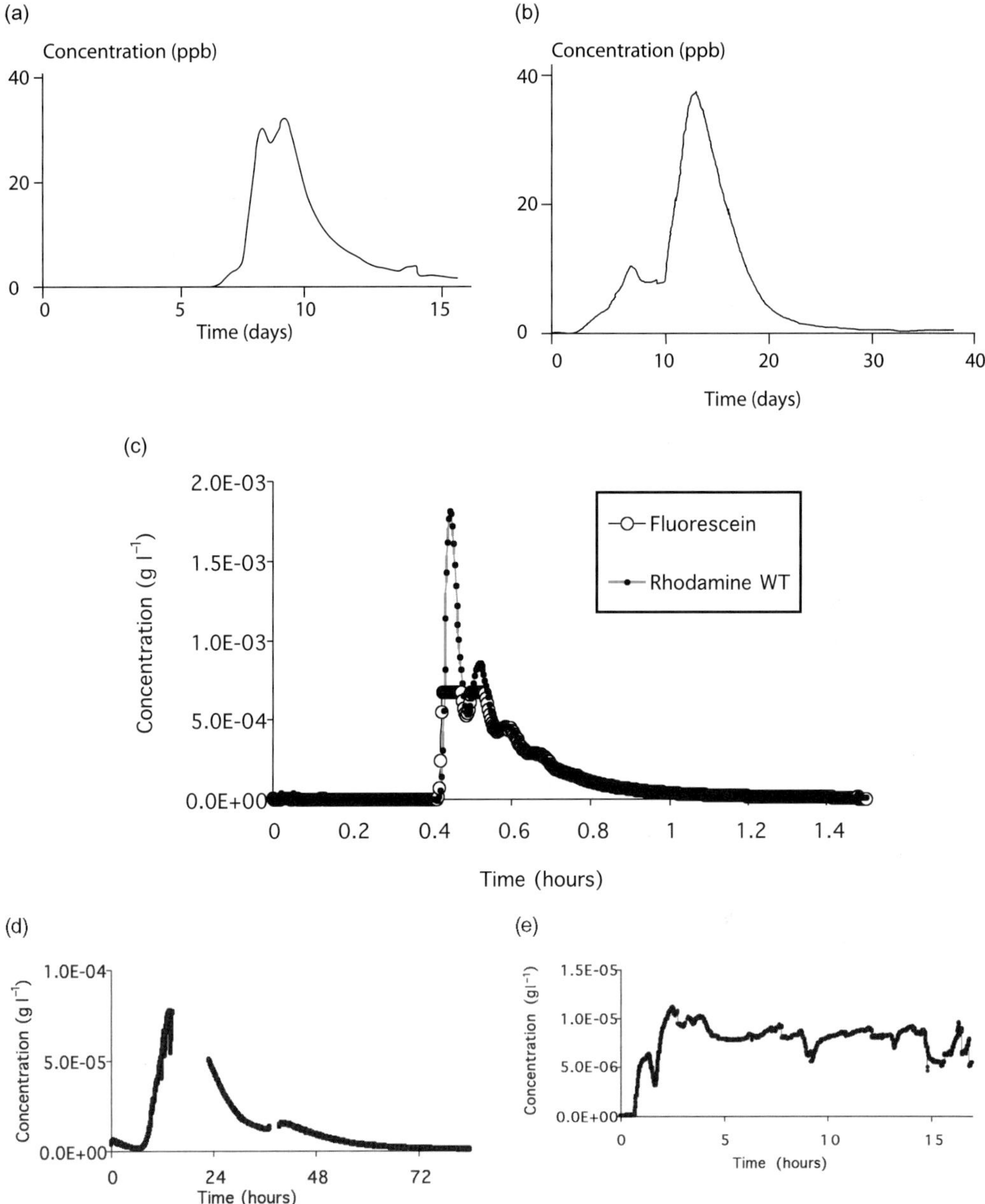

Fig. 12. Example tracer tests at Birmingham University Campus sites. (**a**) & (**b**) Breakthrough over 7 m between packered, apparently unfractured, intervals in the boreholes described by Bouch *et al.* (2006) (Joyce *et al.* 2006). (**c**) & (**d**) Breakthrough in fractured and developed intervals over similar distances between pairs of boreholes approximately 100 m distant from those of (a) & (b) (Sauer in prep). Note that pumping rates/interval thickness ratios for the tests vary considerably. (**e**) Breakthrough at the site of (c) & (d) resulting from forcing the tracer to flow vertically through the profile across mudstone beds (Sauer in prep.).

expected. Modelling work by Stafford *et al.* (1998) shows that a small number of higher permeability truncated fractures in an aquifer with a low-permeability matrix could explain the unusually large spreading of plumes in a fractured permeable-matrix aquifer.

The third tracer test listed by Ward *et al.* (1998) was undertaken by Barker *et al.* (1998)

by injecting fluorescein into an observation borehole and recording the breakthrough at inflows along a tunnel in Liverpool (Fig. 2). Breakthroughs were seen in seven of the eight inflow points into the tunnel, the fastest being at a straight-line rate of approximately 140 m day^{-1} over about 280 m. This is far faster than expected from average groundwater flow rates in this system, again very suggestive of fracture flow. Note that this test differs from conventional tests in that the observation point for breakthrough was not defined before the test began, and hence the solute pathway was much less constrained compared with the other tests described here: although the head gradients in this test were much greater than in most non-test situations, the fact that the pathway was only poorly constrained may be more typical of many systems. The pathway may correspond with either the large subvertical N–S fractures, or the large bedding-plane fractures identified by Hitchmough et al. (submitted) (Table 3).

Taken together, the tracer test results summarized here confirm the importance of fracture flow near wells, and indicate that some fracture flow can occur over distances of up to at least hundreds of metres, but that often a significant percentage of tracer is not recovered or only very slowly recovered.

Given the effect of layering on breakthrough seen in the laboratory, a similar phenomenon might be expected in the field, at least over short longitudinal distances. However, diffusional effects will be less marked as transverse distances are significant: characteristic diffusion times for a range of typical bed thicknesses are given in Figure 13. Although detailed chemical

profiles from piezometer and core pore-water studies often show variations on metre or even submetre scale (e.g. Tellam 1995a; Kinniburgh et al. 2006), method error estimates are difficult to quantify, and are rarely available.

However, pore-water and piezometer profiles through plumes do confirm that transverse dispersion is often limited (<1–2 m) for the distance–time scales of even large pollution plumes (Lerner et al. 2000; Thornton et al. 2001b; Taylor et al. submitted). On larger space and timescales, considerable lateral dispersion can develop: for example, the dispersion profile associated with a fresh–saline groundwater interface in an aquifer in NW England is many tens of metres thick, despite the fact that most of these profiles are in poorly fractured, deeper (100–300 m) parts of the aquifer (Tellam et al. 1986; Brassington et al. 1992; Tellam 1995a; Tellam & Lloyd 1997). In this case, the role of mechanical dispersion and diffusion warrants further investigation (cf. Rubin & Buddemeier 1996). Deep pore-water profiles provide an insight into depth of 'active' flow, but such data are very site specific: for example, in the case of the profiles in NW England, the interface between the fresher and more saline water occurs from less than 50 m to approximately 250 m below ground level depending on location within the regional flow system (Tellam et al. 1986; Brassington et al. 1992). Example studies of pollution penetration depths include Stagg & Tellam (1998), Cronin et al. (2003) and Taylor et al. (submitted).

There are some interesting insights emerging from work on particulates that have relevance for dissolved solute movement. Viable human

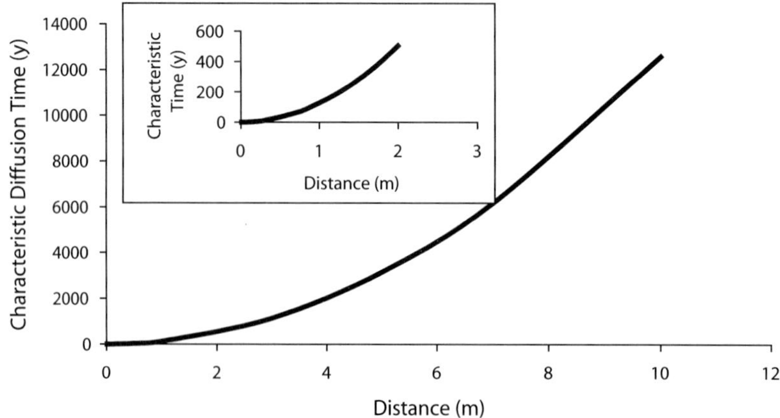

Fig. 13. Characteristic diffusion times [=distance^2/(2 × diffusion coefficient)] over distances equivalent to bed thickness ranges.

enteric viruses have been found in pumped well waters in Birmingham and Nottingham by Powell *et al.* (2000, 2001), and subsequently recovered in piezometers to at least 50 m below ground level (Cronin *et al.* 2003; Powell *et al.* 2003). This is significant given that survival times for such viruses are usually considered to be in terms of months. Some piezometer samples even appear to show concentration seasonality in phase with that in the human gut (Powell *et al.* 2003; Taylor *et al.* 2004), although this needs confirmation. All field detections are at concentrations up to at least eight orders of magnitude below probable source concentrations – virus detection is possible over very considerable concentration ranges. It is unclear whether the presence of the virus particles suggests an extremely small proportion of fast pathways through the sandstones or greater survival times than previously thought for viable viruses; confirmation of seasonality is therefore very important. Recent bacteriophage tracer tests suggest breakthrough pathways are indeed rare, even over lateral distances of 7 m (Joyce *et al.* 2006; same site as Bouch *et al.* 2006, and Fig. 12a, b).

There is other evidence for fast pathways between ground surface and pumped wells. Modelling studies by Tellam & Thomas (2002) and Butcher *et al.* (2006) indicate that chemical steady state following uniform addition of a pollutant at ground surface is only achieved for typical sandstone wells after a few hundred years; both studies were able to show that time to breakthrough could explain the observed wide range of pumped well pollutant concentrations. However, both studies also found that predicted first breakthrough times were sometimes too long. Butcher *et al.* (2006) found that even finite-length fractures extending horizontally from a well could reduce first breakthrough times considerably.

Furlong (2002) takes the argument further. He applies time-variant 3D regional-scale modelling to determine ages of groundwaters at the time of sampling, and is able to demonstrate that pumped well waters can show very great age differentials not only with depth but also with the direction of inflow into the well (Fig. 14). Although his analysis does not take into account the local heterogeneities around the sampled wells, the implied potential for mixing is qualitatively consistent with the mixed isotope ages seen in many pumped well samples, including those he modelled (Tellam 1994; see also Bath *et al.* 1987).

Geophysics

Geophysical techniques may be used to characterize the extent of groundwater contamination, with the application of rapid EM (electromagnetic) mapping techniques being popular in cases where the plume concentrations are significantly greater than the background groundwater. Vertical detail may be provided by resistivity imaging (e.g. Jones *et al.* 1998), but heterogeneity of sandstone properties can sometimes obscure some of the finer detail. Time-lapse imaging offers a potentially powerful tool for monitoring movement of contamination, particularly where concentrations are expected to change quickly (Barker & Moore 1998). In this case, features that are too small to be directly detected can sometimes be detected because they affect flow significantly (cf. Hatzichristodulu *et al.* 2002), and the examination of resistivity changes rather than absolute values can help circumvent the problem of sandstone property heterogeneity.

Regional-scale investigations

Using groundwater sampling survey data, a few studies have found significant changes in chemistry across faults, confirming their effects on solute transport at the regional scale: for example in Birmingham, central England (Jackson & Lloyd 1983) and Merseyside, NW England (Barker *et al.* 1998; Seymour *et al.* 2006).

Few appropriate, long-term, detailed chemical data sets exist that show the breakthrough of a significant change in water chemistry. An exception is the data set from estuary water intrusion in Widnes, NW England (Fig. 15) (Carlyle *et al.* 2004). Here an apparently well-behaved breakthrough occurred over a scale of *c.* 2 km/40 years that can be self-consistently modelled using an equivalent porous medium approach with observed head gradients, realistic matrix permeability/porosity ratios and dispersivities within a normally acceptable range for the distances travelled (10–100 m, depending on assumptions). This is interesting given the model's gross oversimplification of a system that is undoubtedly fractured, faulted and 3D in geometry. However, the data show some scatter and, as most wells do not approach complete breakthrough, some detail in the breakthrough curves may have been missed.

Regional patterns are often discernable in pumped water sample chemical concentrations.

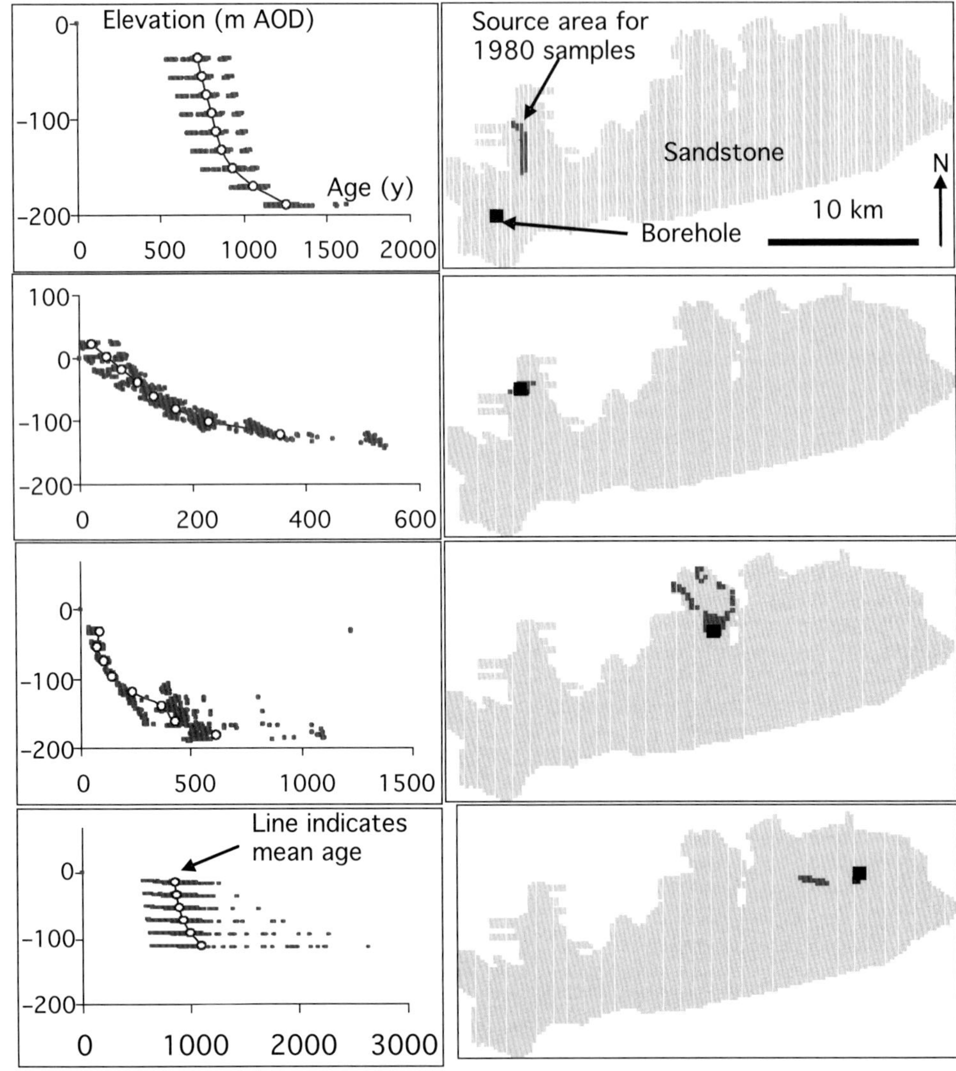

Fig. 14. Ages of water entering example pumping wells in the Lower Mersey Basin aquifer at the time of sampling in 1980, and locations of recharge, as predicted by 3D transient modelling (Furlong 2002). Area marked as sandstone is bordered to the north by underlying low-permeability Carboniferous, and bordered to the south by the overlying Mercia Mudstone Group: Quaternary deposits (sands, clays and peat) cover most of the sandstone area.

The overlying sequence usually exerts the major control by its effect on flow distributions (and ages), be it the Triassic mudstones (e.g. Edmunds *et al.* 1982; Jackson & Lloyd 1983; Rivett *et al.* 1990; Ford & Tellam 1994) or the Quaternary deposits (e.g. Sage & Lloyd 1978; Parker *et al.* 1985; Tellam 1994). There has even been some success at using simple rule-based systems for describing regional distributions (Tellam 1996). However, as indicated in the previous section and implied in well-catchment models (e.g. Papatolios & Lerner 1993; Evers & Lerner 1998; Trowsdale & Lerner 2003), water arriving at a well may be from a wide range of flow paths (Fig. 14), and the regional patterning should be considered as patterning of well-water chemistry rather than groundwater chemistry. Even quite simple descriptions of 3D

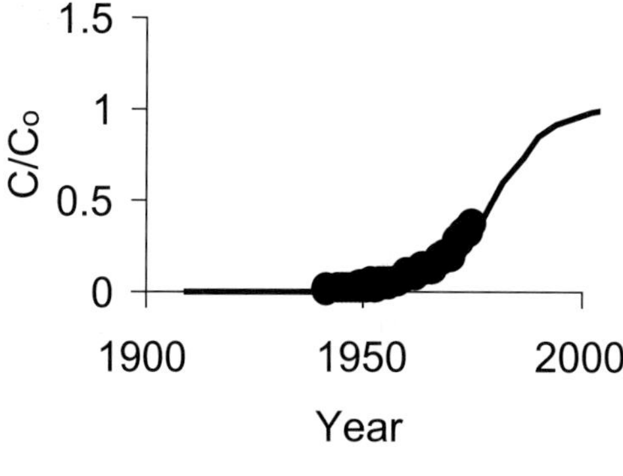
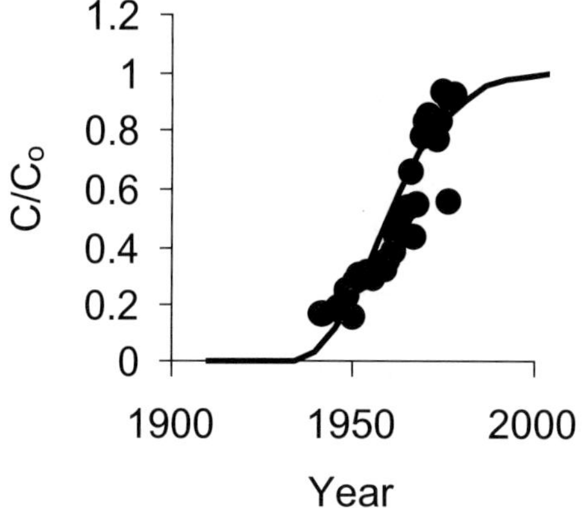
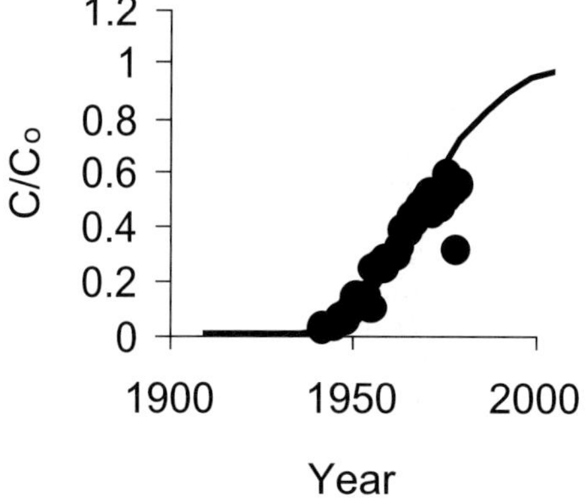

Fig. 15. Example breakthrough curves for estuary water intrusion in response to a regional head gradient, Widnes, Lower Mersey Basin (Carlyle *et al.* 2004; data from Widnes Water Users Association). One-dimensional Ogata & Banks (1961)-style breakthrough curves also indicated.

transient flow systems imply a considerable complexity in the distribution of groundwater ages, as the work of Toth (1963) onwards has indicated. Furlong's (2002) model for the Lower

Mersey Basin indicates a very complex pattern of ages reflecting recharge zones, hydraulic 'traps' associated with discharge zones and low-permeability faulting, and the regional effects of abstraction.

There is much chemical and isotopic evidence that very old water is present in some of the sandstone aquifers (Bath *et al.* 1979; Edmunds *et al.* 1996, 1997; Darling *et al.* 1997, 2003). In some cases these Devensian or older 'palaeowaters' occur because discharge is slowed by the presence of the overlying Mercia Mudstone Group, for example in the case of the East Midlands aquifer (Andrews & Lee 1979; Edmunds *et al.* 1982; Andrews & Kay 1983; Bath *et al.* 1987) or the Birmingham/West Midlands aquifer (Lloyd 1976; Jackson & Lloyd 1983). The saline waters at the base of the sandstone in NW England have ^{18}O, ^{2}H and U series isotopic signatures indicating significant dilution by Devesian (?subglacier) (Tellam 1995*a*) and post-glacial/pre-industrial (Ivanovich *et al.* 1992) recharge: although these waters are not confined by the Mercia Mudstone Group, they have been prevented from being flushed during low glacial base levels by low-permeability faults (see Seymour *et al.* 2006) and these faults also contribute to the present relatively ineffi-cient flushing by regional flow (cf. Furlong 2002). Sparse data suggest the deepest saline water has a heavier isotopic signature.

Summary

The evidence reviewed above suggests that:

- in laboratory column experiments, solute movement is multiregion, with diffusion playing an important role;
- in approximately 10 m borehole-to-borehole forced-gradient tracer tests, fast pathways are important, with overall dispersion being large and breakthroughs often multimodal (and incomplete); this is consistent with the experience of borehole hydraulic tests and with the heavily skewed distribution of fracture trace lengths (Fig. 7c), but the presence also of fast inter-granular pathways cannot be ruled out;
- fast pathways appear to exist from ground surface to pumped borehole and unpumped piezometer intakes at up to at least 50 m depth;
- in one forced-gradient borehole-to-tunnel tracer test over about 200 m, single frac-tures (possibly members of Set 1 or Set 6 in Table 3) were important;
- at the regional scale, fracture flow becomes

less obvious, and in the only long-term data set of which we are aware apparent dispersivity appears to be much as expected.

Note that it may be possible for different results to occur at the same scale when different methods are used (e.g. comparing tests where the sampling points are defined with those where they are not).

Movement of reacting solutes

Introduction

Most of the available chemical studies attempt to explain rather than predict reacting solute movement, an indication of the inherent difficulties. Four reaction classes – dissolution–precipitation, reduction–oxidation, acid–base and sorption–desorption are considered in turn below. Discussion will be limited to sandstone-groundwater interactions, despite the importance Quaternary deposits can have on recharge water chemistry (e.g. Spears & Reeves 1975; Sage & Lloyd 1978).

Dissolution–precipitation reactions

Of the minerals commonly found in the sandstones (Table 5), a few will dissolve in a simple and quantitatively predictable way (gypsum, barite and fluorite): in some cases kinetic data are even available, for dissolution if not for precipitation.

However, because such minerals are often sparsely distributed, and although locally the *in situ* groundwater may be saturated, groundwater samples, being averaged over a larger scale, are often not at saturation (Atteia *et al.* 2005). For example, high [SO_4] but gypsum-undersaturated groundwaters from the confined zone of the Birmingham (Fig. 2) aquifer are interpreted by Jackson & Lloyd (1983) and Hughes *et al.* (1999) to derive from dissolution of patchy gypsum cement. Edmunds *et al.* (1982), Smedley & Brewerton (1997), Shand *et al.* (1997) and Edmunds & Smedley (2000) also deduce dissolution sources for Li, Cs, K and Sr in the East Midlands and Cumbrian aquifers (Fig. 2): with distance down flow gradient, the concentrations increase, again suggesting a source limitation.

In principle, for a particular scale of measurement, such apparently kinetic dissolution should be quantifiable, given adequate data on mineral distributions and their relationships with hydraulic properties. Alternatively, the observed chemical data could be used to investigate the latter. Considerable amounts of data exist on mineral distributions as a function of depth and sometimes of lithology (e.g. Jones *et al.* 1999), but such chemostratigraphic data appear not to have been examined for this purpose (and the role of potentially protective oxide coatings is

Table 5. *Main mineral components of the sandstones, and suggested main water–rock interactions*

Mineral	Main origin	Role
Quartz	Detrital/framework (authigenic)	Source of Si
K-Feldspar	Detrital/framework (authigenic)	Source of Si, trace determinands
Plagioclase	Detrital/framework	Source of Si, trace determinands
Lithics	Detrital/framework	Source of Si, trace. ?removal of O_2 in long term by FeII release
Mica	Detrital	?removal of O_2 in long term by Fe II release. Source of trace determinands
Clays	Detrital, authigenic	Sorption
Calcite	Authigenic (/detrital)	Source of Ca, CO_3, minor Sr, Mg. pH control. ?removal of O_2 in long term by Fe II release. Possibly some sorption
Dolomite	Authigenic (/detrital)	Source of Ca, Mg, CO_3, minor Sr. pH control. ?removal of O_2 in long term by Fe II release
Gypsum	Authigenic	Source of Ca, SO_4. Effect on pH via Ca-carbonate equilibria
Fe ox/hyd	Authigenic	pH; cation (/anion) sorption. Oxidant (?small %). ?catalysis of redox reactions. pH control in absence of carbonates
Mn oxides	Authigenic	Sorption of cations (anions). Oxidant. ?catalysis of redox reactions
Organics	Detrital	Sorption of organics
Pyrite	Authigenic	Source of Fe, SO_4. Reductant
Flourite	Authigenic	Source of F
Heavy	Detrital	Source of trace
Barite	Authigenic	Source of Ba

unclear). At a site adjacent to that described by Bouch *et al.* (2006), Mitchener (2003) could establish no correlations between mineral composition and permeability or porosity, with the exception, curiously, of a positive correlation between permeability and feldspar content. Except at interbasin scale, knowledge of mineral distributions in horizontal directions is less well documented. Clearly, this distribution issue is common to any groundwater reaction involving a solid phase.

As in many aquifers, especially deep ones, apparent equilibrium states in pumped water samples are affected by mixing within a well profile. Examples are shown in Figure 16: here groundwater is saturated throughout the depth profile with barite, but shallower waters have lower [Ba]/[SO_4] ratios; due to the 'algebraic' effect (Plummer 1975), the pumped water is oversaturated. In theory, with different compositions, undersaturation also can similarly occur. Such issues obscure the geochemical interpretation of pumped samples.

Complexation by dissolved organic matter may increase the apparent solubility (especially for those species where acid–base and/or reduction–oxidation are involved in the dissolution process): such effects appear not to have been studied to any great extent in the specific context of the sandstones, although dissolved organic carbon concentrations are low (see below). Colloidal transport can also increase apparent solubility. Even in pumped well samples, total colloid concentrations appear to be less than a few mg l^{-1} (*c.* 10^{11} particles l^{-1}) (Stagg *et al.* 1997, 1998; Stagg & Tellam 1998; Stagg 2000): most of the colloids are dominantly composed of silicates, with significant but secondary amounts of Fe- and, at shallow depths, organic-rich particles (Stagg *et al.* 1998*a*). Metal-rich particles have been identified in polluted urban groundwaters, but not at concentrations that approach water quality standards. However, the number of sites that have been examined is limited, and little is known of situations such as landfill leachate plume migration or recharge through peat deposits where much higher concentrations of particulate matter might be expected.

Reduction–oxidation reactions

The reduction capability of the aquifer, and redox sequences. Classical redox sequences (e.g. Champ *et al.* 1979) have been mapped within the sandstones at a regional scale: Edmunds

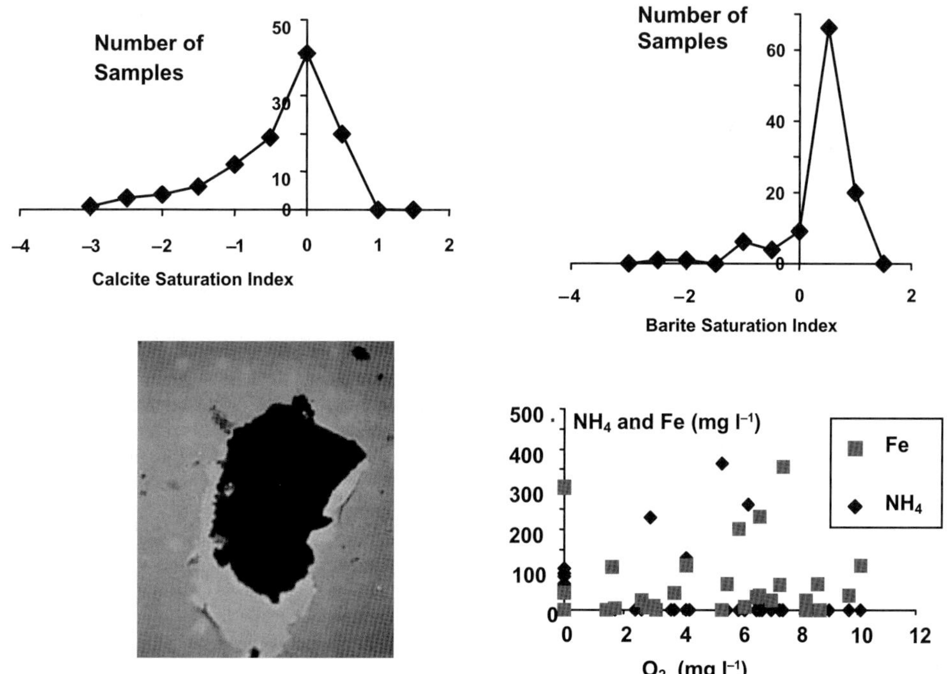

Fig. 16. Disequilibrium in pumped well samples from the Birmingham aquifer: calcite and barite saturation indices, and NH_4 and O_2 concentrations. Barite colloid particle (<1 μm) also shown.

et al. (1982, 1984), for example, describe the sequential removal of O_2, NO_3 and SO_4 with distance downflow in the East Midlands aquifer as the sandstone dips below and continues under the overlying Triassic mudstones. This pattern is also recognized in aquifers with more complex, Quaternary deposit-influenced, flow systems in northern England being particularly noticeable in the distributions of Fe II, NO_3 and O_2 (e.g. Parker *et al.* 1985; Tellam 1994). In general, detailed bacterial or H_2 studies have not yet been undertaken except in research on ?one contaminant plume (Thornton *et al.* 2001*b*).

Given that the sandstone is usually coated in ferric minerals, what are the electron donors? In some cases the overlying Quaternary deposits supply the reductants, usually in the form of organic matter or sulphides. Where Quaternary deposits are missing, some dissolved and particulate organic matter will enter with recharge. Buss *et al.* (2005) give a sandstone average dissolved organic carbon (DOC) concentration of approximately 2.8 mg l⁻¹, but two discrete depth studies on sites with limited Quaternary cover (Edmunds *et al.* 1992: 0–5 m depth; Stagg *et al.* 1998: 0–50 m) suggest that in this case DOC concentrations may be lower, falling rapidly with depth to less than *c.* 1 mg l⁻¹. Some of this decrease may result from O_2 reduction, but some may also result from reduction of Fe III and Mn (III) IV oxides and hydroxides (or possibly from filtration if some of the load is not truly dissolved) (e.g. Banwart 1999; Banerjee & Nesbitt 1999; Klewicki & Morgan 1999). Edmunds *et al.* (1982) suggest that a long-term (approximately several thousands of years: Smedley & Edmunds 2002) sink for O_2 may be Fe II release as ferroan carbonate is dissolved (and ferro-magnesian minerals, continuing the diagenetic process of red-bed production?). Petrographic and redox titration (Jones 2001) data indicate that small amounts of reactive reduced mineral phases, other than those mentioned above, do sometimes occur even within the non-bleached red sandstones.

Once O_2 has been removed Fe concentrations usually rise, although the mechanism again has not in general been quantified: in particular the electron donor is usually uncertain. Smedley & Edmunds (2002) suggest that for the East Midlands aquifer, where O_2 removal appears to have been predominantly from water–sandstone interaction over long periods, the rise in Fe II concentrations is modest because either the solubility of the bulk of the Fe oxides is limited, these being predominantly crystalline haematite, or because of lack of reductant. It is possible that high Fe concentrations suggest in general a source of reductant outside the sandstone: often this will be either from Quaternary deposits or from pollutant sources (e.g. Tellam 1996; Torstensson *et al.* 1998).

Another sink for dissolved Fe II is reduction of Mn oxides (Thornton *et al.* 2000*b*). Mn oxides are likely also to play a significant role in a number of redox transformation processes involving metals (e.g. Cr) and possibly N compounds (e.g. Guha *et al.* 2001), although studies specific to the sandstones appear as yet to be lacking.

Nitrate reduction (*sensu lato*) is suspected to occur in many of the sandstone aquifers, but evidence is often equivocal. However, in some cases denitrification has been positively identified using N_2/Ar and/or isotope data and/or bacterial data (Wilson *et al.* 1994, in only two of 23 samples; Parker *et al.* 1988; Spence *et al.* 2001*a, b*; Williams *et al.* 2001; Fukada *et al.* 2004; Buss *et al.* 2005). In some cases – especially where the nitrate source is also organic-rich – there will be enough organic matter to facilitate NO_3 reduction, but elsewhere there may be other, as yet uninvestigated, processes occurring involving Fe II or Mn, for example (e.g. Staub *et al.* 1996; Luther *et al.* 1997). Records of human pathogenic bacteria at depth (e.g. Barrett *et al.* 1999; Cronin *et al.* 2003; Taylor *et al.* 2004), the presence of NO_3 and SO_4 reducers (e.g. Pickup *et al.* 2001), and the positive identification of NO_3 and SO_4 reduction (see above) suggest that the mobility of bacteria within the sandstones is not limiting. Sandstone median pore diameters are typically tens of microns (Bloomfield *et al.* 2001), although mudstone units may provide more substantial barriers: the dimensions of micro-redox-environments, if they exist as such, are unknown. Overall the capacity for NO_3 reduction is almost certainly limited (Smedley & Brewerton 1997; Buss *et al.* 2005).

Sulphate reduction has been detected in a number of aquifers (e.g. Edmunds *et al.* 1982; Tellam 1994), but sulphides are almost always at low concentrations, and even ³⁴S results are sometimes equivocal (Barker 1996). Sulphide sinks include aqueous Fe II–solid phase Fe III (cf. Mayer *et al.* 2001). Apart from cases of pollution (including by drilling fluid polymers: Barker 1996), there are few examples where SO_4 has been substantially removed within the aquifer by reduction, the obvious limitation again being the lack of organic matter. Where reduction is occurring, the mechanisms in general have not received detailed study.

Methane has been detected in a number of locations (Tellam 1994; Gooddy & Darling

2005). Possible allochthanous sources include overlying Quaternary peat and underlying coal-bearing sequences. The role of CH_4 in sulphate reduction is unknown.

From H_2 and other data from a detailed study of a large phenol plume in the West Midlands, Thornton *et al.* (2001*b*) concluded that under some circumstances, in their case high concentrations of phenol inhibiting biodegradation (Pickup *et al.* 2001), the classical redox sequence may not be followed: at their site, they suggest that phenol oxidation may be possible via NO_3, SO_4, Fe and Mn pathways simultaneously. Unfortunately, lack of mineralogical, thermodynamic and ATP threshold data prevented fuller quantitative analysis, but a method was suggested. This situation is unlikely to be unique to organic contaminants. Such analysis can only be attempted in cases where detailed small-scale sampling has been undertaken: most chemical data available are from pumping wells, and as such mixing will always reduce the resolution, often producing disequilibrium effects not actually present in the groundwater (Fig. 16).

The oxidation capability of the aquifer. The presence of a few per cent Fe III in the sandstones suggests considerable oxidation capacity. However, as the previous section implies, this apparent potential capacity is often not fully realized (see also Thornton *et al.* 1995, 2000*a*, *b*, 2005). In the detailed field and laboratory investigation of the phenol plume mentioned above, mass balance and microcosm studies strongly suggested that only a very small amount, less than 0.15%, of the apparent Fe III and Mn IV capacity of the aquifer was used in oxidizing the organic load (Harrison *et al.* 2001; Thornton *et al.* 2001*a*, *b*). Smedley & Edmunds (2002) report oxalate-extractable Fe, i.e. an indication of 'amorphous' Fe, of <500 mg kg^{-1}, which is less than 2% total Fe. In the case of the phenol plume, over 90% of the observed degradation used dissolved O_2 and NO_3 supplied by dispersive processes in the 2 m-wide dispersion zones at the edges of the plume (Lerner *et al.* 2000).

Whether this oxidizing capacity is typical is uncertain, and capacity will vary with reaction system. In some systems, oxidative capacity may be underutilized because of other limitations. Of the two, MnO_2 is more labile than Fe III oxides/hydroxides in the aquifer. Accessible MnO_2 is finite, perhaps 50–100 mg kg^{-1} (not much less than oxalate-extractable concentrations determined by Smedley & Edmunds 2002) (Thornton *et al.* 1995; Kinniburgh *et al.* 2006),

but may be greater near the surface where authigenesis in the unsaturated zone is commonly observed.

When oxides are dissolved, their sorbed loads are also released into solution as seen, for example, by Thornton *et al.* (2000*b*) in the laboratory and Lewin *et al.* (1994) in the field: there is no information as to whether the oxides most prone to reductive dissolution are also those that have the greatest sorption capacity.

Fossil evidence. 'Fossil flow features' also occur in the sandstone sequence where reducing fluids have passed through removing or reducing the iron oxides/hydroxides and bleached the rock (e.g. Rowe & Burley 1997; cf. Parry *et al.* 2004), turning it from red- to grey- or buff-coloured. The distribution of bleaching often can be seen to be influenced by changes in permeability associated with lamination, mudstones, granulation seams or slip surfaces (Fig. 17). Sometimes bleaching can be traced laterally for over 100 m, often running parallel with bedding, despite being only a few centimetres in thickness, suggesting large volumes of flow and limited transverse dispersion. In some locations, the bleaching affects almost all the rock, and in other places complex patches suggest tortuous 3D flow paths. The presence of occasional oil seeps in the sandstones suggests that even flushing, albeit natural, over millions of years does not necessarily remove all 'pollution' in all cases.

Acid–base reactions

If it is present, dissolution of carbonate will usually buffer the pH to around neutral in recently recharged groundwaters. Incongruent dissolution, especially of dolomite, can subsequently increase the pH (and ^{13}C) considerably (e.g. Edmunds *et al.* 1982; Bath *et al.* 1987), as can processes that rejuvenate carbonate dissolution, for example Ca removal by ion exchange (e.g. Tellam 1994). In the case of polluted recharge water of low pH, buffering by the aquifer is, of course, dependent on the composition of the intruding solution and its existing carbonate saturation state (e.g. Thornton *et al.* 2000*b*, 2001*a*).

In many of the outcrop areas, carbonates have been removed from the rock by previous dissolution, and pHs of less than 5 are common (e.g. Greenwood & Travis 1915; Edmunds & Kinniburgh 1986; Edmunds *et al.* 1992; Moss & Edmunds 1992; Tellam 1995*b*; Shand *et al.* 1997). In other cases, certain formations are apparently carbonate-free, or nearly so (Edmunds &

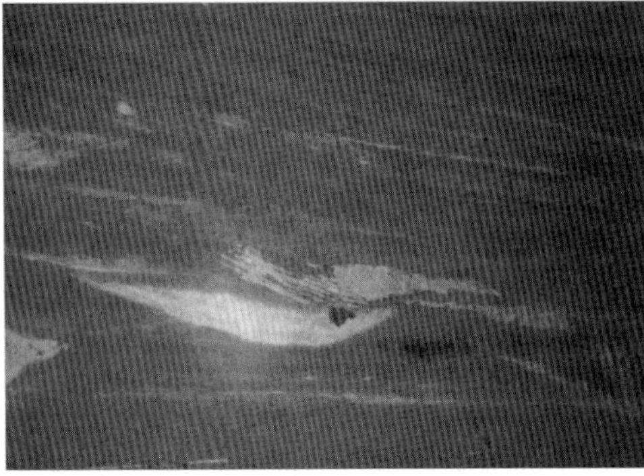

Fig. 17. Three examples of bleached zone outcrops, NW England, showing respectively: effects of faulting; flow controlled by cross-lamination; and 3D sinuous pathways.

Morgan-Jones 1976; Walton 1981). In these systems, there is often still a considerable buffering capacity due to sorption–exchange and dissolution–precipitation processes. The processes are as often suggested for acid soils (e.g. McBride 1994), including Al^{3+} hydrolysis, base cation/H^+/Al^{3+}/Fe/Mn exchange, and dissolution (Moss & Edmunds 1992; Thornton *et al.* 2000*b*). If the intruding water is reducing (e.g. a landfill leachate, as for Thornton *et al.* 2000*a*, *b*), further complications arise through reductive dissolution, even if limited, of the oxides. In addition, if the pH is being buffered at a low value, silicate dissolution will occur to a significant extent; although not as quick as most sorption reactions, saturation with simpler silicates appears to occur within hydrogeologically small travel distances (e.g. Haines & Lloyd 1985; Edmunds *et al.* 1992). Buss (1999) has presented a quantitative model of the interaction of H^+ with samples of the sandstone over short timescales, but, in common with similar aquifers elsewhere, a generic useable quantitative description of the complex array of processes involved has yet to be developed.

Sorption–desorption reactions

Cation-exchange capacities (CECs) for the sandstone are often in the range 1–20 meq/100 g dry mass (Gillespie *et al.* 2001; Mitchener 2003; Carlyle *et al.* 2004). Although most frequently at the lower end of this range, and therefore modest, there is abundant laboratory (e.g. El-Ghonemy 1997, 1998; Thornton *et al.* 2000*b*) and field evidence (e.g. Tellam & Lloyd 1986; Lucey 1987; Tellam 1994) that exchange can modify concentrations significantly. The main exchange phase appears to be provided by clay minerals, but Mitchener (2003) found that for the sequence described by Bouch *et al.* (2006), at pHs of 6.5–7.5, approximately 50% was supplied by oxides, in particular Mn oxides. Mitchener (2003) also found a complex relationship between permeability, porosity, CEC and MnO_2, implying that when MnO_2 is present it increases the CEC significantly which then decreases the significance of the (negative) permeability/CEC correlation: however, a (negative) porosity/CEC correlation remains significant whatever the MnO_2 content.

Taylor & Barker (2006) show that cation-exchange capacity is correlated in the sandstones with matrix conduction, opening up a possible approach for assessing CEC. Scott & Barker (2005) found that surface area is an important influence on the quadrature conductivity in spectral induced polarization measurements.

At a regional scale, in common with many sand aquifers, there is evidence that the Gaines–Thomas convention provides an acceptable approximation for predicting major cation-exchange reactions (Carlyle *et al.* 2004): estuary water intrusion over a period of about 40 years was predicted successfully using laboratory determined exchange parameters. However, further detailed laboratory investigations suggest that an even better description may be afforded by the Rothmund–Kornfeld power function relationship (El-Ghonemy 1997, 1998; Tellam *et al.* 2002), an empirical approach which appears to produce constants independent of sorbed-site composition. Parker *et al.* (2000) and Parker (2005) show that exchange on reference samples of the dominant clays present in the sandstone, and mixtures thereof, can also be described well using the power function approach: similar consideration of the other possible exchange phases has yet to be undertaken. Preliminary work suggests that linear up-scaling may be appropriate in cases where the range of property variation is modest (Parker 2005). Furlong (2002) suggests that ion exchange may allow groundwater age (up to about 15 pore volumes) to be estimated in certain circumstances in the sandstones using such quantification.

The laboratory experiments of Thornton *et al.* (1995, 2000*b*, 2005) suggest that NH_4^+ exchange behaves very similarly, as might be expected, to K^+ exchange, and this seems to be the case in the field too (Lewin *et al.* 1994). Usually a K_d approach is applied when predicting NH_4 migration (Erskine 2000; Gillespie *et al.* 2000; Jones 2001; Adey 2004; Buss *et al.* 2004), and the experiments of Thornton *et al.* (2000*b*) show that breakthroughs for NH_4^+ (and K^+) can indeed be described by retardation factors: geochemical modelling work generally supports this conclusion, but also suggests that the appropriate K_d value is very difficult to predict/measure without extensive knowledge of the system (Clarke 2005).

Sorption of metals/metalloids to the sandstone has not been investigated systematically. The results of some empirical studies are available (e.g. Mimides & Lloyd 1987; Al-Hosni 2001; Thornton *et al.* 2000*b*), but few attempts using recent advances in surface modelling appear to have been published (Read *et al.* 1993). Field studies usually conclude that pH-dependent sorption on the oxide fraction is particularly important (e.g. Spears 1986; Edmunds *et al.* 1989; Ford *et al.* 1992; Ford & Tellam 1994; Smedley & Edmunds 2002; Kinniburgh *et al.* 2006; Shepherd *et al.* 2006). Much has yet to be learnt of the sorption behaviour of the oxide/hydroxide fractions, and how they correspond with those fractions also actively involved with redox reactions. Where MnO_2 reductive dissolution occurs, sorption capacity will be decreased at the same time, but does this also occur with some of the Fe III oxides? The sorption capacity is far from dominating: even natural metal concentrations can exceed water quality standards (Smedley & Edmunds 2002; Kinniburgh *et al.* 2006), and

occasionally pollutant concentrations rise to tens or, even, hundreds of mg l^{-1} (Ford & Tellam 1994).

Again, little is known of anion sorption. Limited work has been done on F^- (Gresswell 2005), but very little apparently on phosphate: neither is commonly at high concentration in the sandstone groundwaters, although there is evidence that F^- distributions are pH-dependent as might be expected.

Sorption of organic pollutants to the sandstone has been the subject of a few studies, including those of Williamson (1994), Barrett (1995) and Shepherd (2003). The fraction of inorganic carbon within the sandstones is limited, being between 0.001 and 0.15% (Williamson 1994; Barrett 1995; and Thornton *et al.* 2000*a*; Steventon-Barnes 2001; Shepherd 2003). The role of organic coatings, if they exist, on the rock surface is unknown, as is the distribution of the detrital organic matter. Mitchener (2003) could find no significant relationship between f_{oc} (fraction of organic carbon) and permeability or porosity for a core from the Wildmoor Sandstone Formation (Bouch *et al.* 2006). Farris (1999) found for sandstones from the Midlands and the Central North Sea that organic matter was correlated with the finer deposits, and completely missing from more than 90% of the sequence. Interestingly, as was found for MnO_2 contents as mentioned above, there is a correlation between organic matter content and colour.

Sorption of chlorinated aliphatics, in particular tetrachloroethene, has been investigated by Shepherd (2003). He found approximately linear isotherms with slopes that vary with lithology; for the more hydrophobic compounds, he concluded that sorption to inorganic phases was not important, despite the low fractions of organic carbon in the sandstones. Thornton *et al.* (2000*a*), who investigated migration in landfill leachates, found less partitioning to the sandstone than Shepherd (2003). Barrett (1995) found that co-solutes increased the sorption of xylene.

Synthesis

The following attempted synthesis is broad and based directly on the evidence summarized above: it is hence biased towards the shallow English succession, for which most evidence is available. Table 6 provides more detail. Scale is stressed. However, it should be noted that the scales of heterogeneity, measurement and interest are not necessarily independent of each other: for example, measurement devices often

Table 6. *Summary of evidence and conclusions (?, limited evidence; ??, very limited evidence; α_L, longitudinal dispersivity; α_T, transverse dispersivity; ϕ_k/ϕ_Σ, = ratio of kinematic porosity to total porosity; BTCs = breakthrough curves)*

Feature	Scale <30 m	30–few 100s of metres	Few 100s of metres–few km	More than a few km
Breakthrough	Multimodal and/or highly dispersive	Usually unimodal?	Unimodal?	Unimodal??
Lateral fracture pathways	Often obvious, becoming less so with greater distance	Rare	None??	None?
Antifractures	Least likely to encounter, but, when present, can be dominant	More likely to encounter	May encounter	Likely to encounter, at least in some regions
Vertical penetration	Small amount of rapid penetration from ground surface	10s of metres?	Many 10s of metres?	Penetration to at least 250 m in places
Matrix barriers	Major effect, but does not completely stop solute migration	Lesser effect	Lesser effect	Overall seen as anisotropy
α_L	Large given the scale (but not a good measure given multimodal BTCs)	Moderate given scale?	Moderate given scale?	Moderate given scale??
α_T	High overall, but limited at edges?	Can be high overall, but limited at plume edges?	??	Can result in mixing over many 10s of metres?
ϕ_k/ϕ_Σ	Lowest	→ 1?	→ 1?	→ 1??
Lithofacies /lithofacies association importance	'Microcycles' significant for flow and chemistry	'Microcycles' becoming less significant	Microscale much less significant; 'mio/macro'-sequences increasingly significant	'Microcycles' insignificant; macroscale potentially significant
Chemical controls	Greater chance of low pH, high O_2 if small lateral distances are associated with shallow depths			Greater chance of high pH, anaerobic conditions
Predictability	Low	Moderate	Moderate?	Moderate?

are different when monitoring at the regional scale than when monitoring at the site scale, and the evidence summarized in the preceding sections is not directly comparable at all scales of interest.

The sandstone sequence is layered at scales from 1 mm to over 100 m, and this layering often shows some cyclicity. With some notable exceptions, these nested cyclic structures have not been mapped.

Over small distances, when flow is parallel with layering, solute transport is multipermeability in character, with diffusive exchange between higher and lower permeability layers (Fig. 11 for small-scale example; Fig. 13). Flow perpendicular to layers does not experience this

effect to the same degree, thus resulting in an anisotropic dispersive behaviour. The lower permeability units will, however, tend to encourage layer-parallel flow: if flow passes around the lower permeability layers, dispersion will increase further. Anisotropy will also be present within the plane of the bedform (i.e. in the approximately horizontal plane) as a result of cross-stratification (Fig. 3). Given that the rock fabrics are present on various scales, there are likely to be several minimum representative volumes (MRVs) for hydraulic properties for matrix flow, and these will be directionally-dependent: values have yet to be established. Analyses of idealized versions of such systems have been available for a long time (e.g.

Mercado 1967; Gelhar *et al.* 1979; Dagan 1989): it is found that dispersivity eventually approaches a constant value dependent on the statistical properties of the sequence, but up to this point non-Fickian behaviour is seen. There is little evidence of systematic change in matrix permeability with depth in most aquifers, although dissolution would be expected to increase permeability near the surface in originally carbonate-containing units.

The layering will often result in multimodal breakthrough up to distances equal to the lateral continuity of the layer (Fig. 12). As the lateral distance travelled increases beyond this, the effects of layering become less obvious as the migrating solution encounters repeated changes in lithofacies or cementation. The distance over which this happens is uncertain, but possibly up to the order of tens of metres (30 m in Table 6). Given the log-normal permeability distributions, unusually high-permeability pathways may occasionally connect up, forming fast solute-conducting pathways (cf. Gutmanis *et al.* 1998).

Superimposed on the lithological variations are structural elements – discontinuities (Figs 4, 7 & 8; Table 3). A small proportion (<10%?) will be of high permeability as a result of permeable fill and/or wide apertures. Exponential fracture length distributions mean that over short distances fracture flow is more common than over longer distances (Fig. 7c). However, when present, permeable long fractures (Table 3) will have a disproportionate effect on flows. Because of their size and frequency, bedding-plane fractures can impart a strong anisotropy to the fracture network system; they are also important in connecting fracture systems vertically. If the rare long subvertical factures are ignored, the permeability MRV of the fracture network is of the order of a few tens of metres in each direction. Rapid, fracture-flow tracer breakthrough can occur over at least 100 m, but as distance increases the chances of this occurring diminish rapidly. At the kilometre scale, breakthrough is effectively single-modal as far as can be discerned (Fig. 15). A relationship between lithology and fracture properties is suspected but not proved. Fractures reduce in frequency with depth, especially below 100–200 m.

Discontinuities can also be of lower permeability than the host sandstone. Such 'antifractures' include faults/granulation seams and veins (Figs 4 & 8). Some faults have enhanced permeability parallel to strike, and reduced permeability in directions perpendicular to strike. When intersecting mudstones, antifractures may form compartments within the sandstone where flow is restricted (Fig. 17). Fault structures will often have a predominant regional alignment, and compartments may sometimes not be completely closed, or they may be partly closed to flows in one direction only. Compartment size will vary from semiregional to subhand specimen size. Often fractures cut through granulation seams, and some degree of funnelling therefore will occur (Fig. 7a). However, kinematic porosity for large-scale flows appears to be similar to total porosity.

Relationships between the distributions of fractures, antifractures, matrix permeability and lithofacies are poorly known. However, it is likely that such relationships exist (cf. Fig. 1). The presence of each of these features will modify the effects of the others.

Groundwater pH is often buffered at neutral to slightly alkaline when carbonates are present, but at 4–6 when absent (Table 5). Highest pH values are normally found when significant incongruent dissolution or ion exchange has occurred, and can be up to at least 9. Carbonates are absent in some formations, and are often removed from the shallower parts of other formations. The reducing capacity of the aquifer is limited, but over extended periods dissolved oxygen is completely removed, possibly in part through Fe II release during incongruent carbonate and silicate dissolution. The oxidizing capacity is also limited. Mn oxides are relatively redox-reactive, but are present in finite amounts. In contrast, Fe oxides are abundant but only a very small proportion is bioavailable in the short term. Sorption capacity is moderate, with cation-exchange capacities of 1–20 meq/100 g dry wt, mostly supplied by clays and Mn oxides, the latter being vulnerable to reductive dissolution. pH-dependent sorption on Fe oxides is also important for metal and some anion sorption. f_{oc} values are typically <0.1%, but nevertheless appear to control the sorption of at least the more hydrophobic organic pollutants. In regions of where palaeoflows have been limited, some soluble minerals (e.g. gypsum) are still present. Distributions of reactive mineral components within the sandstones, and especially in relation to the hydraulic properties, are poorly known. Limited evidence suggests that the organic carbon is more associated with the finer grained beds, and that the CEC is inversely related to the matrix permeability, albeit weakly and only when Mn oxides contents are low. Chemical properties of the fracture fills are unknown, except in a few cases, and thus the effects on solute compositions of flow across or along filled fractures are unknown. Up-scaling procedures for using laboratory geochemical parameter determinations for field prediction cannot be justified without collection and analysis of more mineralogical and hydraulic data on the same sequences: there is an important role here for the use of borehole geophysics in hydraulic and cation-exchange capacity determination at least.

How much of the often fairly limited natural attenuation capacity of the sandstones is available to a migrating solute plume and how readily it is retained during flushing depends on the flow system, the layering and the presence of potential flow barriers (mudstones, antifractures) (Figs 4 & 5). Examination of bleached zones and consideration of the results of numerical modelling suggests that pockets of sandstone are sometimes left unaffected (Fig. 17). However, flow around compartments may actually increase the attenuating capacity by extending flow paths and dividing plumes, thus increasing transverse dispersion. The tortuosity of the matrix and the macroscopic fracture network is limited (actual/straight-line path length is <2).

Concluding comments

The UK Permo-Triassic sandstone is a continental red-bed sequence unremarkable in many ways in a worldwide context, displaying hydrogeological characteristics which are seen in many other fractured permeable-matrix aquifers.

As expected, on a local scale solute movement rates are extremely variable, affected as they are by matrix, fracture and geochemical heterogeneity. Breakthroughs in tracer tests are generally multimodal, with extreme cases ranging from almost karst-like to no breakthrough at all depending on local fracture and 'antifracture' geometry. Precise prediction is difficult. Low tracer test recoveries in general suggest remediation may often be difficult to complete.

At the subregional scale of perhaps a few kilometres, breakthrough appears to be much less complex because the scale of measurement is much greater than the minimum representative volumes for both the hydraulic and chemical systems. At the regional scale, groundwater age distributions can be complicated due to the effects of varying and sometimes geometrically intricate boundary conditions, and chemical variations are further obscured by the usual practice of using pumping wells for monitoring purposes. Nevertheless, there is structure in regional distributions of pumped well-sample chemistry, implying that quantitative analysis of apparent solute movement at this scale is possible in some sense.

The point at which 'local' becomes 'subregional' is uncertain, and will be fuzzy in both the sense of being different at different locations and being gradual at any one location. However, it has been argued that a possible transition range may be 30–100 m or so. This is an important scale range for many point-source pollution problems, including remediation. To improve prediction in this distance range especially, it will be necessary to develop appropriate stochastic descriptions. As is often attempted in oil reservoir engineering contexts, these should be conditioned by the probable lithofacies-controlled structure of the matrix permeability distribution: lithofacies is also suspected to affect fracture/antifracture development and geochemical distributions. The local- to regional-scale transition is very much determined by the measurement device used, and direct modelling of measurement devices may be needed.

To confirm and subsequently use these suspected relationships between geological properties and the required hydraulic/chemical properties, it will be necessary to improve the characterization procedures significantly, despite the wealth of data already available. Much could be achieved from adapting techniques that are more commonly used in other geological disciplines, and in particular petroleum geology, especially in geological characterization and modelling, as advocated strongly by de Marsily *et al.* (2005) in a wider context. Existing, largely geophysical, methods could be used to retrieve geological data using the existing extensive borehole network, but new developments in, for example, resistivity/spectral IP and magnetic resonance sounding could enable hydraulic and even some geochemical properties to be evaluated more directly. Potentially, geophysics has a very significant role to play, but, in addition, direct rock characterization of cores and outcrop is also needed.

Perhaps the main fundamental uncertainties in the geochemical behaviour of the sandstone are: (i) the role of the Fe and Mn oxides and hydroxides in both redox and sorption reactions; and (ii) the distributions of the geochemical components. Distributions may be related to lithofacies, and over short distances the effects of correlations between hydraulic and geochemical parameters may give rise to further (interesting) complexity.

This article is in part a review, drawing heavily on the work of many people. We would like to acknowledge the input of all, whether they are cited or not. We would also like to thank funders with whom we have been involved, especially the UK Natural Environment and Engineering and Physical Sciences Research Councils, and the Environment Agency. We are very grateful to M. Riley, P. Turner and R. Mackay for commenting on parts of the drafts, and to P. Turner for permission to use his photographs.

References

ADEY, J. 2004. *Development of the Partition Coefficient (K_d) Test Method for Use in Environmental Risk Assessments.* (UK) Environment Agency R&D Technical Report, **P1–500/4/TR**.

AIGNER, T., HEINZ, J., HORNUNG, J. & ASPRION, U. 1999. A hierarchical process-approach to reservoir heterogeneity: Examples from outcrop analogues. *Bulletin du Centre de Recherches Elf Exploration Production*, **22**, 1–11.

AIGNER, T., SCHAUER, M., JUNGHANS, W.D. & REINHARDT, L. 1995. Outcrop gamma-ray logging and its applications: Examples from the German Triassic. *Sedimentary Geology*, **100**, 47–61.

AKHURST, M.C., BALL, D.F. *ET AL.* 2006. Towards understanding the Dumfries Basin aquifer, SW Scotland. *In*: BARKER, R.D. & TELLAM, J.H. (eds) *Fluid Flow and Solute Movement in Sandstones: The Offshore UK Permo-Triassic Red Bed Sequence.* Geological Society, London, Special Publications, **263**, 187–198.

ALEXANDER, J., BLACK, J.H. & BRIGHTMAN, M.A. 1987. The role of low-permeability rocks in regional flow. *In*: GOFF, J.C. & WILLIAMS, B.J.P. (eds) *Fluid Flow in Sedimentary Basins and Aquifers.* Geological Society, London, Special Publications, **34**, 173–183.

AL-HOSNI, T. 2001. *Attenuation of cadmium by UK Triassic sandstone aquifer materials.* MSc Project Report, Earth Sciences, University of Birmingham.

ALLEN, D.J., BREWERTON, L.J. *ET AL.* 1997. *The Physical Properties of Major Aquifers in England and Wales.* British Geological Survey Technical Report, **WD/97/34**. Environment Agency R&D Publication, **8**.

ALLEN, D.J., BLOOMFIELD, J.P., GIBBS, B.R. & WAGSTAFF, S.J. 1998. *Fracturing and the Hydrogeology of the Permo-Triassic Sandstones in England and Wales.* British Geological Survey Technical Report WD/98/1, and Environment Agency R&D Technical Report W164.

ALLEN-KING, R.M., HALKET, R.M., GAYLORD, D.R. & ROBIN, M.J.L. 1998. Characterizing the heterogeneity and correlation of perchloroethene sorption and hydraulic conductivity using a facies-based approach. *Water Resources Research*, **34**, 385–396.

ANDREWS, J.N. & KAY, R.L.F. 1983. The U contents and 234U/238U activity ratios of dissolved uranium in groundwaters from some Triassic sandstones in England. *Isotope Geoscience*, **1**, 101–117.

ANDREWS, J.N. & LEE, D.J. 1979. Inert gases in groundwater from the Bunter Sandstone of England as indicators of age and palaeoclimatic trends. *Journal of Hydrology*, **41**, 233–252.

ANDREWS, J.N., EDMUNDS, W.M., SMEDLEY, P.L., FONTES, J.-C.H., FIELD, L.K. & ALLAN, G.L. 1994. Chlorine-36 in groundwater as a palaeoclimatic indicator: the East Midlands Triassic aquifer (UK). *Earth and Planetary Science Letters*, **122**, 159–172.

ANTONELLINI, M. & AYDIN, A. 1994. Effect of faulting on fluid flow in porous sandstones. *AAPG Bulletin*, **78**, 355–377.

ANTONELLINI, M., AYDIN, A. & POLLARD, D.D. 1994. Microstructure of deformation bands in porous sandstones at Arches National Park, Utah. *Journal of Structural Geology*, **16**, 941–959.

ARCHIE, G.E. 1942. The electrical resistivity log as an aid in determining some reservoir characteristics. *Transactions of the American Institute of Mining and Metallurgical Engineers*, **146**, 54–62.

ARCHIE, G.E. 1950. Introduction to petrophysics of reservoir rocks. *AAPG Bulletin*, **34**, 943–961.

ATTEIA, O., ANDRE, L., DUPUY., A. & FRANCESCHI, M. 2005. Contributions of diffusion, dissolution, ion exchange, and leakage from low-permeability layers to confined aquifers. *Water Resources Research*, **41**, WO9412, doi: 10.1029/2003WR002593/.

AUDLEY-CHARLES, M.G. 1970. Stratigraphical correlation of the Triassic rocks of the British Isles. *Quarterly Journal of the Geological Society, London*, **126**, 19–47.

AYDIN, A. 1978. Small faults formed as deformation bands in sandstones. *Pure and Applied Geophysics*, **116**, 913–930.

AYDIN, A. & JOHNSON, A.M. 1978. Development of faults as zones of deformation bands and as slip surfaces in sandstone. *Pure and Applied Geophysics*, **116**, 931–942.

AYDIN, A. & JOHNSON, A.M. 1983. Analysis of faulting in porous sandstones. *Journal of Structural Geology*, **5**, 19–31.

BALLARD, M. 2000. *The role of granulation seams in pollutant transport in the U.K. Triassic Sandstone.* MSc Project Report, University of Birmingham.

BANERJEE, D. & NESBITT, H.W. 1999. XPS study of reductive dissolution of birnessite by oxalate: Rates and mechanistic aspects of dissolution and redox processes. *Geochimica et Cosmochimica Acta*, **63**, 3025–3038.

BANWART, S.A. 1999. Reduction of iron (III) minerals by natural organic matter in groundwater. *Geochimica et Cosmochimica Acta*, **63**, 2919–2928.

BARKER, A.P. 1996. *Isotopic studies of groundwater.* PhD thesis, University of Leeds.

BARKER, A.P., NEWTON, R.J., BOTTRELL, S.H. & TELLAM, J.H. 1998. Processes affecting groundwater chemistry in a zone of saline intrusion into an urban aquifer. *Applied Geochemistry*, **6**, 735–750.

BARKER, J.A. 1988. A generalized flow model for hydrualic tests in fractured rock. *Water Resources Research*, **24**, 1796–1804.

BARKER, R.D. 1974. Interpretation of induced-polarization sounding curves in the frequency domain. *Geophysical Prospecting*, **22**, 610–626.

BARKER, R.D. 1990. Investigation of groundwater salinity by geophysical methods. *In*: WARD, S.H. (eds) *Geotechnical and Environmental Geophysics. Volume 2: Environmental and Groundwater*. Society of Exploration Geophysics, Houston, Texas, USA, 201–212.

BARKER, R.D. & MOORE, J. 1998. The application of time-lapse electrical tomography in groundwater studies. *The Leading Edge*, **17**, 1454–1458.

BARKER, R.D. & WORTHINGTON, P.F. 1973*a*. The hydrological and electrical anisotropy of the Bunter Sandstone of Northwest Lancashire. *Quarterly Journal of Engineering Geology and Hydrogeology*, **6**, 169–175.

BARKER, R.D. & WORTHINGTON, P.F. 1973*b*. Some hydrogeophysical properties of the Bunter Sandstone of northwest England. *Geoexploration*, **11**, 151–170.

BARNES, R.P., DEE, S.J., SANDERSON, D.J. & BOWDEN, R.A. 1998. Interpretation of structural domains in discontinuity data from Nirex deep boreholes at Sellafield. *Proceedings of the Yorkshire Geological Society*, **52**, 177–187.

BARRETT, M.H. 1995. *The transport of mixtures of organic contaminants in groundwater*. PhD thesis, University of Birmingham, UK.

BARRETT, M.H., HISCOCK, K.M., PEDLEY, S.J., LERNER, D.N., TELLAM, J.H. & FRENCH, M.J. 1999. Marker species for identifying urban groundwater recharge sources; the Nottingham case study. *Water Research*, **33**, 3083–3097.

BASHAR, K. 1997. *Developing a conceptual model of intergranular conservative solute transport processes for water flow through laboratory scale samples of the U.K. Triassic Sandstones*. PhD thesis, University of Birmingham, UK.

BASHAR, K. & TELLAM, J.H. 2006. Non-reactive solute movement through saturated laboratory samples of undisturbed stratified sandstone. *In*: BARKER, R.D. & TELLAM, J.H. (eds) *Fluid Flow and Solute Movement in Sandstones: The Offshore UK Permo-Triassic Red Bed Sequence*. Geological Society, London, Special Publications, **263**, 233–251.

BATH, A.H., EDMUNDS, W.M. & ANDREWS, J.N. 1979. Palaeoclimatic trends deduced from the hydrochemistry of a Triassic sandstone aquifer. *International Symposium on Isotope Hydrology, Vienna*, **II, IAEA-SM-228/27**, 545–568.

BATH, A.H., McCARTNEY, R.A., RICHARDS, H.G., METCALF, R. & CRAWFORD, M.B. 1996. Groundwater chemistry in the Sellafield area: a preliminary interpretation. *Quarterly Journal of Engineering Geology and Hydrogeology*, **29**, Supplement 1, S39–S58.

BATH, A.H., MILOWDOWSKI, A.E. & STRONG, G.E. 1987. Fluid flow and diagenesis in the East Midlands Triassic sandstone aquifer. *In*: GOFF, J.C.W. & WILLIAMS, B.J.P. (eds) *Fluid Flow in Sedimentary Basins and Aquifers*. Geological Society, London, Special Publications, **34**, 127–140.

BENTON, M.J., COOK, E. & TURNER, P. 2002. *Permian and Triassic Red Beds and the Penarth Group of Great Britain*. Joint Nature Conservation Committee, Peterborough.

BIVER, P., HAAS, A. & BACQUET, C. 2002. Uncertainties in facies proportion estimation II: Application to geostatistical simulation of facies and assessment of volumetric uncertainties. *Mathematical Geology*, **34**, 703–714.

BLOOMFIELD, J., GOODDY, D.C., BRIGHT, M.I. & WILLIAMS, P.J. 2001. Pore-throat size distributions in Permo-Triassic sandstones from the United Kingdom and some implications for contaminant hydrogeology. *Hydrogeology Journal*, **9**, 219–230.

BLOOMFIELD, J., MOREAU, M.F. & NEWELL, A.J. 2006. Characterization of permeability distributions in six lithofacies from the Helsby and Wilmslow Sandstone formations of the Cheshire Basin, UK. *In*: BARKER, R.D. & TELLAM, J.H. (eds) *Fluid Flow and Solute Movement in Sandstones: The Offshore UK Permo-Triassic Red Bed Sequence*. Geological Society, London, Special Publications, **263**, 83–101.

BOTTRELL, S.H., WEST, L.J. & YOSHIDA, K. 2006. Combined isotopic and modelling approach to determine the source of saline groundwaters in the Selby Triassic Sandstone aquifer, UK. *In*: BARKER, R.D. & TELLAM, J.H. (eds) *Fluid Flow and Solute Movement in Sandstones: The Offshore UK Permo-Triassic Red Bed Sequence*. Geological Society, London, Special Publications, **263**, 325–338.

BOUCH, J.E., HOUGH, E., KEMP, S.J., McKERVEY, J.A., WILLIAMS, G.M. & GRESWELL, R.B. 2006. Sedimentary and diagenetic environments of the Wildmoor Sandstone Formation (UK): implications for groundwater and contaminant transport, and sand production. *In*: BARKER, R.D. & TELLAM, J.H. (eds) *Fluid Flow and Solute Movement in Sandstones: The Offshore UK Permo-Triassic Red Bed Sequence*. Geological Society, London, Special Publications, **263**, 129–153.

BRANEY, M.C., GILLING, D., JEFFERIES, N.L., LINEHAM, T.R. & STONE, N.S. 2001. *Laboratory Measurements of the Mass Transfer Properties of Samples of Permo-Triassic Sedimentary Strata From Cumbria*. UK Nirex Ltd, Report, **NSS/R278**.

BRASSINGTON, F. 1992. Measurements of head variations within observation boreholes and their implications for groundwater monitoring. *Journal of the Institute of Water and Environmental Management*, **6**, 91–100.

BRASSINGTON, F.C. & WALTHALL, S. 1985. Field techniques using borehole packers in hydrogeological investiations. *Quarterly Journal of Engineering Geology*, **18**, 181–193.

BRASSINGTON, F.C., LUCEY, P.A. & PEACOCK, A.J. 1992. The use of down-hole focussed electric logs to investigate saline groundwaters. *Quarterly Journal of Engineering Geology*, **25**, 343–349.

BRERETON, N.R. & SKINNER, A.C. 1974. Groundwater flow characteristics in the Triassic Sandstone in the Fylde area of Lancashire. *Water Services*, **78**, 275–279.

BROHOLM, M.M., JONES, I., TORSTENSSON, D. & ARVIN, E. 1998. Groundwater contamination from a coal carbonization plant. *In*: LERNER, D.N. & WALTON, N.R.G. (eds) *Contaminated Land and Groundwater: Future Directions*. Geological Society,

London, Engineering Geology, Special Publications, **14**, 159–165.

BRYANT, I., CARR, D., CIRILLI, P., DRINKWATER, N., MCCORMICK, D., TILKE, P. & THURMOND, J. 2000. Use of 3D digital analogues as templates in reservoir modelling. *Petroleum Geoscience*, **6**, 195–201.

BUNCH, M.A. 2000. *Spectral IP signatures of granulation seams in Triassic sandstone*. MSc Project Report, University of Birmingham.

BUNCH, M.A. 2006. *A model for simulating dryland alluvial fan development*. PhD thesis, University of Birmingham, UK.

BUNCH, M.A., MACKAY, R., TELLAM, J.H. & TURNER, P. 2004. A model for simulating the deposition of water-lain sediments in dryland environments. *Hydrology and Earth Systems Science*, **8**, 122–134.

BURLEY, S.D. 1984. Patterns of diagenesis in the Sherwood Sandstone Group (Triassic), United Kingdom. *Clay Minerals*, **19**, 403–440.

BUSS, S.R. 1999. *Attenuation of strong acids in the Birmingham sandstone aquifer*. PhD thesis, University of Birmingham.

BUSS, S.R., HERBERT, A.W., MORGAN, P., THORNTON, S.F. & SMITH, J.W.N. 2004. A review of ammonium attenuation in soil and groundwater. *Quarterly Journal of Engineering Geology and Hydrogeology*, **37**, 347–359.

BUSS, S.R., RIVETT, M.O., MORGAN, P. & BEMMENT, C.D. 2005. *Attenuation of Nitrate in the Sub-surface Environment*, (UK) Environment Agency Science Report SC030155/SR2.

BUTCHER, A., LAWRENCE, A., JACKSON, C., CULLIS, E., CUNNINGHAM, J., HASAN, K. & INGRAM, J. 2006. Investigating rising nitrate concentrations in groundwater in the Permo-Triassic aquifer, Eden Valley, Cumbria, UK. *In*: BARKER, R.D. & TELLAM, J.H. (eds) *Fluid Flow and Solute Movement in Sandstones: The Offshore UK Permo-Triassic Red Bed Sequence*. Geological Society, London, Special Publications, **263**, 285–296.

CAMPBELL, J.E. 1982. *Permeability characteristics of the Permo-Triassic sandstones of the Lower Mersey Basin*. MSc thesis, University of Birmingham.

CAMPBELL, J.E. 1986. *West Cheshire Saline Groundwater Investigation (Phase 3)*. North West Water Authority, Warrington, Cheshire, UK.

CARLYLE, H.F., TELLAM, J.H. & PARKER, K.E. 2004. The use of laboratory-determined ion exchange parameters in the prediction of field-scale major cation migration over a 40 year period. *Journal of Contaminant Hydrology*, **68**, 55–81.

CHADWICK, R.A. 1997. Fault analysis of the Cheshire Basin, NW England. *In*: MEADOWS, N.S., TRUEBLOOD, S.P., HARDMAN, M. & COWAN, G. (eds) *Petroleum Geology of the Irish Sea and Adjacent Areas*. Geological Society, London, Special Publications, **124**, 297–313.

CHAMP, D.R., GULENS, J. & JACKSON, R.E. 1979. Oxidation reduction sequences in groundwater flow systems. *Canadian Journal of Earth Sciences*, **16**, 12–23.

CLARKE, L.E. 2005. *Under what circumstances can the Kd approach (i.e. linear sorption) be used to predict attenuation of ammonium in aquifers?* MSc Project Report, University of Birmingham, School of Geography, Earth and Environmental Sciences.

CLEMMENSEN, L.B., ØXNEVAD, I.E.I. & DE BOER, P.L. 1994. Climatic controls on ancient desert sedimentation: some late Palaeozoic and Mesozoic examples from NW Europe and the Western Interior of the USA. *In*: DE BOER, P.L. & D.G.S. (eds) *Orbital Forcing and Cyclic Sequences*. International Association of Sedimentologists, Special Publications, **19**, 439–457.

COLEBY, L.M. 1996. *Heath House Tracer Test Data Report*. British Geological Survey Technical Report, **WD/96/74**.

COLLAR, F.A., 1974. A geophysical interpretation of the structure of the Vale of Clwyd, North Wales. *Geological Journal*, **9**, 65–76.

COLLAR, F.A. & GRIFFITHS, D.H. 1976. A laboratory study of the relationships between induced polarization, permeability and matrix electrical conductivity in Bunter Sandstones. *Quarterly Journal of Engineering Geology and Hydrogeology*, **9**, 57–63.

CRONIN, A., TAYLOR, R.G., POWELL, K.L., BARRETT, M.H., TROWSDALE, S.A. & LERNER, D.N. 2003. Temporal variations in the depth-specific hydrochemistry and sewage-related microbiology of an urban sandstone aquifer, Nottingham, United Kingdom. *Hydrogeology Journal*, **11**, 205–216.

DAGAN, G., 1989. *Flow and Transport in Porous Formations*. Springer, Berlin, 465.

DALLA, E., CASSIANI, G., BROVELLI, A. & PITEA, D., 2004. Electrical conductivity of unsaturated porous media: pore-scale model and comparison with laboratory data. *Geophysical Research Letters*, **31**, Article No. L05609.

DARLING, W.G., EDMUNDS, W.M. & SMEDLEY, P.L. 1997. Isotopic evidence for palaeowaters in the British Isles. *Applied Geochemistry*, **12**, 813–829.

DARLING, W.G., BATH, A.H. & TALBOT, J.C. 2003. The O and H stable isotopic composition of fresh waters in the British Isles: 2. Surface waters and groundwaters. *Hydrology and Earth Systems Science*, **7**, 183–195.

DAVIES, R.K. & HANDSCHY, J.W. 2003. Introduction to the *AAPG Bulletin* thematic issue on fault seals. *AAPG Bulletin*, **87**, 377–380.

DE MARSILY, G., DELAY F., GONCALVES, J., RENARD, P., TELES, V. & VIOLETTE, S. 2005. Dealing with spatial heterogeneity. *Hydrogeology Journal*, **13**, 161–183.

DIGGES LA TOUCHE, S.V. 1998. *Unsaturated flow in the Triassic Sandstones of the United Kingdom*. PhD thesis, University of Birmingham.

EDMUNDS, W.M. 1986. Geochemistry of geothermal waters in the U.K. *In*: DOWNING, R.A. & GRAY, D.A. (eds) *Geothermal Energy: The Potential in the United Kingdom*. HMSO, London, 111–122.

EDMUNDS, W.M. 1996. *Indicators of Rapid Environmental Change in the Groundwater Environment*. Balkema, Rotterdam, 135–150.

EDMUNDS, W.M. & KINNIBURGH, D.G. 1986. The susceptibility of UK groundwaters to acidic deposition. *Journal of Geological Society, London*, **143**, 707–720.

EDMUNDS, W.M. & MORGAN-JONES, M. 1976. Geochemistry of groundwaters in British Triassic Sandstones: the Wolverhampton–East Shropshire area. *Quarterly Journal of Engineering Geology and Hydrogeology*, **9**, 73–101.

EDMUNDS, W.M. & SMEDLEY, P.L. 2000. Residence Time Indicators in Groundwater: The East Midlands Triassic Sandstone Aquifer. *Applied Geochemistry*, **15**, 737–752.

EDMUNDS, W.M., BATH, A.H. & MILES, D.L. 1982. Hydrochemical evolution of the East Midlands Triassic sandstone aquifer, England. *Geochimica et Cosmochimica Acta*, **46**, 2069–2081.

EDMUNDS, W.M., BREWERTON, L.J., SHAND, P. & SMEDLEY, P.L. 1997. *The Natural (Baseline) Quality of Groundwater in England and Wales. Part 1: A Guide to the Natural (Baseline) Quality Study.* British Geological Survey Technical Report, **WD/97/54**.

EDMUNDS, W.M., BREWERTON, L.J. & SMEDLEY, P.L. 1996. *Natural Quality of Groundwaters in Aquifers in England and Wales: A Scoping Study.* British Geological Survey Technical Report, **WDI96146R**.

EDMUNDS, W.M., COOK, J.M., KINNIBURGH, D.G., MILES, D.L. & TRAFFORD, J.M. 1989. *Trace Element Occurrences in British Groundwaters.* British Geological Survey, Research Report, **SD/89/3**.

EDMUNDS, W.M., KINNIBURGH, D.G. & MOSS, P.D. 1992. Trace metals in interstitial waters from sandstones: acidic inputs to shallow groundwaters. *Environmental Pollution*, **77**, 129–141.

EDMUNDS, W.M., MILES, D.L. & COOK, J.M. 1984. A comparative study of sequential redox processes in three British aquifers. *In*: ERIKSSON, E. (ed.) *Hydrochemical Balances of Freshwater Systems.* IAHS Publication, **150**, 55–70.

EDWARDS, H.E., BECKER, A.D. & HOWELL, J.A. 1993. Compartmentalization of an aeolian sandstone by structural heterogeneities: Permo-Triassic Sandstone, Moray Firth, Scotland. *In*: NORTH, C.P. & PROSSER, D.J. (eds) *Characterization of Fluvial and Aeolian Reservoirs.* Geological Society, London, Special Publications, **73**, 339–365.

EL-GHONEMY, H. 1997. *Laboratory experiments for quantifying and describing cation exchange in UK Triassic Sandstones.* PhD thesis, University of Birmingham.

EL-GHONEMY, H. 1998. *Estimation of Solid Phase Activity Coefficients for Non-ideal Cation Exchange Behaviour in Laboratory Samples of UK Triassic Sandstones.* Geological Society, London, Engineering Geology Special Publications, **14**, 213–217.

ERSKINE, A. 2000. Transport of ammonium in aquifers: retardation and degradation. *Quarterly Journal of Engineering Geology and Hydrogeology*, **33**, 161–170.

EVERS, S. & LERNER, D.N. 1998. How uncertain is our estimate of a wellhead protection zone. *Ground Water*, **36**, 49–57.

FARRIS, M. 1999. *Palynomorph taphonomy in continental red-bed facies, UK West Midlands and Central North Sea.* PhD thesis, University of Keele, UK.

FISHER, Q.L. & KNIPE, R.J. 1998. Fault sealing processes in in siliclastic sediments. *In*: JONES, G. *ET AL.* (eds) *Faulting and Fault Sealing in Hydrocarbon Reservoirs.* Geological Society, London, Special Publications, **147**, 117–134.

FITCH, F.J., MILLER, J.A. & THOMPSON, D.B. 1966. The palaeogeographic significance of isotopic age determination on detrital micas from the Triassic of the Stockport–Macclesfield district, Cheshire. *Palaeogeography, Palaeoclimatology, Palaeoecology*, **2**, 281–312.

FLETCHER, S.W. 1989. *Groundwater Resource Development in the Region Between Wolverhampton and Bridgnorth.* Unpublished report, Severn Trent Water Authority, Birmingham, UK.

FORD, M. & TELLAM, J.H. 1994. Source, type of extent of inorganic contamination within the Birmingham urban aquifer system, UK. *Journal of Hydrology*, **156**, 101–135.

FORD, M., TELLAM, J.H. & HUGHES, M. 1992. Pollution-related acidification in the urban aquifer, Birmingham, UK. *Journal of Hydrology*, **140**, 297–312.

FOWLES, J. & BURLEY, S. 1994. Textural and permeability characteristics of faulted, high porosity sandstones. *Marine and Petroleum Geology*, **11**, 608–623.

FUKADA, T., HISCOCK, K.M. & DENNIS, P.F. 2004. A dual isotope approach to the nitrogen hydrochemistry of an uban aquifer. *Applied Geochemistry*, **19**, 709–719.

FURLONG, B.V. 2002. *Regional scale solute transport in the Permo-Triassic Sandstone aquifer of the Lower Mersey Basin, north west England.* PhD thesis, University of Birmingham.

GABRIELSEN, R.H., AARLAND, R.-K. & ALSAKER, E. 1998. Identification and spatial distribution of fractures in porous, siliclastic sediments. *In*: COWARD, M.P., DALTABAN, T.S. & JOHNSON, H. (eds) *Structural Geology in Reservoir Characterization.* Geological Society, London, Special Publications, **127**, 49–64.

GELHAR, L.W., GUTJAHR, A.L. & NAFF, R.L. 1979. Stocastic analysis of macrodispersion in a stratified aquifer. *Water Resources Research*, **15**, 1387–1397.

GIBSON, R.G. 1998. Physical character and fluid-flow properties of sandstone-derived fault zones. *In*: COWARD, M.P., DALTABAN, T.S. & JOHNSON, H. (eds) *Structural Geology in Reservoir Characterization.* Geological Society, London, Special Publications, **127**, 83–97.

GIBSON, R.G. & BENTHAM, P.A. 2003. Use of fault-seal analysis in understanding petroleum migration in a complexly faulted anticlinal trap, Columbus Basin, offshore Trinidad. *AAPG Bulletin*, **87**, 465–478.

GILLESPIE, M.R., KEMP, S.J., VICKERS, B.P., WATERS, C. & GOWING, C.J. 2001. *Cation Exchange Capacity (CEC) of Selected Lithologies from England, Wales and Scotland.* (UK) Environment Agency R&D Technical Report, **P2-222/TR**.

GILLESPIE, M.R., LEADER, R.U. *ET AL.* 2000. *CEC and K_d Determination in Landfill Performance Evaulation: A Review of Methodologies and Preparation* *of Standard Materials for Laboratory Analysis.* (UK) Environment Agency R&D Project Record, **P1/254/01**.

GOODDY, D.C. & BLOOMFIELD, J.P. 2006. Controls on dense non-aqueous phase liquid transport in Permo-Triassic Sandstones, UK. *In*: BARKER, R.D. & TELLAM, J.H. (eds) *Fluid Flow and Solute Movement in Sandstones: The Offshore UK Permo-Triassic Red Bed Sequence.* Geological Society, London, Special Publications, **263**, 253–264.

GOODDY, D.C. & DARLING, W.G. 2005. The potential for methane emissions from groundwaters in the UK. *Science of the Total Environment,* **339**, 117–126.

GOUZE, P., TENCHINE, S. *ET AL.* 2003. A study of the dispersion in a single thin fracture using Positron Emission Projected Imaging. *In*: KRÁSNÝ, J., HRKAL, Z. & BRUTHANS, J. (eds) *Groundwater in Fractured Rock, IHP–VI. Series on Groundwater,* **7**, 407–408.

GREENWOOD, H.W. & TRAVIS, C.B. 1915. The mineralogical and chemical constitution of the Triassic rocks of Wirral, Part II. *Proceedings of the Liverpool Geological Society,* **12**, 161–188.

GRESSWELL, R.A. 2005. *Quantitative characterisation of fluoride sorption in the UK Triassic sandstone and the development of a fluoride reactive transport model.* MSc Project Report, University of Birmingham.

GRESWELL, R.B., TELLAM, J.H., LLOYD, J.W. & PARKER, D. 1998. Positron emission tomography studies of solute movement through rock. *In*: BRAHANA, J., ECKSTEIN, Y., ONGLEY, L.K., SCHNEIDER, R. & MOORE, J.E. (eds) *Gambling with Groundwater, Proceedings of the IAH Conference, Las Vegas,* 561–566.

GRIFFITHS, D.H., BARKER, R.D. & FINCH, J.W. 1981. Recent applications of the electrical resistivity and induced-polarisation methods to hydrogeological problems. *In*: *A Survey of British Hydrogeology 1980.* The Royal Society, London.

GUADAGNINI, A. & WINTER, C.L. 2004. Special Issue: Stochastic models of flow and transport in multiple-scale heterogeneous porous media. *Journal of Hydrology,* **294**, 1–212.

GUHA, H., SAIERS, J.E., BROOKS, S., JARDINE, P. & KRISHNASWAMY JAYACHANDRAN. 2001. Chromium transport, oxidation, and adsorption in manganese-coated sand. *Journal of Contaminant Hydrology,* **49**, 311–334.

GUTMANIS, J.C., LANYON, G.W., WYNN, T.J. & WATSON, C.R. 1998. Fluid flow in faults: a study of fault hydrogeology in Triassic sandstone and Ordovician volcaniclastic rocks at Sellafield, north-west England. *Proceedings of the Yorkshire Geological Society,* **52**, 159–175.

HAINES, T.S. & LLOYD, J.W. 1985. Controls on silica in groundwater environments in the United Kingdom. *Journal of Hydrology,* **81**, 277–295.

HARRISON, I., WILLIAMS, G.M., HIGGO, J.J.W., LEADER, R.U., KIM, A.W. & NOY, D.J. 2001. Microcosm studies of microbial degradation in a coal tar distillate plume. *Journal of Contaminant Hydrology,* **53**, 319–340.

HARTLEY, L.T. 1998. *NAPSAC (Release 4.1) Technical Summary Document.* AEA Technology Report, **AEA-D&R-0271**.

HATZICHRISTODULU, V.C., BARKER, R.D. & TELLAM, J.H. 2002. *High Resolution Electrical Monitoring of Fluid Flow Through the Unsaturated Zone.* Environment Agency Technical Report, **P2-042/TR1**.

HEATHCOTE, J.A., JONES, M.A. & HERBERT, A.W. 1996. Modelling groundwater flow in the Sellafield area. *Quarterly Journal of Engineering Geology and Hydrogeology,* **29**, Supplement 1, S59–S82.

HERBERT, A.W. 1992. *Groundwater flow and transport in fractured rock.* PhD thesis, University of Bath.

HEWITT, W. 1898. Notes on some sections exposed by excavations on the site of the new technical schools, Byrom Street, Liverpool. *Proceedings of the Liverpool Geological Society Session the Thirty-ninth,* **8**, 268–273.

HITCHMOUGH, A.M., RILEY, M.S., HERBERT, A.W. & TELLAM, J.H. Submitted. Estimating the hydraulic properties of the fracture network in a sandstone aquifer. *Journal of Contaminant Hydrology.*

HOUGH, E., PEARCE, J.M., KEMP, S.J., MCKERVEY, J. & OATES, D. 2001. *Regional Survey and Sampling of Sediment Filled Fractures and Evaluation of Prospective Research Sites: A Report Based on Fieldwork Undertaken in 2000–2001.* British Geological Survey Technical Report, **CR/01/14**.

HOUSTON, J. 2004. High-resolution sequence stratigraphy as a tool in hydrogeological exploration in the Atacama Desert. *Quarterly Journal of Engineering Geology and Hydrogeology,* **37**, 7–17.

HOWARD, K.W.F. 1988. Beneficial aspects of sea-water intrusion. *Ground Water,* **17**, 250–257.

HOWELL, J. & MOUNTNEY, N. 1997. Climate cyclicity and accommodation space in arid to semi-arid depositional systems: an example from the Roliegend Group of the UK southern North Sea. *In*: ZIEGLER, K., TURNER, P. & DAINES, S.R. (eds) *Petroleum Geology of the Southern North Sea: Future Potential.* Geological Society, London, Special Publications, **123**, 63–86.

HUGHES, A.J., TELLAM, J.H., BOTTRELL, S.H., LLOYD, J.W., BARKER, A.P., STAGG, K.A. & BARRETT, M.H. 1999. Sulphate Isotope signatures in borehole waters from three urban Triassic Sandstone aquifers, UK. *In*: ELLIS, J.B. (ed.) *Impacts of Urban Growth on Surface Water and Groundwater Quality.* IAHS Publication, **259**, 143–149.

HUNTER-WILLIAMS, N. 1995. *Investigation of the response of the Triassic Sandstone aquifer Nurton to extended abstraction and of the controls on groundwater flow.* MSc Project Report, University of Birmingham.

HUNTOON, P.W. 1986. Incredible tale of Texasgulf Well 7 and fracture permeability, Paradox Basin, Utah. *Ground Water,* **24**, 643–653.

IVANOVICH, M., TELLAM, J.H., LONGWORTH, G. & MONAGHAN, J.J. 1992. Rock/water interaction timescales involving U and Th isotopes in a Permo-Triassic sandstone. *Radiochimica Acta,* **58/59**, 423–432.

IXER, R.A. & VAUGHAN, D.J. 1982. The primary ore

mineralogy of the Alderley Edge deposit, Cheshire. *Mineralogical Magazine*, **46**, 485–492.

JACKSON, D. & LLOYD, J.W. 1983. Groundwater chemistry of the Birmingham Triassic sandstone aquifer and its relation to structure. *Quarterly Journal of Engineering Geology*, **16**, 135–142.

JAMISON, W.R. & STEARNS, D.W. 1982. Tectonic deformation of the Wingate sandstone, Colorado National Monument. *AAPG Bulletin*, **66**, 2584–2608.

JEFFCOAT, A.M. 2002. *Exploring the hydraulic properties of discontinuity geometry in the UK Triassic Sandstones*. PhD thesis, University of Birmingham.

JENSEN, J.L., CORBETT, P.W.M., PICKUP, G.E. & RINGROSE, P.S. 1996. Permeability semivariograms, geological structure and flow performance. *Mathematical Geology*, **28**, 419–435.

JONES, D.G., MORTON, A.C., LENG, M.J., HASLAM, H.W., MILOWDOWSKI, A.E., STRONG, G.E. & KEMP, S.J. 1999. Provenance of basin fill material. *In*: PLANT, J.A., JONES, D.G. & HASLAM, H.W. (eds) *The Cheshire Basin: Basin Evolution, Fluid Movement and Mineral Resources in a Permo-Triassic Rift Setting*. British Geological Survey, Keyworth, 90–124.

JONES, I. 2001. *Processes controlling the transport and attenuation of contamination from a coking plant in a sandstone aquifer*. PhD thesis, University of Bradford.

JONES, I., DAVISON, R.M. & LERNER, D.N. 1998. The importance of understanding groundwater flow history in assessing present-day groundwater contamination patterns: a case study. *In*: LERNER, D.N. & WALTON, N.R.G. (eds) *Contaminated Land and Groundwater: Future Directions*. Geological Society, London, Engineering Geology, Special Publications, **14**, 137–148.

JONES, I. & LEARNER, D.N. 1995. Level-determined sampling in an uncased borehole. *Journal of Hydrology*, **171**, 291–317.

JOYCE, E., RUEEDI, J., CRONIN, A.A., PEDLEY, S., GRESWELL, R.B. & TELLAM, J.H. 2006. *Fate and transport of phage and viruses in UK sandstone aquifers*. Final report on Environment Agency Research Project, **NC/02/47**.

KELLER, G.V. 1966. Electrical properties of rocks and minerals. *In*: CLARK, S.P. (eds) *Handbook of Physical Constants. Geological Society of America, Memoir*, **97**, 553–578.

KEMP, S.J., PEARCE, J.M. & OATES, D. 2003. In situ sampling of sediment-filled fractures. *Geotechnique*, **53**, 665–668.

KINNIBURGH, D.G., NEWELL, A.J., DAVIES, J., SMEDLEY, P.L., MILODOWSKI, A.E., INGRAM, J.A. & MERRIN, P.D. 2006. The arsenic concentration in groundwater from the Abbey Arms Wood observation borehole, Delamere, Cheshire, UK. *In*: BARKER, R.D. & TELLAM, J.H. (eds) *Fluid Flow and Solute Movement in Sandstones: The Offshore UK Permo-Triassic Red Bed Sequence*. Geological Society, London, Special Publications, **263**, 265–284.

KLEWICKI, J.K. & MORGAN, J.J. 1999. Dissolution of β-MnOOH particles by ligands: Pyrophosphate, ethylenediaminetetratacetic acid, and citrate. *Geochimica et Cosmochimica Acta*, **63**, 3017–3024.

KNIPE, C., LLOYD, J.W., LERNER, D.N. & GRESWELL, R.B. 1993. *Rising Groundwater Levels in Birmingham, and Their Engineering Significance*. ed.; CIRIA (Construction Industry Research and Information Association), London, Special Publications, **92**, 114.

KNOTT, S.D. 1994. Fault thickness versus displacement in the Permo-Triassic sandstones of NW England. *Journal of the Geological Society, London*, **151**, 17–25.

KOLTERMAN, C.E. & GORELICK, S.M. 1996. Heterogeneity in sedimentary deposits: a review of structure-imitating, process-imitating, and descriptive approaches. *Water Resources Research*, **32**, 2617–2658.

LERNER, D.N., THORNTON, S.F. *ET AL*. 2000. Ineffective natural attenuation of degradable organic compounds in a phenol-contaminated aquifer. *Ground Water*, **38**, 922–928.

LEVEILLE, G.P., KNIPE, R. *ET AL*. 1997. Compartmentalization of Rotliegendes gas reservois by sealing faults, Jupiter Fields area, southern North Sea. *In*: ZIEGLER, K., TURNER, P. & DAINES, S.R. (eds) *Petroleum Geology of the Southern North Sea: Future Potential*. Geological Society, London, Special Publications, **123**, 87–104.

LEWIN, K., YOUNG, C.P., BRADSHAW, K., FLEET, M. & BLAKEY, N.C. 1994. *Landfill Monitoring Investigations at Burntstump Landfill. Sherwood Sandstone, Nottingham 1978–1993* (ENV 9003), Final Report of the Department of the Environment CWMD35194: DOE3388/1.

LIU, N. & OLIVER, D.S. 2004. Automatic history matching of geologic facies. *Journal of Petroleum Engineers*, **9**, 429–436.

LIU, N. & OLIVER, D.S. 2005. Ensemble Kalman filter for automatic history matching of geologic facies. *Journal of Petroleum Science and Engineering*, **47**, 147–161.

LIWANG, J. 2005. Introducing geological processes in reservoir models. *Geo ExPro*, **2**, 28–32.

LLOYD, J.W. 1976. Hydrochemistry and groundwater flow patterns in the vicinity of Stratford-upon-Avon. *Quarterly Journal of Engineering Geology*, **9**, 315–326.

LOVELL, M.A., JACKSON, P.D., HARVEY, P.K. & FLINT, R.C. 2006. High-resolution petrophysical characterization of samples from an aeolian sandstone: the Permian Penrith Sandstone of NW England. *In*: BARKER, R.D. & TELLAM, J.H. (eds) *Fluid Flow and Solute Movement in Sandstones: The Offshore UK Permo-Triassic Red Bed Sequence*. Geological Society, London, Special Publications, **263**, 49–63.

LOVELOCK, P.E.R. 1977. Aquifer Properties of the Permo-Triassic Sandstones in the United Kingdom. *Bulletin of the Geological Survey of Great Britain*, **56**.

LUBCZYNSKI, M. & ROY, J. 2005. MRS contribution to hydrogeological system parametrization. *Near Surface Geophysics*, **3**, 131–139.

LUCEY, P.A. 1987. *The hydrochemistry of groundwater*

in the West Cheshire aquifer. MPhil, University of Lancaster.

LUTHER, G.W., SUNDBY, B., LEWIS, B.L., BRENDEL, P.J. & SILVERBERG, N. 1997. Interactions of manganese with the nitrogen cycle: Alternative pathways to denitrogen. *Geochimica et Cosmochimica Acta,* **61**, 4043–4052.

MAIN, I., MAIR, K., OHMYOUNG KWON, ELPHICK, S. & NGWENYA, B. 2001. Experimental constraints on the mechanical and hydraulic properties of deformation bands in porous sandstones: a review. *In:* HOLDSWORTH, R.E., STRACHAN, R.A., MAGLOUGHLIN, J.F. & KNIPE, R.J. (eds) *The Nature and Tectonic Significance of Fault Zone Weakening.* Geological Society, London, Special Publications, **186**, 43–63.

MANZOCCHI, T., RINGROSE, P.S. & UNDERHILL, J.R. 1998. Flow through fault systems in high-porosity sandstones. *In:* COWARD, M.P., DALTABAN, T.S. & JOHNSON, H. (eds) *Structural Geology in Reservoir Characterization.* Geological Society, London, Special Publications, **127**, 65–82.

MARTINIUS, A.W. & NAESS, A. 2005. Uncertainty analysis of fluvial outcrop data for stochastic reservoir modelling. *Petroleum Geoscience,* **11**, 203–214.

MATHERON, G., BEUCHER, H., DE FOUQUET, C., GALLI, A., GUERILLOT, D. & RAVENNE, C. 1987. Conditional simulation of the geometry of fluvio-deltaic reservoirs. *In: Conditional Simulation of the Geometry of Fluvio-deltaic Reservoirs,* SPE Annual Technical Conference and Exhibition, Society of Petroleum Engineers, Dallas, TX, paper 16753-MS.

MAYER, K.U., BENNER, S.G., FRIND, E.O., THORNTON, S.F. & LERNER, D.N. 2001. Reactive transport modeling of processes controlling the distribution and natural attenuation of phenolic compounds in a deep sandstone aquifer. *Journal of Contaminant Hydrology,* **53**, 341–368.

MCBRIDE, M.B. 1994. *Environmental Chemistry of Soils.* Oxford University Press, Oxford, 406.

MERCADO, A. 1967. The spreading of injected waters in a permeable stratified aquifer. *In: Artificial Recharge and Management of Aquifers.* IAHS Publication, **72**, 23–36.

METCALF, R., ROCHELLE, C.A., SAVAGE, D. & HIGGO, J.W. 1994. Fluid–rock interactions during continental red bed diagenesis: implications for theoretical models of mineralization in sedimentary basins. *In:* PARNELL, J. (ed.) *Geofluids: Origin, Migration and Evolution of Fluids in Sedimentary Basins.* Geological Society, London, Special Publications, **78**, 301–324.

MIALL, A.D. 1990. *Principles of Sedimentary Basin Analysis.* Springer, New York.

MILODOWSKI, A.E. & GILLESPIE, M.R. 1997. *The Petrology of Fractures and Fracture Mineralisation in Sellafield BH3,* British Geological Survey Report CC945/527/CF-I-A.

MILODOWSKI, A.E., GILLESPIE, M.R. & KEMP, S.J. 1994. *A Review of the Fracture Mineralogy and Mineral Distribution in the Sellafield Boreholes,* British Geological Survey, Technical Report WG-194137.

MILODOWSKI, A.E., GILLESPIE, M.R., NADEN, J., FORTEY, N.J., SHEPHERD, T.J., PEARCE, J.M. &

METCALF, R. 1998. The petrology and paragenesis of fracture mineralization in the Sellafield area, west Cumbria. *Proceedings of the Yorkshire Geological Society,* **52**, 215–241.

MILODOWSKI, A.E., STRONG, G.E. *ET AL.* 1999. Diagenesis of the Permo-Triassic rocks. *In:* PLANT, J.A., JONES, D.G. & HASLAM, H.W. (eds) *The Cheshire Basin.* British Geological Survey, Keyworth, 125–175.

MIMIDES, T. & LLOYD, J.W, 1987. Toxic metal adsorption in the Triassic Sandstone aquifer of the English East Midlands. *Environmental Geology and Water Science,* **10**, 135–140.

MITCHENER, R.G.R. 2003. *Hydraulic and chemical property correlations of the Triassic sandstone of Birmingham.* PhD thesis, University of Birmingham, UK.

MOORE, C.C. 1898. The chemical examination of sandstones from Prenton Hill and Bidston Hill. *Proceedings of the Liverpool Geological Society, Session the thirty-ninth,* **8**, 241–267.

MOSS, P.D. & EDMUNDS, W.M. 1992. Processes controlling acid attenuation in the unsaturated zone of a Triassic Sandstone aquifer (UK), in the absence of carbonate minerals. *Applied Geochemistry,* **7**, 573–583.

MOUNTNEY, N.P.T. & THOMPSON, D.B. 2002. Stratigraphic evolution and preservation of aeolian dune and damp/wet interdune strata: an example from the Triassic Helsby Sandstone Formation, Cheshire Basin, UK. *Sedimentology,* **49**, 805–833.

MOYSEY, S., CAERS, J., KNIGHT, R. & ALLEN-KING, R.M. 2003. Stochastic estimation of facies using ground penetrating radar data. *Stochastic Environmental Research and Risk Assessment,* **17**, 306–318.

NARR, W. & SUPPE, J. 1991. Joint spacing in sedimentary rocks. *Journal of Structural Geology,* **13**, 63–72.

NAYLOR, H., TURNER, P., VAUGHAN, D.J., BOTCE, A.J. & FALLICK, A.E. 1989. Genetic studies of red bed mineralization in the Triassic of the Cheshire Basin, northwest England. *Journal of the Geological Society, London,* **146**, 685–699.

NEWELL, A.J. 2006. Calcrete as a source of heterogeneity in Triassic fluvial sandstone aquifers (Otter Sandstone Formation, SW England). *In:* BARKER, R.D. & TELLAM, J.H. (eds) *Fluid Flow and Solute Movement in Sandstones: The Offshore UK Permo-Triassic Red Bed Sequence.* Geological Society, London, Special Publications, **263**, 119–127.

ODLING, N.E. & RODEN, J.E. 1997. Contaminant transport in fractured rocks with significant matrix permeability, using natural fracture geometries. *Journal of Contaminant Hydrology,* **27**, 263–283.

OGATA, A. & BANKS, R.B. 1961. *A Solution of the Differential Equation of Longitudinal Dispersion in Porous Media.* US Geological Survey, Professional Paper, **411-A**, A1–A9.

OLORUNFEMI, M.O. & GRIFFITHS, D.H. 1985. A laboratory investigation of the induced polarization of the Triassic Sherwood Sandstone of Lancashire and its hydrogeological applications. *Geophysical Prospecting,* **33**, 110–127.

PAPATOLIOS, K.T. & LERNER, D.N. 1993. Defining a borehole capture zone in a complex sandstone aquifer: a modelling case study from Shropshire, UK. *Quarterly Journal of Engineering Geology and Hydrogeology*, **26**, 193–204.

PARKER, J., FOSTER, SS..D., SHERRATT, R. & ALDRICK, J. 1985. *Diffuse Pollution and Groundwater Quality in the Triassic Sandstone Aquifer in Southern Yorkshire*. British Geological Survey Report, **17**, No. 5.

PARKER, J.M., MASON, J. & KELLY, D.P. 1988. *A Study of Denitrification in the Triassic Sandstone Aquifer in Yorkshire, England*. British Geological Survey Hydrogeology Report, **WD/88/23**.

PARKER, K.E. 2005. *Calculating ion exchange parameters for pure clay assemblages*. PhD thesis, University of Birmingham.

PARKER, K.E., TELLAM, J.H. & CLIFF., M.I. 2000. Predicting ion exchange parameters for Triassic Sandstone aquifers. *In*: SILILO, O.T.N. (ed.) *Groundwater: Past Achievements and Future Challenges*. Balkema, Rotterdam, 581–586.

PARRY, W.T., CHAN, M.A. & BEITLER, B. 2004. Chemical bleaching indicates episodes of fluid flow in deformation bands in sandstone. *AAPG Bulletin*, **88**, 175–191.

PATNODE, H.W. & WYLLIE, M.R.J. 1950. The presence of conductive solids in reservoir rocks as a factor in electric log interpretation. *Transactions of the American Institute of Mining and Metallurgical Engineers*, **189**, 47–52.

PICKUP, G.E., RINGROSE, P.S., JENSON, J.L. & SORBIE, K.S. 1994. Permeability tensors for sedimentary structures. *Mathematical Geology*, **26**, 227–250.

PICKUP, R.W., RHODES, G., ALAMILLO, M.L., MALLINSON, H.E.H., THORNTON, S.F. & LERNER, D.N. 2001. Microbiological analysis of multi-level borehole samples from a contaminated groundwater system. *Journal of Contaminant Hydrology*, **53**, 269–284.

PITTMAN, E.D. 1981. Effects of fault-related granulation on porosity and permeability of quartz sandstones. *AAPG Bulletin*, **65**, 2381–2387.

PLANT, J.A., JONES, D.G. & HASLAM, H.W. 1999. *The Cheshire Basin: Basin Evolution, Fluid Movement and Mineral Resources in a Permo-Triassic Rift Setting*. British Geological Survey: Keyworth, 263.

PLUMMER, L.N. 1975. Mixing of sea water with calcium carbonate groundwater. *Geological Society of America, Memoir*, **42**, 219–236.

POKAR, M., WEST, L.J. & ODLING, N.E. 2006. Petrophysical characterization of the Sherwood Sandstone from East Yorkshire, UK. *In*: BARKER, R.D. & TELLAM, J.H. (eds) *Fluid Flow and Solute Movement in Sandstones: The Offshore UK Permo-Triassic Red Bed Sequence*. Geological Society, London, Special Publications, **263**, 103–118.

POOLE, E.G. & WHITEMAN, A.J. 1955. Variations in the thickness of the Collyhurst Sandstone in the Manchester area. *Bulletin of the Geological Survey of Great Britain*, **9**, 33–41.

POOLE, E.G. & WHITEMAN, A.J. 1966. *The Geology of the Country Around Nantwich and Whitchurch*. Memoir of the Geological Survey of Great Britain.

POWELL, K.L., BARRETT, M.H. *ET AL.* 2001. Virus-colloid interactions: implications for groundwater monitoring. *In*: SEILER, K.-P. & WOHNLICH, S. (eds) *New Approaches Characterizing Groundwater Flow*, **1**, Balkema, Rotterdam, 169–174.

POWELL, K.L., BARRETT, M.H., PEDLEY, S., TELLAM, J.H., STAGG, K.A., GRESWELL, R.B. & RIVETT, M. 2000. Enteric virus detection in groundwater using glass wool. *In*: SILILO, O.T.N. (ed.) *Groundwater: Past Achievements and Future Challenges*. Balkema, Rotterdam, 813–816.

POWELL, K.L., TAYLOR, R.G. *ET AL.* 2003. Microbial contamination of two urban sandstone aquifers in the UK. *Water Research*, **37**, 339–352.

PRICE, M.M., MORRIS, B.L. & ROBERTSON, A.S. 1982. A study of intergranular and fissure permeability in Chalk and Permian aquifers, using double packer injection testing. *Journal of Hydrology*, **54**, 401–423.

PRINGLE, J.K., WESTERMAN, A.R., CLARK, J.D., DRINKWATER, N.J. & GARDINER, A.R. 2004. 3D high resolution digital models of outcrop analogue study sites to constrain rservoir model uncertainty: an example from Alport, Derbyshire, UK. *Petroleum Geoscience*, **10**, 343–352.

PRIVETT, K.D. 2006. The capillary characteristic model of petroleum hydrocarbon saturation in the Permo-Triassic sandstone and its implications for remediation. *In*: BARKER, R.D. & TELLAM, J.H. (eds) *Fluid Flow and Solute Movement in Sandstones: The Offshore UK Permo-Triassic Red Bed Sequence*. Geological Society, London, Special Publications, **263**, 297–309.

PROSSER, D.J. & MASKALL, R. 1993. Permeability variation within aeolian sandstone: a case study using core cut sub-parallel to slipface bedding, the Auk Field, Central North Sea, UK. *In*: NORTH, C.P. & PROSSER, D.J. (eds) *Characterization of Fluvial and Aeolian Reservoirs*. Geological Society, London, Special Publications, **73**, 377–397.

READ, D., LAWLESS, T.A., SIMS, R.J. & BUTTER, K.R. 1993. Uranium migration through intact sandstone cores. *Journal of Contaminant Hydrology*, **13**, 277–289.

REES, S.B. 2006. Investigation and management of a kerosene leakage into a Permo-Triassic sandstone aquifer in the UK. *In*: BARKER, R.D. & TELLAM, J.H. (eds) *Fluid Flow and Solute Movement in Sandstones: The Offshore UK Permo-Triassic Red Bed Sequence*. Geological Society, London, Special Publications, **263**, 311–324.

REEVES, M.J., SKINNER, A.C. & WILKINSON, W.B. 1975. The relevance of aquifer-flow mechanisms to exploration and development of groundwater resources. *Journal of Hydrology*, **25**, 1–21.

RIMSTIDT, J.D. & VAUGHAN, D.J. 2003. Pyrite oxidation: a state-of-the-art assessment of the reaction mechanism. *Geochimica et Cosmochimica Acta*, **67**, 873–880.

RINGROSE, P.S. & CORBETT, P.W.M. 1994. Controls on two-phase fluid flow in heterogeneous sandstones.

In: PARNELL, J. (eds) *Geofluids: Origin, Migration and Evolution of Fluids in Sedimentary Basins.* Geological Society, London, Special Publications, **78**, 141–150.

RIVETT, M.O., LERNER, D.N., LLOYD, J.W. & CLARK, L. 1990. Organic contamination of the Birmingham aquifer, UK. *Journal of Hydrology*, **113**, 307–323.

ROBERTS, I. 1869. On the wells and water of Liverpool. *Proceedings of the Liverpool Geological Society*, **1**, 84–97.

ROWE, J.R. & BURLEY, S.D. 1997. Faulting and porosity modification in the Sherwood Sandstone at Alderley Edge, northern Cheshire: an exhumed example of fault-related diagenesis. *In*: MEADOWS, N.S., TRUEBLOOD, S.P., HARDMAN, M. & COWAN, G. (eds) *Petroleum Geology of the Irish Sea and Adjacent Areas.* Geological Society, London, Special Publications, **124**, 325–352.

RUBIN, H. & BUDDEMEIER, R.W. 1996. Transverse dispersion of contaminants in fractured permable formations. *Journal of Contaminant Hydrology*, **176**, 133–151.

RUBIN, H., SOLIMAN, A.M., BIRKHOLZER, J. & ROUVE, G. 1996. Transport of a tracer slug in a fractured permeable formation. *Journal of Hydrology*, **176**, 133–151.

RUSHTON, K.R. & SALMON, S. 1993. Significance of vertical flow through low-conductivity zones in Bromsgrove Sandstone aquifer. *Journal of Hydrology*, **152**, 131–152.

SAGE, R.C. & LLOYD, J.W. 1978. Drift deposit influences on the Triassic Sandstone aquifer of N.W. Lancashire as inferred by hydrochemistry. *Quarterly Journal of Engineering Geology*, **11**, 209–218.

SAUER, R.I. In preparation. *Solute transport around pumping wells in a sandstone aquifer.* PhD thesis, University of Birmingham, UK.

SCOTT, J.B.T. & BARKER, R.D. 2003. Determining pore-throat size in Permo-Triassic sandstones from low-frequency electrical spectroscopy. *Geophysical Research Letters*, **30**, 3-1–3-4.

SCOTT, J.B.T. & BARKER, R.D. 2005. Characterization of sandstone by electrical spectroscopy for stratigraphical and hydrogeological investigations. *Quarterly Journal of Engineering Geology and Hydrogeology*, **38**, 143–154.

SCOTT, J.B.T. & BARKER, R.D. 2006. Pore geometry of Permo-Triassic sandstone from measurements of electrical spectroscopy. *In*: BARKER, R.D. & TELLAM, J.H. (eds) *Fluid Flow and Solute Movement in Sandstones: The Offshore UK Permo-Triassic Red Bed Sequence.* Geological Society, London, Special Publications, **263**, 65–81.

SEGAR, D.A. 1993. *The effect of open boreholes on groundwater flows and chemistry.* PhD thesis, University of Birmingham, UK.

SEIF EL-DEIN, S.A.E.-B. 1983. *Diagenesis and aquifer characteristics of the Sherwood Sandstone Group in north Nottinghamshire.*

SEYMOUR, K.J., INGRAM, J.A. & GEBBETT, S.J. 2006. Structural controls on groundwater flow in the Permo-Triassic sandstones of NW England. *In*: BARKER, R.D. & TELLAM, J.H. (eds) *Fluid Flow and Solute Movement in Sandstones: The Offshore*

UK Permo-Triassic Red Bed Sequence. Geological Society, London, Special Publications, **263**, 169–186.

SHAND, P., HARGREAVES, R. & BREWERTON, L.J. 1997. *The Natural (Baseline) Quality of Groundwaters in England and Wales: The Permo-Triassic Sandstones of Cumbria, North-west England.* (UK) Environment Agency R&D Project Record, **W6/i722/3**, Birmingham, UK.

SHEPHERD, K.A. 2003. *Contamination and groundwater quality in the Birmingham aquifer.* PhD thesis, University of Birmingham.

SHEPHERD, K.A., ELLIS, P.A. & RIVETT, M.O. 2006. Integrated understanding of urban land, groundwater, baseflow and surface-water quality – The City of Birmingham, UK. *Science of the Total Environment*, **360**, 180–195.

SMEDLEY, P.L. & BREWERTON, L.J. 1997. *The Natural (Baseline) Quality of Groundwaters in England and Wales: The Triassic Sherwood Sandstone of the East Midlands and South Yorkshire.* (UK) Environment Agency R&D Project Record, **W6/i722/1**.

SMEDLEY, P.L. & EDMUNDS, W.M. 2002. Redox patterns and trace-element behavior in the East Midlands Triassic Sandstone Aquifer, U.K. *Ground Water*, **40**, 44–58.

SMITH, R. & MOLLER, N. 2003. Sedimentology and reservoir modelling of the Ormen Lange field, mid Norway. *Marine and Petroleum Geology*, **20**, 601–613.

SPEARS, D. 1986. Pollution investigation of a Triassic sandstone aquifer: the role of mineralogy. *In*: CRIPPS, J., BELL, F.G. & CULSHAW, MG. (eds) *Groundwater in Engineering Geology.* Geological Society, London, Engineering Geology, Special Publications, **3**, 211–218.

SPEARS, D.A. & REEVES, M.J. 1975. The influence of drift deposits on groundwater quality in the Vae of York. *Quarterly Journal of Engineering Geology*, **8**, 255–269.

SPENCE, M.J., BOTTRELL, S.H., HIGGO, J.J.W., HARRISON, I. & FALLICK, A.E. 2001a. Denitrification and phenol degradation in a contaminated aquifer. *Journal of Contaminant Hydrology*, **53**, 305–318.

SPENCE, M.J., BOTTRELL, S.H., HIGGO, J.J.W., THORNTON, S.F. & LERNER, D.N. 2001b. Isotopic modelling of the significance of bacterial sulphate reduction for phenol attenuation in a contamnated aquifer. *Journal of Contaminant Hydrology*, **53**, 285–304.

STAFFORD, P., TORAN, L. & MACKAY, L. 1998. Influence of fracture truncation on dispersion: A dual permeability model. *Journal of Contaminant Hydrology*, **30**, 79–100.

STAGG, K.A. 2000. *An investigation into colloids in Triassic Sandstone groundwaters, UK.* PhD thesis, University of Birmingham.

STAGG, K.A. & TELLAM, J.H. 1998. The use of image analysis in interpreting electron microscope photographs of groundwater colloids. *In*: BRAHANA, J., ECKSTEIN, Y., ONGLEY, L.K., SCHNEIDER, R. & MOORE, J.E. (eds) *Gambling with*

Groundwater. Proc IAH Conference, Las Vegas. IAH, Smyrna, GA, 533–539.

STAGG, K.A., KLEINERT, U.O., TELLAM, J.H. & LLOYD, J.W. 1997. Colloidal populations in urban and rural groundwaters, UK. *In*: CHILTON, P.J. *ET AL*. (eds) *Urban groundwater. Proc IAH Conference, Nottingham, UK*. Balkema, Rotterdam, 187–192.

STAGG, K.A., TELLAM, J.H., BARRETT, M.H. & LERNER, D.N. 1998. Hydrochemical variations with depth in a major UK aquifer: the fractured, high permeability Triassic Sandstone. *In*: BRAHANA, J., ECKSTEIN, Y., ONGLEY, L.K., SCHNEIDER, R. & MOORE, J.E. (eds) *Gambling with Groundwater. Proc IAH Conference, Las Vegas*. IAH, Smyrna, GA, 53–58.

STANISTREET, I.G. & STOLLHOFEN, H. 2002. Hoanib River flood deposits of Namib Desert interdunes as analogues for thin permeability barrier mudstone layers in aeolianite reservoirs. *Sedimentology*, **49**, 719–736.

STAUB, K.L., BENZ, M., SCHINK, B. & WIDDEL, F. 1996. Anaeorobic, nitrate-dependent microbial oxidation of ferrous iron. *Applied and Evironmental Microbiology*, **62**, 1458–1460.

STEEL, R.J. & THOMPSON, D.B. 1983. Structures and textures in Triassic braided stream conglomerates ('Bunter Pebble Beds') in the Sherwood Sandstone Group, north Staffordshire, England. *Sedimentology*, **30**, 341–367.

STEELE, A. & LERNER, D.N. 2001. Predictive modelling of NAPL injection tests in variable aperture spatially correlated fractures. *Journal of Contaminant Hydrology*, **49**, 287–310.

STEVENTON-BARNES, H. 2001. *Solid organic carbon in UK aquifers: its role in sorption of organic contaminants*. PhD thesis, University of London.

STREETLY, H., HAMILTON, C.A.L., BETTS, C., TELLAM, J.H. & HERBERT, A.W. 2002. Reconnaissance tracer tests in the Triassic Sandstone aquifer north of Liverpool, UK. *Quarterly Journal of Engineering Geology and Hydrogeology*, **35**, 167–178.

STREETLY, M., CHAKRABARTY, C. & MCLEOD, R. 2000. Interpretation of pumping tests in the Sherwood Sandstone Group, Sellafield, Cumbria, UK. *Quarterly Journal of Engineering Geology and Hydrogeology*, **33**, 281–299.

STREETLY, M.J., HEATHCOTE, J.A. & DEGNAN, P.J. 2006. Estimation of vertical diffusivity from seasonal fluctuations in groundwater pressures in deep boreholes near Sellafield, NW England. *In*: BARKER, R.D. & TELLAM, J.H. (eds) *Fluid Flow and Solute Movement in Sandstones: The Offshore UK Permo-Triassic Red Bed Sequence*. Geological Society, London, Special Publications, **263**, 155–167.

STRONG, G.E., MILOWDOWSKI, A.E., PEARCE, J.M., KEMP, S.J., PRIOR, S.V. & MORTON, A.C. 1994. The petrology and diagenesis of Permo-Triassic rocks of the Sellafield area, Cumbria. *Proceedings of the Yorkshire Geological Society*, **50**, 77–89.

SZURLIES, M., BACHMANN, G.H., MENNING, M., NOWACZYK, N.R. & KADING, K.-C. 2003. Magnetostratigraphy and high-resolution lithostratigraphy of the Permian–Triassic boundary interval in Central Germany. *Earth and Planetary Science Letters*, **212**, 263–278.

TAYLOR, R.G., CRONIN, A.A., LERNER, D.N., TELLAM, J.H., BOTTRELL, S.H., RUEEDI, J. & BARRETT, M.H. Submitted. Penetration of anthropogenic solutes in sandstone aquifers underlying two mature cities in the UK. *Hydrogeology Journal*.

TAYLOR, R.G., CRONIN, A.A., PEDLEY, S., ATKINSON, T.C. & BARKER, J.A. 2004. The implications of groundwater velocity variations on microbial transport and wellhead protection: review of field evidence. *FEMS Microbiology and Ecology*, **49**, 17–26.

TAYLOR, R.G., CRONIN, A.A., TROWSDALE, S.A., BAINES, O.P., BARRETT, M.H. & LERNER, D.N. 2003. Vertical groundwater flow in Permo-Triassic sediments underlying two cities in the Trent River Basin (UK). *Journal of Hydrology*, **284**, 92–113.

TAYLOR, S. & BARKER, R.D. 2002. Resistivity of partially saturated Triassic Sandstone. *Geophysical Prospecting*, **50**, 603–613.

TAYLOR, S.B. & BARKER, R.D. 2006. DC electrical properties of Permo-Triassic sandstone. *In*: BARKER, R.D. & TELLAM, J.H. (eds) *Fluid Flow and Solute Movement in Sandstones: The Offshore UK Permo-Triassic Red Bed Sequence*. Geological Society, London, Special Publications, **263**, 199–217.

TELES, V., DELAY, F. & DE MARSILY, G. 2004. Comparison of genesis and geostatistical methods for characterizing the heterogeneity of alluvial media: Groundwater flow and transport. *Journal of Hydrology*, **294**, 103–121.

TELLAM, J.H. 1994. The groundwater chemistry of the Lower Mersey Basin Permo-Triassic Sandstone aquifer system, UK: 1980 and pre-industrialisation/urbanisation. *Journal of Hydrology*, **161**, 287–325.

TELLAM, J.H. 1995a. Hydrochemistry of the saline groundwaters of the lower Mersey Basin Permo-Triassic sandstone aquifer, UK. *Journal of Hydrology*, **165**, 45–84.

TELLAM, J.H. 1995b. Urban groundwater pollution in the Birmingham Triassic Sandstone aquifer. *In*: *4th Annual IBC Conference on Groundwater Pollution*. IBC, London, 10.

TELLAM, J.H. 1996. Interpreting the borehole water chemistry of the Permo-Triassic sandstone aquifer of the Liverpool area, UK. *Geological Journal*, **31**, 61–87.

TELLAM, J.H. 2004. 19th century studies of the hydrogeology of the Permo-Triassic Sandstones of the northern Cheshire Basin, England. *In*: MATHER, J. (ed.) *200 Years of British Hydrogeology*. Geological Society, London, Special Publications, **225**, 89–105.

TELLAM, J.H. & LLOYD, J.W. 1986. Problems in the recognition of seawater intrusion by chemical means: an example of apparent chemical equivalence. *Quarterly Journal of Engineering Geology and Hydrogeology*, **19**, 389–398.

TELLAM, J.H. & LLOYD, J.W. 1997. The flushing of saline groundwater from a Permo-Triassic Sandstone aquifer system in NW England. *In*: FEI JIN

& KROTHE, N.C. (eds) *Proceedings of the 30th International Geological Congress (Beijing, China, Aug 1996)*, **22**, VSB, Utrect, 177–184.

TELLAM, J.H. & THOMAS, A. 2002. Well water quality and pollutant source distributions in an urban aquifer. *In*: HOWARD, K.W.F. & ISRAFILOV, R.G. (eds) *Current Problems in Hydrogeology in Urban Areas, Urban Agglomerates and Industrial Centres*. NATO Science Series, IV. Earth and Environmental Sciences, **8**, 139–158.

TELLAM, J.H., LLOYD, J.W. & WALTERS, M. 1986. The morphology of a saline groundwater body; its investigation, description and possible explanation. *Journal of Hydrology*, **83**, 1–21.

TELLAM, J.H., CARLYLE, H.F., EL-GHONEMY, H., PARKER, K.E. & MITCHENER, R.G.R. 2002. Towards quantification of ion exchange in a sandstone aquifer., *In*: THANGARAJAN, M., RAI, S.N. & SINGH, V.S. (eds) *Towards Quantification of Ion Exchange in a Sandstone Aquifer. Sustainable Development and Management of Groundwater Resources in Semi-arid Regions with Special Reference to Hard Rocks*. Oxford & IBH Publishing, New Delhi, 205–214.

TELLAM, J.H., THORNTON, S.F., LERNER, D.N., KLEINERT, U.O. & HARRIS, R.C. 1997. *Leachate Migration Through the Triassic Sandstones at Burntstump Landfill and Elsewhere*. Environment Agency R&D Report P183, (UK) Environment Agency, Bristol.

THOMPSON, D.B. 1970*a*. Sedimentation of the Triassic (Scythian) red pebbly sandstones in the Cheshire Basin and its margins. *Geological Journal*, **7**, 183–216.

THOMPSON, D.B. 1970*b*. The stratigraphy of the so-called Keuper Sandstone Formation (Scythian–?Anisian) in the Permo-Triassic Cheshire Basin. *Quarterly Journal of the Geological Society, London*, **126**, 151–181.

THORNTON, S.F., TELLAM, J.H. & LERNER, D.N. 2000*b*. Attenuation of landfill leachate by Triassic sandstone aquifer materials, 1. Fate of inorganic pollutants in laboratory columns. *Journal of Contaminant Hydrology*, **43**, 327–354.

THORNTON, S.F., BRIGHT, M.I., LERNER, D.N. & TELLAM, J.H. 2000*a*. Attenuation of landfill leachate by Triassic sandstone aquifer materials, 2. Sorption and degradation of organic pollutants in laboratory columns. *Journal of Contaminant Hydrology*, **43**, 355–383.

THORNTON, S.F., QUIGLEY, S., SPENCE, M.J., BANWART, S.A., BOTTRELL, S. & LERNER, D.N. 2001*b*. Processes controlling the distribution and natural attenuation of dissolved phenolic compounds in a deep sandstone aquifer. *Journal of Contaminant Hydrology*, **53**, 233–267.

THORNTON, S.F., LERNER, D.N. & BANWART, S.A. 2001*a*. Assessing the natural attenuation of organic contaminants in aquifers using plume-scale electron and carbon balances: model development with analysis of uncertainty and parameter sensitivity. *Journal of Contaminant Hydrology*, **53**, 199–232.

THORNTON, S.F., TELLAM, J.H. & LERNER, D.N. 2005.

Experimental and modelling approaches for the assessment of chemical impacts of leachate migration from landfills: A case study of a site on the Triassic sandstone aquifer in the UK East Midlands. *Geotechnical and Geological Engineering*, **23**, 811–829.

THORNTON, S.F., LERNER, D.N. & TELLAM, J.H. 1995. *Laboratory Studies of Landfill Leachate–Triassic Sandstone Interactions*. Wastes Technical Division, Department of the Environment Report, **CWM/035A/94**.

TORSTENSSON, D., THORNTON, S.F., BROHOLM, M.M. & LERNER, D.N. 1998. Hydrochemistry of pollutant attenuation in groundwater contaminated by coal tar wastes. *In*: LERNER, D.N. & WALTON, N.R.G. (eds) *Contaminated Land and Groundwater: Future Directions*. Geological Society, London, Engineering Geology, Special Publications, **14**, 149–157.

TOTH, J. 1963. A theoretical analysis of groundwater flow in a small drainage basin. *Journal of Geophysical Research*, **68**, 4795–4812.

TRAVIS, C.B. & GREENWOOD, H.W. 1911. The mineralogical and chemical constitution of the Triassic rocks of Wirral, Part I. *Proceedings of the Liverpool Geological Society*, **11**, 116–139.

TROWSDALE, S.A. & LERNER, D.N. 2003. Implications of flow patterns in the sandstone aquifer beneath the mature conurbation of Nottingham (UK) for source protection. *Quarterly Journal of Engineering Geology and Hydrogeology*, **36**, 197–206.

TURNER, P., PILLING, D., WALKER, D., EXTON, J., BINNIE, J. & SABAOU, N. 2001. Sequence stratigraphy and sedimentology of the late Triassic TAG-I (Blocks 401/402, Berkine Basin, Algeria). *Marine and Petroleum Geology*, **18**, 959–981.

UNDERHILL, J.R. & WOODCOCK, N.H. 1987. Faulting mechanisms in high-porosity sandstones: New Red Sandstone, Arran, Scotland. *In*: JONES, M.E. & PRESTON, R.M.F. (eds) *Deformation of Sediments and Sedimentary Rocks*. Geological Society, London, Special Publications, **29**, 91–105.

UNIVERSITY OF BIRMINGHAM/NWWA. 1981. *Lower Mersey Basin Saline Groundwater Study: Final Summary Report*. North West Water Authority, Warrington, Cheshire, UK.

VAIL, P.R., AUDEMARD, F., BOWMAN, S.A., EISNER, P.N. & PEREZ-CRUZ, C. 1991. The stratigraphic signatures of tectonics, eustacy, and sedimentology – an overview. *In*: EINSELE, G., RICKEN, W. & SEILACHER, A. (eds) *Cycles and Events in Stratigraphy*. Springer, Berlin, 617–657.

VAUGHAN, A., HANSEN, T., CARDON, H. & RADCLIFFE, N. 1999. Litho-flow facies prediction in an alluvial fan/fluvial system, Central North Sea. *Petroleum Geoscience*, **5**, 409–419.

WALSH, J.J., WATTERSON, J., HEATH, A.E. & CHILDS, C. 1998. Representation and scaling of faults in fluid flow models. *Petroleum Geoscience*, **4**, 241–251.

WALTHALL, S. & CAMPBELL, J.E. 1986. The measurement, interpretation and use of permeability values with specific reference to fissured aquifers. *In*: CRIPPS, J.C., BELL, F.G. & CULSHAW, M.G. (eds) *Groundwater in Engineering Geology*. Geological

Society, London, Engineering Geology, Special Publications, **3**, 273–278.

WALTHALL, S. & INGRAM, J.A. 1984. The investigation of aquifer parameters using multiple piezometers. *Ground Water*, **22**, 25–30.

WALTON, N.R.G. 1981. The origin of high iron concentrations in groundwater from the Trassic aquifer in East Devon. *Journal of the Institute of Water Engineers*, **36**, 63–77.

WARD, R.S., WILLIAMS, A.T., BARKER, J.A., BREWERTON, L.J. & GALE, I.N. 1998. *Groundwater Tracer Tests: A Review and Guidelines for Their Use in British Aquifers*. British Geological Survey, Report, **WD/98/19**.

WARRINGTON, G., AUDLEY-CHARLES, M.G., ELLIOTT, R.E., IVIMEY-COOK, H.C., ROBINSON, P.L., SHOTTON, F.W. & TAYLOR, F.M. 1980. *A Correlation of the Triassic Rocks in the British Isles*. Geological Society, London, Special Report, **13**, 78.

WEALTHALL, G.P., STEELE, A., BLOOMFIELD, J.P., MOSS, R.H. & LERNER, D.N. 2001. Sediment filled fractures in the Permo-Triassic sandstones of the Cheshire Basin: observations and implications for pollutant transport. *Journal of Contaminant Hydrology*, **50**, 41–51.

WHITE, K.E. 1986. Water tracers and investigatory techniques. *In*: BRANDON, T.W. (ed.) *Groundwater: Occurrence, Development and Protection*. The Institution of Water Engineers and Scientists, London, 353–383.

WILLIAMS, G.M., PICKUP, R.W., THORNTON, S.F., LERNER, D.N., MALLINSON, H.E.H., MOORE, Y. & WHITE, C. 2001. Biogeochemical characterisation of a coal tar distillate plume. *Journal of Contaminant Hydrology*, **53**, 175–197.

WILLIAMSON, D.J. 1994. *Controls on the transport of organic pollutants through Permo-Triassic sandstone aquifers: low concentration sorption of selected species on mineral substrates*. PhD thesis, University of Birmingham, UK.

WILLIS, B.J. & WHITE, C.D. 2000. Quantitative outcrop data from flow simulation. *Journal of Sediment Research*, **70**, 788–802.

WILLS, L.J. 1970. The Triassic succession in the central Midlands in its regional setting. *Quarterly Journal of the Geological Society, London*, **126**, 225–283.

WILLS, L.J. 1976. *The Trias of Worcestershire and Warwickshire*. Institute of Geological Sciences, Report No. 76/2.

WILSON, G., ANDREWS, J.N. & BATH, A.H. 1994. The nitrogen isotope composition of groundwater nitrates from the East Midlands Triassic Sandstone aquifer, England. *Journal of Hydrology*, **157**, 35–46.

WINSAUER, W.O., SHEARIN, H.M., MASSON, P.H. & WILLIAMS, M. 1952. Resistivity of brine-saturated sands in relation to pore geometry. *AAPG Bulletin*, **36**, 253–277.

WINSHIP, P., BINLEY, A. & GOMEZ, D. 2006. Flow and transport in the unsaturated Sherwood Sandstone: characterization using cross-borehole geophysical methods. *In*: BARKER, R.D. & TELLAM, J.H. (eds) *Fluid Flow and Solute Movement in Sandstones: The Offshore UK Permo-Triassic Red Bed Sequence*. Geological Society, London, Special Publications, **263**, 219–231.

WONG, P.M., TAMHANE, D. & WANG, L. 1997. A neural-network approach to knowledge-based well interpolation: A case study of a fluvial sandstone reservoir. *Journal of Petroleum Geology*, **20**, 363–372.

WORTHINGTON, P.F. 2004. The effect of scale on the petrophysical estimation of intergranular permeability. *Petrophysics*, **45**, 59–72.

WORTHINGTON, P.F. & BARKER, R.D. 1972. Methods for the calculation of true formation factors in the Bunter Sandstone of northwest England. *Engineering Geology*, **6**, 213–228.

WORTHINGTON, P.F. & GRIFFITHS, D.H. 1975. The application of geophysical methods in the exploration and development of sandstone aquifers. *Quarterly Journal of Engineering Geology and Hydrogeology*, **8**, 73–102.

WORTHINGTON, P.H. 1973. Estimation of the permeability of a Bunter Sandstone aquifer from laboratory investigations and borehole resistivity measurements. *Water and Water Engineering*, **77**, 251–257.

WYLLIE, M.R.J. & GREGORY, G.H.F. 1953. Formation factors of unconsolidated porous media: influence of particle shape and effect of cementation. *Transactions of the American Institute of Mining and Metallurgical and Petroleum Engineers*, **198**, 103–110.

YANG, C.S. & BAUMFALK, Y.A. 1997. Application of high-frequency cycle analysis in high-resolution sequence stratigraphy. *In*: ZIEGLER, K., TURNER, P. & DAINES, S.R. (eds) *Petroleum Geology of the Southern North Sea: Future Potential*. Geological Society, London, Special Publications, **123**, 181–203.

YANG, Y., LERNER, D.N., BARRETT, M.H. & TELLAM, J.H. 1999. Quantification of groundwater recharge in the city of Nottingham, UK. *Environmental Geology*, **38**, 183–198.

YEO, I.W., DE FREITAS, M.H. & ZIMMERMAN, R.W. 1998. Effect of shear displacement on the aperture and permeability of a rock fracture. *International Journal of Rock Mechanics and Mining Sciences*, **35**, 1051–1070.

High-resolution petrophysical characterization of samples from an aeolian sandstone: the Permian Penrith sandstone of NW England

M. A. LOVELL[1], P. D. JACKSON[2], P. K. HARVEY[1] & R. C. FLINT[2,3]

[1]*Department of Geology, University of Leicester, Leicester LE1 7RH, UK
(e-mail: mtl@leicester.ac.uk)*

[2]*British Geological Survey, Keyworth, Nottingham NG12 5GG, UK*

[3]*Present address: Department of Aeronautical and Automotive Engineering,
Loughborough University, Leicestershire LE11 3TU, UK*

Abstract: The Penrith Sandstone is an orange/red, mainly homogeneous, friable rock made up of well-rounded, highly spherical quartz grains, often showing euhedral overgrowths of quartz. Sandstone samples from Stoneraise Quarry, NW England, exhibit a remarkable degree of rounding and very high sphericity, along with frosted textures typical of aeolian deposits. Chemically, the rock is predominantly SiO_2 (>95%), with no evidence of carbonate cements. Quartz predominates with a small proportion (10%) of feldspar. The grain size across heterogeneous zones varies from very fine (100 μm) to coarse sand (700 μm). There is no evidence of the presence of clay minerals. Petrophysically, based on the measurements made in this study, the Penrith Sandstone is a typical clean sandstone characterized by moderate porosity (12%) and core-plug permeability (10^{-14}–10^{-12} m^2), and Archie 'm' exponents between 1.90 and 1.91, suggesting a reasonably clean 'Archie' rock with no excess conductivity associated with clays or bound water. Capillary pressure curves for four samples demonstrate unimodal pore-size distributions with a single modal range that varies between 25–50 and 70–80 μm. Because of the relative simplicity of its petrophysics, the sandstone is thus potentially very useful in fundamental studies, and also in the trialling of new techniques.

We use imaging techniques to investigate the degree of heterogeneity and the fabric of the Penrith Sandstone. Conventional optical images are complemented by electrical resistivity, porosity and mini-permeametry images. These two-dimensional maps of resolution of approximately 5 mm show a spatial similarity determined by the rock fabric. The detailed images show a wider degree of variation and heterogeneity than the plug-averaged values. The success of the resistivity imaging method suggests that the technique could be used in deriving correlations that could be used to interpret borehole resistivity imaging logs. However, in the present study, correlations of property values derived from the imaging do show considerable scatter: this suggests that heterogeneity even below the scale of the imaging is also important, a conclusion supported by thin-section and electron-microscope data.

The Penrith Sandstone is of Lower Permian age and outcrops in NW England (Fig. 1). An aeolian sandstone, often with good porosity and permeability, it is used for petrophysical studies because of its relative simplicity and similarity to some reservoir rocks such as the Rotleigendes of the southern North Sea. Macchi (1990) describes the Penrith Sandstone as a classic aeolian deposit in which component 'millet seed' grains of quartz display a remarkable degree of rounding and very high sphericity. Although somewhat coarser than many modern inland dune sands, the grains exhibit the frosted textures formerly regarded as conclusive evidence of an aeolian mode of origin. A general description is given by Waugh (1970).

As part of a programme of work concerned with the development of electrical core imaging, samples of the sandstone were collected from Stoneraise Quarry, which is located about 5 km NE of Penrith (Fig. 1). This quarry was, until the early 1990s, working a single barchan sand dune; work at the time of sampling had extended into an adjacent dune set. A number of blocks of sandstone were collected to provide a range of grain sizes, with material varying from homogeneous to distinctly heterogeneous. In hand specimen, the sandstone appears as an orange/red, mainly homogeneous, friable rock. In places, visible laminations are seen interspersed with areas of more uniform or massive beds. The rock is made up predominantly of

From: BARKER, R. D. & TELLAM, J. H. (eds) 2006. *Fluid Flow and Solute Movement in Sandstones: The Onshore UK Permo-Triassic Red Bed Sequence.* Geological Society, London, Special Publications, **263**, 49–63.

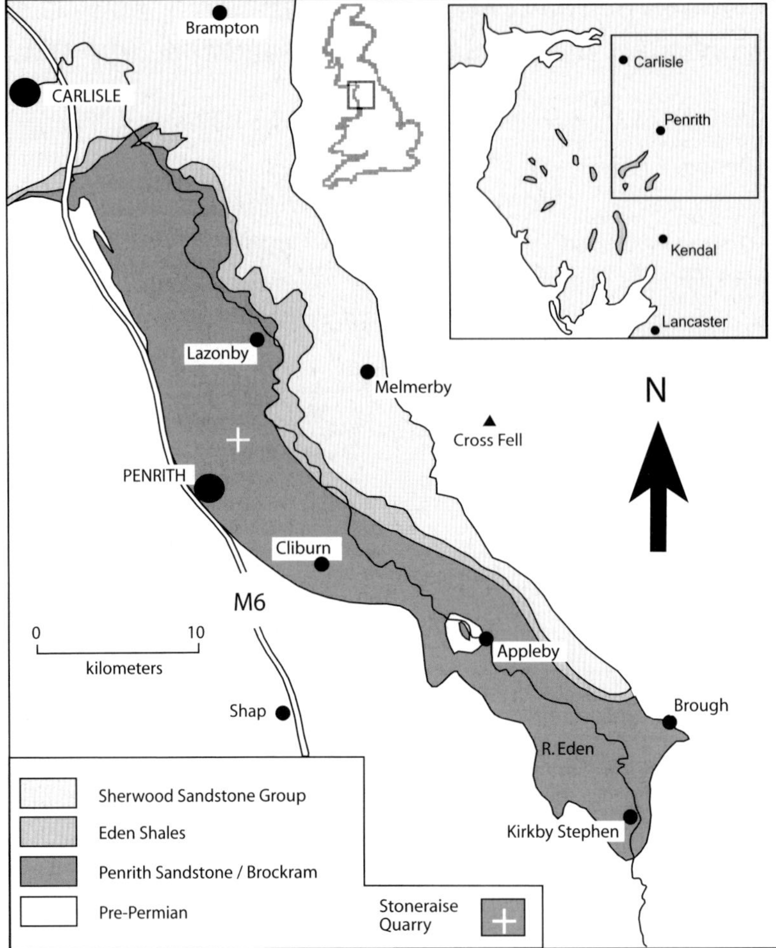

Fig. 1. Map showing location of Penrith Sandstone and the sampling location (Stoneraise Quarry). The inset figure shows the far NW coastline of England, with the Irish Sea to the west.

well-rounded, highly spherical quartz grains, often showing euhedral overgrowths of quartz. The well-rounded grains are covered with a red, dust-like, coating of iron oxide or hydroxide on their surfaces: in the case of the grains showing overgrowths, the overgrowths have formed over the oxide coatings. Examination of polished sections in reflected light suggests that goethite is the most likely iron mineral present (Harvey *et al.* 1995).

The aims of the work reported here are to examine the small-scale spatial distribution of porosity, permeability and electrical resistivity, and to evaluate the relationships between these properties, using, where appropriate, petrographical and geochemical evidence. For this work a number of the Stoneraise Quarry

samples were cut into blocks 45 mm wide, 40 mm deep, and between 200 and 260 mm in length: the long axis of the blocks was approximately perpendicular to the bedding. Where possible, contiguous blocks were chosen so that one could be imaged for porosity, permeability and electrical resistivity, whist the adjoining block could be plugged for conventional petrophysical measurements, and also sampled for mineralogical, geochemical, petrographic and X-ray characterization.

Mineralogy, chemistry and fabric

Examination of polished thin sections of these rocks shows that while quartz is by far the most abundant mineral present, quartz grains exhibit two distinct types: single grains, which make up the majority, and a much smaller number of grains made up of multiple subgrains (Harvey *et al.* 1995).

Within the less homogeneous zones characterized by the presence of thin laminations, quartz makes up approximately 90% of the total minerals, the remainder being potassium feldspar (orthoclase). The grain-size variation within these heterogeneous zones is extensive, ranging from less than 100 µm to about 700 µm, very fine–coarse sand. Figure 2a shows a good example of this type of heterogeneity.

Within the coarser and generally more homogeneous beds, the quartz content increases to 95% by volume or more, with a corresponding decrease in the potassium feldspar (orthoclase) to about 5%. The grain-size variation in these coarser beds is much more restricted. This feature can be clearly seen in Figure 2b, where the grain size ranges from 600 to 1000 µm, coarse–very coarse sand, and shows the presence of two or more different grain-size populations with cementation playing an important role.

The rock is cemented throughout by quartz overgrowths that are in optical continuity, and hence crystallographic continuity, with the quartz grains on which they grew. The original well-rounded grains can be clearly seen in Figure 2b, being picked out by the iron oxide/hydroxide coatings, with the quartz overgrowths being superimposed on these.

Examination of the sandstone by scanning electron microscopy (SEM) highlights some of the features that control this rock's porosity and permeability. Figure 3a shows the extent of the development of the overgrowths and its effect on the overall porosity. The two grains in the centre of the figure show the two extremes of the development of the overgrowths. The grain showing good crystal faces is a quartz overgrowth that has had little impedance to growth. In places, grain–grain cementation has occurred as the void between the grains has been filled by the overgrowths. The voids between the triple grain contacts appear to be filled to a lesser extent, so that these zones therefore make up the majority of the porosity. Figure 3b shows one of these triple grain contacts in which a void is partially filled by a feldspar grain that is

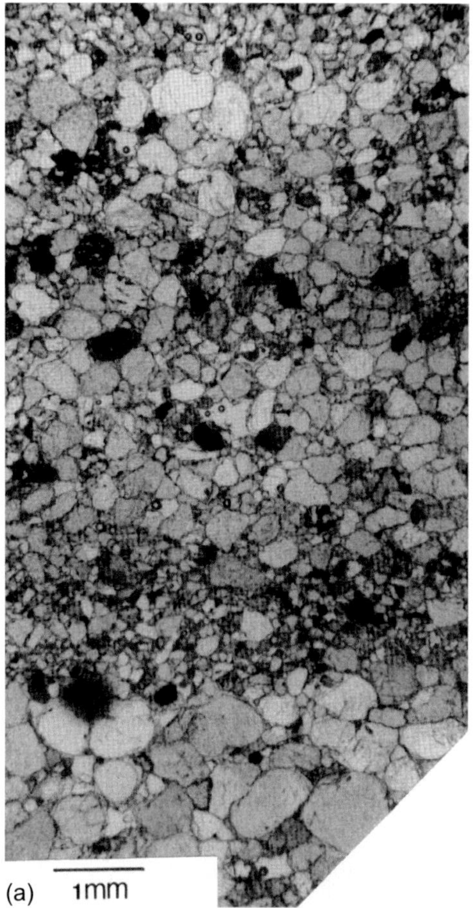

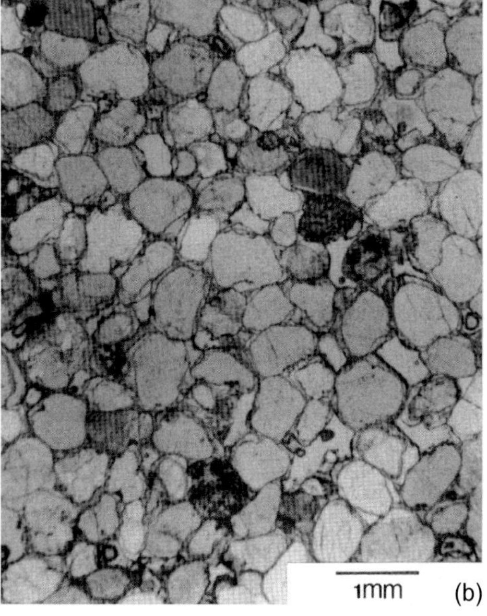

Fig. 2. Thin section images of the Penrith Sandstone. (**a**) Heterogeneous sample with fine-grained laminae with more pervasive cementation. (**b**) Syntaxial quartz overgrowths act as the cementing agent, but leave high porosity in this coarse lamina.

beginning to break down. Even so, the void is still relatively free from blockage. Figure 3c & 3d shows a more open packing of the grains and therefore a more developed porosity. Again, the cementation of the grains is by quartz over-growths, and in Figure 3d the void is partially filled by small quartz crystals.

No clays or micas were found in the thin sections or SEM samples examined, an important mineralogical characteristic that is further

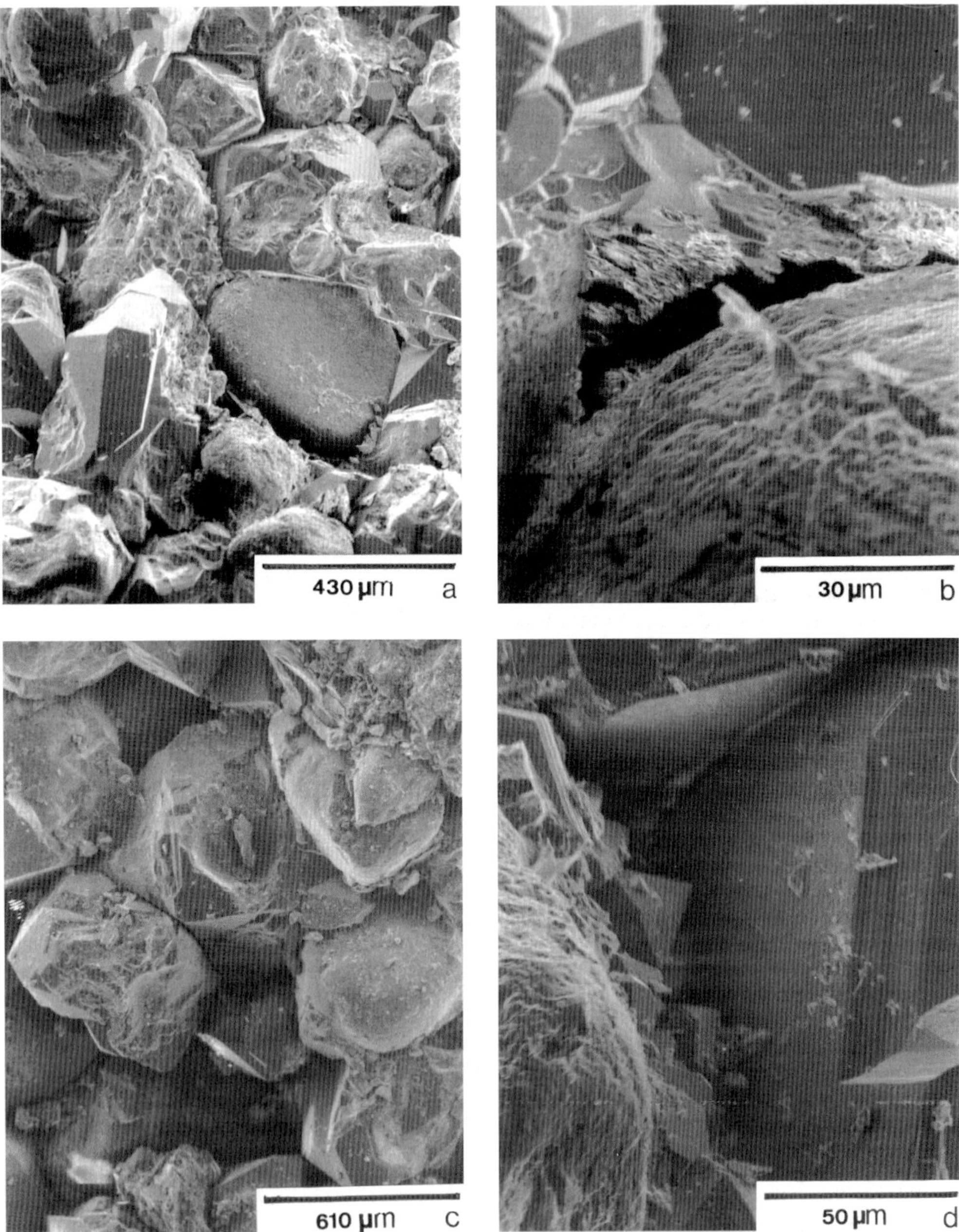

Fig. 3. Scanning electron microscope images of the Penrith Sandstone. (a), (b) & (d) Quartz overgrowths impinging on pore space with developed pyramidal crystal faces. (a) & (c) Overgrowths in contact with each other and partially or completely occluding the pore space.

backed up by chemical analysis of representative samples of the sandstone. This is also supported by results presented later in this paper where electrical measurements are made for a range of fluid salinities. The major element chemistry of four samples from the same block was obtained by X-ray fluorescence analysis using a fusion technique, and the modal mineralogy calculated in terms of quartz, potassium feldspar and iron oxide. These results are summarized in Table 1.

The chemistry is entirely compatible with the polished section observations in that the least homogeneous sample (Sample 14) has significantly higher alumina and potash, and lower silica, than the other three. In terms of the calculated mineralogy, Sample 14 has nearly 10% potassium feldspar, about twice the figure for the rest of the block. CaO, Na_2O and MnO are below their detection limits of 0.01 wt per cent (wt%) in all cases. LOI is the loss on ignition at 1025 °C, and is a measure of volatiles (water and carbon dioxide mainly) in the sample together with some adjustment for the oxidation of any divalent iron that was originally present. The latter has a negligible effect in these samples, and the very low LOI figures are a further confirmation of the absence of micas and clay minerals.

A grain density calculated from the estimated modal mineral values [ρ(calculated)] is in good agreement with that directly determined [ρ(measured)] during the petrophysical characterization (Table 1).

Overall, therefore, the rock chosen for this study is a very clean sandstone composed essentially of well-rounded quartz grains together with a small percentage of potash feldspar and a dusting of iron hydroxide. The rock is cemented by the overgrowth of quartz, often forming good crystal habit as growth takes place into the void space, on the original detrital grains. The most obvious heterogeneity in the rock is caused by a variation in grain size, with thin laminae of fine-grained sandstone occurring within the generally coarser rock. There is an approximately equal thickness of overgrowth in both the coarser and finer grain layers (Fig. 2). The pore connections or throats in the coarser grained material should hence be larger and more open than in the finer grained laminations. SEM observations confirm this and show a generally smaller pore size with few openings between pores in the finer grained laminae. The latter may therefore be expected to have a distinctly lower permeability and higher resistivity.

Petrophysical characteristics

Conventional analyses: porosity–permeability–capillary pressure curves

Four core plugs were taken for the following petrophysical analyses as follows: helium

Table 1. *Chemical analysis of the Penrith Sandstone*

	Sample 14	Sample 16	Sample 17	Sample 18
SiO_2 (% by weight)	95.88	98.06	97.28	98.01
Al_2O_3 (% by weight)	2.25	1.08	1.36	1.10
TiO_2 (% by weight)	0.07	0.03	0.04	0.04
Fe_2O_3 (% by weight)	0.63	0.36	0.40	0.37
MgO (% by weight)	0.11	0.09	0.11	0.10
CaO (% by weight)	<0.01	<0.01	<0.01	<0.01
Na_2O (% by weight)	<0.01	<0.01	<0.01	<0.01
K_2O (% by weight)	1.25	0.47	0.69	0.46
MnO (% by weight)	<0.01	<0.01	<0.01	<0.01
P_2O_5 (% by weight)	0.02	0.02	0.02	0.01
LOI* (% by weight)	0.36	0.23	0.29	0.23
Total (% by weight)	100.57	100.34	100.19	100.32
Calculated modal composition				
Quartz (% by volume)	89.45	95.18	93.36	95.10
Orthoclase (% by volume)	9.90	4.44	6.12	4.49
Fe-oxide (% by volume)	0.61	0.36	0.44	0.37
Density (kg m^{-3}) ρ(calculated)	2.652	2.652	2.651	2.652
Density (kg m^{-3}) ρ(measured)	2.648	2.646	2.649	2.647
Difference in density ρ(calculated–measured) %	0.15	0.23	0.06	0.19

* Loss on ignition.

porosity, gas permeability, four-electrode electrical resistivity and mercury injection capillary pressure curves.

The helium porosity measurements provide an estimate of the connected porosity of the samples. Gas permeabilities were uncorrected for slippage at the pore wall (Klinkenberg effect) but still provide a relative indication of the gas permeability of each sample. Porosity and permeability measurements on the same samples (Table 2) show that while the porosity of the samples remains consistent at around 12%, the gas permeability ranges from 1.6×10^{-14} to 8.0×10^{-13} m^2. This variability in permeability includes considerably lower values than those generated through mini-permeameter measurements shown on the image in Figure 6 later. These low-permeability values (samples 16 and 17) may be a result of the development of more pervasive cementation through quartz overgrowths, as described in Figures 2 and 3, reducing the flow at the plug scale, while not impinging on the localized mini-permeameter measurement. The core plugs were drilled perpendicular to the mini-permeameter measurement surface (i.e. the gas permeameter and mini-permeameter values were measured along the same axes).

Archie m values were calculated from measurements of electrical resistivity and porosity on block samples using a standard four-electrode core-analysis technique. It was assumed there was no excess surface conduction, i.e. $a = 1$ in Archie's equation:

$$\rho_o = a\rho_w\phi^{-m} \qquad (1)$$

where ρ is resistivity, subscripts o and w refer to the water-saturated rock and the saturating water, and ϕ is fractional porosity.

The values of m calculated from these measurements range from 1.90 to 1.91 (Table 3), suggesting a reasonably clean 'Archie' rock with no excess conductivity associated with clays or bound water (Worthington 1982).

For each of the four samples of Penrith Sandstone from this location, mercury injection capillary pressure curves (Figs 4 & 5) demonstrate that each sample has a unimodal pore-size distribution with a single modal range which varies between 25–50 and 70–80 μm. Examination of the capillary pressure curves for these samples suggests the lower permeabilities correspond with an increase in the proportion of smaller pore spaces.

These results are supported by examination of thin sections of the sandstone, the latter clearly demonstrating the extent of the development of quartz overgrowths, effectively in-filling the voids between the grains.

Images

Introduction. In addition to the conventional analyses, detailed images were also obtained using optical photography, high-resolution electrical resistivity measurements, surface porosity from image analysis and mini-permeametry.

Porosity and permeability images. Surface porosity was obtained by taking a greyscale image of the upper surface that had been previously surface ground and then filled with talc. Pores and grains are thus assigned a discrete greyscale contrast enabling quantification of the pore-space proportion. Edge effects, where pores appear to overlap the sample boundary, are minimized because the sample size is large compared with the pore size: while variations in apparent v. true pore size may be important locally (for example in calculating Archie m values from porosity and resistivity data at the

Table 2. *Porosity and permeability values from core-plug measurements*

Sample No.	Porosity (%)	Permeability (m^2)
14	13.0	8.0×10^{-13}
16	11.7	1.6×10^{-14}
17	12.3	6.8×10^{-14}
18	13.0	3.0×10^{-13}

Table 3. *Archie 'm' values for the block shown in Figures 6–9*

Porosity (%)	Measured		Calculated from smoothed image	
	Formation Factor	Archie 'm' value	Formation Factor	Archie 'm' value
20.1	20.96	1.90	20.62	1.89
20.2	20.85	1.90	19.01	1.84
19.4	22.75	1.91	23.29	1.92

pore scale), they may be assumed to provide a representative view at the larger scale.

Permeability images were obtained at a similar surface resolution (5 × 5 mm), the minipermeameter measurements (Bourke 1991, 1993; Bourke *et al.* 1993) being located at the centre of the sampling zone for the porosity measurement (Fig. 6). A more detailed consideration of these separate measurements and the different volumes of rock interrogated is given by Lovell *et al.* (1998*b*).

Plug-sample porosity and permeability data show reasonable agreement with the images presented in Figure 6 given the different measurement methods used, although the detailed images inevitably show a wider degree of variation than the plug-averaged values.

Figure 7 shows logs derived from the images in Figure 6 by averaging pixel values across the sample. The visual similarity in the responses can again be seen, and the core can be zoned through application of cut-off values (see below).

Electrical images. Researchers have previously obtained porosity and permeability images or maps; electrical images of core are, however, a novel concept. Consequently, we discuss here their origin and principle.

Electrical resistivity measurements play a central role in both laboratory and downhole petrophysics. Extensive experimental and theoretical studies are well documented, whilst laboratory measurements provide essential control parameters for calibrating and evaluating *in situ* data. Over the past two decades downhole electrical resistance images have been produced by specially designed tools (e.g. Ekstrom *et al.* 1986; Boyeldieu & Jeffreys 1988). These downhole images of the borehole wall represent the highest resolution currently available using resistivity, with a vertical measurement interval of approximately 2.5 mm. The resulting images are well reported in the literature, and can show bedding in sedimentary sequences, fractures, foliations and property contrasts, all of which may be compared with visual inspection of the core (e.g. Luthi 1990; Lovell *et al.* 1998*a*). At present these images are particularly useful in aiding our understanding of geological processes, mapping structural discontinuities and assessing possible petrophysical changes qualitatively. Indeed, in mud-filled boreholes the downhole electrical images are the closest yet obtainable to a traditional geological log. Acoustic images are also possible, and while these are especially useful for determining borehole shape (and stress) and surface texture, they are generally not as useful for sedimentological purposes.

These downhole imaging techniques use multiple single-point resistance measurements to assess fine changes in electrical resistivity at the surface of the borehole wall. The resistance measurements are essentially relative and cannot be converted to resistivities with great confidence. Furthermore, each individual button electrode is compensated by equalization techniques that produce superb images, but further complicate quantification. Whilst contrasts can thus be identified and utilized to highlight features of interest, the resistivity of a sealed fracture cannot, however, be assessed with sufficient confidence to predict its porosity. The need to convert downhole images to quantitative resistivities is thus of considerable importance. Attempts at using conventional downhole logs to calibrate the electrical images have been made using reasonably high-resolution logs, such as the microspherically focused tools. The difficulty is in obtaining a quantitative resistivity measurement at a similar scale to the resolution of the imaging device. In thickly bedded, homogeneous sequences the constraints are such that calibration may be feasible (Hansen & Parkinson 1999). Unfortunately, in thinly bedded or heterogeneous zones the devices sample different volumes and thus comparison and calibration is less reliable. This sample-scale problem is complicated further by differences in measurement, processing and display characteristics, as described above.

Laboratory investigations of electrical resistivity have been made for many decades and require relatively simple measurement techniques. The simplicity is, however, misleading in that there is no universally accepted method. Whilst the objective has normally been to obtain an average value of the gross resistivity of core samples and plugs, this has the disadvantage of ignoring the smaller scale features and can be degraded by electrode effects. Jackson *et al.* (1995) describe theoretical and practical experiments that have resulted in the development of a high-resolution electrical resistivity imaging device for use on core samples in the laboratory (Jackson *et al.* 1990, 1991). Potentially, such images could provide the quantitative resistivity data for calibrating downhole images.

In porous rocks the electrical flow is generally considered to be dominated by electrolytic conduction through the pore fluid. Sundberg (1932) defined the 'formation factor' (FF) as an

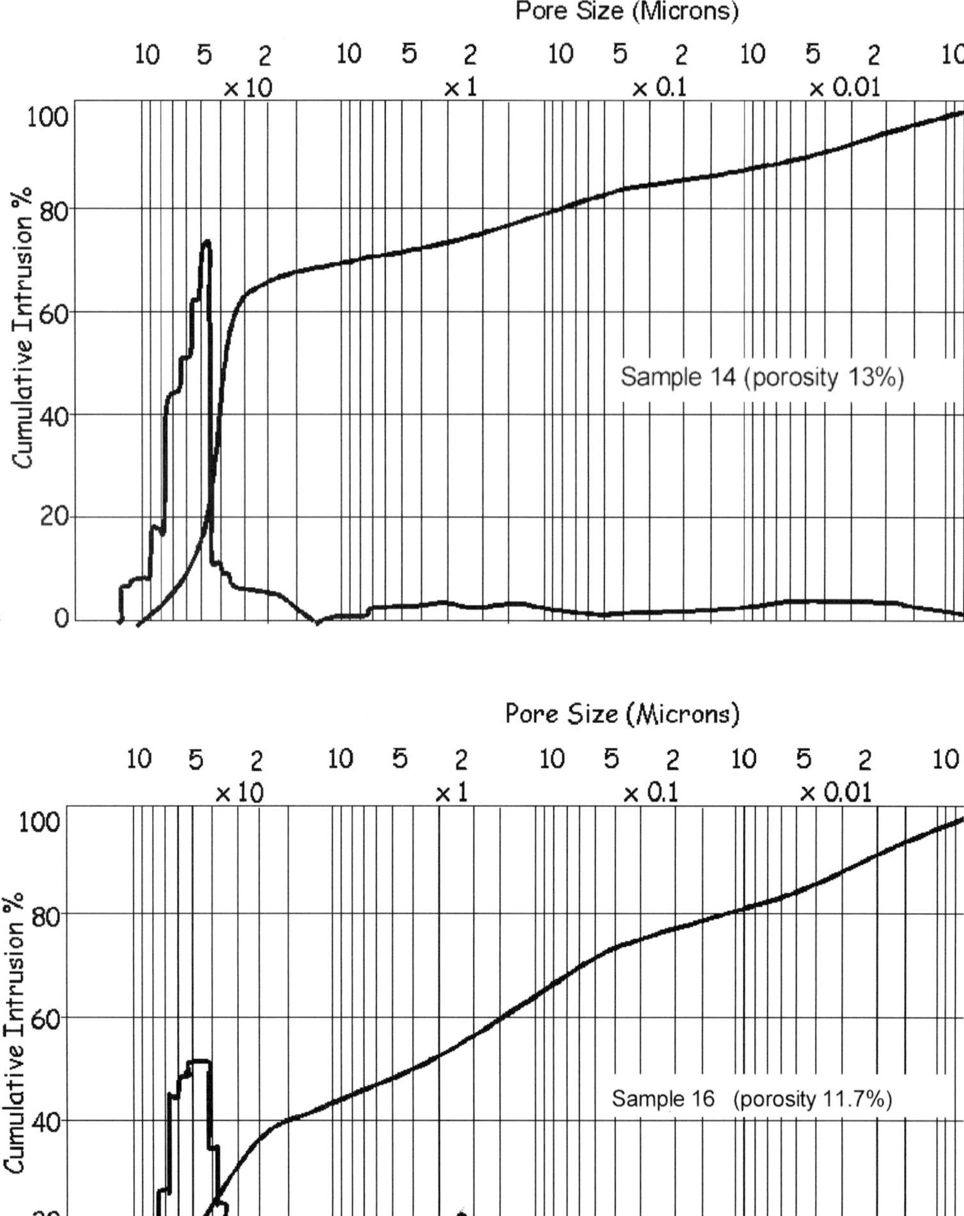

Fig. 4. Mercury injection capillary pressure curves for four samples of Penrith Sandstone.

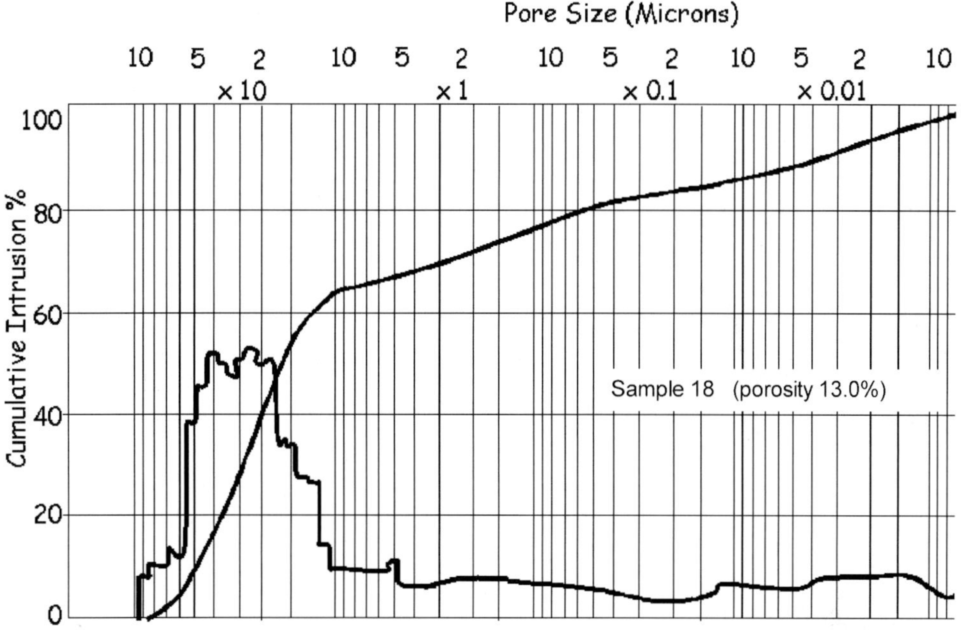
Pore Size (Microns)
10 5 2 10 5 2 10 5 2 10 5 2 10
× 10 × 1 × 0.1 × 0.01
Cumulative Intrusion %
100
80
60
40
20
0
Sample 18 (porosity 13.0%)

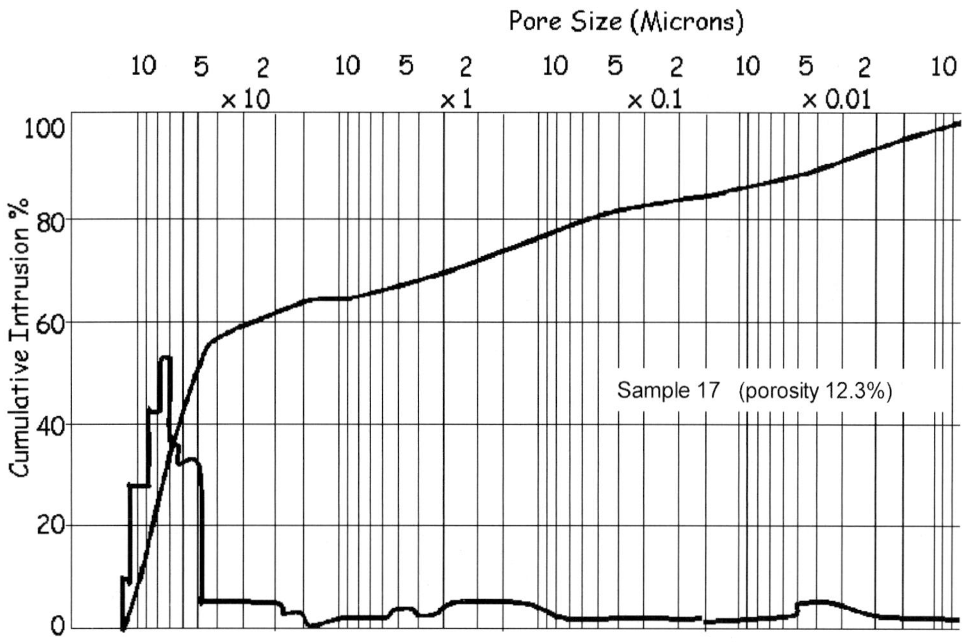
Pore Size (Microns)
10 5 2 10 5 2 10 5 2 10 5 2 10
× 10 × 1 × 0.1 × 0.01
Cumulative Intrusion %
100
80
60
40
20
0
Sample 17 (porosity 12.3%)

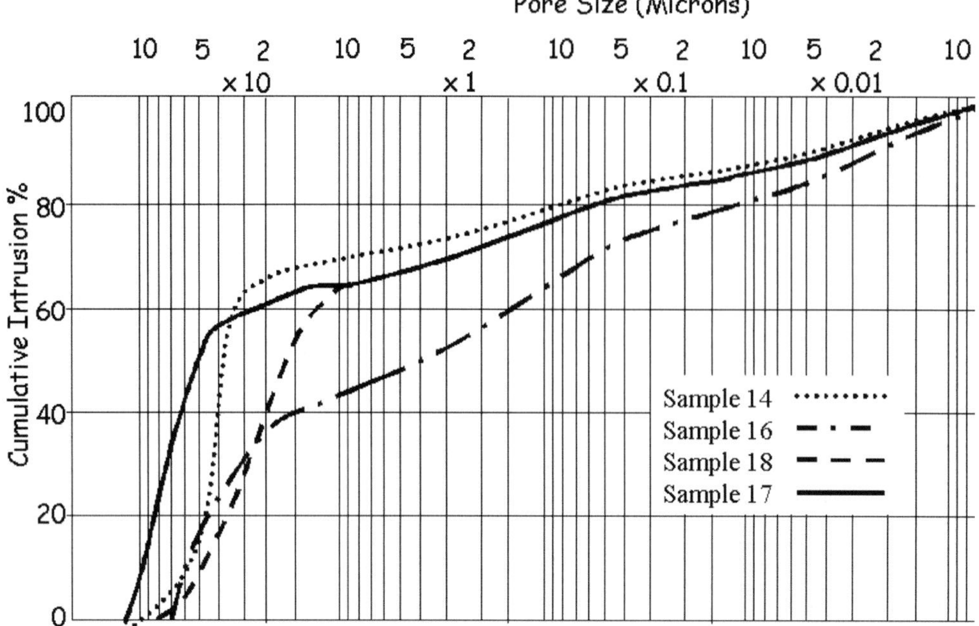

Fig. 5. Mercury injection capillary pressure curves compared for four samples of Penrith Sandstone.

intrinsic property of a fully saturated rock, independent of the nature of the pore fluid

$$FF = \frac{\rho_o}{\rho_w}. \tag{2}$$

This was shown by Archie (1942, 1950) to be related to the porosity (ϕ) (see equation 1).

The electrical images presented here (Figs 6, 8 & 9) use a four-electrode measurement technique, with current passing through electrodes separate from those that measure the potential developed. It can be shown that the resistance can be converted to a resistivity by multiplying by a factor that is related to the geometry of the electrodes; thus, the measurements are quantitative resistivities. Furthermore, experiments with calibration samples demonstrate both high precision and accuracy, together with excellent repeatability (Lovell *et al.* 1998*b*).

Resaturation effects. The samples were resaturated using a large volume of fluid under vacuum that was introduced at one end, thus displacing any gas. Two weeks were allowed for any diffusion to occur in the $50 \times 80 \times 260$ mm sample. Figure 8 shows an electrical formation factor image at ambient conditions and under pressure; the visual similarity between the images infers there is no significant gas present in the larger pores and that the resaturation process has been accomplished.

Shale effects. Figure 9 shows two electrical formation factor images of the same sample of rock saturated with different salinity fluids with fluid resistivities of 0.165 and 0.54 Ωm. The consistency of formation factor values between the two images supports the interpretation of the Penrith Sandstone as a clean sandstone. This conclusion, supported by the chemical and optical evidence discussed above, is important when considering the electrical properties of water-saturated rocks. At low salinities clay particles provide an excess conductivity that reduces the formation factor. Thus, in these circumstances the formation factor is no longer a constant but depends on the salinity of the saturating fluid. The results here demonstrate that this is not a problem with these samples of Penrith Sandstone.

Image interpretation. A collection of images for a single Penrith Sandstone block are presented in Figure 6. These show a photograph of the upper surface of the sandstone block, together with the porosity map and permeability map of the same surface. Figure 7 allows an even clearer, but averaged, comparison to be made. It

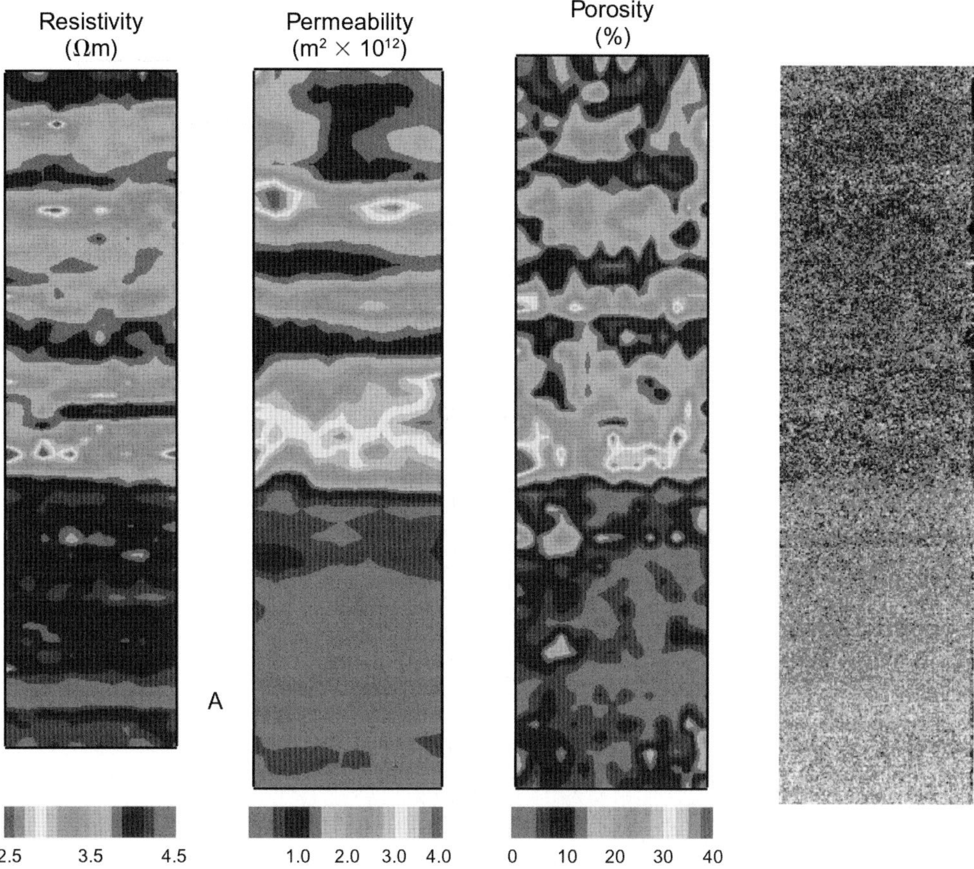

Fig. 6. Electrical, permeability, porosity and optical images. The length of the block in each case is 26 cm. The lower part of the block has a resistivity of more than $c.$ 4 Ωm, a permeability of less than $c.$ 1 × 10^{-12} m², and a porosity of less than $c.$ 10%. The upper part of the block, in the lighter coloured bands, has a resistivity mainly between 2.5 and 3.5 Ωm, a permeability of 1.5 to 4 × 10^{-12} m², and a porosity of more than 15%: the darker coloured bands have properties similar to the upper part of the block.

can be seen that increasing porosity and permeability correspond both with each other and with increasing electrolytic conductivity (i.e. decreased resistivity); effectively more pore space in the rock (increasing porosity) enables easier fluid flow (higher permeability) and easier electrical conduction (more conductive and better connected fluid).

The images demonstrate an overall similarity with increased porosity, permeability and conductivity towards the top half of the mapped core. These highly porous and permeable volumes in the upper part of the core correspond in turn with more homogeneous and coarser grained zones where the grain size is well sorted and where there are relatively few quartz overgrowths. In contrast, the lower half of the mapped core is characterized by lower porosity, permeability and conductivity. In addition, there is a linear horizontal feature visible cutting across the core, most clearly seen in the conductivity image as a discrete narrow zone of low conductivity (or high resistivity) (A, Fig. 6). In the permeability map this feature is also visible but as a more diffuse zone; this is not surprising given the different physics of the measurements and the fact that the gas flow for the mini-permeameter invades a zone of finite size. In contrast, the porosity map appears to have very poor resolution. The acquisition of this porosity image is based on separation of pores from grains using image analysis of the talc-impregnated ground surface. This surface map has minimal depth of penetration and the

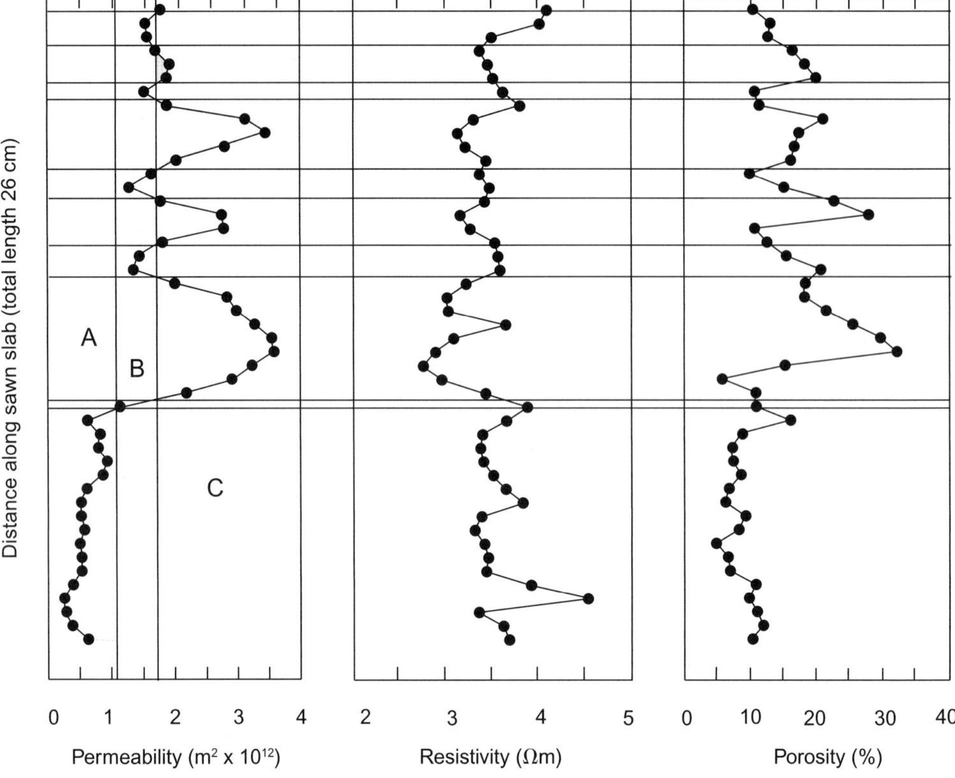

Fig. 7. Core logs derived from images in Figure 5. An example classification based on permeability values is indicated (see text).

(disturbed) surface examined may not be representative.

The images, however, correlate well and suggest that simple relationships may exist between the porosity, permeability and the electrical conductivity (resistivity). Figure 10 shows histograms for both the porosity and the permeability; while the porosity distribution is unimodal, the permeability distribution seems to be tri- or even quadramodal: the logs of Figure 7, for example, can be interpreted on the basis of three classes (labelled A, B and C). Zoning of core on the basis of porosity and permeability values seems to indicate that fine-scale fabric changes are important in determining the values of these properties. A cross-plot of porosity against permeability (Fig. 11) shows considerable scatter, suggesting that the gross relationship at the large sample size is not as simple as initially suggested, and that fine-scale sedimentological fabric related to deposition and overgrowths is important at the pore scale.

Conclusions

The Penrith Sandstone is a clean aeolian sandstone, dominated by quartz grains, with quartz overgrowths. Chemically, the rock consists predominantly of SiO_2 (>95%), with no evidence in the analyses presented here of the presence of carbonate cements or clay minerals. The grain size varies from very fine sand (100 μm) to coarse or very coarse sand (>700 μm).

Petrophysically, the Penrith Sandstone is a typical clean sandstone characterized by relatively high porosity and permeability. Detailed imaging of core samples demonstrates that, while the formation is well sorted at the larger scale, there is fine-scale variability resulting from thin laminations of finer grained material together with quartz overgrowths that tend to infill larger voids, thus reducing permeability. Capillary pressure curves for four samples demonstrate a unimodal pore-size distribution with a modal range that varies between 25–50 and 70–80 μm. Porosity measurements are

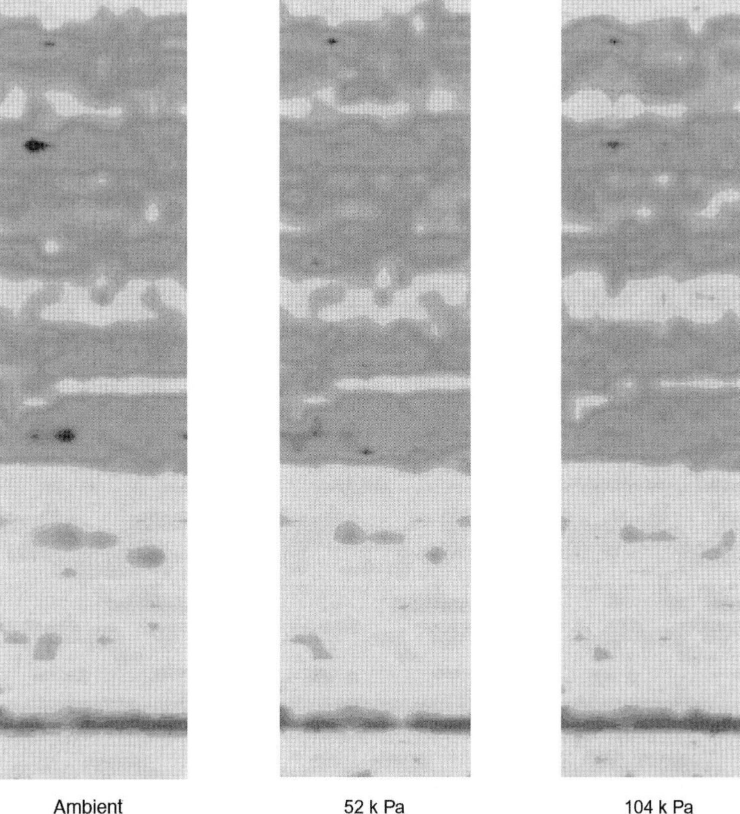

Fig. 8. Formation factor distributions at different water pressures. The similarity of the results indicates that full resaturation of the sample was achieved. In all three images, the darker shades of the upper part of the block represent formation factors in the range 13–19, and the lighter shades of the lower part factors between 17 and 22. The dark band near the base represents a value of 25–26.

consistent values at around 12%, with the permeability ranging from 1.6×10^{-14} to 8.0×10^{-13} m^2. Examination of the capillary pressure curves suggests lower permeabilities correspond with an increase in the proportion of smaller pore spaces. This agrees with examination of thin sections of the sandstone. Calculations of Archie m exponents range from 1.90 to 1.91, suggesting a reasonably clean 'Archie' rock (i.e. non-shaly) with no excess conductivity associated with clays or bound water. Because of its relative simplicity, the Penrith Sandstone is potentially a very useful rock for petrophysical investigations of new techniques and fundamental issues. In this study novel imaging techniques undertaken on blocks of the sandstone provided insights into the issues of scale.

Two-dimensional optical, electrical resistivity, porosity and mini-permeametry images of blocks of the sandstone indicated that the distributions of property values are affected by subtle variations in the fabric of the rock. The success of the novel resistively imagining method suggests that this technique may be an appropriate way to quantify borehole resistivity imaging logs. In the present investigation, although the resistivity, porosity and permeability distributions are clearly all related to rock fabric, there is still substantial uncertainty in the relationships between the properties. This appears to be due to variations in fabric at scales even below the resolution of the quantitative methods used here, but which are, nevertheless, indicated by thin section and electron microscope images also obtained.

The authors thank the Natural Environment Research Council (Ocean Drilling Program Special Topic Initiative) for support in developing electrical resistivity core imaging, a collaborative research programme

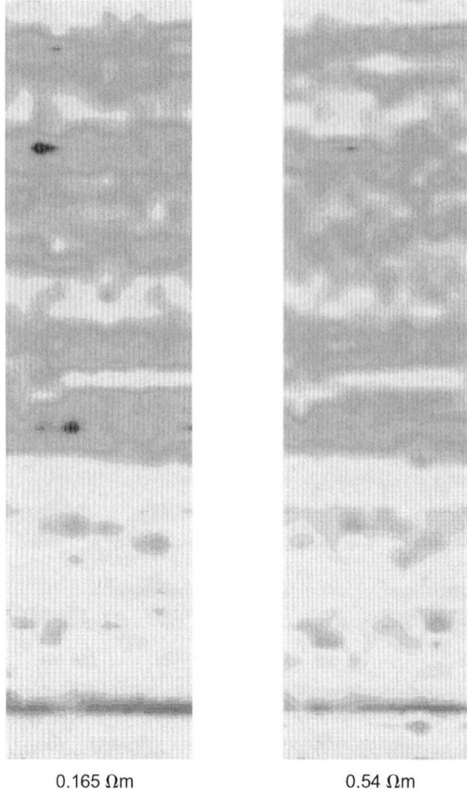

Fig. 9. Formation factor distributions at different salinities (0.165 and 0.54 Ωm). The similarity of the images suggests that excess conduction due to clays is not significant. In both images the darker shades of the upper part of the block represent formation factors in the range 13–19, and the lighter shades of the lower part factors between 17 and 22. The dark band near the base represents a value of 25–26.

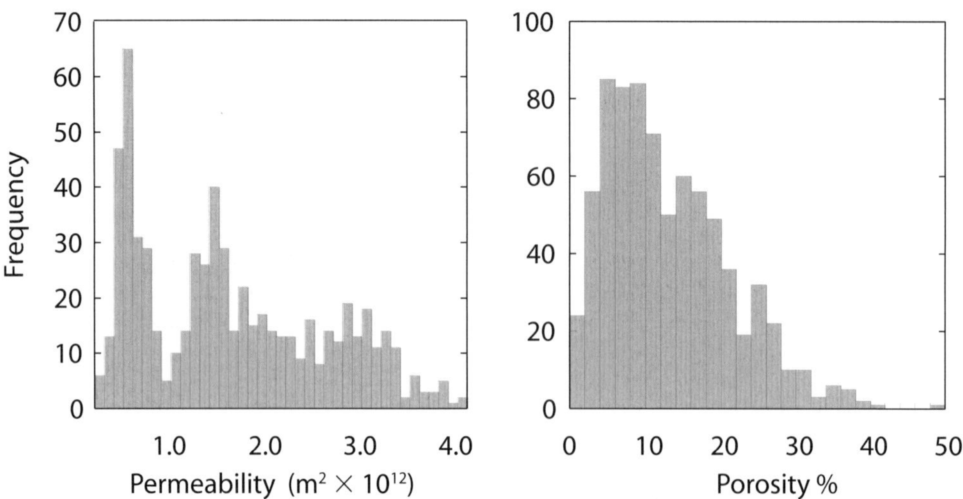

Fig. 10. Porosity and permeability frequency distributions derived from micrograph image analysis and mini-permeameter data.

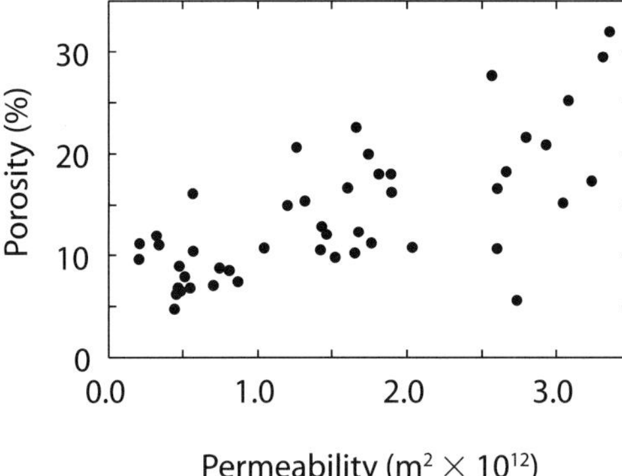

Fig. 11. Porosity and permeability plot for porosity data derived by image analysis and permeabilities derived from mini-permeameter experiments.

between the University of Leicester and the British Geological Survey. Enterprise Oil Ltd, LASMO North Sea plc, Shell UK Ltd, Mobil North Sea Ltd and the Offshore Supplies Office of the UK Department of Trade and Industry are thanked for their support under the LAMBDA programme (Laboratory Measurements and Borehole Data Analysis). P. D. Jackson and R. C. Flint acknowledge this paper is published with the permission of the Director, BGS.

References

ARCHIE, G.E. 1942. The electrical resistivity log as an aid in determining some reservoir characteristics. *Transactions of the American Institute of Mining and Metallurgical Engineers*, **146**, 54–62.

ARCHIE, G.E. 1950. Introduction to petrophysics of reservoir rocks. *AAPG Bulletin*, **34**, 943–961.

BOURKE, L.T. 1991. Permeability imaging for detailed reservoir characterization. *In*: WORTHINGTON, P.F. & LONGERON, D. (eds) *Advances in core evaluation II – Reservoir Appraisal*. Gordon & Breach, Philadelphia, PA, 407–427.

BOURKE, L.T. 1993. Core permeability imaging – its relevance to conventional core characterization and potential application to wireline measurement. *Marine and Petroleum Geology*, **10**, 297–408.

BOURKE, L.T., CORBIN, N., BUCK, S.G. & HUDSON, G. 1993. Permeability images – a new technique for enhanced reservoir characterization. *In*: ASHTON, M. (ed.) *Advances in Reservoir Geology*. Geological Society, London, Special Publications, **69**, 219–232.

BOYELDIEU, C. & JEFFREYS, P. 1988, Formation microscanner: new developments. *In*: *11th European Formation Evaluation Symposium Transactions*. Society of Professional Well Log Analysts, Oslo, Norway, paper WW.

EKSTROM, M., DAHAN, C.A., CHEN, M., LLOYD, P.M. &

ROSSI, D.J. 1986. Formation imaging with micro-electrical scanning arrays. *In*: *27th Annual Logging Symposium Transactions*. Society of Professional Well Log Analysts, Houston, TX, paper BB.

HANSEN, T. & PARKINSON, D.N. 1999. Insights from simultaneous acoustic and resistivity imaging. *In*: LOVELL, M.A., WILLIAMSON, G. & HARVEY, P.K. (eds) *Borehole Imaging – Applications and Case Histories*. Geological Society, London, Special Publications, **159**, 191–202.

HARVEY, P.K., LOVELL, M.A. *ET AL.* 1995. Electrical resistivity core imaging III: Characterisation of an aeolian sandstone. *Scientific Drilling*, **5**, 165–176.

JACKSON, P.D., LOVELL, M.A. *ET AL.* 1995. Electrical resistivity core imaging I: A new technology for high resolution investigation of petrophysical properties. *Scientific Drilling*, **5**, 139–152.

JACKSON, P.D. LOVELL, M.A., PITCHER, C.A., GREEN, C.A., EVANS, C.J., FLINT, R. & FORSTER, A. 1990. Electrical resistivity imaging of core samples. *Transactions of the European Core Analysis Symposium (EUROCAS I), London, May*, Gordon & Breach Science.

JACKSON, P.D. & ODP LEG 133 SHIPBOARD PARTY. 1991. Electrical resistivity core scanning: a new aid to the evaluation of fine scale sedimentary structure in sedimentary cores. *Scientific Drilling*, **2**, 41–54.

LOVELL, M.A., HARVEY, P.K., BREWER, T.S., WILLIAMS, C.G., JACKSON, P.D. & WILLIAMSON, G. 1998*a*. Application of FMS images in the Ocean Drilling Program: an overview. *In*: CRAMP, A., MACLEOD, C.J., LEE, S.V. & JONES, E.J.W. (eds) *Geological Evolution of Ocean Basins: Results From the Ocean Drilling Program*. Geological Society, London, Special Publications, **131**, 287–303.

LOVELL, M.A., HARVEY, P.K., JACKSON, P.D., BREWER, T.S., WILLIAMSON, G. & WILLIAMS, C.G. 1998*b*. Interpretation of core and log data – integration or calibration? *In*: HARVEY, P.K. & LOVELL, M.A. (eds) *Core-log Integration*. Geological Society, London, Special Publications, **136**, 39–51.

LUTHI, S.M. 1990. Sedimentary structures of clastic rocks identified from electrical borehole images. *In*: HURST, A., LOVELL, M.A. & MORTON, A.C. (eds) *Geological Applications of Wireline Logs*. Geological Society, London, Special Publications, **48**, 3–10.

MACCHI, L.A. 1990. *Field Guide to Continental Permo-Triassic Rocks of Cumbria and of North West Cheshire*. Liverpool Geological Society.

SUNDBERG, K. 1932. Effect of impregnating waters on electrical conductivity of soils and rocks. *Transactions of the American Institute of Mining and Metallurgical Engineers*, **79**, 367–391.

WAUGH, B. 1970. Petrology, provenance and silica diagenesis of the Penrith Sandstone (Lower Permian) of Northwest England. *Journal of Sedimentary Petrology*, **40**, 1226–1240.

WORTHINGTON, P.F. 1982. The influence of shale effects upon the electrical resistivity of reservoir rocks. *Geophysical Prospecting*, **30**, 673–687.

Pore geometry of Permo-Triassic sandstone from measurements of electrical spectroscopy

JULIAN B. T. SCOTT & R. D. BARKER

*School of Geography, Earth and Environmental Sciences, University of Birmingham,
Birmingham B15 2TT, UK*

Abstract: In order to provide an indirect method of estimating the hydraulic properties of sandstone aquifer rocks, electrical spectroscopy measurements were made on Permo-Triassic Sandstone over the frequency range 0.0001–1000 Hz. Samples from several boreholes across Britain are shown to exhibit a phenomenon called electrical relaxation, which we have modelled using a generalized Cole–Cole equation. The Cole–Cole parameters correlate well with other hydrogeologically important parameters determined using mercury injection capillary pressure and nitrogen adsorption techniques. Of greatest importance is that the relaxation time, τ, is strongly related to the dominant pore-throat size from which intergranular permeability may be estimated. A normalized version of the chargeability appears to be related to the surface conductivity and this gives the possibility of estimating the pore surface area to volume ratio, a value that is important in determining both permeability and sorption. The potential uses of electrical spectroscopy in hydrogeology and many other fields, including geology and petroleum petrophysics, are only now becoming apparent, and further advances are certain.

Low-frequency electrical spectroscopy is the measurement and analysis of the variation of electrical resistivity with frequency of the applied current. In geophysics it is often called spectral induced polarization (SIP) and has been used for many years, mainly for mineral resource location and discrimination (Van Voorhis *et al.* 1973; Pelton *et al.* 1978). Work over the last 20 years has shown that electrical spectroscopy can also provide useful information on non-mineralized sedimentary rocks and soils, from characterization of internal structure to determination of pore-fluid chemistry and contamination (Olhoeft 1985; Börner *et al.* 1993; Slater & Lesmes 2002). However, in the past, the variation in properties over the lowest frequency range 0.0001–1000 Hz has often been ignored due to either limited measurement capability or because studies focused on rock types with little frequency variation, but it is in this frequency range where the most interesting and useful information is now being obtained (Lesmes & Mòrgan 2001; Scott & Barker 2003).

The Permo-Triassic sandstone is the second most important aquifer in England and Wales, providing much of the groundwater abstraction for major conurbations including Birmingham and Manchester (Allen *et al.* 1997). A knowledge of its small-scale internal structure and hydraulic properties is important in understanding fluid flow through the sandstone and in water resource and contaminant studies. It is possible that this information may be obtained indirectly from measurements of spectral induced polarization. Here we present the first detailed study of spectral induced polarization of Permo-Triassic sandstone from the British Isles. The results are compared with hydrogeologically important parameters, such as cation-exchange capacity, pore surface area and pore-throat size distributions, determined using a variety of other techniques. We provide details of the measurements and develop some explanations for the observations.

Electrical spectroscopy

Conventional measurement of resistivity is achieved by passing a direct electrical current through a material while monitoring the voltage. The resistance is computed from the ratio of the voltage to current, and the material property, i.e. its resistivity, is determined by multiplying the resistance by some factor that accounts for the size of the sample and the geometry of the measuring contact points (electrodes).

When an alternating current is employed, the observed voltage will often be slightly out of phase with the applied current. The phase shift (φ) as well as the resistivity magnitude (ρ) are normally measured. These quantities generally vary with frequency of the applied current and

From: BARKER, R. D. & TELLAM, J. H. (eds) 2006. *Fluid Flow and Solute Movement in Sandstones: The Onshore UK Permo-Triassic Red Bed Sequence.* Geological Society, London, Special Publications, **263**, 65–81.
0305–8719/06/$15 © The Geological Society of London 2006.

are therefore measured at a number of different frequencies, hence the term electrical spectroscopy. Often the in-phase (real) and out-of-phase (quadrature) components of resistivity (ρ', ρ'') or conductivity (σ', σ'') are used and these can be related to the phase shift φ by

$$\tan \varphi = \frac{\rho''}{\rho'} = -\frac{\sigma''}{\sigma'} \qquad (1)$$

and the resistivity and conductivity magnitudes ($\rho*$ and $\sigma*$) by

$$\rho* = \sqrt{\rho'^2 + \rho''^2} \text{ and } \sigma* = \sqrt{\sigma'^2 + \sigma''^2}. \qquad (2)$$

The complex resistivity and conductivity are related to each other and to the dielectric permittivity (ε^*) by

$$\sigma* = \frac{1}{\rho*} = i\omega\varepsilon*, \text{where } i = \sqrt{-1}. \qquad (3)$$

The frequency-dependent phase shift is the result of a phenomenon known as polarization, which is caused by a relative shift of positive and negative charges (Jonscher 1983, 1996). Going from small to large scale, some of the main polarization causes are the shifting of electric fields around individual atoms, the realignment of dipolar molecules and ionic diffusion. These different processes act at different speeds when an electric field is applied. It takes a finite time (the relaxation time) after the application of an electric field for the charges to realign or re-distribute themselves and the same length of time to return to equilibrium after cessation of that field. When an alternating sinusoidal field is applied with a period much longer than a particular relaxation time, that process will seem to happen almost instantaneously and not contribute to the out-of-phase (quadrature) response. When the period is much shorter than the relaxation time, there will be no significant charge movement and therefore no contribution to the quadrature response. When the period is the same as the relaxation time there will be a maximum quadrature response.

Electrical spectroscopy is used for a wide range of scientific applications. These include investigating the physics of liquids and monitoring biological systems. Complex electrical measurements for these and other purposes are regularly made over the frequency range of 10^{-6}–10^{12} Hz (Kremer 2002). Electrical relaxation phenomena are very important in electrical spectroscopy, occurring in many different materials at all frequencies, and have a wide variety of causes, although these are often poorly understood.

The aim of the work presented here is an analysis of the polarization response observed in the Permo-Triassic sandstone over relatively low frequencies (0.0001–1000 Hz), below the point at which inductive charge transfer becomes significant.

Laboratory measurements on Permo-Triassic sandstone

Sandstone samples

Twenty-eight samples of Permo-Triassic sandstone were taken from eight boreholes distributed across the sandstone outcrop in England (Fig. 1), the exception being the Winterbourne Kingston samples which were obtained from a deep oil exploration borehole. The samples were chosen to provide a wide variety of sandstone types from within the UK Permo-Triassic. Their petrological characteristics, determined from thin section and optical and scanning electron microscope studies, are summarized in Table 1. In addition all sandstones had a small percentage of iron oxide and clay (illite–smectite) coating the grains.

Electrical measurements

For the electrical measurements, all samples were cut into cylindrical cores each measuring

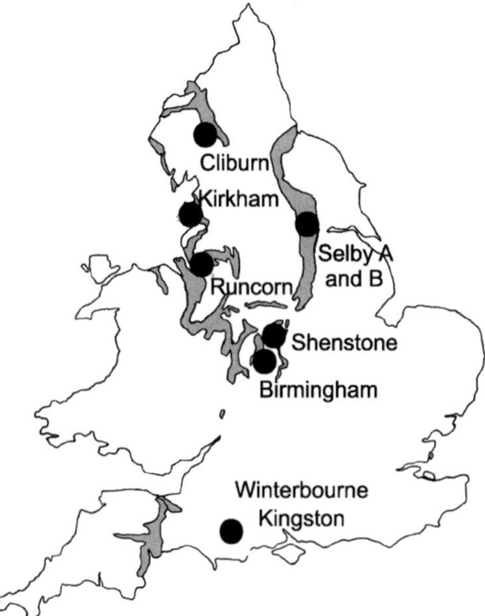

Fig. 1. Map of the boreholes sampled in this study. The main Permo-Triassic sandstone areas of England and Wales are shaded grey.

Table 1. *Sandstone sample descriptions*

Site	Samples	Formation	Description
Birmingham	Ba4, Ba10-12, Ba18, Ba21, Ba23, Ba31, Ba33, Ba34A/B, Ba37, Ba40, Ba42	Wildmoor	Variable, red-brown soft, poorly-cemented, sublitharenitic sandstone (B40 bleached grey, B18 contains 1 mm thin marl parting)
Cliburn	C7, C8	Penrith Sandstone	Dark red, fluvial sandstone with some feldspar
Kirkham	K7, K9, K10	Helsby Sandstone	Coarse-grained, well-sorted, aeolian sandstones with quartz overgrowths (K7 fluvial)
Runcorn	R5, R7, R16, R18	Helsby Sandstone	Coarse-grained, aeolian, arkosic sandstones
Selby A	Sa7	Wildmoor	Red, poorly-cemented fluvial sandstone
Selby B	Sb3	Wildmoor	Red, poorly-cemented fluvial sandstone
Shenstone	SH4	Kidderminster	Poorly-sorted, red, medium- to coarse-grained litharenite
Winterbourne Kingston	W2, W9	Otter Sandstone	Fine to coarse, pale red sandstones with up to 30% feldspar, carbonate (largely dolomite) cement, much unconnected pore space

35 mm in diameter and 30–70 mm in length. The samples were dried at 80 °C for 48 h. Tests confirmed that electrical results were the same as when samples are dried at lower temperatures for longer times (Scott *et al.* 2003). The samples were then vacuum saturated with a synthetic groundwater containing 60 mg l^{-1} Ca^{2+}, 30 mg l^{-1} Mg^{2+}, 34 mg l^{-1} Na^+, 142 mg l^{-1} Cl^-, 120 mg l^{-1} SO_4^{2-} and 30 mg l^{-1} HCO_3^-, a chemistry that closely matched the cation concentrations in the Triassic sandstone aquifer measured at the University of Birmingham borehole (Mitchener 2003). The same water was used for all the samples, irrespective of origin, in order to provide a firm basis for comparison. The measured synthetic groundwater resistivity varied slightly between samples, from 11.5 to 14 Ωm, due to difficulty in accurate reproduction of the ionic balance.

Measurements of the resistivity magnitude and the phase angle were made at several frequencies over the range of 10^{-3}–10^3 Hz, using a SIP–Fuchs spectral IP system, a commercially available system from Radic Research. This passes a sinusoidal current through the sample at set frequencies, starting at 1500 Hz, then decreasing in steps of a half down to 1.43 mHz, giving a total of 21 readings. Lower frequencies were needed for the measurement of sample Ba12, in order to fully characterize its phase spectrum, and this was done using a standard impedance analyser and software specifically written for this purpose at the Department of Electrical Engineering, University of Birmingham.

Over this low-frequency range it is important to use a four-electrode measurement technique and for this the sample cell of Taylor & Barker (2002) was used along with Ag–AgCl non-polarizing electrodes. The measurement temperature, was the ambient laboratory temperature which was monitored to ensure stability at 20 ± 1 °C. Errors due to the experimental system were checked by repeating measurements for a number of samples on a system using circular potential electrodes at the Technical University of Clausthal (TUC), Germany (Scott *et al.* 2003; no significant differences were found.

The electrical results are presented in the form of resistivity and phase spectra. The phase shift is normally negative and so is conventionally plotted with the negative phase scale up. In this way the phase spectra generally show a relaxation 'peak' (Fig. 2), which appears at different frequencies for different samples and occurs over the range of 10^{-4}–10 Hz. This demonstrates that different relaxation times are dominant for different samples. In order to determine the dominant relaxation times and other important induced polarization parameters, the data were modelled using the generalized Cole–Cole equation (Klein & Sill 1982; Dias 2000):

$$\rho^* = \rho_0 \left[1 - m \left(1 - \frac{1}{[1 + (i\omega\tau)^c]^k} \right) \right] \qquad (4)$$

where $0 \leq (c;k) \leq 1$, ρ^* is the complex resistivity and ρ_0 is the resistivity at the limit as ω tends to 0. The parameter, m, gives the magnitude of

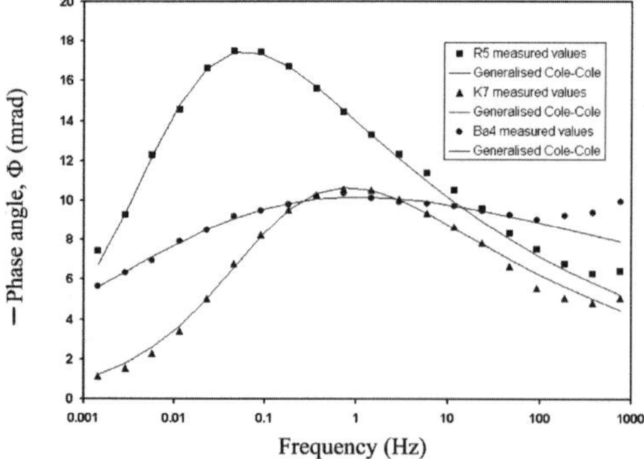

Fig. 2. Electrical phase-angle spectra with generalized Cole–Cole model parameter fitting at frequencies of less than 100 Hz.

the polarization and is roughly equivalent to the chargeability as defined by Seigel (1959). The model parameters c and k give the symmetrical and asymmetrical distribution of relaxation times around a central relaxation time of τ.

The model fitting was carried out for at least 12 data points around the phase peak region using a least-squares minimization routine. Points at frequencies over 100 Hz were not included in the fitting process in order to avoid any influence of electromagnetic coupling across the sample, the effects of which start to appear at higher frequencies. In all cases it was possible to obtain a fit to within experimental error, although with spectra that had low phase angles or a widely spread relaxation peak a wider range of parameters could provide a good fit to the spectra. For this reason the model fitting was repeated several times to find the range of each parameter that could give a good fit to the data. The mean value for each parameter is given in Table 2, along with the range of values fitted and the parameter errors. For the low-frequency resistivity, ρ_0, the value used was the resistivity at 5 mHz ($\rho_{5\,\text{mHz}}$), chosen to avoid the occasional measurement errors occurring at very low frequencies and which were due to the long duration of individual measurements. The generalized Cole–Cole model fits to three of the samples (Fig. 2) demonstrate how well the asymmetric data can be modelled. The slight increase in phase angle above 100 Hz is largely due to electromagnetic coupling across the sample holder and is therefore not a property of the sandstone and so was not included in the modelling.

The chargeability, m, determined from the generalized Cole–Cole modelling, is a measure

of the polarization. However, the magnitude of m will also depend on the in-phase conductivity of the bulk pore fluid. In order to isolate the effects of the pore surface it is necessary to normalize m by dividing it by the in-phase resistivity (Lesmes & Frye 2001), so that

$$m_\text{n} = m\sigma_0 = \frac{m}{\rho_0}. \qquad (5)$$

This normalized chargeability m_n can also be compared to σ'', which is considered to be closely linked to the surface conduction in the rock matrix (Vinegar & Waxman 1984).

Cation-exchange capacity (CEC)

CEC is a measure of the ability of the rock to exchange cations between the pore surface and the saturating solution. As exchange sites occur mainly on pore-lining clays and oxyhydroxides, the CEC is one measure of the quantity of pore-lining material present in the rock. The standard way to measure CEC is to saturate the exchange sites with a cation that preferentially bonds to these sites and does not occur in the groundwater chemistry. The method used was first to disaggregate the rock with a pestle and mortar, and then saturate it with ammonium chloride, in order to replace all the cation-exchange sites with ammonium ions. The released cations in solution are then detected using ICP-AES (inductively coupled plasma-atomic emission spectrophotometry) and the CEC of the rock determined. A correction is made for the increase in carbonate due to some cations being released along with the carbonate cement and not coming from the exchange sites, by using a standard carbonate alkalinity titration. Rather than use the CEC in petrophysical applications, it is more common to use the cation exchange to pore volume ratio (Q_v) (Vinegar & Waxman 1984; Sen *et al.* 1990) given by

$$Q_v = \frac{1}{100} \times \text{CEC} \left[\frac{1-\phi}{\phi} \right] d_\text{m} \qquad (6)$$

where ϕ is the fractional porosity, d_m is the matrix density in g cm^{-3}, CEC is in meq 100 g^{-1} and Q_v is in units of meq ml^{-1}.

Surface area analysis

Surface area analysis is commonly carried out using the nitrogen adsorption method. This has an advantage over mercury injection in that it is not affected by bottleneck pores (Lowell & Shields 1991) that occur in sandstones and could

Table 2. *Generalized Cole–Cole parameters for Permo-Triassic sandstone samples*

Sample	$\rho_{5\ mHz}$ (Ωm) error = 2%	m	$\pm m$ error	τ (s)	τ error (s)	c	$\pm c$ error	k	$\pm k$ error	m_n (S m^{-1})	Frequency range fitted (Hz)
Ba4	98.9	0.159	0.013	25	10	0.35	0.03	0.20	0.05	0.00161	0.001–100
Ba10	73.7	0.071	0.010	17	6	0.72	0.05	0.20	0.07	0.000963	0.001–3
Ba11	86.3	0.070	0.001	35	25	0.24	0.04	0.91	0.09	0.000811	0.001–3
Ba12	90.8	0.153	0.001	335	10	0.35	0.01	0.99	0.02	0.00169	0.0003–0.3
Ba18	51.9	0.088	0.002	37	7	0.57	0.04	0.38	0.06	0.00170	0.001–3
Ba21	32.0	0.148	0.066	1.4	0.8	0.37	0.04	0.15	0.08	0.00463	0.04–100
Ba23	78.4	0.088	0.005	18	4	0.59	0.04	0.21	0.02	0.00170	0.001–3
Ba31	62.0	0.074	0.001	4.7	0	0.58	0.00	0.61	0.00	0.00119	0.001–3
Ba33	50.6	0.067	0.001	3.3	0.3	0.61	0.01	0.33	0.02	0.00132	0.002–11
Ba34A	53.0	0.094	0.003	2.8	0.2	0.64	0.02	0.30	0.03	0.00177	0.005–50
Ba34B	60.4	0.084	0.002	2.8	0.4	0.58	0.02	0.35	0.04	0.00139	0.005–50
Ba37	64.9	0.075	0.001	3.8	2.4	0.58	0.01	0.34	0.03	0.00116	0.005–11
Ba40	37.0	0.093	0.001	1.4	0.1	0.53	0.01	0.29	0.02	0.00251	0.01–100
Ba42	78.7	0.061	0.003	14	2	0.65	0.08	0.17	0.04	0.000775	0.01–100
C7	88.0	0.091	0.002	11	1	0.65	0.02	0.35	0.03	0.00103	0.001–3
C8	74.6	0.145	0.002	11	1	0.57	0.02	0.37	0.03	0.00194	0.001–6
K7	88.8	0.074	0.001	1.4	0.2	0.61	0.01	0.29	0.01	0.000833	0.01–100
K9	213	0.057	0.003	33	22	0.30	0.03	0.29	0.03	0.000268	0.001–3
K10	217	0.058	0.002	50	30	0.29	0.02	0.53	0.12	0.000267	0.001–3
R5	99.5	0.102	0.002	19	2	0.70	0.03	0.26	0.03	0.00103	0.001–3
R7	106	0.130	0.002	18	3	0.51	0.02	0.35	0.04	0.00123	0.002–25
R16	69.0	0.125	0.003	6.9	1.5	0.62	0.04	0.27	0.02	0.00181	0.002–100
R18	83.5	0.115	0.007	21	0	0.58	0.03	0.32	0.01	0.00138	0.001–12
Sa7	32.8	0.097	0.007	0.74	0.37	0.42	0.02	0.56	0.07	0.00296	0.01–100
Sb3	49.2	0.111	0.011	11	1	0.63	0.04	0.25	0.02	0.00226	0.01–6
SH4	152	0.119	0.002	15	5	0.37	0.02	0.57	0.08	0.000743	0.001–3
W2	127	0.041	0.019	9.8	5.2	0.56	0.06	0.06	0.04	0.000323	0.002–25
W9	109	0.044	0.003	33	28	0.27	0.08	0.35	0.20	0.000404	0.002–25
Max.	217	0.153		335		0.72		0.99		0.00463	
Min.	32.0	0.041		0.74		0.24		0.06		0.000267	
Mean	87	0.09		27		0.5		0.4		0.00141	
SD	46	0.03		62		0.1		0.2		0.00091	

give artificially high values for surface area. Surface area is fractal in nature and the smaller the unit of area the measurement tool is capable of measuring, the larger the value of surface area obtained. Nitrogen adsorption measures pore space down to the scale of the size of a nitrogen molecule (about 3.5 Å) and is therefore capable of measuring much of the surface roughness provided by the pore-lining material. There are several methods for calculating the surface area from nitrogen adsorption, the most common method being the Brunauer, Emmett and Teller (BET) method (Lowell & Shields 1991).

Nitrogen adsorption was carried out using the Micromeritics ASAP 2010 chemisorption equipment. Samples used for nitrogen adsorption analysis were taken from the end cuts of the cylindrical samples used for the electrical measurements. Each sample was broken off carefully, in order to avoid unnecessary destruction of the pore space, and the optimum mass of sample to use was determined by trial and error. Enough sample was needed to provide the minimum surface area required for the measurement accuracy of the equipment and the maximum limit on sample quantity was the size of the glass bulb in which the sample is contained. The mass used varied between 2 and 8.5 g, with an average mass of 5.6 g. Samples were dried initially in an oven at 80 °C for 48 h. Then the sample was dried at 150 °C under vacuum by the analysis equipment. This high level of drying was likely to have affected the clay structure but was unavoidable for nitrogen adsorption measurements where a stable high vacuum was required. Even at this temperature a drying time of 8 h was needed to achieve the vacuum required by the equipment. Collapse of the clay structure due to the sample being dry and heated could result in a reduction of surface area from the fully saturated sandstone *in situ*. However, the surface area should still give some guide as to the quantity of pore-lining material. The surface area was calculated using a five-point BET method. The pore surface-area-to-volume ratio (S_{POR}) (Sen *et al.* 1990; Börner *et al.* 1996) is given by

$$S_{POR} = \left[\frac{1-\phi}{\phi}\right] S_m d_m \qquad (7)$$

where d_m is the matrix density in g cm^{-3}, S_m the specific surface area in m^2 g^{-1} and S_{POR} is in units of µm^{-1}.

Mercury injection capillary pressure (MICP)

Background. Mercury injection porosimetry is used to measure the size of pores and the distribution of different pore sizes in a material (Lowell & Shields 1991). A sample of the material is first evacuated, surrounded by mercury and then subjected to increasing pressure to force the mercury into the sample. With each step increase in pressure, the volume of mercury intruded into the sample is recorded. The greater the pressure, the smaller the pores the mercury will intrude into, enabling a pore-size distribution to be determined. In geological studies the technique is often referred to as mercury injection capillary pressure (MICP).

In a sandstone the mercury must pass through the interconnecting spaces between the pores (i.e. the pore throats) in order to fill the pores that are shielded by these spaces. Therefore, importantly, MICP actually measures the pressures needed to intrude mercury through the pore throats. The volume intruded will be entirely dependent on the volume of pore space that is accessed through these pore throats. Bloomfield *et al.* (2001) have made measurements on pore-throat size distributions of Permo-Triassic sandstones from the UK and link them closely with the hydraulic permeability of the rock. Assuming cylindrical pores, mercury porosimetry is capable of measuring pore diameters in the range from 0.3 mm to 30 Å, which is equivalent to a pressure range of 3.5 kPa–414 MPa.

Measurement technique. Mercury injection porosimetry was carried out, using the Micromeritics Autopore III mercury injection porosimeter, on samples taken from the end cuts of the cylindrical cores used for the electrical measurements. Care was taken in sampling in order to maintain the integrity of the pore structure of the sandstone. Samples were then dried at 80 °C for 48 h so that the clay structure was altered as little as possible, although the clays would inevitably be partially dehydrated, causing some change in their structure. An optimum sample mass of between 1.7 and 3.1 g was found by trial and error to use the correct volume of mercury for accurate measurement. Pressure was increased in 54 steps from 3.52 kPa to 228 MPa and the pore-throat diameter, D (in m), at each pressure was given by the Washburn equation (Bloomfield *et al.* 2001)

$$D = \frac{\alpha(-4\cos\theta)}{P_c} \qquad (8)$$

where the contact angle θ is taken as $140°$, P_c is the capillary pressure (in N m^{-2}) and α is the interfacial tension taken as 0.485 N m^{-1}.

Model fitting. The results of the MICP are plotted as cumulative volume of intruded mercury against pressure or pore-throat size (Fig. 3). The results for our samples are similar to those presented in a recent study of MICP results for 153 samples of UK Permo-Triassic sandstone by Bloomfield *et al.* (2001).

In Figure 3 mercury is the non-wetting-phase liquid and the assumption is made that there will be some residual water left within the mainly evacuated pores. The 'entry pressure' corresponds to the pressure required to invade the largest pores in the sandstone (Bloomfield *et al.* 2001). The 'breakthrough pressure' corresponds to the formation of a continuum of a non-wetting phase through the pore network and the 'dominant intrusion pressure' is the pressure at which there is maximum intrusion of the mercury. These pressures all correspond to equivalent pore-throat sizes, so that, for example, the dominant intrusion pressure is associated with the dominant pore-throat size.

Here we have modelled the pore-throat distribution using a van Genuchten type equation that has been found to fit the results well (van Genuchten 1980; Bloomfield *et al.* 2001). This is particularly useful because the van Genuchten equation is well known in hydrogeology (Fetter 1999) and the parameters calculated from MICP can then be used to estimate other properties such as entry pressure for contaminants (Gooddy *et al.* 2002) and permeability (van Genuchten 1980; Fetter 1999; Kamath 1992). Before fitting the van Genuchten model

it was first necessary to modify the MICP data to remove the effect of surface conformance (Bloomfield *et al.* 2001), which is the lowest pressure required to cause the mercury to take on the shape of the rough surface of the sample. This value is subtracted from the subsequent measurements. The van Genuchten model (Bloomfield *et al.* 2001) is given by

$$P_c = P_0 \left(S_e^{-1/m_{VG}} - 1 \right)^{1-m_{VG}} \qquad (9)$$

where P_c is the capillary pressure and P_0 is a characteristic capillary pressure for the medium; Bloomfield *et al.* (2001) consider P_0 to be roughly equivalent to the breakthrough pressure. The parameter, m_{VG}, is a pore-size distribution index. S_e is the normalized wetting-fluid saturation also known as the effective saturation (Fetter 1999) and is given by

$$S_e = \frac{S_w - S_r}{S_m - S_r} \qquad (10)$$

where S_w is the wetting fluid saturation, S_r is the residual fluid saturation, in this case the residual water saturation, and S_m is the fluid content at natural saturation, which is taken as 0 as the samples have been dried. The three parameters, P_0, m_{VG} and S_r were fitted to the results using a least-squares minimization with the standard settings for the Solver routine in Microsoft Excel. The breakthrough pore-throat size (D_{P0}) is given by the Washburn equation (equation 8) from the value of P_0.

Three examples of measured MICP curves are given in Figure 4 along with the van Genuchten fits to these data. It is seen that the van Genuchten parameters provide acceptable models for the curves. The most useful parameters are the breakthrough pore-throat sizes (D_{P0}) and m_{VG}. A small value of m_{VG} indicates a greater spread in the distribution of pore-throat sizes. The dominant intrusion occurs at the pressure at which the rate of mercury entering the sample is highest. This is generally close to P_0 and can be calculated by solving,

$$\frac{d^2 P_c}{dS_w^2} = 0 \qquad (11)$$

to find the dominant intrusion pressure and then from this the dominant pore-throat diameter ($D_{dominant}$) can be calculated from the Washburn equation (equation 8).

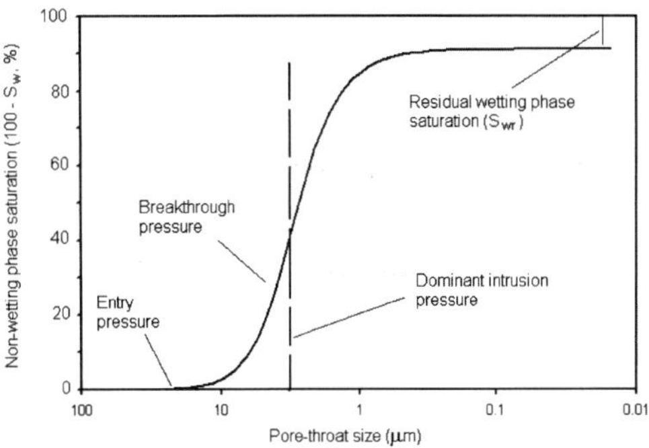

Fig. 3. Pore-throat diameter, from mercury injection capillary pressure curves, against saturation, where mercury is the non-wetting phase.

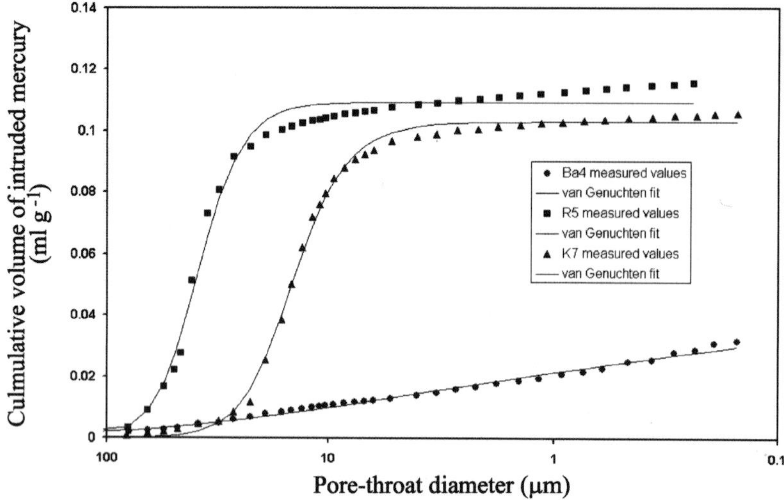

Fig. 4. Mercury injection curves with van Genuchten model parameter fitting (see Table 2 for parameters).

Results

Sandstone properties

The effective porosity, Q_v, S_{POR}, and the van Genuchten parameters for the 28 samples studied are summarized in Table 3 and Figure 5. Porosity generally varies between 15 and 30%, with a mean of 23.6%, which is close to the results of a study of more than 10 000 samples of UK Permo-Triassic sandstone carried out by Allen *et al.* (1997), who measured a mean porosity of 23.8%. Q_v varies between 0.05 and 5.0 meq ml^{-1}, and has a logarithmic distribution as shown in Figure 5B. Pore surface area, S_{POR}, is also best shown as a logarithmic distribution (Fig. 5C), with values varying over two orders of magnitude, from 1 to 100 μm^{-1}. The range of dominant pore-throat size, $D_{dominant}$ is shown in Figure 5D to vary from 5 to 50 μm. Samples Ba4 and Ba21 are notable amongst the Birmingham samples for their relatively low porosity and high Q_v, and Ba4 for its high S_{POR}. Both samples have a low van Genuchten distribution index m_{VG} and, because of the wide distribution of the pore-throat sizes in these two samples, it was not possible to give a dominant pore-throat size. Sample Ba4 contains a significant amount of clay-grade material while sample Ba21 is very fine grained, both reasons for high Q_v and high S_{POR} values. The aeolian samples, K9 and K10, both have relatively low porosities and this is likely to be due to the presence of the pore-filling quartz overgrowths that often surround the grains, quartz having a much lower surface area than clay.

The electrical parameters from the generalized Cole–Cole fitting are summarized in Table 2 and Figure 6. The low-frequency resistivity ($\rho_{5\,mHz}$) ranges over an order of magnitude from 30 to 220 Ωm (Fig. 6A). The normalized chargeability distribution is shown in Figure 6B and the relaxation time distribution in Figure 6C. There is a wide range of relaxation times in these samples, with most falling between 0.74 and 50 s, sample Ba12 is exceptional with a relaxation time of over 300 s.

Correlations

The main aim of this study is the investigation and determination of useful relationships between the frequency-dependent electrical parameters and other important parameters such as porosity, Q_v, S_{POR} and pore-throat size.

The most widely used model for representing the relationship between resistivity and porosity of fully saturated sandstones is probably the following derived by Archie (1942) for a clean sandstone containing no electrically conductive solid material

$$\rho_o/\rho_w = F = a/\phi^m \qquad (12)$$

where F is the formation factor, ρ_o is bulk resistivity, ρ_w the pore-water resistivity, ϕ is the fractional porosity, and a and m are formation constants. The coefficient a is related to

Table 3. *Porosity, cation-exchange capacity, pore surface area and van Genuchten parameters for Permo-Triassic sandstone samples*

Sample	Porosity (%)	Q_v (meq ml^{-1})	S_{POR} (μm^{-1})	van Genuchten parameters from MICP		
				m_{VG}	D_{PO} (μm)	$D_{dominant}$ (μm)
Ba4	12.4	3.60	93.98	0.42	39	
Ba10	28.7	0.32	7.11	0.83	40	36
Ba11	24.8	1.06	15.35	0.78	46	39
Ba12	22.7	0.89	13.81	0.78	56	47
Ba18	21.3	0.87	73.92	0.67	18	12
Ba21	18.8	1.42		0.43	3.1	
Ba23	25.2	1.12	14.95			
Ba31	26.6	0.30	9.65	0.86	34	32
Ba33	27.8	0.70	11.98			
Ba34A	26.9	0.29	25.13	0.83	23	21
Ba34B	26.5	0.30	25.65	0.83	23	21
Ba37	27.7	0.08	12.10		35	33
Ba40	23.4	0.54	58.64	0.80	9.9	8.6
Ba42	27.8	0.80		0.88	43	41
C7	25.8		17.30	0.81	30	27
C8	26.0			0.83	16	15
K7	26.4	0.10	18.77	0.78	16	14
K9	16.3		1.70	0.79	29	25
K10	16.1			0.79	39	33
R5	24.9		11.35	0.82	42	38
R7	23.3			0.77	45	37
R16	24.1	0.15		0.76	24	20
R18	23.6			0.82	34	30
Sa7	27.3	0.67	39.10	0.76	12	9.5
Sb3	31.9	0.29	14.26	0.75	40	32
SH4	15.0			0.75	41	33
W2	20.3		2.71			
W9	19.5	0.06		0.74	27	21
Max.	31.9	3.60	93.98	0.88	56	47
Min.	12.4	0.06	1.70	0.42	3.1	8.6
Mean	23.6	0.71	26	0.76	31	27
SD	4.6	0.80	25	0.11	13	11

tortuosity and is often given the value 1, while m may take values between 1.2 and 4.0 (Winsauer *et al.* 1952; Mendelson & Cohen 1982).

It was recognized early on that departures from this relationship occur when the rock matrix includes minerals that provide additional conductive paths (Waxman & Smits 1968; Barker & Worthington 1973) and associated polarization effects. The presence of polarization effects in shaly sandstone has led to the development of complex resistivity survey and logging techniques. Low-frequency SIP techniques have been carried out in the mHz–kHz range and various studies (Vinegar & Waxman 1984; Börner & Schön 1991) have attempted to define the parameter relationships, but these studies were made on sandstones that had no significant variation in phase angle with frequency. This is not the case for the majority of Permo-Triassic sandstones from the UK. Here we outline the correlations found from our studies.

Chargeability. Our measurements show only a very weak correlation of polarization magnitude, or normalized chargeability, m_n, with Q_v (Fig. 7) and it would be meaningless to try to fit a trend to these data. This contradicts the results of Vinegar & Waxman (1984), who found a correlation between the quadrature conductivity σ'', one measure of polarization magnitude, with Q_v, and Börner *et al.* (1996) who found an approximate correlation between σ'' and CEC.

The correlation between m_n and S_{POR} (Fig. 8) is much stronger and fitting the data with a power trend, using a least-squares regression, gives

$$S_{POR} = a(m_n)^b \qquad (13)$$

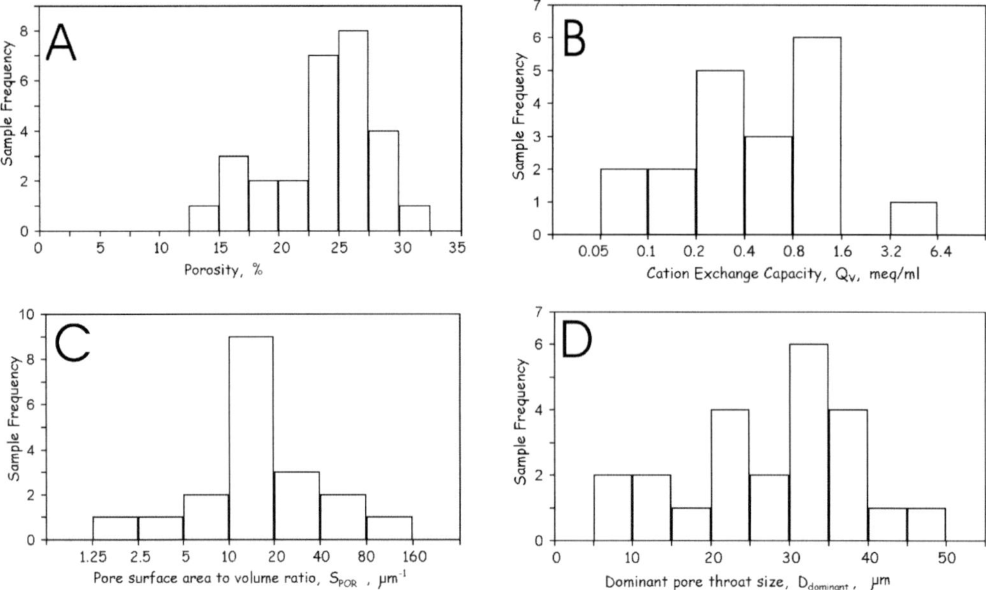

Fig. 5. Histograms of sandstone properties. (**A**) Effective porosity. (**B**) Cation-exchange capacity, Q_v. (**C**) S_{POR} from nitrogen adsorption BET measurement. (**D**) Dominant pore-throat size from mercury injection.

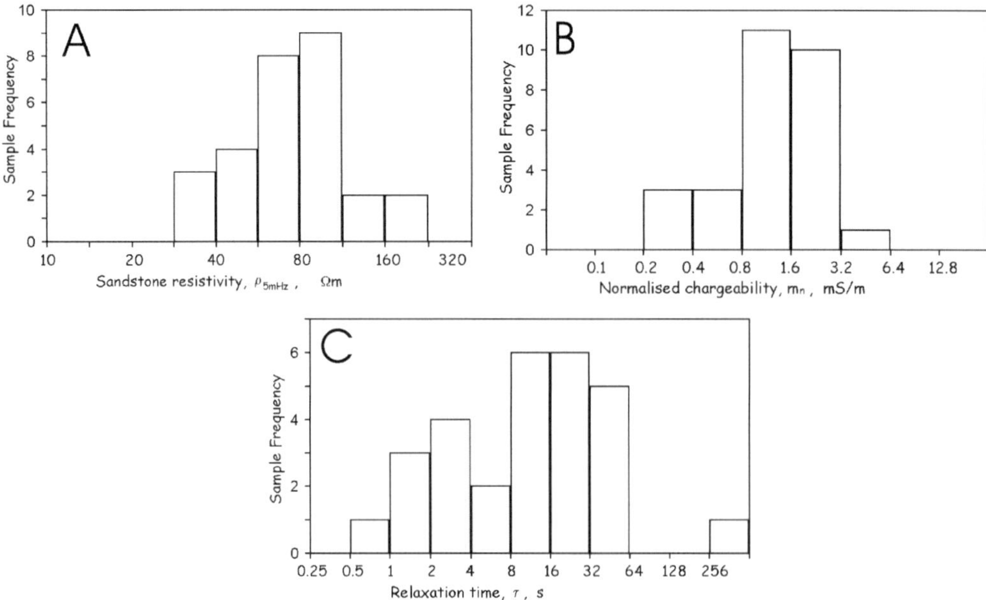

Fig. 6. Histograms of the electrical properties. (**A**) Resistivity at 5 mHz. (**B**) Chargeability from the generalized Cole–Cole model, normalized by the low-frequency resistivity (equation 5). (**C**) Relaxation time from the generalized Cole–Cole model.

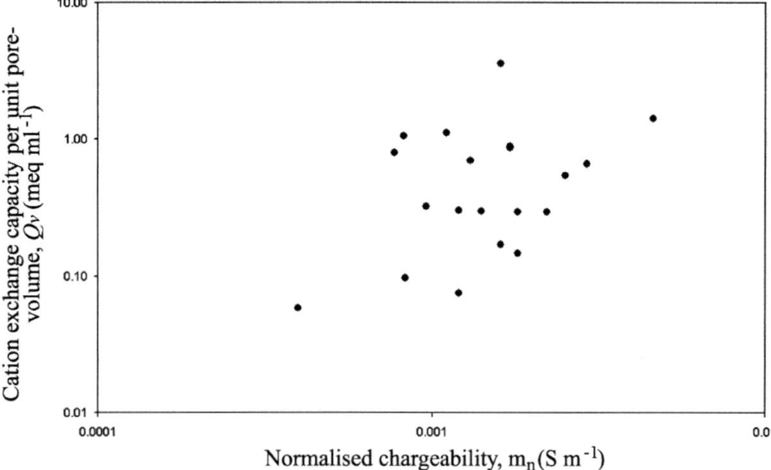

Fig. 7. Comparison of normalized chargeability from the generalized Cole–Cole fit with the cation-exchange capacity to pore volume ratio.

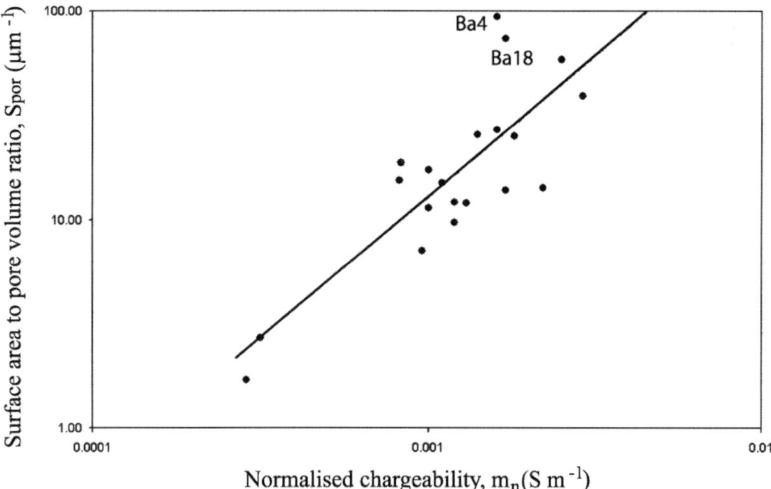

Fig. 8. Comparison of normalized chargeability from the generalized Cole–Cole fit with the surface area to pore volume ratio, calculated using the BET nitrogen adsorption method.

with a correlation coefficient (R^2) of 0.66, where $a = 154\,000$ and $b = 1.36$. The two samples that fall away from the trend (Ba4 and Ba18) have the two highest S_{POR} values and also have a considerable amount of fine and clay-grade material that is not evenly distributed across the grain surface of the sample. Because of this uneven distribution of surface area, these samples can be removed from the trend to give an improved correlation coefficient of 0.76 with $a = 54\,400$ and $b = 1.23$. The correlation is very similar to that demonstrated by Börner *et al.*

(1996) for their σ''_n parameter which is calculated using their constant phase angle model. However, the results on the samples Ba4 and Ba18 suggest that this nearly linear trend is probably only applicable to homogeneous sandstones with an even distribution of clay or oxides covering the grain surface. The results of Börner *et al.* (1996) and the results shown here suggest that these relationships are only good for an order of magnitude estimate of S_{POR} which could then be used to give permeability estimates by a Kozeny–Carmen type relationship

(Börner *et al.* 1996). Although further ambiguities may make the results inaccurate, modified versions of this permeability estimation method have shown some reasonable results for unconsolidated sediments (Slater & Lesmes 2002). However, the Kozeny–Carmen type permeability model is based on samples with low surface area having higher permeability, and this relationship is likely to be inappropriate for these samples because the presence of the quartz overgrowths (e.g. samples K9 and K10) is likely to significantly lower the permeability as well as the surface area.

Relaxation time. The correlation between pore-throat size and relaxation time is shown in Figures 9 and 10 for all samples for which both the van Genuchten and generalized Cole–Cole parameters were determined, except sample Ba18 which has a strong clay banding and is highly inhomogeneous. The error in the τ calculation is the most significant and error bars have been included for all samples (Fig. 9). The samples with an error in τ of less than 50% have been fitted with logarithmic trends, using a least-squares regression. The correlation between D_{P0} and τ (solid line in Fig. 9) is given by

$$D_{P0} = 8.04 \ln \tau + 14.0 \qquad (14)$$

with a correlation coefficient of 0.75, where D_{P0} is in µm and τ is in s. The correlation between $D_{dominant}$ and τ (solid line in Fig. 10) is given by

$$D_{dominant} = 6.87 \ln \tau + 12.4 \qquad (15)$$

with a correlation coefficient of 0.73, where $D_{dominant}$ is in µm and τ is in s. This correlation between $D_{dominant}$ and τ is similar to the results given by Scott & Barker (2003) where the dominant pore-throat diameter, $D_{dominant}$, is shown to correlate with the phase peak frequency, f_{peak}, based on a smaller set of samples. Converting the dominant pore-throat value to µm the fit shown by Scott & Barker (2003) is given by

$$D_{dominant} = 5.9 \ln \tau + 27 \qquad (16)$$

where the relationship between relaxation time and peak frequency is given by

$$\tau = \frac{1}{2\pi f_{peak}}. \qquad (17)$$

The model fitting may be more practical because D_{P0} is easily calculated and is very close in value to $D_{dominant}$, and the discrete increases in mercury pressure make accurate picking of $D_{dominant}$ difficult. It can also be difficult to pick f_{peak} due to the use of discrete frequencies. From the generalized Cole–Cole model τ has been shown to have a potentially large error, particularly when there is a wide spread of the phase peak region, and fitting of this model gives an estimate of the potential range of τ. Therefore the model fitting of the van Genuchten and generalized Cole–Cole parameters, along with an appropriate error analysis, can give more meaningful results on a wider range of samples.

Distribution. Owing to the correlation between D_{P0} and τ, shown above, it should be expected that there will be some correlation between the distribution parameters m_{VG}, c and k. Samples Ba4 and Ba21 both have a widely spread pore-throat size distribution and hence have the lowest values of m_{VG}. They also have extremely low values of c and k, and no clear peak relaxation time. However, samples with low chargeability also have widely spread relaxation times and hence low values of c and k, a notable example being sample W2. This implies that the correlation between the pore-throat size distribution and the spread of the electrical spectra is also dependent on the magnitude of the polarization. Therefore any correlation between the parameters m_{VG}, c and k should either be done only on samples with a high enough polarization or should take into account the magnitude of the polarization. It would be necessary to have a larger data set in order to examine this possible correlation.

Discussion

General findings

The results presented above reinforce the idea of Börner & Schön (1991) that the magnitude of the low-frequency polarization in sandstones is dependent on the pore surface area. However, this is only true for the homogeneous samples that have an even distribution of clays or oxyhydroxides around the grains giving an even distribution of surface area throughout the sandstone. Samples with quartz overgrowths or calcite cement, the aeolian samples from Kirkham and the samples from Winterbourne Kingston have an increasingly lower pore surface area giving rise to a significantly lower polarization than the other samples. Pore surface area is not only important for possible permeability calculation (Börner *et al.* 1996) but also in estimating the potential sorbtion of any contaminants passing through the rock.

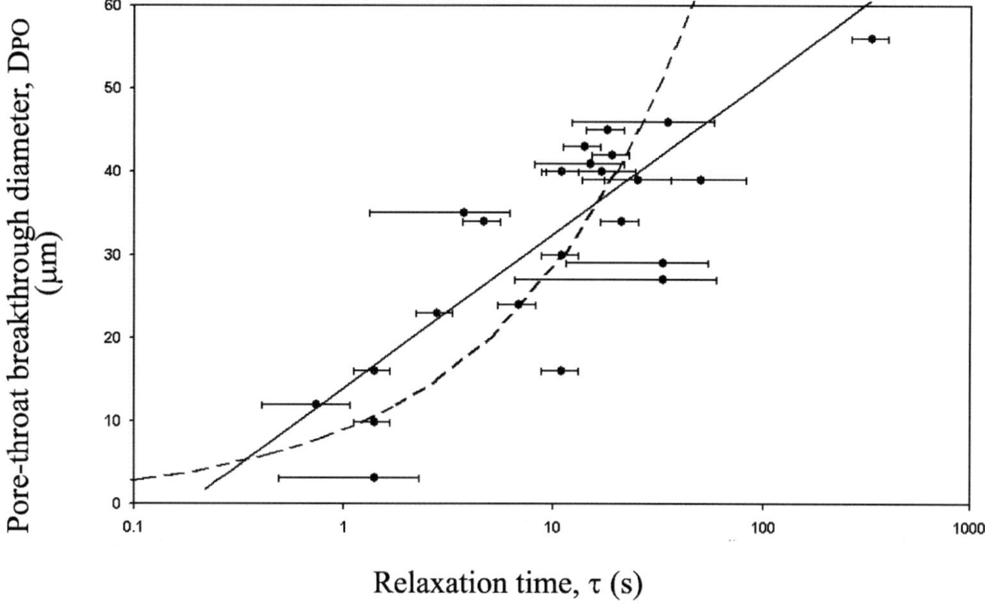

Fig. 9. Comparison of relaxation time from the generalized Cole–Cole fit with the breakthrough pore-throat diameter, from van Genuchten fitting of mercury injection curves. The solid line shows a logarithmic trend fitted to samples with a τ error ≤ 50%. The dashed line shows the best fit of the spherical colloid relaxation model (equation 18) using the pore-throat diameter.

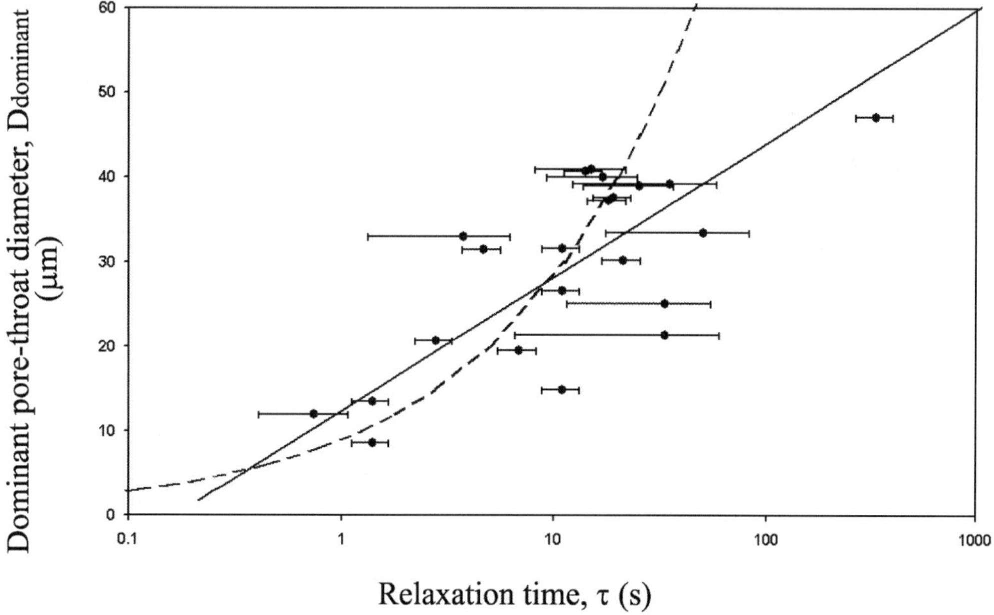

Fig. 10. Comparison of relaxation time from the generalized Cole–Cole fit with the dominant pore-throat diameter, calculated from van Genuchten fitting of mercury injection curves. The solid line shows a logarithmic trend fitted to the results. The dashed line shows the best fit of the spherical colloid relaxation model (equation 18) using the pore-throat diameter.

As the magnitude of the polarization is a measure of the quantity of ions associated with the pore surface, it is expected to show a stronger correlation with Q_v than does the pore surface area. However, this was not the case here.

There is a good correlation between relaxation time and pore-throat size, although it is not possible to be certain from these results exactly what length scale in the sandstone gives rise to the observed relaxation times. Possible relaxation time length scales in the sandstone are pore-throat diameter, pore diameter, length between pore throats or length of pore throat, these in combination with the varying shapes of the pores and pore throats mean that a much higher level of measurement of the pore geometry would be needed to examine which length scale or combination of length scales was the most important. However, the correlation between pore-throat diameter and relaxation time is strong enough to be able to make a calculation of the pore-throat size from the electrical measurements and to use this to estimate the permeability of the sandstone or the entry pressure needed for non-aqueous contaminants or even commercial hydrocarbons.

The amount of information contained in the electrical spectra of a sandstone means that not only can hydrogeologically important parameters be derived but also that the electrical spectra may be an effective way of fingerprinting a particular lithology. This could prove useful in correlating between boreholes.

Modelling

Relaxation models. Although it has been found that the pore geometry is an important factor in the low-frequency relaxation observed in sandstones, there is still some debate as to which ions are the most significant and what is the exact relaxation mechanism. It is generally accepted that the polarization and corresponding low-frequency relaxation in non-metallically mineralized sandstones is caused by ion diffusion in or around the electrochemical double layer (EDL) (Nettelblad & Niklasson 1997; Chelidze & Gueguen 1999; Lesmes & Morgan 2001). The EDL is formed when a liquid containing ions is in contact with the solid rock matrix. Then the rock surface tends to have an associated negative charge. This surface negative charge attracts cations to it that form an EDL. The Stern model of the EDL (Reppert *et al.* 2001) consists of a layer (the Stern layer) where ions are relatively immobile and held rigidly by electrostatic forces. Beyond this there is a diffuse layer that extends into the solution (the Gouy–Chapman diffuse layer).

Ions in and around the EDL are less mobile than those in the pore fluid and the excess of cations in the EDL along with different layer mobilities causes relative movement of charge resulting in polarization. The main contributions to the in-phase and quadrature low-frequency electrical conduction are generally considered to be the movement of ions in the Stern and the diffuse layers. There is also some suggestion that a membrane type polarization occurs, where in zones such as pore throats the cations in the pore fluid are more mobile than the anions due to the blocking effect of the EDL. This type of model was proposed by Marshal & Madden (1959) and has been used recently by Titov *et al.* (2002); in this case both the length of the pore and the length of the pore throat could make a significant contribution to the electrical relaxation. However, it is generally accepted that the excess, in-phase conductivity observed in shaly sandstone is the result of movement of cations in the EDL. Revil & Glover (1998) and Revil (1999) claim that most of this excess conductivity is due to cations moving in the Stern layer. Ions in the Stern layer are free to 'site hop' between exchange sites along the surface, and monovalent ions are generally only weakly bound and free to move tangentially along the grain surface. This excess movement of cations, either in the Stern or diffuse layers, is also most likely to be the cause of the observed polarization and the observed quadrature conductivity. Lesmes & Morgan (2001) suggest that movement of cations in the Stern layer is important in the polarization of sandstone. Relaxation models for these surface type EDL relaxations have generally been developed on colloidal particles. The simplest models of this type in general use are for relaxation around a sphere (Schwarz 1962; Dukhin & Shilov 1974; Fixman 1980). The relaxation time for ions in the Stern layer from these models is given by

$$\tau_s = \frac{R^2}{2D_s} \qquad (18)$$

where τ_s is the relaxation time, R is the radius of the sphere and D_s is the diffusion coefficient of the ions in the Stern layer. Correspondingly, a similar equation can apply to the diffuse layer, where τ_d and D_d are the relaxation time and diffusion coefficient, respectively, for ions in the diffuse layer.

Ion mobility. The ion diffusion coefficient is extremely important as this controls the relaxation time as well as the length scale involved. Lesmes & Morgan (2001) use a bulk pore-fluid diffusion coefficient of $D_{bulk} = 2 \times 10^{-9}$ m^2 s^{-1} for KCl and state that in general $D_s < D_d < D_{bulk}$, where D_{bulk} is the diffusion coefficient away from the EDL. They state that a diffusion coefficient, $D = 0.11 \, D_{bulk}$ best fits their data and so assume this is the diffusion coefficient in the EDL; for this they use the grain radius as the relaxation length, assuming that the sand grains behave like spherical colloidal particles, even though the grains are often tightly packed. An equivalent model is applied to the pores, with the pore size being assumed to be approximately the same as the grain size.

The absolute mobility, μ, of an ion is related to the diffusion coefficient by the Einstein–Smoluchowski relation:

$$D = \frac{\mu kT}{e} \qquad (19)$$

where k is Boltzman's constant, T is the absolute temperature and e is the charge of an electron. Revil (1999) gives some typical mobility values for monovalent ions, quoting a dilute solution mobility for Na$^+$ at 25 °C of 5.19 $\times$ 10^{-8} m^2 s^{-1} V^{-1}. Revil (1999), using the results of Raythatha & Sen (1986) on clay suspensions, estimates a Stern layer mobility for Na$^+$ at 25 °C of 5.14 $\times$ 10^{-9} m^2 s^{-1} V^{-1} claiming this to be independent of clay mineral type. Calculation of the diffusion coefficient for Na$^+$ from the results of Revil (1999) and equation 19 gives a bulk value, D_{bulk}, of 1.31 $\times$ 10^{-9} m^2 s^{-1} and a surface diffusion value, D_s, of 1.31 $\times$ 10^{-10} m^2 s^{-1}. The surface ion mobility, μ_s, and hence the surface diffusion, D_s, can be estimated from the excess in-phase conductivity, σ_s, and Q_v (Revil 1999) using

$$\mu_s \approx \frac{3}{2}\left(\frac{1-\phi}{\phi}\right)\frac{\sigma_s}{Q_v} \qquad (20)$$

where σ_s is calculated from multiple salinity measurements. Using the results of Taylor & Barker (2006) on UK Permo-Triassic sandstone gives a $D_s = 6.58 \times 10^{-11} \pm 2.4 \times 10^{-11}$ m^2 s^{-1} for Na$^+$, which is slightly lower than the values given above. The measurements in this study use a synthetic groundwater containing a concentration ratio of approximately 2:1:1 for Ca^{2+}:Na$^+$:Mg^{2+}. Calcium and magnesium ions are less mobile in bulk solution and being bivalent would be expected to be more strongly bonded in the EDL, further reducing their mobility.

Application to experimental results. The results of Börner *et al.* (1993) and Scott & Barker (2002) show that Na$^+$ gives a higher magnitude polarization than Ca^{2+} and Mg^{2+} so it is likely to have a major contribution to the combined relaxation effect, although Ca^{2+} and Mg^{2+} are preferentially bonded to the exchange sites at the mineral surface making the effects on the double layer complicated. Using a simple spherical grain size relaxation model, the observed relaxation times, in the range of 1–100 s in this study and the range of sandstone grain diameters of 0.1–1 mm (fine–coarse), would require a D_s of 1.25 $\times$ 10^{-9} m^2 s^{-1}, which is equivalent to the bulk diffusion coefficient. This shows that the relaxation length scale is not the grain diameter if this is a correct model to use. The dashed line in Figures 9 and 10 is the best fit of the spherical colloid relaxation model (equation 18) to the pore-throat size and this uses a D_s value of 1 $\times$ 10^{-11} m^2 s^{-1}, which is slower than the diffusion coefficient suggested by the results of Taylor & Barker (2006). Simple relaxation models can only provide estimates, and the uncertainty in relaxation processes and diffusion coefficients makes length scale estimates inaccurate. However, the correlation with pore-throat size in Figures 9 and 10 and equations (14) and (15) are extremely good and indicate the possibility of this being the most important length scale. For example, sample Ba12 does not have a larger grain size than some of the other samples but its relaxation time is much greater. The sample also has the largest pore-throat size, which suggests that the length scale may be related to pore-throat size or possibly pore size, but *not* the grain size. Pore throats are not always simple spheres and the length of the pore throats may also be extremely important; this has already been suggested by Titov *et al.* (2002). A lot of further work is necessary including image analysis for grain-size and pore-size distributions in order to make more informed conclusions.

Another important factor that is often ignored is that the strength of the electrical relaxation will also be affected by the dimensions of the pore or pore throat. Low-frequency conduction is mainly through the bulk pore fluid and in the main pore volume there is a lower surface area to volume ratio than in the pore-throats. With the greater in-phase conduction through this fluid it can be seen from equation 1 that this would reduce the phase angle measured. Therefore the surface conduction in pore throats would be expected to be far more significant than that in the main pore space. It is important to note that the basic relaxation models cannot be applied directly to sandstones because the dimensions and shape of pores and pore throats will vary significantly. Also the pore-throat diameter calculated from mercury injection will also not be accurate because it is based on a simple cylindrical model. Therefore the correlations observed here are extremely good considering the various ambiguities and inaccuracies involved.

Conclusions

The low-frequency electrical spectra of Permo-Triassic sandstone from the UK are often characterized by a relaxation phenomenon that can be seen clearly as a negative peak in the phase spectra. The relaxation peaks when present are usually asymmetric and can be modelled by a generalized Cole–Cole equation (Klein & Sill 1982; Dias 2000). This model gives five parameters that describe the spectra well over at least three decades of frequency, and sometimes as many as all five decades of frequency between 0.001 and 100 Hz. The parameters all have electrical significance and correlate well with other hydrogeologically important parameters obtained from the techniques of MICP and nitrogen adsorption. Of particular importance is that the relaxation time, τ, is strongly related to the dominant pore-throat size. This could have many hydrogeologically important, uses including permeability estimation based on van Genuchten models (Kamath 1992). A normalized version of the chargeability appears to be related to the surface conductivity and this gives the possibility of estimating the pore surface area to volume ratio, a value that is important both in permeability estimation and in sorption estimation with relation to contaminant transport. There is a correlation between the spread of the relaxation distribution and the spread of the pore-throat distribution, although this is weak due to the interaction of other parameters such as chargeability. Finally, the electrical spectra for shaly sandstone contain so much information that they could potentially provide an effective lithological fingerprint. There is a need for many more extensive studies on sandstones using both real and artificial samples, and development of the associated theory of the electrical relaxation processes taking place. The potential uses of electrical spectroscopy in hydrogeology and many other fields including geology and petroleum petrophysics are only now becoming apparent, and further advances are certain.

The authors wish to thank P. Atkins for assistance with electrical measurements and H. Mills for assistance in nitrogen adsorption and mercury injection measurements. Our thanks also to the British Geological Survey, Wallingford for providing sandstone samples for this study. A number of people have assisted with the laboratory work and we would like to thank P. Hands, R. Livesey and J. Harris. Finally, we would like to acknowledge J. Tellam and S. Taylor for their helpful discussions.

References

ALLEN, D.J., BREWERTON, L.J. ET AL. 1997. *The Physical Properties of Major Aquifers in England and Wales*. British Geological Survey, Technical Report, **WD/97/34**. Environment Agency R&D Report Publication, **8**.

ARCHIE, G.E. 1942. The electrical resistivity log as an aid in determining some reservoir characteristics. *Transactions of the American Institute of Mining, Metallurgical and Petroleum Engineers*, **146**, 54–62.

BARKER, R.D. & WORTHINGTON, P.F. 1973. Some hydrogeophysical properties of the Bunter sandstone of northwest England. *Geoexploration*, **11**, 151–170.

BLOOMFIELD, J.P., GOODDY, D.C., BRIGHT, M.I. & WILLIAMS, P.J. 2001. Pore-throat size distributions in Permo-Triassic sandstones from the United Kingdom and some implications for contaminant hydrogeology. *Hydrogeology Journal*, **9**, 219–230.

BÖRNER, F.D. & SCHÖN, J.H. 1991. A relation between the quadrature component of electrical conductivity and the specific surface area of sedimentary rocks. *The Log Analyst*, **32**, 612–613.

BÖRNER, F.D., GRUHNE, M. & SCHÖN, J.H. 1993. Contamination indications derived from electrical properties in the low-frequency range. *Geophysical Prospecting*, **41**, 83–98.

BÖRNER, F.D., SCHOPPER, J.R. & WELLER, A. 1996. Evaluation of transport and storage properties in the soil and groundwater zone from induced polarization measurements. *Geophysical Prospecting*, **44**, 583–601.

CHELIDZE, T.L. & GUEGUEN, Y. 1999. Electrical spectroscopy of porous rocks: a review – I. Theoretical models. *Geophysical Journal International*, **137**, 1–15.

DIAS, C.A. 2000. Developments in a model to describe low-frequency electrical polarisation of rocks. *Geophysics*, **65**, 437–451.

DUKHIN, S.S. & SHILOV, V.N. 1974. *Dielectric Phenomena and the Double Layer in Disperse Systems and Polyelectrolytes*. Wiley, New York.

FETTER, C.W. 1999. *Contaminant Hydrogeology*, 2nd edn. Prentice-Hall, Upper Saddle River, NJ.

FIXMAN, M. 1980. Charged macromolecules in external fields, I, the sphere. *Journal of Chemical Physics*, **72**, 5177–5186.

GOODDY, D.C., BLOOMFIELD, J.P., HAROLD, G. & LEHARNE, S.A. 2002. Towards a better understanding of trichloroethene entry pressure in the matrix of Permo-Triassic sandstones. *Journal of Contaminant Hydrology*, **59**, 247–265.

JONSCHER, A.K. 1983. *Dielectric Relaxation in Solids*. Chelsea Dielectrics Press, London.

JONSCHER, A.K. 1996. *Universal Relaxation Law*. Chelsea Dielectrics Press, London.

KAMATH, J. 1992. Evaluation of accuracy of estimating air permeability from mercury injection data. *SPE Formation Evaluation*, **4**, 304–310.

KLEIN, J.D. & SILL, W.R. 1982. Electrical properties of artificial clay-bearing sandstone. *Geophysics*, **47**, 1593–1605.

KREMER, F. 2002. Dielectric spectroscopy – yesterday, today and tomorrow. *Journal of Non-crystaline Solids*, **305**, 1–9.

LESMES, D.P. & FRYE, K.M. 2001. Influence of pore fluid chemistry on the complex conductivity and induced polarization responses of Berea sandstone. *Journal of Geophysical Research*, **106**, 4079–4090.

LESMES, D.P. & MORGAN, F.D. 2001. Dielectric spectroscopy of sedimentary rocks. *Journal of Geophysical Research*, **106**, 13 329–13 346.

LOWELL, S. & SHIELDS, J.E. 1991. *Powder Surface Area and Porosity*, 3rd edn. Chapman & Hall, London.

MARSHALL, D.J. & MADDEN, T.R. 1959. Induced polarization, a study of its causes. *Geophysics*, **24**, 790–816.

MENDELSON, K.S. & COHEN, M.H. 1982. The effect of grain anisotropy on the electrical properties of sedimentary rocks. *Geophysics*, **47**, 257–263.

MITCHENER, R.G.R. 2003. *Hydraulic and chemical property correlations of the Triassic Sandstone of Birmingham*. PhD thesis, University of Birmingham.

NETTELBLAD, B. & NIKLASSON, G.A. 1997. Dielectric relaxations in liquid-impregnated porous solids. *Journal of Materials Science*, **32**, 3783–3800.

OLHOEFT, G.R. 1985. Low-frequency electrical properties. *Geophysics*, **50**, 2492–2503.

PELTON, W.H., WARD, S.H., HALLOF, P.G., SILL, W.R. & NELSON, P.H. 1978. Mineral discrimination and removal of inductive coupling with multifrequency IP. *Geophysics*, **43**, 588–609.

RAYTHATHA, R. & SEN, P.N. 1986. Dielectric-properties of clay suspensions in MHz to GHz range. *Journal of Colloid and Interface Science*, **109**, 301–309.

REPPERT, P.M., MORGAN, F.D., LESMES, D.P. & JOUNIAUX, L. 2001. Frequency-dependent streaming potentials. *Journal of Colloid and Interface Science*, **234**, 194–203.

REVIL, A. 1999. Ionic diffusivity, electrical conductivity, membrane and thermoelectric potentials in colloids and granular porous media: A unified model. *Journal of Colloid and Interface Science*, **212**, 503–522.

REVIL, A. & GLOVER, P.W.J. 1998. Nature of surface electrical conductivity in natural sands, sandstones and clays. *Geophysical Research Letters*, **25**, 691–694.

SCHWARZ, G. 1962. A theory for the low-frequency dielectric dispersion of colloidal particles in electrolyte solution. *Journal of Physical Chemistry*, **66**, 2636–2642.

SCOTT, J. & BARKER, R. 2002. The spectral induced polarisation response of Triassic sandstone from the United Kingdom. *In: Proceedings of the 8th Meeting of the Environmental and Engineering Geophysical Society – European Section, Aviero*, EEGS, 363–366.

SCOTT, J.B.T. & BARKER, R.D. 2003. Determining pore-throat size in Permo-Triassic sandstones from low-frequency electrical spectroscopy. *Geophysical Research Letters*, **30**, 1450.

SCOTT, J., SCHLEIFER, N., WELLER, A. & BARKER, R. 2003. The spectral induced polarisation of groundwater saturated sandstones. *In: Proceedings of the 9th European Meeting of Environmental and Engineering Geophysics, Prague, EEGS*, 265–268.

SEIGEL, H.O. 1959. A theory for induced polarization effects (for step excitation function). *In*: WAIT, J.R. (ed.) *Overvoltage Research and Geophysical Applications*. Pergamon Press, New York, 4–21.

SEN, P.N., STRALEY, C., KENYON, W.E. & WHITTINGHAM, M.S. 1990. Surface-to-volume ratio, charge density, nuclear magnetic relaxation, and permeability in clay-bearing sandstones. *Geophysics*, **55**, 61–69.

SLATER, L. & LESMES, D.P. 2002. Electrical-hydraulic relationships observed for unconsolidated sediments. *Water Resources Research*, **38**, 1213.

TAYLOR, S. & BARKER, R. 2002. Resistivity of partially saturated Triassic sandstone. *Geophysical Prospecting*, **50**, 603–613.

TAYLOR, S. & BARKER, R.D. 2006. Modelling the DC electrical response of fully and partially saturated Permo-Triassic sandstone. *Geophysical Prospecting* (in press).

TITOV, K., KOMAROV, V., TARASOV, V. & LEVITSKI, A. 2002. Theoretical and experimental study of time domain-induced polarization in water-saturated sands. *Journal of Applied Geophysics*, **50**, 417–433.

VAN GENUCHTEN, M.TH. 1980. A closed-form equation for predicting the hydraulic conductivity of unsaturated soils. *Soil Science Society of America Journal*, **44**, 892–898.

VAN VOORHIS, G.D., NELSON, P.H. & DRAKE, T.L. 1973. Complex resistivity spectra of porphyry copper mineralization. *Geophysics*, **38**, 49–60.

VINEGAR, H.J. & WAXMAN, M.H. 1984. Induced polarisation of shaly sands. *Geophysics*, **49**, 1267–1287.

WAXMAN, M.H. & SMITS, L.J.M. 1968. Electrical conductivities in oil-bearing shaly sands. *Society of Petroleum Engineers Journal*, **243**, 107–122.

WINSAUER, W.O., SHEARIN, H.M., JR., MASSON, P.H. & WILLIAMS, M. 1952. Resistivity of brine-saturated sands in relation to pore-geometry. *AAPG Bulletin*, **36**, 253–277.

WORTHINGTON, P.F. 1976. Hydrogeophysical properties of parts of the British Trias. *Geophysical Prospecting*, **24**, 672–695.

Characterization of permeability distributions in six lithofacies from the Helsby and Wilmslow sandstone formations of the Cheshire Basin, UK

J. P. BLOOMFIELD, M. F. MOREAU & A. J. NEWELL

British Geological Survey, Maclean Building, Crowmarsh Gifford, Wallingford, Oxfordshire OX10 8BB, UK (e-mail: jpb@bgs.ac.uk)

Abstract: The lithofacies and permeability structure of the Abbey Wood borehole, near Delamere, Cheshire, UK, have been investigated so that permeability distributions can be established for characteristic sandstone lithofacies in the Permo-Triassic sandstone aquifer. The borehole penetrates the Helsby Sandstone and Wilmslow Sandstone formations and six major facies (coarse- and fine-grained fluvial channel fill sandstones, sandy sabkha, sand sheet and dune aeolian sandstones, and a massive sandstone facies) and one minor lithofacies, a mudstone facies, have been identified. The permeability distributions for each lithofacies show subtle systematic differences. The two fluvial sandstone facies and the massive sandstone facies have systematically lower permeabilities than the aeolian sandstones, and within the aeolian sandstones the sand sheet and dune sandstones appear to be systematically more permeable than the sabkha facies. Statistical tests on the permeability distributions of each facies indicate that the distributions are statistically distinct and so support the lithofacies classification.

The Permo-Triassic sandstone aquifer is the second most important aquifer in England and Wales, supplying about a quarter of all licensed groundwater abstractions, and a number of large cities, such as Birmingham in the Midlands and Manchester and Liverpool in NW England, obtain their water supplies from this aquifer. In NW England the water is used for industrial and brewery supplies. A characteristic feature of the aquifer is that it is heterogeneous and that it has a wide range of hydraulic conductivities. Consequently, there is a need to understand this heterogeneity if groundwater flow and solute migration are to be predicted.

The Permo-Triassic sandstone aquifer is a dual-permeability aquifer, with both matrix and fracture components contributing to the hydraulic conductivity (Allen *et al.* 1997). The relative contributions of matrix and fracture permeability vary geographically, with depth, and as a function of scale. Despite extensive studies, the influence of fracturing on regional-scale hydraulic conductivity distribution is still poorly understood, although in the upper 200 m of the unconfined aquifer it may have a significant contribution (Price *et al.* 1982). The present study focuses on the matrix component of the hydraulic conductivity distribution and, in particular, on the geological controls on the hydraulic conductivity of the matrix.

The main hydraulic characteristics of the matrix of the Permo-Triassic sandstone aquifer are similar to many other sandstone formations (Jensen *et al.* 1996), i.e. a wide range of hydraulic conductivities, layered heterogeneity in conductivity at the bed scale and hydraulic anisotropy within individual beds (Bloomfield *et al.* 2001). Based on over 9000 measurements, Allen *et al.* (1997) have reported that the hydraulic conductivity of the matrix varies by over seven orders of magnitude from 10^{-6} to 20 m day^{-1} (equivalent to a permeability range of 10^{-17}–10^{-11} m^2 at 10 °C) with a median value of 0.56 m day^{-1}. The true range of *in situ* matrix hydraulic conductivities is likely to be even greater, since unconsolidated sands with very high hydraulic conductivities and clays and marls with very low hydraulic conductivities will not have been tested (Allen *et al.* 1997). There is a need for both a better understanding of the internal sedimentary architecture of the sandstone aquifer and of the distribution of lithofacies, and for better-constrained parameterization of the hydraulic properties of individual sandstone lithofacies. This paper is a contribution towards the latter task.

The heterogeneous physical characteristics of the matrix of the aquifer are a function of the wide range of lithologies present and are related to variations in grain size, sorting, and the extent and nature of the cement, pore-filling and pore-lining phases (Strong & Milodowski 1987; Strong 1993; Bouch *et al.* 2006; Newell 2006). It is the consequent variation in the pore-throat

From: BARKER, R. D. & TELLAM, J. H. (eds) 2006. *Fluid Flow and Solute Movement in Sandstones: The Onshore UK Permo-Triassic Red Bed Sequence.* Geological Society, London, Special Publications, **263**, 83–101.

size distributions that controls the diverse hydraulic characteristics of the matrix (Bloomfield *et al.* 2001). The sandstones are composed of predominantly fluvial siliciclastic sediments with some aeolian sands, and contain local mudstones and marls and finer grained and, occasionally, well-cemented layers. These latter lithologies may act as confining or partially confining layers (Allen *et al.* 1997).

Correlations between sedimentary facies and hydraulic conductivity distributions have proved to be particularly useful in stochastic models of flow and transport in sandstone hydrocarbon reservoirs (some more recent examples include Aigner *et al.* 1996; Hern & Stell 1997; Anderson *et al.* 1999; Hornung & Aigner 1999; Bahar & Kelkar 2000). This modelling approach generally consists of identifying volumes of rock within a reservoir with specific facies characteristics, determining hydraulic conductivity distributions for each facies on the basis of either direct or indirect measurements or on the basis of prior knowledge, and populating the model volumes with the appropriate distributions of hydraulic conductivity. Such models can then be used as the basis for flow or transport studies. This approach offers great potential for the investigation of the onshore Permo-Triassic aquifer, but, to date, has not been used due to the cost of obtaining sufficient data to condition stochastic models and due to the computational effort that is involved in such models. As computational speed increases in the future, data availability is likely to become the limiting factor.

Although an understanding of facies distributions and sedimentary architecture gained from studies of hydrocarbon reservoirs may be applicable to onshore sandstone aquifers, it may be inappropriate to apply permeability data derived from hydrocarbon reservoir studies in the onshore hydrogeological context as the permeability distributions in sandstone aquifers may be different to permeability distributions in sedimentologically similar hydrocarbon reservoirs due to removal of cement phases and modification of pore-lining mineral assemblages during recent groundwater circulation (Strong & Milodowski 1987). Consequently, if sedimentary facies models and stochastic models are to become useful tools for the study of the Permo-Triassic sandstone aquifer, a first step is to establish representative permeability distributions for sandstone facies in onshore sandstone aquifers.

This paper describes the results of a detailed investigation of the lithological and physical features of the Helsby and Wilmslow sandstone formations at Abbey Arms Wood, Delamere, as seen in a continuously cored 150 m borehole. It presents and discusses correlations between sedimentary facies and hydraulic conductivity distributions, as well as additional parameters such as matrix porosity, solid-phase geochemistry and borehole gamma logs. The work described in this paper is part of a larger study, undertaken in conjunction with the Environment Agency and United Utilities, into the hydrogeological controls on groundwater quality and, in particular, into the occurrence of arsenic in groundwater in the Cheshire Basin. A companion paper (Kinniburgh *et al.* 2006) describing the groundwater chemistry of the Abbey Arms Wood borehole is also presented in this volume.

Site description, and geological and hydrogeological context

Borehole location and description

The Abbey Arms Wood borehole is situated near Delamere, Cheshire, UK [NGR 356418 368133] adjacent to road number B4162 (Fig. 1). The borehole was drilled using the rotary airflush technique with the upper section of the borehole drilled using a rock roller bit to 9.55 m below ground level (m bGL). Below this depth the sandstone was cored using a double core barrel with rigid plastic core-lining tubes to a total depth of 150 m bGL. The borehole was reamed out to 200 mm diameter and was flushed for 5 h. Permanent steel casing was installed and grouted to a depth of 9.55 m bGL to screen off overlying Quaternary sand and gravel deposits. The borehole was completed as a groundwater level observation borehole and equipped with a data-logger to record the water level.

Regional structural setting

The Abbey Arms Wood borehole is located in the northern part of the Cheshire Basin (Fig. 1). The Cheshire Basin is one of the largest onshore basins in the UK. It is a post-Variscan sedimentary basin that contains up to 4.5 km of Permian and Triassic strata (Chadwick 1997), and is a broadly asymmetrical half-graben that deepens slightly towards the Wem–Red Rock Fault system that forms the SE margin of the basin. The Permo-Triassic sequence has a regional dip toward the ESE and thins gradually toward the mainly unfaulted western basin margin (Plant *et al.* 1999).

The Abbey Arms Wood borehole, located in

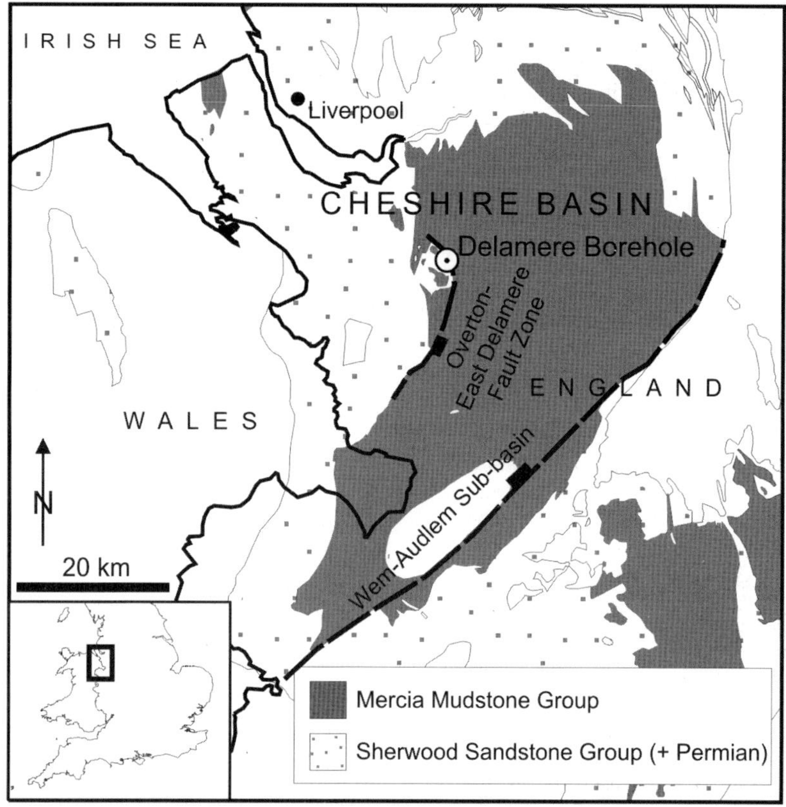

Fig. 1. Map illustrating the location of the Abbey Arms Wood borehole. SSG, Sherwood Sandstone Group; MMG, Mercia Mudstone Group. The Helsby and Wilmslow sandstone formations are units within the SSG.

an important structural position on the 'Western Slope' of the Cheshire Basin, is less than 1 km west of the 'Overton–East Delamere Fault Zone', part of a major structural belt that defines the western margin of the Wem–Audlem Sub-basin (Plant *et al.* 1999). This structural belt is some 70 km in length, runs northward from the Wem Fault in the south to north of Liverpool, and has an easterly downthrow that locally exceeds 1000 m at the top of the pre-Permian basement. The structural belt has an important control on the geological outcrop pattern, with the Sherwood Sandstone Group generally occurring to the west of the structural belt and the Mercia Mudstone Group to the east. The structural belt broadly forms the eastern margin of the 'Mid-Cheshire Ridge'.

Stratigraphy of the Delamere area

The Abbey Arms Wood borehole penetrates the Helsby Sandstone Formation and the Wilmslow Sandstone Formation, the two uppermost units of the Sherwood Sandstone Group, a thick succession of Triassic sandstones, mudstones and conglomerates deposited in arid continental fluvial, aeolian and lacustrine environments. Figure 2 shows a simplified lithological log of the borehole and includes an indication of sedimentary structures encountered. For reference, the high-resolution gamma log is shown on the left-hand side of Figure 2.

In general terms, the Helsby Sandstone Formation comprises reddish brown, well-cemented, sometimes pebbly, fluvial sandstones interstratified with friable aeolian sandstones. An attempt has been made to subdivide the Helsby Sandstone into formal members (Delamere, Thuraston and Frodsham members) based on the proportion of aeolian and fluvial facies (Thompson 1970). However, this scheme has generally proven difficult to apply because of complex lateral and vertical facies change (Warrington *et al.* 1980).

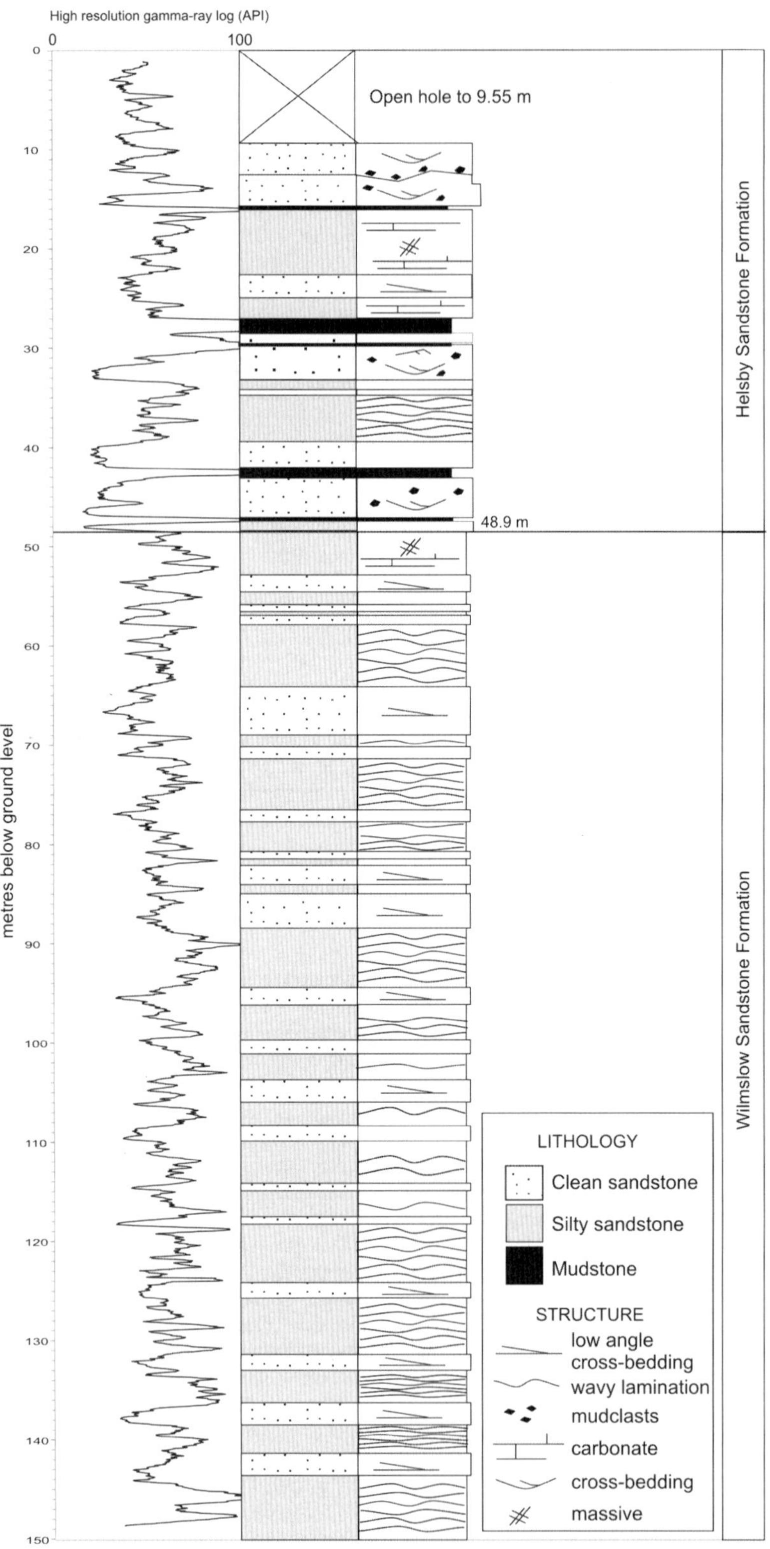

High resolution gamma-ray log (API)
0
100
metres below ground level
Open hole to 9.55 m
48.9 m
Helsby Sandstone Formation
Wilmslow Sandstone Formation
LITHOLOGY
Clean sandstone
Silty sandstone
Mudstone
STRUCTURE
low angle cross-bedding
wavy lamination
mudclasts
carbonate
cross-bedding
massive

The Helsby Sandstone Formation is generally thought to rest with a sharp erosive unconformity on the underlying Wilmslow Sandstone Formation and, in the basin centre, reaches a thickness of up to 230 m (Evans *et al.* 1993). The Wilmslow Sandstone comprises predominantly red, fine-grained, argillaceous cross-bedded sandstones with some interbedded siltstones and mudstones. The Wilmslow Sandstone Formation is 920 m thick in the Knutsford borehole in the centre of the basin (Evans *et al.* 1993). The contact between the Helsby Sandstone and Wilmslow Sandstone is relatively well defined in the Abbey Arms Wood borehole, occurring at a depth of 48.9 m bGL and marked by a change from relatively silty, fine- to medium-grained flat-bedded sandstones to clean cross-bedded sandstones and mudstones (Fig. 2).

Lithofacies and depositional environments encountered in the Delamere borehole

A lithofacies, or facies, is a rock unit with a distinctive set of characteristics, such as grain size and sedimentary structure, and is generally produced by a particular process or depositional environment. Table 1 provides a description and interpretation of the principal lithofacies of the Helsby and Wilmslow sandstone formations that are the focus of the present study. The Helsby and Wilmslow sandstones are each characterized by a distinct association of lithofacies that provide the key to the identification of depositional environment.

The Helsby Sandstone is composed predominantly of coarse-grained fluvial channel-fill sandstones interbedded with subordinate mudstones, aeolian sandstones and massive sandstones with zones of carbonate cementation. The interpretation of sandstones as fluvial is based on the presence of erosion surfaces (scoured channel bases), rounded mudclasts, and the association with fine-grained micaceous sandstones and mudstone. An important feature of the fluvial sandstones is that most appear to be composed of reworked aeolian sand grains, characterized by a high degree of rounding and a frosted surface texture. The mature composition and lack of clay matrix probably accounts for the relatively low gamma-ray values of the

fluvial sandstones (Fig. 2). Overall, the association of facies is consistent with previous depositional models of the Helsby Sandstone that envisage a mixed fluvial–aeolian environment (Thompson 1970).

The Wilmslow Sandstone differs significantly from the Helsby Sandstone. It is primarily a monotonous succession of wavy and irregularly stratified fine- to medium-grained silty sandstones. These wavy bedded units are characterized by undulose, wavy crinkly and irregular stratification defined by preferentially cemented thin very fine-grained and silty laminae a few mm to 1–2 cm apart. Analogous facies are developed in the Sherwood Sandstone Group in the East Irish Sea Basin, where they were previously interpreted as fluvial sheetflood deposits (Meadows & Beach 1993). However, recent work has demonstrated that these irregularly laminated silty sandstones are the product of deposition on an inland aeolian sandy sabkha, where deposition results from the trapping and adhesion of sand on and under thin surface salt crusts (Mountney & Thompson 2002).

Direct evidence for evaporitic conditions is generally absent from the rock record of this environment because the salts are dissolved under conditions of sediment aggradation and rising water table. The sandy sabkha deposits are interbedded with thin units of aeolian sandsheet–dune sandstones. These sandstones are relatively 'clean' (free from clay and silt) compared to the sabkha deposits and have a lower gamma-ray response. They are also more poorly cemented and friable than the other lithologies, and are often represented by zones of core loss. The gamma-ray curve (Fig. 2) highlights the arrangement of clean aeolian sandstones with sabkha deposits in cleaning-upwards cycles. These cycles may reflect periodic fluctuations in the palaeo-water table that allowed the encroachment and inundation of the sabkha by 'dry' aeolian deposits.

Large-scale geometry of the lithofacies

Outcrop studies of the Helsby Sandstone Formation in the Cheshire Basin show that fluvial sandstones and mudstones are often channel fills that have limited (tens of metres) lateral extent (Mountney & Thompson 2002). In contrast, in the East Irish Sea Basin, aeolian

---

Fig. 2. Stratigraphic log of the Abbey Arms Wood borehole with a high-resolution gamma-ray log for reference (the units of the high-resolution gamma-ray log are standard American Petroleum Institutes gamma-ray units).

Table 1. *Principal lithofacies of the Helsby and Wilmslow sandstone formations in the Abbey Arms Wood borehole*

Facies	Facies code	Facies description	Facies interpretation
Aeolian sandy sabkha	Aw	Sandstone, generally dark reddish brown, fine–medium grained, poorly sorted, moderately cemented, distinctive wavy lamination highlighted by thin silty laminae and irregular lenses of sand	Deposition of windblown sand, silt and clay on a siliciclastic sabka by the adhesion and trapping of sediment on and under salt crusts
Aeolian sand sheet	Al	Sandstone, generally reddish brown or pale reddish brown, fine–coarse grained, moderately sorted, moderately cemented to friable, distinctive low-angle/horizontal pinstripe (wind-ripple) lamination	Deposition of wind-blown sand as a low-relief sand sheet where wind-ripples form the dominant sedimentary structure and dune slip-faces are absent
Aeolian dune	Ax	Sandstone, generally reddish brown or pale reddish brown, fine–coarse grained, moderately sorted, moderately cemented to friable, high-angle cross-lamination	Deposition of wind-blown sand on a slip-faced aeolian dune
Coarse-grained fluvial channel fill	Fx	Sandstone, generally reddish brown or orange brown, fine–coarse grained, moderately to well sorted, cross-bedded, contains mudclasts often lining erosion surfaces	Deposition of sand in a fluvial channel environment by the lateral and downstream migration of dunes and bars
Fine-grained fluvial channel fill	Fl	Sandstone, generally reddish brown or orange brown, fine-grained, moderately to well sorted, laminated or ripple cross-laminated, commonly micaceous	Deposition in a fluvial channel, generally fine-grained bar-top deposits
Mudstone	M	Mudstone, dark reddish brown, generally massive	Deposition from suspension in standing water body, e.g. channel plug, shallow lake
Massive sandstone	Sm	Sandstone, generally reddish brown or dark reddish brown, fine–coarse grained, moderately to well sorted, can include carbonate cemented zones generally as bedding parallel layers and nodules	Pedogenically destratified sands of fluvial or aeolian origin. Carbonates probably represent early digenetic calcretes precipitated from Ca-rich groundwaters in the near-surface zone

dune and sandy sabka deposits have been correlated over several kilometres (Herries & Cowan 1997).

Hydrogeological context

The water level in the Abbey Arms Wood borehole was 40.06 m bGL in April 2002. The regional flow in the Cheshire Basin is towards the north and NE, although it is probable that this is distorted locally by pumping from public supply boreholes, in particular the Delamere Pumping Station [NGR 356050 367700]. Local groundwater flow in the Delamere area is believed to be towards the west. Groundwater residence times in the Sherwood Sandstone at shallow depths are relatively short, as shown by modern ^{14}C dates (Kinniburgh *et al.* 2006). Pumping tests in the West Cheshire region, closer to the Abbey Arms Wood borehole, indicate transmissivities typically ranging from 100 to 500 m^2 day^{-1}.

Approximately 300 m to the east of the borehole the N–S-trending normal East Delamere Fault brings the Helsby Sandstone and underlying formations into contact with the Mercia Mudstone Group, a throw of around 400 m (Fig. 1). This fault is likely to constitute a hydraulic boundary with very limited groundwater flow into or out of the relatively low-permeability Mercia Mudstone.

Sample selection, testing programme and methods

Sample selection and preparation

The core was sampled at approximately 2.5 m intervals for material for physical properties testing. Sampling intervals were selected to be as representative as possible of the full range of lithologies encountered in the core, although it was not possible to sample the few mudstones in the Helsby Sandstone Formation (due to local core loss associated with the mudstones and because when core was recovered the highly fissile nature of the mudstone made the cutting of core plugs impossible) or the fine-grained fluvial channel-fill deposits, which are a very minor component of the overall sequence. At each sampling interval, wherever possible, both a horizontal and a vertical core plug were prepared. Horizontal samples were cut sub-parallel to bedding and vertical samples were cut subperpendicular to bedding. The core plugs were cut as right-cylindrical plugs of about 24.5 mm diameter and about 27.5 mm length. All samples were then oven-dried at 60 °C for a minimum of 24 h prior to any physical properties tests.

Physical properties testing programme

A total of 103 core plugs from 51 sampling intervals were cut and tested for permeability. Fifty-one of these were horizontally oriented plugs and 52 were vertically oriented plugs. The 51 horizontally oriented core plugs were also tested for porosity, bulk density and grain density.

In addition to the core plugs, the whole core was tested using a mini-permeameter. This was used to provide a more detailed permeability profile along the length of the whole core. A total of 365 mini-permeameter measurements were made over a drilled length of core of 139.4 m. Given the small amount of core loss during drilling, this approximates to one mini-permeameter measurement every 0.39 m, i.e. almost an order of magnitude better sampling resolution than the core-plug sampling. The mini-permeameter was also used on the 103 core plugs. This enabled the mini-permeameter measurements on the whole core to be calibrated against the results of the standard permeability tests on the core plugs.

Methods overview

A standard liquid resaturation method was used to determine effective porosity, bulk and grain density (Bloomfield *et al.* 1995). Permeability was determined using nitrogen under steady-state conditions. Mini-permeametry was performed using nitrogen as the permeant under steady-state flow conditions and assuming radial flow geometry. Lithofacies were identified on the basis of inspection of hand specimens. Brief details of each of the methods is given below. In addition, data for calcium ion concentrations for the whole rock phase are briefly presented and discussed. These data are taken from Kinniburgh *et al.* (2006) and a full description of the analysis methods used to obtain the rock chemistry data for the sandstones is given in that reference.

Porosity. Effective porosity, bulk density and grain density were measured using a liquid resaturation method based on the Archimedes' principle. A sample to be tested is dried, weighed and then placed in a resaturation jar. The jar is evacuated then flooded with propanol. Propanol is used as it is relatively inert with respect to the core and, in particular, reduces the potential for swelling clays to modify the porosity during testing. The sample is allowed to saturate for at least 24 h. The saturated sample is then weighed, first immersed in the propanol and then, still saturated with propanol, in air. For each sample, a record is made of dry weight w [M], propanol saturated weight in air S_1 [M] and saturated weight immersed in propanol S_2 [M]. The density of the propanol ρ_f [ML^{-3}] is also noted. From these values, sample dry bulk density ρ_b [ML^{-3}], grain density ρ_g [ML^{-3}] and effective porosity ϕ [–] can be calculated as follows:

$$\rho_b = (w\rho_f)/(S_1 - S_2) \tag{1}$$

$$\rho_g = (w\rho_f)/(w - S_2) \tag{2}$$

$$\phi = (S_1 - w)/(S_1 - S_2). \tag{3}$$

The effective errors on the porosity measurements are approximately ±0.5 porosity %; for example, for a measured porosity of 23.7% the true porosity would lie in the range 23.2–24.2%.

Core-plug permeametry. Permeability tests were performed on samples under steady-state conditions. A full description of the methodology can be found in Bloomfield & Williams (1995). Samples were constrained in a core holder and a pressure-regulated supply of nitrogen gas was applied to one end of the sample (the downstream end of the sample was held at atmospheric pressure). A soap-film flow meter was used to measure the outflow of nitrogen from the downstream end of the sample.

Permeability was calculated using the measured sample dimensions, differential pressure and the steady-state gas flow rate as follows:

$$k = \mu\, Q\, L\, P_o / [A\, (P_i^2 - P_o^2)] \qquad (4)$$

where k is permeability [L^2], μ is gas viscosity [MT^{-1}L^{-1}], Q is the volumetric gas flow rate measured at atmospheric pressure, L and A are the sample length [L] and area [L^2], respectively, P_o is the pressure downstream of the sample [ML^{-1}T^{-1}] and P_i is the pressure upstream of the sample [ML^{-1}T^{-1}]). The effective errors associated with the permeability measurements are about $\pm 2.5\%$ of measured sample permeability.

Mini-permeametry. Probe or mini-permeameters have become increasingly popular in recent years for characterizing permeability distributions within individual lithofacies and for quantifying the effects of small-scale heterogeneities on permeability (Hurst & Goggin 1995). The mini-permeameter samples a small volume of rock by forcing nitrogen gas through an injection tip that is sealed against an otherwise unconfined dry sample. The gas flows radially through the tested region of the sandstone and out through the free surface. Permeability, k_m [L^2], is calculated using a modified form of Darcy's law from the known injection pressure and flow rate, and the geometry of the tip and the rock as follows:

$$k_m = \frac{2\mu Q_b P_b T_{act}}{a G_0 (P_1^2 - P_2^2) T_{ref}} \qquad (5)$$

where G_0 is a geometrical shape factor [–], μ is viscosity [MT^{-1}L^{-1}], Q_b is volumetric flow rate [L^{3}T^{-1}], P_b is a reference pressure [ML^{-1}T^{-1}] taken to be 1×10^5 Pa, T_{act}/T_{ref} [–] is the ratio of the temperature of the flowing gas, T_{act}, to a reference temperature, T_{ref}, of 21 °C, a is the internal radius of the tip of the permeameter [L], and P_1 and P_2 are the upstream and downstream pressures, respectively [ML^{-1}T^{-1}].

A Temco MP-402 mini-permeameter was used. Measurements were made on the cleaned surface of the whole core at about 30–40 cm intervals along its length. Consequently, the measurements are analogous to horizontal permeability measurements on core plugs. Mini-permeametry measurements were only made on unfractured consolidated core of at least 4 cm in length. Between 5 and 10 measurements were made at each sampling point, and an average value was calculated. Measurements were attempted at 396 sampling points. However, at 31 sampling points the permeability was below the effective resolution of the mini-permeameter (about 10^{-15} m^2) and these data have not been included in the subsequent data analysis. Note that in addition to the limited censoring of the permeability distribution due to the resolution of the mini-permeameter, there is also a sampling bias. The core recovery, although good, was incomplete, some of the core was too friable to be tested with the mini-permeameter, and some of the core intervals were judged (on the basis of visual inspection) to be too impermeable and were not tested.

In addition, the mini-permeameter was used on the ends on unconstrained horizontal plug samples so that the mini-permeameter readings could be calibrated against the core-plug measurements. Up to 10 spot mini-permeameter measurements were made on the ends of each horizontally oriented core plug. The repeatability of these measurements was good. For example, for one relatively high-permeability sample permeabilities were in the range 1.46×10^{-12}–1.50×10^{-12} m^2 with an average permeability of 1.48×10^{-12} m^2 and a standard deviation of 1.58×10^{-14} m^2 (based on 10 spot measurements), and for one relatively low-permeability sample permeabilities were in the range 1.50×10^{-13}–1.52×10^{-13} m^2 with an average permeability of 1.51×10^{-13} m^2 and a standard deviation of 9.87×10^{-16} m^2 (based on seven spot measurements). Consequently, the average values are considered representative of the spot measurements on the plugs and are presented below.

Identification of lithofacies. Lithofacies were defined on the basis of the lithology and the main sedimentary structures present within each bed. In general, the distinction of aeolian and fluvial facies on simple textural criteria such as sorting or the presence of frosted wind-blown grains was not possible because, for example, aeolian sands could be reworked into fluvial deposits. Three main lithologies were recognized, clean sandstones, silty sandstones and mudstones. Fluvial facies were identified as clean, cross-bedded or cross-laminated sandstones that contained reworked mudstone clasts and erosion surfaces. The fluvial sandstones are associated with mudstones that were deposited from suspension during periods of low flow within the channel. Aeolian dune and sandsheet facies were identified as clean sandstones that showed moderate- to high-angle cross-stratified grain-flow strata or low- to moderate-angle cross-stratified translatent strata that lack internal ripple-form lamination (Hunter 1977). Aeolian sandy sabkha deposits were identified as silty sandstones that showed a diffuse, irregular wavy lamination. Comparable facies have been identified from the Lower Triassic Ormskirk Sandstone Formation of the East Irish Sea Basin where they were thought to originate under conditions of rising water table as wind-blown sand and silt adheres to the damp substrate or is trapped by algal and evaporitic topography (Herries & Cowan 1997).

Results

The core-plug data are listed in full in Table 2, and are summarized in Tables 3 and 4. Table 3 summarizes the porosity and density data. Table 4 summarizes the permeability data for the core plugs. It also summarizes the mini-permeametry measurements on the core plugs, and the mini-permeametry measurements made on the whole core corrected using the correlation between standard and mini-permeametry measurements on the sore plugs (Fig. 4, equation 6). Table 5 summarizes the mini-permeametry data as a function of sedimentary facies (note that eight mini-permeametry measurements were made in the Fl facies, but that no core-plug measurements were obtained for this facies, Table 2). Figure 3 is a cross-plot of permeability as a function of porosity, and Figure 4 shows the correlation between permeability measured using the mini-permeameter and standard methods on the core-plug samples. On the basis of this cross-plot, a calibration has been established for the mini-permeametry measurements on the whole core. Figure 5 shows permeability against depth, variation in porosity and whole-rock calcium ion concentration, and also gives a comparison of the gamma-log and core-plug permeability. Figure 6 is an illustration of variation of permeability and the gamma log in two short representative depth intervals, one from the Helsby Sandstone Formation and one from the Wilmslow Sandstone Formation. Figure 7 is a plot of core-plug permeability as a function of whole-rock calcium ion concentration. Figure 8 is a probability plot of the corrected mini-permeameter data for the different facies.

Porosity results

Table 3 shows that porosity, bulk density and grain density are in the ranges 5.9–26.6%, 1.94–2.51 g cm^{-3}, and 2.63–2.67 g cm^{-3}, respectively, with a mean porosity of 20.3% and mean bulk and grain densities of 2.11 and 2.65 g cm^{-3}. The porosity values are broadly consistent with those previously reported for the Triassic sandstones of the central and NW England (Allen *et al.* 1997 reported porosity varying in the range 2–35% with a median value of 26%). The porosity data are approximately normally distributed, but there is only a weak correlation between porosity and horizontal permeability based on measurements made on the core plugs (Fig. 3). This indicates that porosity alone is a poor predictor of hydraulic conductivity. In addition, Figure 3 shows that there are no clear correlations between porosity and sedimentary facies. Porosity is generally more variable in the top 50 m bGL, but no significant reduction in porosity with depth in the borehole can be recognized (Fig. 5). The density results also appear to be reliable as the average grain density of 2.65 g cm^{-3} is consistent with the density of quartz, the major mineral component of the sandstones.

Permeability results

Table 4 shows that core-plug permeability varies over almost five orders of magnitude, between 1.4×10^{-16} and 7.8×10^{-12} m^2, with a geometric mean of 1.85×10^{-13} m^2. These values are broadly consistent with permeabilities for the Permo-Triassic sandstones. Allen *et al.* (1997) reported permeability varying by just over six orders of magnitude from 1×10^{-17} to about 2×10^{-11} m^2 for the sandstones, with a median permeability of the order of 1×10^{-12} m^2. Table 3 also shows that, based on core-plug measurements, the vertical permeabilities are on average only slightly lower than the horizontal permeabilities (the geometric mean horizontal permeability is 3.1×10^{-13} m^2 while the corresponding vertical permeability is 1.1×10^{-13} m^2). From this it is inferred that the sandstones intercepted by the Abbey Arms Wood borehole only exhibit moderate–weak hydraulic anisotropy at the core-plug scale.

Table 4 shows that the mini-permeameter measurements on the horizontal core-plug samples correlate well with permeability determined by standard permeability tests, with respective geometric mean permeabilities of 2.8×10^{-3} and 3.1×10^{-13} m^2. This good cor–relation between the two measurement

Table 2. *Table of results for tests on core samples. V and H denote vertically and horizontally oriented cores, respectively. Fx is coarse-grained fluvial channel fill, Sm is massive sandstones, Al is aeolian sand sheets, Aw is aeolian sandy sabkha and Ax is aeolian dune facies. The first column of permeability results are those obtained using standard core-plug methods, the second using the mini-permeameter on the core plugs*

Core orientation	Depth (m bGL)	Facies	Permeability (standard) (m^2)	Permeability (mini-perm.) (m^2)	Porosity (%)	Bulk density (g cm^{-3})	Grain density (g cm^{-3})
H	11.00	Fx	3.1×10^{-12}	1.3×10^{-12}	20.0	2.122	2.652
V	11.00	Fx	2.6×10^{-12}	1.5×10^{-12}	–	–	–
H	13.61	Fx	5.4×10^{-13}	2.6×10^{-13}	15.4	2.247	2.657
V	13.61	Fx	3.9×10^{-13}	2.6×10^{-13}	–	–	–
H	16.25	Fx	2.7×10^{-13}	1.5×10^{-13}	19.0	2.156	2.662
V	16.25	Fx	1.7×10^{-13}	5.6×10^{-14}	–	–	–
H	20.59	Fx	7.8×10^{-12}	5.1×10^{-12}	26.6	1.942	2.647
V	20.59	Fx	7.5×10^{-12}	4.4×10^{-12}	–	–	–
H	23.21	Fx	3.5×10^{-12}	2.4×10^{-12}	22.3	2.059	2.650
V	23.21	Fx	1.6×10^{-12}	2.7×10^{-12}	–	–	–
H	26.44	Fx	3.9×10^{-13}	1.3×10^{-14}	17.4	2.195	2.657
V	26.44	Fx	3.9×10^{-15}	1.2×10^{-13}	–	–	–
V	26.44	Fx	3.2×10^{-13}	5.3×10^{-13}	–	–	–
H	28.67	Fx	9.6×10^{-13}	5.1×10^{-13}	18.9	2.157	2.660
V	28.67	Fx	8.0×10^{-13}	3.9×10^{-13}	–	–	–
H	32.17	Fx	4.0×10^{-12}	1.2×10^{-12}	20.2	2.116	2.652
V	32.17	Fx	2.2×10^{-12}	1.7×10^{-12}	–	–	–
H	34.85	Fx	5.1×10^{-13}	2.2×10^{-13}	19.9	2.121	2.648
V	34.85	Fx	1.1×10^{-13}	2.1×10^{-13}	–	–	–
H	37.79	Fx	9.1×10^{-13}	8.7×10^{-13}	20.8	2.102	2.654
V	37.79	Fx	3.3×10^{-13}	2.2×10^{-13}	–	–	–
H	41.08	Fx	2.1×10^{-12}	1.1×10^{-12}	21.7	2.076	2.652
V	41.08	Fx	2.2×10^{-12}	2.3×10^{-12}	–	–	–
H	43.66	Fx	8.6×10^{-15}	7.8×10^{-13}	19.7	2.129	2.650
V	43.66	Fx	4.7×10^{-15}	1.2×10^{-12}	–	–	–
H	46.07	Fx	9.8×10^{-15}	9.1×10^{-13}	17.8	2.181	2.653
V	46.07	Fx	1.0×10^{-14}	1.1×10^{-12}	–	–	–
H	48.13	Fx	9.7×10^{-15}	9.0×10^{-14}	16.6	2.210	2.650
V	48.13	Fx	5.3×10^{-15}	7.1×10^{-14}	–	–	–
H	50.96	Al	1.2×10^{-13}	2.4×10^{-13}	15.0	2.258	2.655
V	50.96	Al	8.0×10^{-16}	2.9×10^{-13}	–	–	–
H	53.62	Ax	2.3×10^{-12}	2.0×10^{-12}	24.3	2.001	2.645
V	53.62	Ax	1.9×10^{-12}	1.1×10^{-12}	–	–	–
H	55.59	Al	3.6×10^{-13}	4.2×10^{-13}	21.0	2.096	2.654
V	55.59	Al	2.7×10^{-14}	1.5×10^{-13}	–	–	–
H	58.79	Aw	7.2×10^{-13}	6.0×10^{-13}	21.7	2.079	2.654
V	58.79	Aw	4.0×10^{-13}	6.7×10^{-14}	–	–	–
H	60.32	Aw	1.6×10^{-12}	4.6×10^{-13}	23.9	2.015	2.650
V	60.32	Aw	2.2×10^{-13}	3.6×10^{-13}	–	–	–
H	63.79	Ax	2.7×10^{-12}	1.8×10^{-12}	23.6	2.017	2.640
V	63.79	Ax	4.4×10^{-13}	7.0×10^{-13}	–	–	–
H	65.55	Aw	2.1×10^{-12}	9.3×10^{-13}	23.3	2.023	2.637
V	65.55	Aw	1.3×10^{-12}	1.1×10^{-12}	–	–	–
H	68.66	Al	1.1×10^{-12}	9.9×10^{-13}	23.5	2.020	2.640
V	68.66	Al	1.6×10^{-12}	2.7×10^{-12}	–	–	–
H	71.21	Al	3.1×10^{-13}	1.1×10^{-13}	21.0	2.088	2.645
V	71.21	Al	1.1×10^{-13}	1.4×10^{-13}	–	–	–
H	74.73	Aw	6.6×10^{-13}	6.2×10^{-13}	21.2	2.082	2.642
V	74.73	Aw	2.3×10^{-14}	2.6×10^{-13}	–	–	–
H	77.76	Al	1.7×10^{-12}	1.2×10^{-12}	20.8	2.091	2.640
V	77.76	Al	3.2×10^{-13}	1.2×10^{-12}	–	–	–
H	80.72	Aw	2.7×10^{-13}	2.1×10^{-13}	20.0	2.116	2.644
V	80.72	Aw	1.0×10^{-13}	9.0×10^{-14}	–	–	–
H	84.03	Aw	1.3×10^{-13}	1.6×10^{-13}	19.3	2.137	2.647

continued

Table 2. *Continued*

Core orientation	Depth (m bGL)	Facies	Permeability (standard) (m^2)	Permeability (mini-perm.) (m^2)	Porosity (%)	Bulk density ($g\,cm^{-3}$)	Grain density ($g\,cm^{-3}$)
H	84.03	Aw	1.3×10^{-13}	1.6×10^{-13}	19.3	2.137	2.647
V	84.03	Aw	8.1×10^{-14}	1.5×10^{-13}	–	–	–
H	86.69	Al	2.0×10^{-12}	1.6×10^{-12}	26.1	1.944	2.631
V	86.69	Al	1.8×10^{-12}	3.4×10^{-12}	–	–	–
H	88.32	Al	1.1×10^{-12}	1.2×10^{-12}	22.8	2.036	2.637
V	88.32	Al	4.5×10^{-13}	5.1×10^{-13}	–	–	–
H	89.49	Aw	2.4×10^{-13}	3.1×10^{-15}	19.6	2.135	2.656
V	89.49	Aw	3.4×10^{-15}	2.2×10^{-14}	–	–	–
H	91.43	Aw	5.0×10^{-13}	1.7×10^{-13}	23.1	2.032	2.643
V	91.43	Aw	9.5×10^{-15}	1.8×10^{-13}	–	–	–
H	94.51	Ax	5.3×10^{-13}	2.5×10^{-13}	20.5	2.092	2.633
V	94.51	Ax	4.0×10^{-13}	1.2×10^{-12}	–	–	–
H	97.98	Aw	3.5×10^{-13}	5.8×10^{-13}	23.0	2.039	2.647
V	97.98	Aw	2.2×10^{-13}	4.8×10^{-13}	–	–	–
H	101.61	Aw	1.7×10^{-13}	1.4×10^{-13}	21.9	2.068	2.646
V	101.61	Aw	1.2×10^{-13}	1.7×10^{-13}	–	–	–
H	104.53	Al	6.6×10^{-13}	9.0×10^{-13}	22.7	2.041	2.641
V	104.53	Al	5.3×10^{-13}	1.3×10^{-12}	–	–	–
H	105.95	Aw	4.0×10^{-14}	5.0×10^{-14}	19.2	2.147	2.659
V	105.95	Aw	2.0×10^{-14}	1.3×10^{-13}	–	–	–
H	109.86	Al	1.5×10^{-13}	1.1×10^{-13}	22.7	2.051	2.655
V	109.86	Al	4.7×10^{-14}	1.3×10^{-14}	–	–	–
H	113.73	Aw	5.5×10^{-13}	1.1×10^{-13}	21.9	2.062	2.641
V	113.73	Aw	6.9×10^{-13}	1.0×10^{-12}	–	–	–
H	117.14	Sm	2.9×10^{-13}	2.9×10^{-13}	21.6	2.072	2.643
V	117.14	Sm	2.5×10^{-13}	4.4×10^{-13}	–	–	–
H	119.54	Al	1.3×10^{-14}	7.1×10^{-15}	14.5	2.278	2.663
V	119.54	Al	1.0×10^{-14}	1.2×10^{-13}	–	–	–
H	123.21	Aw	2.3×10^{-15}	1.5×10^{-14}	10.2	2.398	2.669
V	123.21	Aw	1.4×10^{-16}	6.4×10^{-14}	–	–	–
H	129.14	Aw	9.8×10^{-14}	4.9×10^{-13}	19.0	2.154	2.658
V	129.14	Aw	1.0×10^{-15}	2.2×10^{-12}	–	–	–
H	132.48	Aw	7.9×10^{-16}	1.4×10^{-15}	5.9	2.506	2.662
V	132.48	Aw	5.4×10^{-14}	2.6×10^{-14}	–	–	–
H	135.48	Al	2.5×10^{-14}	1.0×10^{-13}	19.2	2.140	2.649
V	135.48	Al	1.7×10^{-14}	1.0×10^{-13}	–	–	–
H	136.58	Sm	1.5×10^{-13}	1.0×10^{-13}	21.4	2.072	2.636
V	136.58	Sm	1.0×10^{-13}	4.2×10^{-13}	–	–	–
H	140.27	Aw	1.2×10^{-13}	1.2×10^{-13}	21.1	2.088	2.647
V	140.27	Aw	3.9×10^{-14}	4.4×10^{-14}	–	–	–
H	142.00	Al	1.6×10^{-13}	3.2×10^{-13}	20.1	2.110	2.641
V	142.00	Al	7.3×10^{-14}	3.6×10^{-14}	–	–	–
H	144.12	Ax	1.5×10^{-12}	8.5×10^{-13}	20.7	2.085	2.630
V	144.12	Ax	8.6×10^{-13}	9.2×10^{-13}	–	–	–
H	145.72	Aw	7.0×10^{-14}	9.9×10^{-14}	19.9	2.117	2.643
V	145.72	Aw	2.2×10^{-14}	1.1×10^{-13}	–	–	–
H	147.74	Aw	4.5×10^{-13}	4.6×10^{-13}	23.3	2.018	2.633
V	147.74	Aw	3.3×10^{-13}	5.2×10^{-13}	–	–	–
H	148.98	Al	2.0×10^{-12}	1.5×10^{-12}	21.3	2.071	2.633
V	148.98	Al	3.7×10^{-13}	1.4×10^{-12}	–	–	–

Table 3. *Summary statistics for the porosity and density data*

Statistic	Porosity (%)	Bulk density (g cm^{-3})	Grain density (g cm^{-3})
n	51	51	51
Minimum	5.9	1.94	2.63
Maximum	26.6	2.51	2.67
Arithmetic mean	20.3	2.11	2.65
Geometric mean	19.9	2.11	2.65
SD	3.5	0.10	0.01
Skewness	−1.7	1.7	0.03

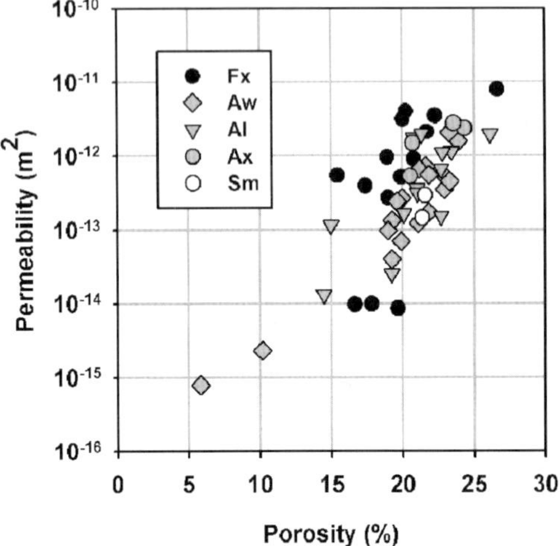

Fig. 3. Cross-plot of matrix permeability as a function of porosity and sedimentary facies. Fx is coarse-grained fluvial channel fill; SM is massive sandstones; Al is aeolian sand sheets; Aw is aeolian sandy sabkha; and Ax is aeolian dune facies.

techniques is illustrated in Figure 4, which shows the correlation between core-plug permeability using the standard method and results on the same horizontally oriented samples obtained using mini-permeametry. Figure 4 also shows that permeability measurements are systematically higher using the standard method on the whole core plugs compared with the mini-permeameter measurements. This is due to the flow geometry associated with the mini-permeameter measurements. In standard permeability tests on horizontally oriented core plugs the gas flows through the entire sample parallel to the predominantly subhorizontal fabric. In mini-permeameter tests on the ends of horizontally oriented core plugs, however, gas flows radially from the tip of the mini-permeameter through the rock to the free surface of the core plug. Even small anisotropies in the permeability structure of the core will reduce the measured permeability. It is inferred from Figure 4 that most core plugs tested exhibit a slight permeability anisotropy associated with grain-scale fabrics (consistent with the observations above that there is a slight permeability anisotropy at the core-plug scale).

A linear regression of the standard permeability data, k, on the mini-permeameter data for horizontally oriented plug samples, k_{m}, has the form:

$$k = 1.475k_{m} \ \mathrm{m}^2 \qquad (6)$$

with an r^2 value of 0.85. This correlation has been used to correct the mini-permeameter measurements (made in a horizontal direction) on the whole core (Table 4, right-hand column).

Figure 5 shows depth trends in horizontal and

Table 4. *Summary statistics for the permeability data. The first three columns summarize the results of standard tests on the core plugs, the fourth column summarizes the results of the mini-permeametry on the same core-plug samples, and the last column summarizes the mini-permeametry measurements made on the whole core corrected using the correlation between standard and mini-permeametry measurements on the sore plugs (Fig. 4 and equation 6)*

Statistic	Core-plug permeability			Mini-permeametry	
	All samples	Horizontal samples	Vertical samples	On core plugs Horizontal samples	Corrected permeability on whole core
n	103	51	51	51	365
Minimum (m^2)	1.4×10^{-16}	7.9×10^{-16}	1.4×10^{-16}	1.4×10^{-15}	5.9×10^{-15}
Maximum (m^2)	7.8×10^{-12}	7.8×10^{-12}	7.5×10^{-12}	5.1×10^{-12}	9.4×10^{-12}
Arithmetic mean (m^2)	7.8×10^{-13}	9.7×10^{-13}	6.0×10^{-13}	6.7×10^{-13}	1.3×10^{-12}
Geometric mean (m^2)	1.8×10^{-13}	3.1×10^{-13}	1.1×10^{-13}	2.8×10^{-13}	1.7×10^{-12}
SD (m^2)	1.3×10^{-13}	1.4×10^{-12}	1.2×10^{-12}	8.6×10^{-13}	1.6×10^{-12}
Skewness	3.4	2.9	4.2	3.1	2.0

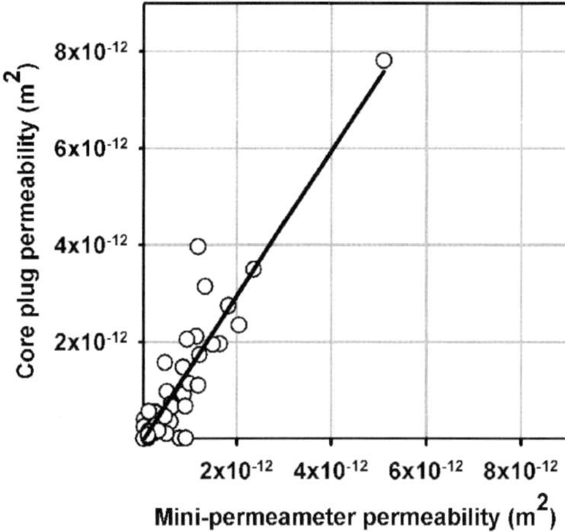

Fig. 4. Correlation between permeability measured using the mini-permeameter and standard methods on the core-plug samples. The line shows a linear regression through the data.

vertical permeability based on plug measurements, and the corrected mini-permeametry measurements for the whole core. Overall, there does not appear to be a significant depth trend in matrix permeability or any significant systematic variation in matrix anisotropy with depth. Cyclic trends in permeability can, however, be seen. As previously noted, the Helsby Sandstone Formation shows a number of fining-up cycles. These can be seen in the permeability depth profile in Figure 5, and in the example shown in Figure 6 in the detailed interval between about 33 and 27 m bGL. In contrast, in the Wilmslow Sandstone Formation there are a number of cleaning-upwards cycles that can be seen to be associated with increasing permeabilities in the depth profile in Figure 5, and in the example shown in Figure 6 in the detailed interval between 112 and 106 m bGL.

In addition, the mini-permeametry and core-plug permeability measurements indicate more small-scale variability and a wider range in matrix permeability in the Helsby Sandstone Formation than in the Wilmslow Sandstone Formation (Fig. 5). Towards the base of the borehole some of the aeolian sandstones are relatively friable and less cohesive. In this interval it was difficult to make mini-permeability measurements on the core. Hence, the density of observations is less and any variability in matrix permeability may be less well characterized.

Figure 5 shows a generally good agreement between trends in the calibrated mini-permeability data and the horizontal permeability measurements made on the core plugs. The mini-permeameter profile provides substantially more information on the detail of the fining-upwards sequences in the Helsby Sandstone Formation than the core-plug measurements. The mini-permeametry, however, is unable to characterize the occasional low-permeability units where permeability is less than about 10^{-15} m². These low-permeability intervals are often found near the top of the fining-upwards cycles in the top 50 m, and are commonly associated with small positive spikes on the gamma log indicating elevated clay content in the matrix.

Figure 5 shows relatively low concentrations of calcium in the whole rock. This supports the observation that there is relatively little calcite cement present in the sandstones. Where limited calcite cement is inferred to be present it is principally in the Helsby Sandstone Formation or it is below about 120 m bGL in the Wilmslow Sandstone Formation. Where calcite cement is not inferred to be present permeability is independent of the calcium concentration in the whole rock, and where it is inferred to be present permeability is only very weakly correlated with permeability (Fig. 7).

Comparison of the high-resolution gamma log with core-plug permeability measurements (Fig. 5) shows that there is a good inverse correlation between permeability and the gamma log within the Helsby Sandstone Formation and a weaker correlation within the Wilmslow Sandstone Formation.

Table 4 shows that the average permeabilities of the different sedimentary facies identified in the Abbey Arms Wood borehole vary by up to about one order of magnitude. For example, the geometric mean permeability of the fine-grained fluvial sandstones (Fl), the least permeable of the sandstone facies, is 9.7 × 10^{-14} m², whereas the corresponding mean for the cross-laminated aeolian sandstone (Ax) is 1.7 × 10^{-12} m². Inspection of the normal probability plots of corrected mini-permeameter measurements on the whole core in Figure 8 shows subtle differences in the permeability distributions between the fluvial and aeolian facies and within the aeolian facies (see Table 5). The two fluvial sandstone facies, the fine-grained fluvial sandstones (Fl) and the cross-bedded fluvial sandstones (Fx), and the massive sandstones (Sm) facies have systematically lower permeabilities than the aeolian sandstones. Within the aeolian sandstones the differences in permeability distribution appear to reflect primarily differences in clay content, with the low-angle laminated and cross-bedded sandstone interbeds (Al and Ax) being

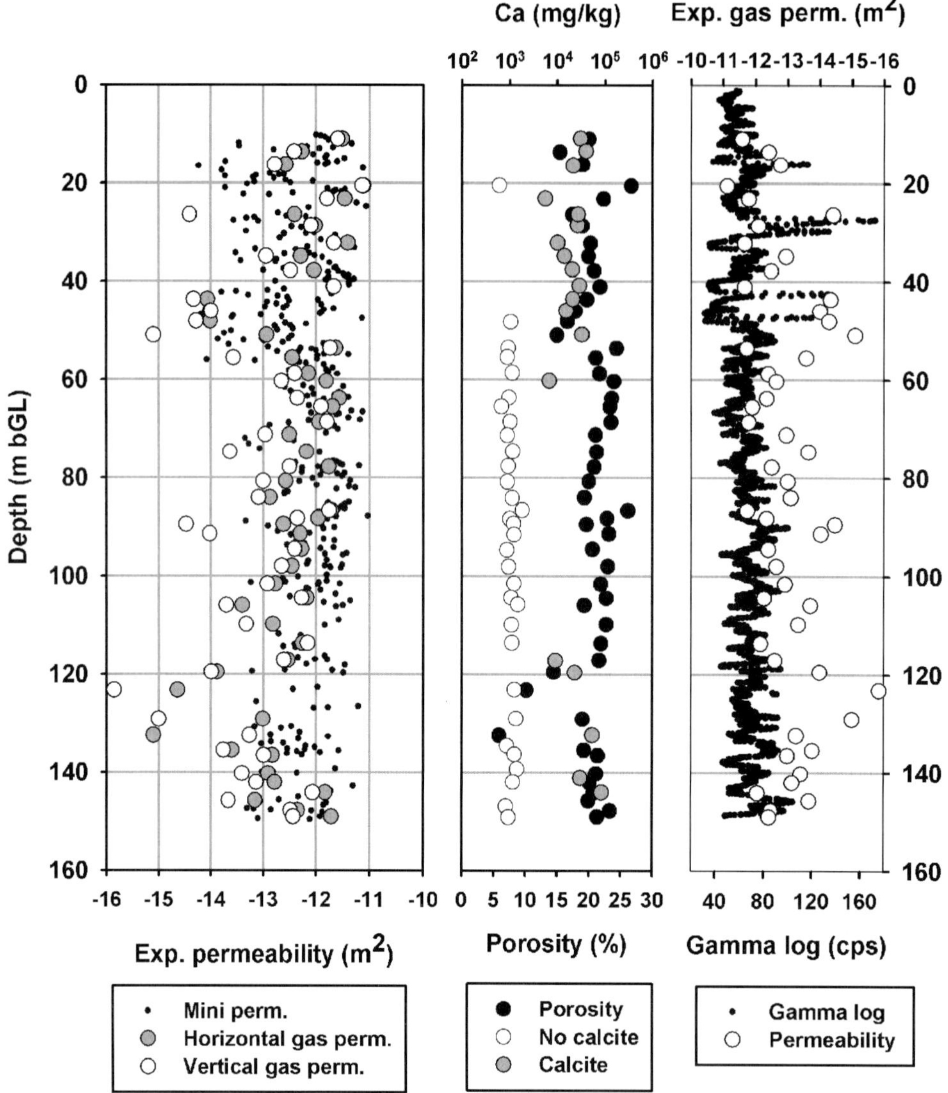

Fig. 5. Summary depth plots of the permeability, porosity, calcium ion whole-rock chemistry and gamma logs for the Abbey Arms Wood borehole. Permeability data include horizontal and vertical permeability based on core-plug measurements and the calibrated mini-permeability measurements on the whole core. See text for description and discussion of data.

systematically more permeable than the aeolian sandy sabkha (Aw) facies. The Al and Ax facies have relatively low clay and silt contents with respect to the sabkha deposits.

Discussion and conclusions

Sandstone lithofacies have been identified on the basis of sedimentological characteristics alone (Table 1). It has been noted that, although they exhibit similar permeability distributions, subtle systematic differences in the permeability distributions can be identified as a function of lithofacies. Are the permeability distributions for each facies statistically distinct? Although the distributions appear to be log-normal (Fig. 8), because the form of the permeability distributions is unknown, parametric tests cannot be used to compare the distributions. Hence, a (two-sided) Kolmogorov–Smirnov

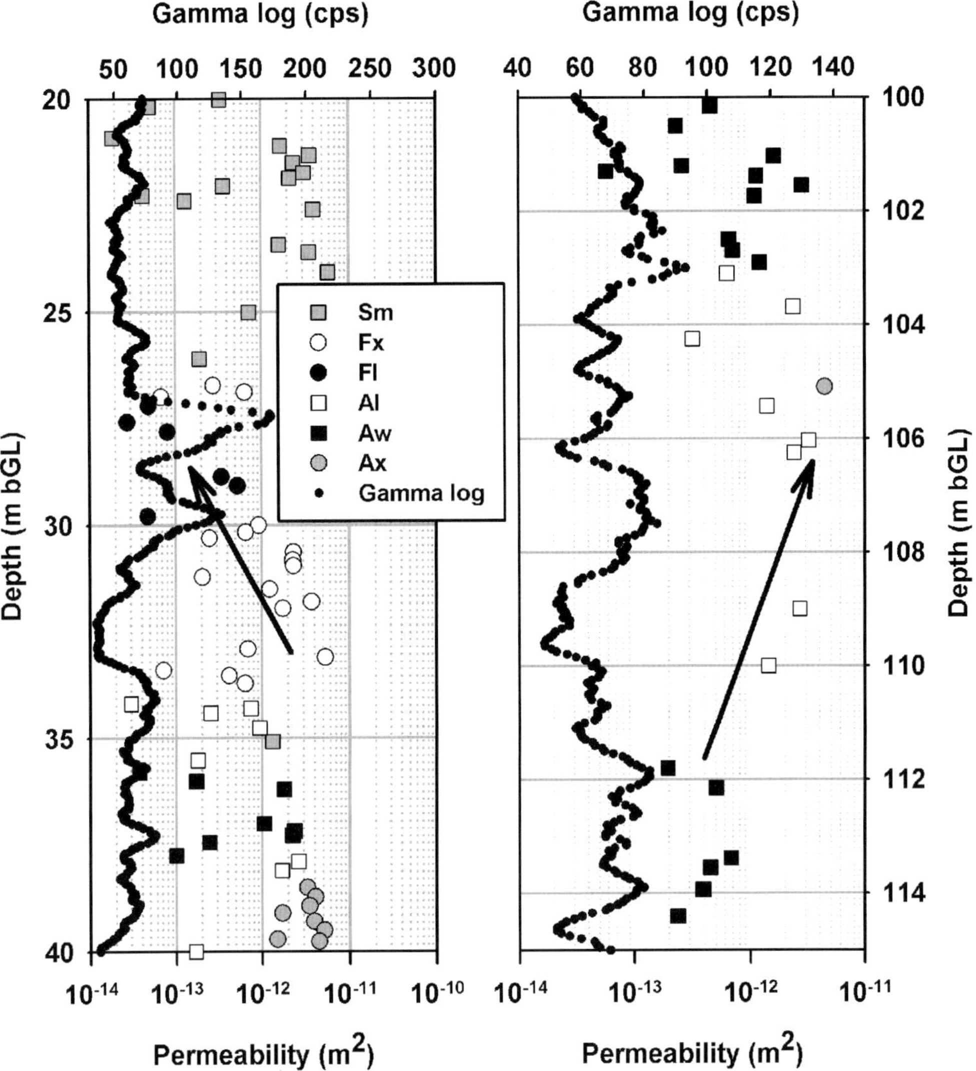

Fig. 6. Illustration of variation in permeability and the gamma-ray log across a fining-upwards sequence in the Helsby Sandstone Formation between about 33 and 27 m bGL, and across a cleaning-upwards cycle in the Wilmslow Sandstone Formation between about 11 and 105 m bGL. The units of the reference gamma-ray logs are in raw counts per s.

(non-parametric) test has been used (Conover 1980). This test determines whether two independent samples come from the same distribution by comparing the two-sample cumulative distribution functions. The test assumes that both samples come from exactly the same distribution. The test, however, is not ideal as each distribution in the test has an unequal sample size and the sample size for the fine-grained fluvial facies (Fl) is very small so the results of the test should only be taken as broadly indica-

tive. The results of the Kolmogorov–Smirnov test (Table 6) suggest that the probabilities that the paired distributions are the same is generally low or very low. Only the paired permeability distributions for the aeolian sabkha (Aw) and aeolian sand sheet (Al) facies and the massive sandstone (Sm) and fine-grained fluvial channel fill sandstone (Fl) facies have probabilities greater than 0.5 and none of paired distributions appear to come from the same distribution at probability levels of 0.9 or

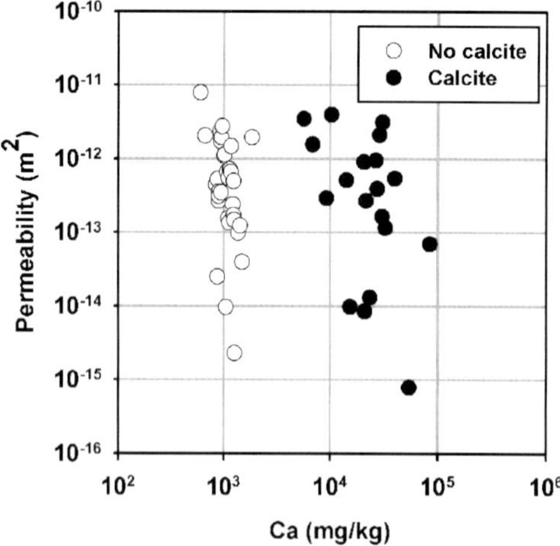

Fig. 7. Illustration of core-plug permeability as a function of calcium ion concentration in the whole rock. The figure shows two populations. One population where calcite cement is inferred to be absent and permeability is independent of Ca concentration, and one population where calcite is inferred to be present and there is a weak correlation with permeability.

greater. Consequently, it is inferred that the permeability distributions for each sedimentary facies are distinct.

A range of factors may control the permeability of the sandstones. There is only a poor correlation between effective porosity and permeability (Fig. 3) as would be expected, as permeability generally shows a better correlation with some measure of pore-size distribution (Dullien 1979). Pore-throat sizes in the sandstones may be significantly modified by clays (particularly affecting pore throats) and by cements (commonly calcite and dolomite) affecting both pores and pore throats. Calcite is present as a cement phase in the Abbey Arms Wood borehole (Kinniburgh *et al.* 2006). Higher values of calcium in the rock may indicate the presence of calcite-cemented horizons, mainly in the Helsby Sandstone Formation (Fig. 5). The calcite cement principally appears to be an early cement preserving an uncompacted grain fabric and is interpreted as a calcrete. Minor weakly ferroan calcite cement also occurs as later diagentic overgrowths. Both types of calcite cement have undergone significant corrosion and dissolution and there is only a very weak correlation between calcite cement, when present, and permeability. Whereas a number of lines of evidence indicate that there is a correlation between clay content and permeability. As has been noted previously, the sandier facies

generally exhibit higher permeabilities while facies containing more clays and silts generally exhibit slightly lower permeabilities. These trends are consistent with the correlation between the gamma log and permeability (Fig. 5), where high gamma-log values are associated with high clay contents and lower permeabilities, particularly in the Helsby Sandstone Formation. Based on these observations it is inferred that there is little remaining calcite cement in or near the pore throats, most having been removed by dissolution, that the cement phase does not appear to have a significant influence on hydraulic conductivity, and that clay content appears to be an important factor influencing the hydraulic conductivity of the matrix particularly in the Helsby Sandstone Formation. The weaker correlation between permeability and the gamma log in the Wilmslow Sandstone Formation may indicate that clay content is less important in controlling permeability in this formation and that grain-size distribution and grain sorting may be more important in influencing the distribution of hydraulic conductivity in this formation. Unfortunately, we have no data on grain size distributions or sorting to support this inference.

Summary and conclusions

- This study has identified seven distinct sedimentary facies within the Helsby and Wilmslow sandstone formations in the Abbey Arms Wood borehole; coarse- and fine-grained fluvial channel-fill facies, a mudstone facies, aeolian sandy sabkha, sand sheet and dune facies, and a facies associated with destratified sands of either fluvial or aeolian origin that gives rise to massive sandstones.

- Each of the two formations can be characterized by a representative assemblage of these facies with the Helsby Sandstone Formation consisting predominantly of fluvial sandstones with minor interstratified friable aeolian sandstones and the Wilmslow Sandstone Formation consisting principally of fine-grained cross-bedded aeolian sandstones. A sequence of fining-upwards cycles is present in the Helsby Sandstone Formation and a sequence of drying-upwards cycles is present in the Wilmslow Sandstone Formation.

- Detailed permeability profiling using a mini-permeameter on the whole core has enabled sufficient measurements to be made so that permeability distributions can be established for each facies.

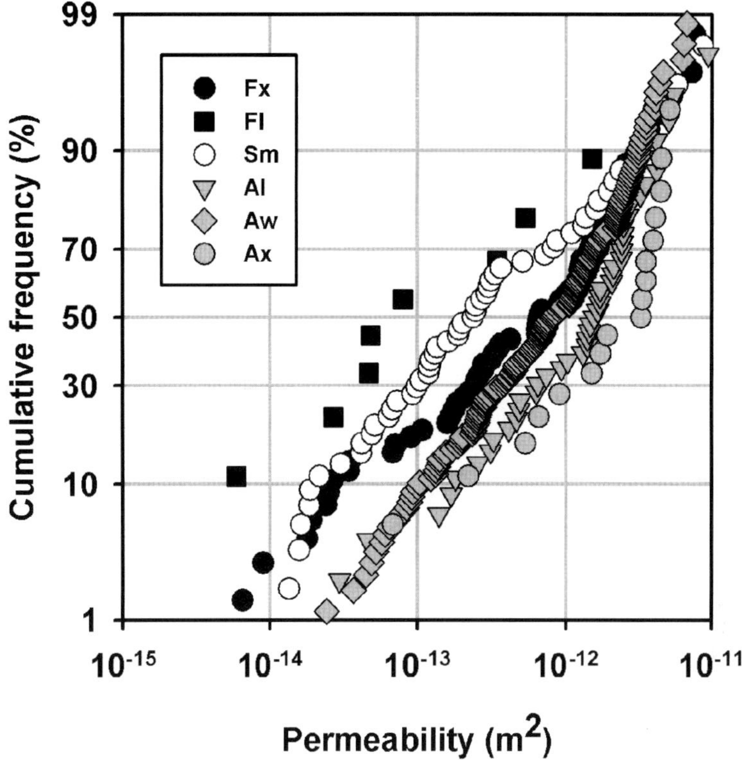

Fig. 8. Normal probability plots of corrected mini-permeametry measurements on the whole core as a function of sedimentary facies. Fx is coarse-grained fluvial channel fill; SM is massive sandstones; Al is aeolian sand sheets; Aw is aeolian sandy sabkha; and Ax is aeolian dune facies. Units given here for permeability are millidarcies where 1 mD = 1 $\times$ 10^{-15} m^2.

Table 5. *Summary of corrected mini-permeametry measurements as a function of sedimentary facies*

Statistic/facies	Fx	Fl	Sm	Al	Aw	Ax
n	66	8	63	45	166	17
Minimum (m^2)	6.5×10^{-15}	5.9×10^{-15}	8.3×10^{-15}	3.0×10^{-14}	2.2×10^{-14}	6.9×10^{-14}
Maximum (m^2)	7.7×10^{-12}	1.5×10^{-12}	8.7×10^{-12}	9.4×10^{-12}	7.5×10^{-12}	5.1×10^{-12}
Arithmetic mean (m^2)	1.2×10^{-12}	3.2×10^{-13}	9.5×10^{-13}	1.9×10^{-12}	1.3×10^{-12}	2.5×10^{-12}
Geometric mean (m^2)	4.8×10^{-13}	9.7×10^{-14}	2.8×10^{-13}	1.2×10^{-12}	7.0×10^{-13}	1.7×10^{-12}
SD (m^2)	1.6×10^{-12}	5.1×10^{-13}	1.6×10^{-12}	1.8×10^{-12}	1.4×10^{-12}	1.7×10^{-12}
Skewness	2.3	2.2	2.9	2.0	1.8	−1.1

Table 6. *Results of a two-sample Kolmogorov–Smirnov tests on corrected mini-permeametry measurements as a function of sedimentary facies. Fx is coarse-grained fluvial channel fill, Sm is massive sandstones, Al is Aeolian sand sheets, Aw is Aeolian sandy sabkha and Ax is Aeolian dune facies*

Probability	Fx	Fl	Sm	Al	Aw
Fx	–	–	–	–	–
Fl	0.073	–	–	–	–
Sm	0.114	0.52	–	–	–
Al	0.016	0.187	0	–	–
Aw	0.318	0.187	0	0.781	–
Ax	0.009	0	0	0.052	0.003

- The mean permeabilities for each facies are within about one order of magnitude; however, the permeability distributions can be differentiated on the basis of cumulative probability plots and the distributions for each facies appear to be statistically distinct and support the sedimentological classification.
- An important control on permeability appears to be clay content, particularly in the Helsby Sandstone Formation, and the limited calcite cement remaining in the sandstone appears to have a negligible effect on the hydraulic conductivity profile.

This pilot study has successfully demonstrated that it is feasible to obtain sufficient permeability measurements to characterize permeability distributions in sedimentary facies in the Permo-Triassic sandstone aquifer. The aim of future studies is to obtain additional information on permeability distributions within the Helsby and Wilmslow formations and other formations in the aquifer. Characteristic spatial correlations in permeability should also be investigated on a facies-by-facies basis. It is noted that if stochastic models of flow and transport models are to be built using data arising out of such studies, then it will also be necessary to develop a better understanding of the internal sedimentary architecture of the sandstone aquifer. It is also noted that the extensive modification of the cement phases by dissolution, inferred to be principally due to recent groundwater circulation, supports the view that if stochastic models of the sandstone aquifer are to be developed they must be based on permeability distributions obtained from measurement of on-shore aquifers. Published core permeability data from off-shore sandstone hydrocarbon reservoirs, where cement phase modification by freshwater circulation is likely to be limited, is likely to be inappropriate for use in stochastic models of on-shore aquifers.

We would like to thank A. Milodowski for his work on the calcite cements. The work described in this paper was undertaken as part of a larger study in conjunction with J. Ingram and R. Ward of the Environment Agency and P. Merrin of United Utilities. This paper is published with the permission of the Executive Director of the British Geological Survey.

References

AIGNER, T., ASPRION, U., HORNUNG, J., JUNGHANS, W.-D. & KOSTREWA, R. 1996. Integrated outcrop analogue studies for Triassic alluvial reservoirs: examples from southern Germany. *Journal of Petroleum Technology*, **19**, 393–406.

ALLEN, D.J., BREWERTON, L.J. ET AL. 1997. *The Physical Properties of Major Aquifers in England and Wales*. British Geological Survey, Technical Report, **WD/97/34**. Environment Agency R&D Report Publication, **8**.

ANDERSON, M.P., AIKEN, J.S., WEBB, E.K. & MICKELSON, D.M. 1999. Sedimentology and hydrogeology of two braided stream deposits. *Sedimentary Geology*, **129**, 187–199.

BAHAR, A. & KELKAR, M. 2000. Journey from well logs/cores to integrated geological and petrophysical properties simulation: a methodology and application. *SPE Reservoir Evaluation & Engineering*, **3**, 444–456.

BLOOMFIELD, J.P. & WILLIAMS, A. 1995. An empirical liquid permeability–gas permeability correlation for use in aquifer properties studies. *Quarterly Journal of Engineering Geology*, **28**, S143–S150.

BLOOMFIELD, J.P., BREWERTON, L.J. & ALLEN, D.J. 1995. Regional trends in matrix porosity and dry density of the Chalk of England. *Quarterly Journal of Engineering Geology*, **28**, S131–S142.

BLOOMFIELD, J.P., GOODDY, D.C., BRIGHT, M.I. & WILLIAMS, P.J. 2001. Pore-throat size distributions in Permo-Triassic sandstones from the United Kingdom and some implications for contaminant hydrogeology. *Hydrogeology Journal*, **9**, 219–230.

BOUCH, J.E., HOUGH, E., KEMP, S.J., MCKERVEY, J.A., WILLIAMS, G.M. & GRESWELL, R.B. 2006. Sedimentary and diagenetic environments of the Wildmoor Sandstone Formation (UK): implications for groundwater and contaminant transport, and sand production. *In*: TELLAM, J.H. & BARKER, R.D. (eds) *Fluid Flow and Solute Movement in Sandstones: The Offshore UK Permo-Triassic Red Bed Sequence*. Geological Society, London, Special Publications, **263**, 129–153.

CHADWICK, R.A. 1997. Fault analysis of the Cheshire Basin, north-west England. *In*: MEADOWS, N.S., TRUEBLOOD, S., HARDMAN, M. & COWAN, G. (eds) *The Petroleum Geology of the Irish Sea and Adjacent Areas*. Geological Society, London, Special Publications, **124**, 297–313.

CONOVER, W.J. 1980. *Practical Nonparametric Statistics*. Wiley, New York.

DULLIEN, F.A.L. 1979. *Porous Media. Fluid Transport and Pore Structure*. Academic Press, New York.

EVANS, D.J., REES, J.G. & HOLLOWAY, S. 1993. The Permian to Jurassic stratigraphy and structural evolution of the Cheshire Basin. *Journal of the Geological Society, London*, **150**, 857–870.

HERN, C.Y. & STELL, N.C.T. 1997. Petrophysical and facies description of the mixed fluvio-aeolian Frobisher reservoir: a genetic unit approach to reservoir characterization. *In*: OAKMAN, C.D., MARTIN, J.H. & CORBETT, P.W.M. (eds) *Cores from the Northwest European Hydrocarbon Province: An Illustration of Geological Applications from Exploration to Development*, Geological Society, London, 187–195.

HERRIES, R.D. & COWAN, G. 1997. Challenging the

'sheetflood' myth: the role of water-table-controlled sabkha deposits in redefining the depositional model for the Ormskirk Sandstone Formation (Lower Triassic), East Irish Sea Basin. *In*: MEADOWS, N.S., TRUEBLOOD, S.P., HARDMAN, M. & COWAN, G. (eds) *Petroleum Geology of the Irish Sea and Adjacent Areas*. Geological Society, London, Special Publications, **124**, 253–276.

HORNUNG, J. & AIGNER, T. 1999. Reservoir and aquifer characterization of fluvial architectural elements: Stubensandstein, Upper Triassic, southwest Germany. *Sedimentary Geology*, **129**, 215–280.

HUNTER, R.E. 1977. Terminology of cross-stratified sedimentary layers and climbing ripple structures. *Journal of Sedimentary Petrology*, **47**, 697–706.

HURST, A. & GOGGIN, D. 1995. Probe permeametery: An overview and bibliography. *AAPG Bulletin*, **79**, 463–473.

JENSEN, J.L., CORBETT, P.W.M., PICKUP, G.E. & RINGROSE, P.S. 1996. Permeability semivariograms, geological structure and flow performance. *Mathematical Geology*, **28**, 419–435.

KINNIBURGH, D.K., NEWELL, A.J., DAVIES, J., SMEDLEY, P.L., MILODOWSKI, A.E., INGRAM, J. & MERRIN, P.D. 2006. The arsenic concentration in groundwater from Abbey Arms Wood observation borehole, Delamere, Cheshire, UK. *In*: TELLAM, J.H. & BARKER, R.D. (eds) *Fluid Flow and Solute Movement in Sandstones: The Offshore UK Permo-Triassic Red Bed Sequence*. Geological Society, London, Special Publications, **263**, 265–284.

MEADOWS, N.S. & BEACH, A. 1993. Structural and climatic controls on facies distribution in a mixed fluvial and aeolian reservoir: the Triassic Sherwood Sandstone in the Irish Sea. *In*: NORTH, C.P. & PROSSER, D.P. (eds) *Characterization of Fluvial and Aeolian Reservoirs*. Geological Society, London, Special Publications, **73**, 247–263.

MOUNTNEY, N.P. & THOMPSON, D.B. 2002. Stratigraphic evolution and preservation of aeolian dune and damp/wet interdune strata: an example from the Triassic Helsby Sandstone Formation, Cheshire Basin, UK. *Sedimentology*, **49**, 805–835.

NEWELL, A.J. 2006. Calcrete as a source of heterogeneity in Triassic fluvial sandstones aquifers (Otter Sandstone Formation, SW England). *In*: TELLAM, J.H. & BARKER, R.D. (eds) *Fluid Flow and Solute Movement in Sandstones: The Offshore UK Permo-Triassic Red Bed Sequence*. Geological Society, London, Special Publications, **263**, 119–127.

PLANT, J.A., JONES, D.G. & HASLAM, H.W. 1999. *The Cheshire Basin: Basin Evolution, Fluid Movement and Mineral Resources in a Permo-Triassic Rift Setting*. British Geological Survey Memoir, Keyworth, Nottingham.

PRICE, M., MORRIS, B. & ROBERTSON, A. 1982. A study of permeability variations in Chalk and Permean aquifers using double packer injection testing. *Journal of Hydrology*, **54**, 401–423.

STRONG, G.E. 1993. Diagenesis of the Triasic Sherwood Sandstone Group Rocks, Preston, Lancashire, UK: a possible evaporitic cement precursor to secondary porosity? *In*: NORTH, C.P. & PROSSER, D.J. (eds) *Characterization of Fluvial and Aeolian Reserviors*. Geological Society, London, Special Publications, **37**, 279–289.

STRONG, G.E. & MILODOWSKI, A.E. 1987. Aspects of the diagenesis of the Sherwood Sandstones of the Wessex Basin and their influence on reservoir characteristics. *In*: MARSHAL, J.D. (ed.) *Diagenesis of Sedimentary Sequences*. Geological Society, London, Special Publications, **36**, 325–337.

THOMPSON, D.B. 1970. The stratigraphy of the so-called Keuper Sandstone Formation (Scythian–Anisian) in the Permo-Triassic Cheshire Basin. *Quarterly Journal of the Geological Society, London*, **126**, 151–181.

WARRINGTON, G., AUDLEY-CHARLES, M.G. ET AL. 1980. *A Correlation of Triassic Rocks in the British Isles*. Geological Society, London, Special Report, **13**.

Petrophysical characterization of the Sherwood Sandstone from East Yorkshire, UK

MAGDELINE POKAR[1], L. J. WEST[2] & N. E. ODLING[2]

[1]*ResearchSEA Ltd, 65 Covent Garden, Willingham, Cambridge CB4 5GD*
(e-mail: m.pokar@researchsea.com)

[2]*School of Earth and Environment, University of Leeds, Leeds LS2 9 JT, UK*

Abstract: Petrophysical tests were conducted on core samples from the unsaturated zone of the Sherwood Sandstone Group in East Yorkshire. Tests were conducted to determine which physical parameters most influenced its hydraulic conductivity values. The main parameters analysed were grain-size distribution, pore-throat size distribution, clay content, mineralogy and porosity. A constant flow rate permeameter was used to measure saturated hydraulic conductivity values in the vertical direction (perpendicular to lamination), K_v, and horizontal direction (parallel to lamination), K_h. Hydraulic conductivity values in the vertical direction, K_v, ranged from 0.004 to 0.12 m day^{-1} while values in the horizontal direction, K_h, ranged from 0.01 to 0.17 m day^{-1}. Hydraulic conductivity anisotropy, K_h/K_v, varied from 0.6 to 35. Scanning electron microscope analysis showed this anisotropy to be caused mainly by millimetre-scale laminations. Representative bulk hydraulic conductivity values were estimated from the core data; bulk horizontal hydraulic conductivity, K_{hb}, was estimated as 0.1 m day^{-1}, and bulk vertical hydraulic conductivity, K_{vb}, as 0.01 m day^{-1}. Principal components analysis and multiple regression analysis were used to determine parameters that affect hydraulic conductivity most. Grain sorting is established to be the most important parameter to influence K_v values; samples with fine laminations have relatively low K_v.

A detailed investigation of the petrophysical factors influencing matrix hydraulic conductivity was undertaken as part of a larger project on shallow groundwater flow in the Permo-Triassic Sandstone of the Selby–Doncaster area, Yorkshire, UK (Binley *et al.* 2001, 2002; Pokar 2002). The aim was to identify the key petrophysical variables that influence matrix hydraulic conductivity, and to estimate bulk (outcrop-scale) hydraulic conductivity values. The approach involved measuring petrophysical properties such as porosity, pore size and grain-size distribution, and hydraulic conductivity (both along and across laminations where these were present) on cores recovered from selected field sites and then conducting statistical analysis on the resulting data. Measurements included hydraulic conductivity, porosity and pore-throat size distribution by mercury intrusion, grain-size distribution, mineralogy by X-ray diffraction (XRD), and clay content, fine-scale texture and porosity by scanning electron microscopy (SEM). Finally, the thickness of lithological units (beds) in cores was used with the matrix hydraulic conductivities to predict bulk hydraulic conductivity in the horizontal and vertical directions.

Field sites

The two field sites investigated are in the Doncaster–Selby area of NE England at Lings Farm smallholding, near Hatfield, Yorkshire, UK [national grid reference SE 653 079], and near Eggborough, Yorkshire, UK [SE 570 232] (Fig. 1). The Sherwood Sandstone aquifer in the study area is characterized by fining-upwards sequences of cross-bedded medium- to fine-grained sandstones, representing in-channel deposits of a braided river system and occasional clay horizons that may represent slack water or overbank deposits. Lithologically, the aquifer consists mainly of red-brown, fine- to medium-grained and fine-grained sandstone with sporadic thin lenses of red-brown to grey-green mudstone and layers of rolled mudstone fragments (Allen *et al.* 1997). The western margin of the southern North Sea Basin, where the field sites for this study are located, underwent shallow burial of about 1 km (Burley 1984), and the sandstones are only very weakly cemented. The Quaternary (superficial) deposits overlying the Permo-Triassic sandstones at the two field sites are relatively thin (<2 m thick), and consist of fluvioglacial sands and gravels.

From: BARKER, R. D. & TELLAM, J. H. (eds) 2006. *Fluid Flow and Solute Movement in Sandstones: The Onshore UK Permo-Triassic Red Bed Sequence*. Geological Society, London, Special Publications, **263**, 103–118.
0305–8719/06/$15 © The Geological Society of London 2006.

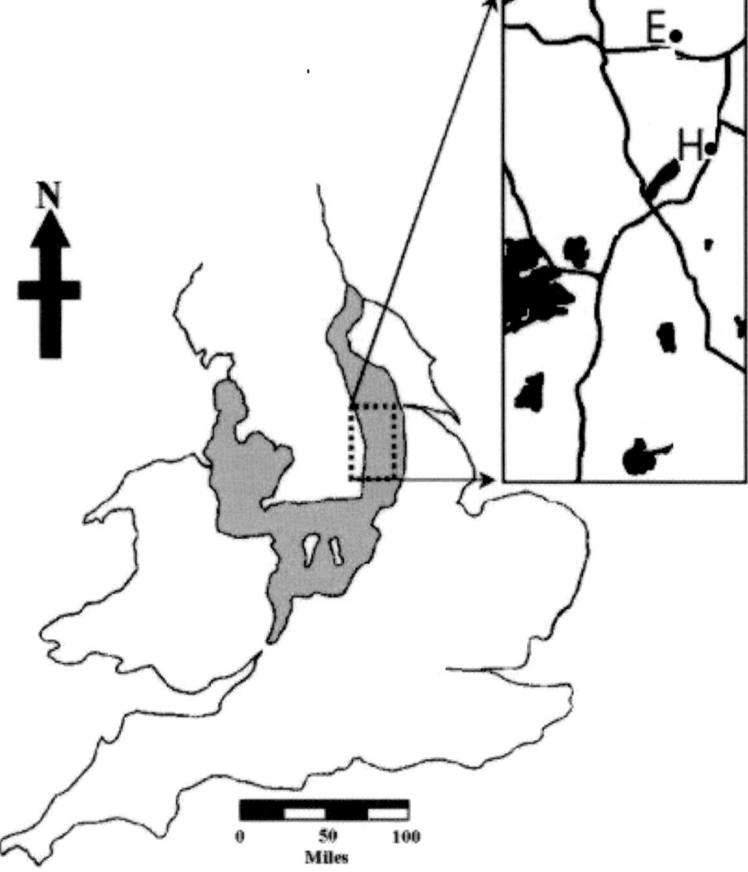

Fig. 1. Location of field sites and Permo-Triassic sandstone outcrop. E, Eggborough field site; H, Hatfield field site.

Materials and methods

Core description

Cores, 102 mm in diameter, from both sites were drilled with a polymer flush rotary drilling method. Core from the Lings Farm site near Hatfield was taken at 30° from vertical in order to identify any vertical fractures. No vertical fractures were intersected by the inclined core, although clay-filled vertical fractures are seen in a nearby quarry exposure (Truss 2004). The core was visually classified into different units (beds) based on grain size (Fig. 2a, d). Core from both sites consisted mainly of medium sandstone units, grading upwards into interlaminated fine and medium sandstone units, and subsequently into fine sandstone units. Medium sandstone units were generally between 0.5 and 2.5 m thick, interlaminated fine and medium sand-stone units were between 0.25 and 0.6 m thick, and fine sandstone units ranged from 0.2 to 1 m thick. The core from the Hatfield site also contained a unit of medium sandstone contain-ing centimetre-scale discontinuous siltstones.

Hydraulic conductivity measurement

A constant flow rate permeameter was used to measure saturated hydraulic conductivity on subsamples that were approximately 1–2 cm³ in volume, and rectilinear in shape (typically 1 cm × 1 × 2 cm). Dental putty was used to seal the gaps around the edges of the samples. Owing to time and logistical constraints, only distilled water was used as the permeant fluid; samples were soaked in distilled de-aired water under vacuum for 24 h before testing. Next, water was forced through the samples at a specified rate;

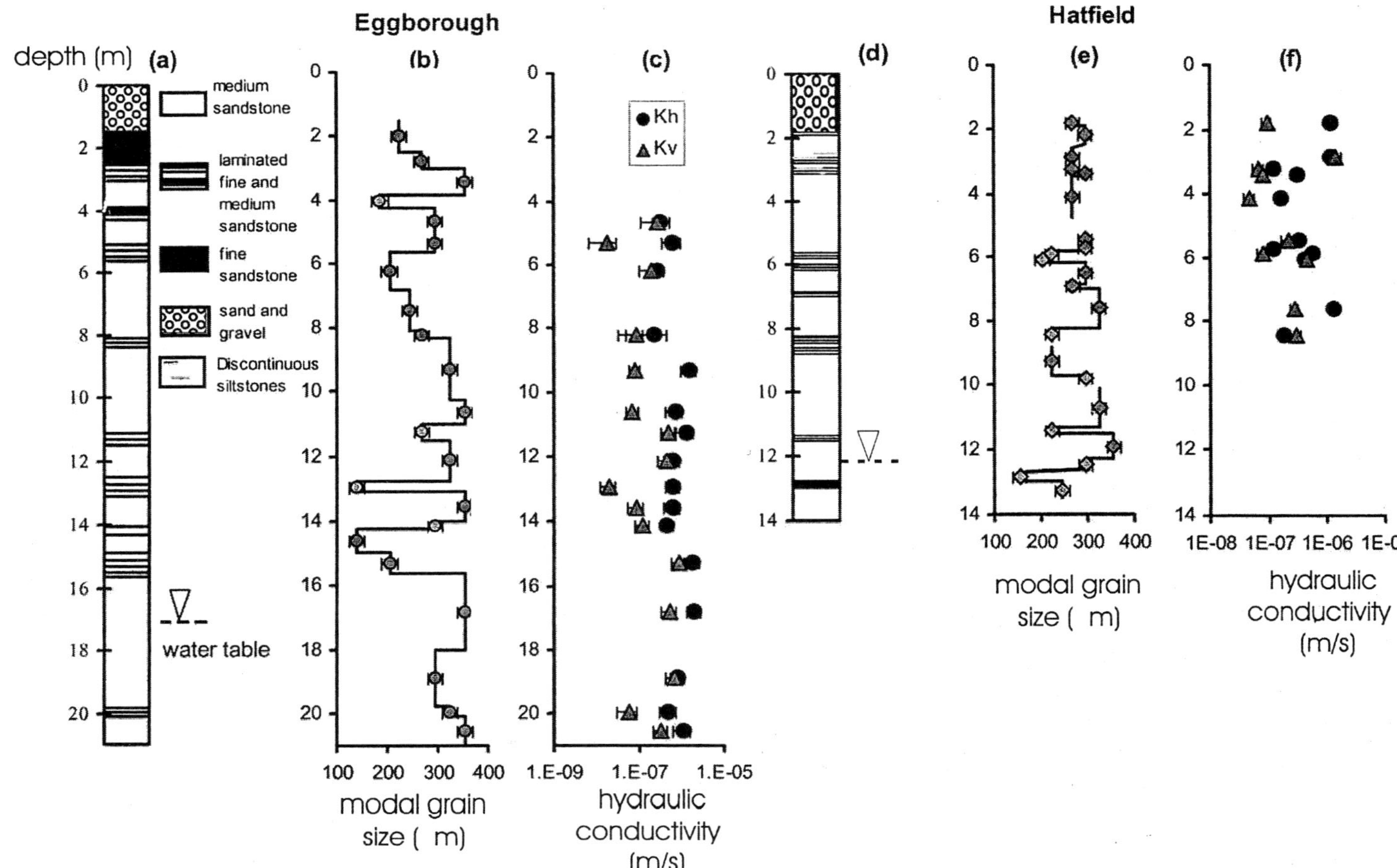

Fig. 2. (**a**) Visual core log from the Eggborough field site. (**b**) Modal grain size from the Eggborough field site. (**c**) Hydraulic conductivity from core retrieved from the Eggborough field site. (**d**) Visual core log from the Hatfield field site. (**e**) Modal grain size from core retrieved from the Hatfield field site. (**f**) Hydraulic conductivity from core retrieved from the Hatfield field site.

here typical flow rates used were 0.3, 0.5, 0.7 and 1 ml min^{-1} (the system is computerized and a flow pump allows a constant flow rate to be set by the user). The pressure difference created across the sample was measured using a differential pressure transducer; this pressure difference was allowed to reach a constant value before the pressure difference was recorded. Pressure difference was then converted into hydraulic head and divided by the sample length to obtain the hydraulic gradient, and Darcy's law was then used to calculate the hydraulic conductivity. After completion of tests in one flow direction (parallel to lamination), the sample was allowed to dry out and prepared for hydraulic conductivity tests in the direction perpendicular to lamination (to measure the amount of permeability anisotropy within the samples). For the Eggborough core, the same samples were then analysed by mercury porosimetry. However, the hydraulic conductivity samples from Hatfield were too poorly cemented to be subjected to further tests and therefore adjacent samples from the original core were used instead.

In order to test for the impact of the permeant on the sandstone pore structure, six consecutive tests were conducted on a medium-grained sandstone sample. The six separate tests gave measured differences of up to 17% from the geometric mean; there was no systematic tendency for the hydraulic conductivity to either increase or decrease with successive tests. However, for well-cemented, clay-free sandstones the repeatability of hydraulic conductivity measurements using this equipment is better than 8% (M. Kay pers. comm.). The wider range of values shown by the relatively friable Sherwood Sandstone sample may reflect processes such as migration of fine particles.

Pore-throat size distribution

The ability and efficiency of a sedimentary rock to transmit fluid is largely dependent on the interconnectedness and size of its pore space. Mercury porosimetry was used to measure the pore-size distribution of core samples. Measurement using mercury injection provides a semi-quantitative measure of pore-throat size distributions (Netto 1993).

The tests were conducted on an Autopore II 9220 machine, which is capable of measuring pore-throat diameters ranging from 0.003 to 360 μm. Samples were dried in the oven for at least 24 h before testing. Details of measurement by mercury porosimetry can be found in Webb & Orr (1997). The pore diameter is calculated using the Washburn equation

$$D = \frac{-4\gamma\cos\theta}{P} \tag{1}$$

where D is the pore diameter [L], γ the surface tension [MT^{-2}] and θ the contact angle. The surface tension for mercury is 0.485 Nm^{-1}, while the contact angle between mercury and solid varies between 112° and 142°, with 130° being the most commonly used (Webb & Orr 1997). P is the fluid pressure applied to the sample [ML^{-1}T^{-2}]. A plot of the pressure (or equivalent pore-throat diameter) v. volume of mercury intruded gives an indication of the pore-size distribution.

Scanning electron microscopy

The total porosity and mineralogy of selected samples were characterized by viewing resin impregnated thin sections in a scanning electron microscope (SEM) in backscattered mode and using energy dispersive spectrometry (EDS). Backscattered images were analysed using Scion image-processing software for porosity, and histograms of X-ray energy from the EDS were used to identify minerals in the sample. Details of the analysis can be found in Pokar (2002). Porosity was estimated from the average of four micrographs. The error in porosity estimated from micrographs is less than 3%.

Particle-size distribution

Hydraulic conductivity is affected by grain size and sorting. A laser diffraction technique was used to obtain a particle-size distribution of the sandstones. The distribution gives information about grain sizes within the sample, clay content and sorting. The rock samples were disaggregated using a pestle and mortar, and further dispersed with Calgon (sodium hexametaphosphate) solution. A Coulter Counter series LS230 was used, which is capable of differentiating particle sizes from 0.04 to 2000 μm. It has an additional system known as PIDS (polarization intensity differential scattering), which is used to distinguish particles of less than 0.4 μm. Clay content was derived from grain-size distributions, where clay is defined as having a grain size of ≤ 2 m.

X-ray diffraction

This method was used to identify clay mineral phases present in the sandstones. Clay minerals are important because some clays (e.g. smectite) can swell and block pore throats, which inhibits water flow. Small amounts of smectite can cause a large reduction in porosity and hydraulic conductivity. Clay phases in the Sherwood Sandstone samples were analysed by X-ray diffraction using a Phillips PW1840 Diffractometer. Identification of clay minerals are based on procedures outlined by Lindholm (1987) and Brindley & Brown (1984), and included glycolation and heat treatment.

Results

Hydraulic conductivity and SEM images

Core samples were classified visually according to grain size into medium sandstone units, interlaminated fine and medium sandstone units, and fine sandstone units (Fig. 2). For the Eggborough field site, samples from 16 units were tested for hydraulic conductivity. Seven were of medium sandstone units, while the rest were of interlaminated fine and medium sandstone. No samples were tested from the fine sandstone units as these samples disintegrated during sample preparation.

Hydraulic conductivity values ranged from 2.3×10^{-7} to 2.0×10^{-6} m s^{-1} (0.02–0.17 m day^{-1}) for the horizontal component (K_h), while the vertical component, K_v, ranged from 8.3×10^{-8} to 8.7×10^{-7} m s^{-1} (0.007–0.075 m day^{-1}) (Table 1 and Fig. 2c). The highest anisotropy was detected in sample A3 with a K_h to K_v ratio of 35. About half of the samples showed much lower anisotropies of between 1 (i.e. isotropic) and 3.

Samples from the Lings Farm site near Hatfield site were weaker than those from the Eggborough site; therefore two separate subsamples had to be used for hydraulic conductivity measurements parallel to, and perpendicular to, the core axis; 10 pairs of samples were tested. Hydraulic conductivity at 90° to the core axis ranged from 1.2×10^{-7} to 1.4×10^{-6} m s^{-1} (0.01–0.12 m day^{-1}), while that parallel to the axis ranged from 4.7×10^{-8} to 1.4×10^{-6} m s^{-1} (0.004–0.12 m day^{-1}) (Table 2 and Fig. 2f). These values are similar to those for K_h and K_v, respectively, from the tests on core from the Eggborough field site. Hydraulic conductivity anisotropy of Hatfield samples ranged from 0.6 to 12.6, i.e. the high end of the range is below that for Eggborough samples. However, this probably resulted from the 30° inclination to the vertical of the Hatfield core (the core from the Eggborough site was vertical). For this reason, the hydraulic conductivity data from the

Hatfield samples were excluded from further analysis.

SEM image analysis shows that samples which have high-permeability anisotropy often show millimetre-scale laminations, parallel or subparallel to bedding, and preferred grain orientation (Fig. 3). Some anisotropic samples like A11 (Fig. 3e, f) have very little clay content within laminations, which are composed of relatively fine-grained quartz and feldspar, while others like A3 (Fig. 3a, b) and A8 (Fig. 3c, d) have a high clay content within laminations.

Grain size, mineralogy, modal pore-throat size

Modal grain size for samples from the Eggborough core ranged from 140 to 356 μm with a geometric mean of 277 μm and a standard deviation of 69 μm. Those for samples from the Hatfield core ranged from 154 to 356 μm, with a geometric mean of 273 (m and a standard deviation of 46 μm. In general, higher clay contents were found in finer grained layers (see Tables 1 and 2).

Grain sorting was quantified from particle-size distribution using its geometric standard deviation. Lower values represent better sorting. Fine sandstones (e.g. unit D1, Table 1) and interlaminated fine and medium sandstones (e.g. A8) are more poorly sorted than medium-grained sandstones (e.g. B10). In general, samples from Hatfield (Table 2) were better sorted than those from Eggborough and sandstones with higher clay content were generally poorly sorted.

SEM image analysis indicates that the sandstones consist mainly of quartz (50–60% by volume), feldspar (10–15%) and clay (1–3%). Based on XRD and EDS measurements, kaolinite was the most common clay phase followed by illite and mixed phases of illite and smectite. Mixed phases of illite and smectite were more common in samples from Eggborough. Figure 4 shows the results of a typical X-ray diffraction measurement for the Sherwood Sandstone samples (sample A5), while Figure 5 shows the backscattered image of a kaolinite pore fill. Other minerals and compounds identified include quartz, potassium feldspar, and small traces of iron oxide, biotite, muscovite, apatite, titanium oxide, rutile, zircon, barium sulphate and chlorite.

The modal pore-throat diameter for samples from Eggborough ranged from 10 to 64 μm, with a mean of 35 μm and a standard deviation of 16 μm. Samples from Hatfield ranged from 16 to

Table 1. *Results from petrophysical measurements on core from the Eggborough field site*

Sample	Top (m)	Base (m)	Hydraulic conductivity K_h (m s^{-1})	Hydraulic conductivity K_v (m s^{-1})	Anisotropy K_h/K_v	Mercury porosity ϕ_{MICP} (%)	SEM porosity ϕ_{SEM} (%)	Modal pore-throat (pt_{mod}) (µm)	Modal grain size d_{mod} (µm)	SD g.s. sd (µm)	Clay content cl_{GS} (%)	Clay content cl_{SEM} (%)
Dl	1.5	2.5				23.9		14.9	223.4	7.8	9.5	
Al	2.5	3.0							269.2	3.5	1.6	
BI	3.0	3.8							356.1	3.9	1.8	
D2/A2	3.8	4.3							185.3	4.5	3.3	
B2	4.3	5.1	3.1×10^{-7}	2.7×10^{-7}	1.2	29.0	37	32.2	295.5	3.8	1.8	3
A3	5.1	5.6	6.2×10^{-7}	1.8×10^{-8}	35.1	27.2	33	18.0	295.5	4.6	2.7	2
A4	5.6	6.8	2.8×10^{-7}	1.9×10^{-7}	1.5	25.3	27	18.6	203.5	4.1	2.5	2
B3	6.8	8.1					34		245.2	4.9	3.2	1
AS	8.1	8.4	2.3×10^{-7}	8.3×10^{-8}	2.8		27		269.2	4.8	3.2	2
B4	8.4	10.3	1.5×10^{-6}	7.9×10^{-8}	18.6	32.6	34	53.1	324.3	3.8	1.4	1
BS	10.3	11.0	7.1×10^{-7}	6.9×10^{-8}	10.4	30.8	34	43.5	356.1	4.5	2.2	3
A6	11.0	11.5	1.3×10^{-6}	4.9×10^{-7}	2.6	28.7	34	31.2	269.2	4.1	1.6	2
A7	11.5	12.8	6.1×10^{-7}	4.3×10^{-7}	1.4	32.7	36	45.8	324.3	3.7	1.4	1
A8	12.8	13.1	6.2×10^{-7}	2.0×10^{-8}	31.6	28.4	33	18.0	140.1	4.9	4.9	3
B6	13.1	14.0	6.0×10^{-7}	8.4×10^{-8}	7.2	32.1	34	63.8	356.1	4.3	2.0	2
A9	14.0	14.3	4.5×10^{-7}	1.2×10^{-7}	3.8	26.4	35	10.8	295.5	3.7	1.6	5
B7	14.3	15.0					33		140.1	4.0	2.2	6
A1O	15.0	15.6	1.8×10^{-6}	8.7×10^{-7}	2.0		33	27.7	203.5	4.5	3.1	6
B8	15.6	18.0	2.0×10^{-6}	5.0×10^{-7}	4.0	31.9	35	52.0	356.1	4.0	1.7	2
B9	18.0	19.8	8.1×10^{-7}	6.6×10^{-7}	1.2	32.0	34	43.5	295.5	3.6	1.3	1
All	19.8	20.1	5.0×10^{-7}	$S.7 \times 10^{-8}$	8.7	29.8	34	21.6	324.3	4.5	2.2	2
B1O	20.1	21.0	1.1×10^{-6}	3.3×10^{-7}	3.3	29.6	31	42.1	356.1	3.8	1.4	1

Data shown are: horizontal hydraulic conductivity (K_h); vertical hydraulic conductivity (K_v); anisotropy (K_h/K_v); porosity from mercury intrusion, ϕ_{MICP}; porosity from SEM image analysis, ϕ_{SEM}; modal pore-throat size, pt_{mod}; modal grain size, d_{mod}; standard deviation from grain-size distribution, sd; clay content from grain-size distribution, cl_{GS}; and clay content from SEM measurements, cl_{SEM}. Unit labelling is as follows: A, interlaminated fine and medium sandstone; B, medium sandstone; C, medium sandstone with discontinuous siltstone; and D, fine-grained sandstone.

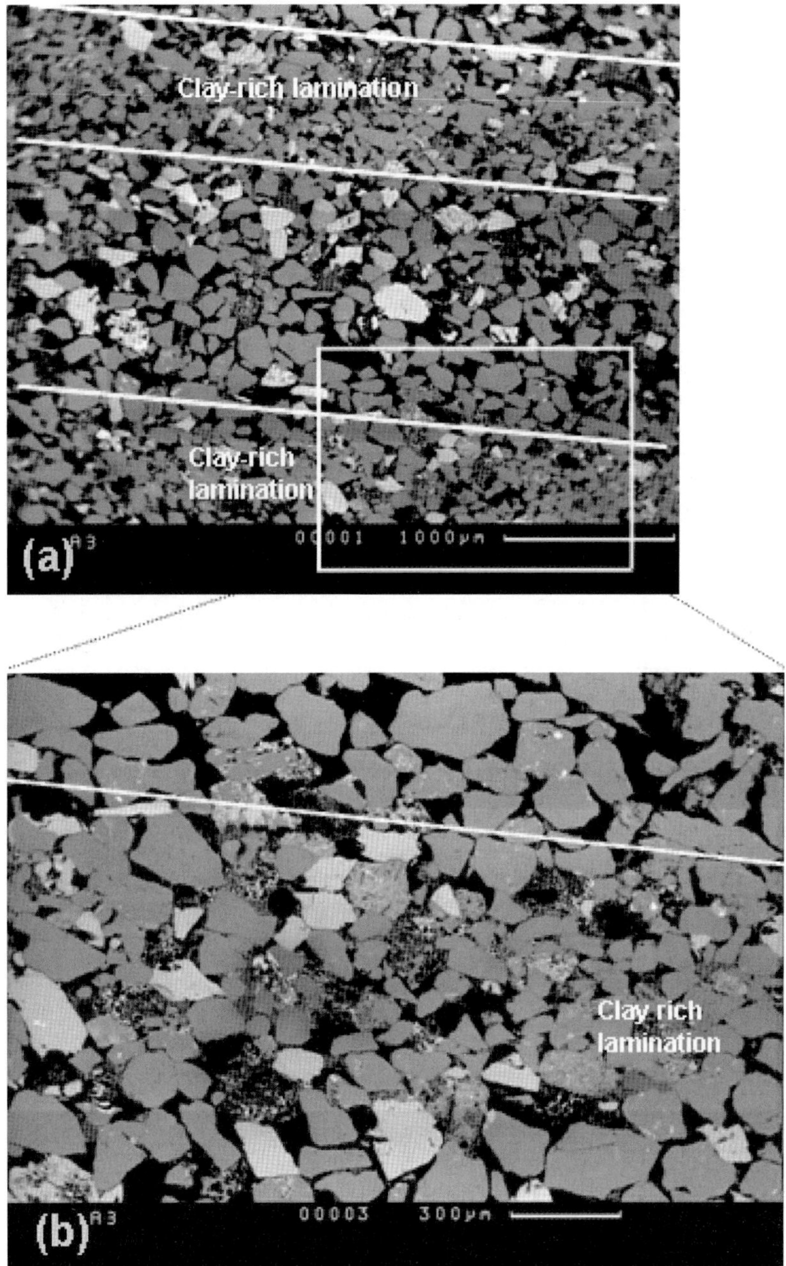

Fig. 3. (**a**) SEM micrograph of sample A3 from the Eggborough field site showing horizontal clay-rich lamination. (**b**) SEM micrograph of sample A3 at a larger scale.

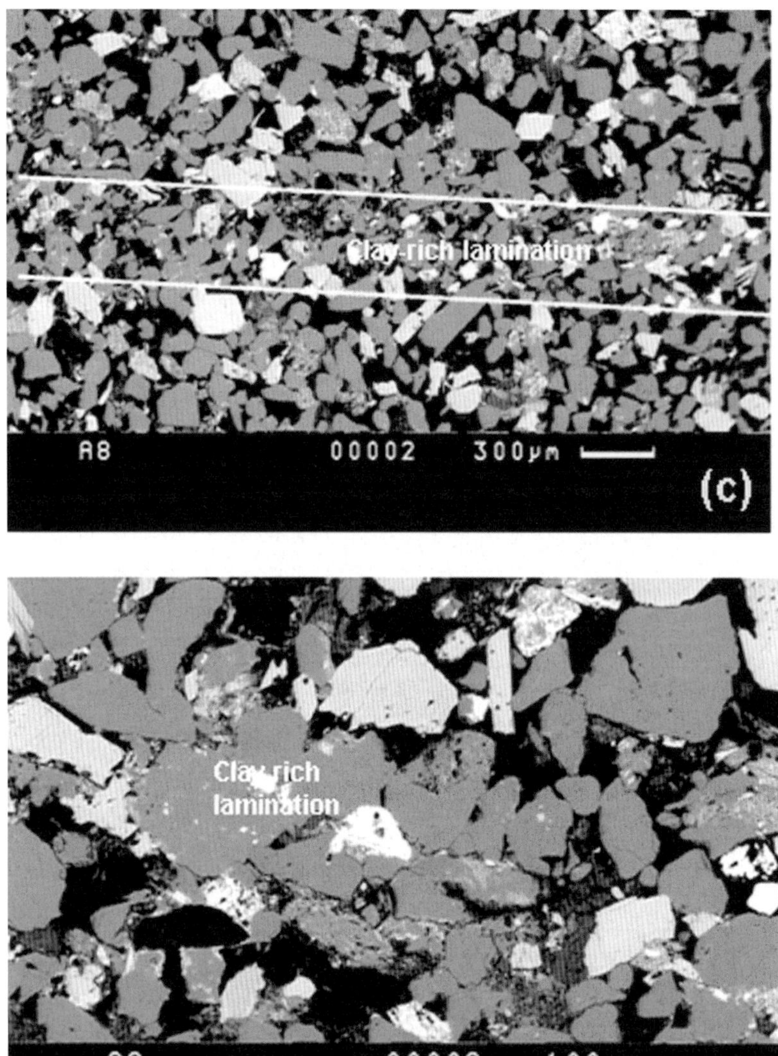

Fig. 3. (**c**) SEM micrograph of sample A8 from the Eggborough field site showing clay-rich lamination. (**d**) SEM micrograph of sample A8 at a larger scale.

62 μm, with a mean of 38 μm and a standard deviation of 11 μm. This shows that pore-throat sizes for samples from Hatfield and Eggborough are similar. Figure 6 shows four examples of cumulative intrusion v. pore-throat diameter curves for Eggborough samples. Fine sandstone sample D1 generally had a lot of very small pore throats; the modal pore-throat diameter was 15 μm. Small pore throats down to 0.01 μm were detected, representing those within clay phases.

In contrast, medium-grained samples like B8 and B5 showed large modal pore-throat diameters of 40–50 μm and much less response in the narrower pore-throat region (<0.1 μm), which indicates low clay content. Sample A6, interlaminated fine and medium sandstone, showed intermediate behaviour.

Porosity calculated from mercury intrusion, ϕ_{MICP}, for samples from Eggborough ranged from 24 to 32%, with a mean of 29% and a

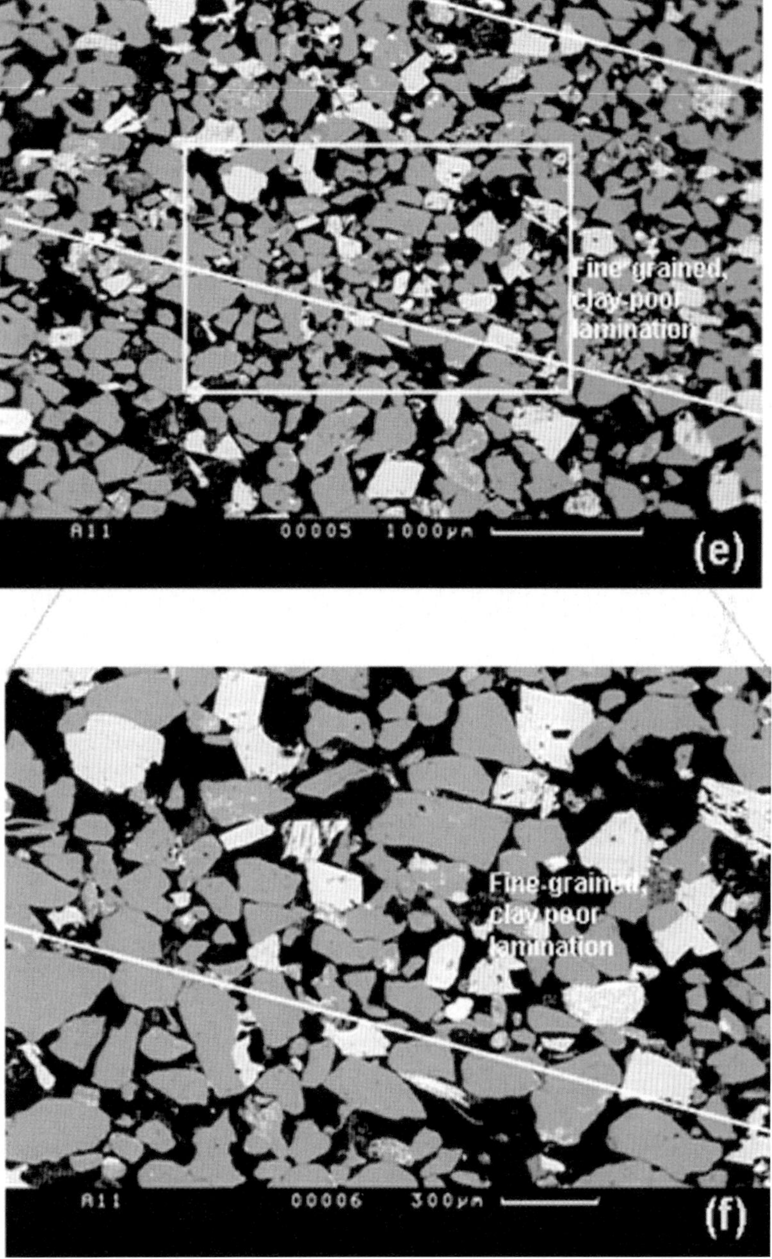

Fig. 3. (e) SEM micrograph of sample A11 from the Eggborough field site showing fine-grained, but clay-poor, lamination. (f) SEM micrograph of sample A11 at a larger scale.

standard deviation of 3% (see Table 1). For Hatfield, the results ranged from 17 to 33%, with a mean of 28% and a standard deviation of 5% (see Table 2). For Eggborough samples, porosity from SEM image analysis, ϕ_{SEM}, ranged from 27 to 37%, with a mean of 34% and a standard deviation of 3%; Hatfield samples ranged from 31 to 40%, with a mean of 36% and a standard deviation of 4%. These data indicate that the porosities at the two sites are similar; SEM analysis gives somewhat higher porosity values than mercury intrusion.

Table 2. *Results from petrophysical measurements on the core from the Hatfield field site*

Sample	Top (m)	Base (m)	Hydraulic conductivity K_h (m s^{-1})	Hydraulic conductivity K_v (m s^{-1})	Anisotropy K_h/K_v	Mercury porosity $\varnothing_{MICP}$ (%)	SEM porosity $\varnothing_{SEM}$ (%)	Modal pore throat V(pt_{mod}) (µm)	Modal pore throat H(pt_{mod}) (µm)	Modal grain size d_{mod} (µm)	SD g.s. sd (µµm)	Clay content cl_{GS} (%)	Clay content cl_{SEM} (%)
A1	1.7	1.9	1.2×10^{-6}	9.4×10^{-8}	12.6	28.5	40		53.75	269.2	3.7	2	2
BI	1.9	2.4								295.5	3.6	1.6	
C1	2.6	3.1	1.2×10^{-6}	1.4×10^{-6}	0.8					269.2	3.9	2.2	
B2	3.1	3.3	1.2×10^{-7}	6.7×10^{-8}	1.8		34			269.2	4.2	2.4	1
C2	3.3	3.5	3.2×10^{-7}	8.3×10^{-8}	3.9		33	39.74		295.5	3.6	1.6	2
B3	3.5	4.8	1.6×10^{-7}	4.7×10^{-8}	3.4	32.6			46.58	269.2	3.3	1.4	
B4	5.3	5.6	3.3×10^{-7}	2.2×10^{-7}	1.5	32.6		46.15		295.5	3.5	1.4	
A2	5.6	5.8	1.2×10^{-7}			27.6	31	43.62		295.5	3.5	1.4	3
B5	5.8	6.0	5.7×10^{-7}	8.1×10^{-8}	7.0	30.9	36	30.32	34.03	223.4	3.9	2.2	1
A3	6.0	6.2	4.2×10^{-7}	4.5×10^{-7}	0.9	29.5		31.28		203.5	3.7	2	
B6	6.2	6.8					40	30.65	34.45	295.5	3.5	1.3	1
A4	6.8	7.0					40	34.56	45.33	269.2	4.1	2.3	1
B7	7.0	8.2	1.4×10^{-6}	2.8×10^{-7}	4.9	27.1			60.68	324.3	3.4	1.2	
A5	8.2	8.7	1.8×10^{-7}	3.0×10^{-7}	0.6	16.5	39	62.2		223.4	4.1	2.4	3
B8	8.8	9.7							37.16	223.4	4.4	3	
B9	9.7	10.0					38		26.49	295.5	3.4	1.3	1
B10	10.1	11.3					40	38.4		324.3	3.2	1.1	2
A6	11.3	11.5					35	16.16	30.09	223.4	4.3	2.6	1
B11	11.5	12.3						35.6	60.79	356.1	3.6	1.5	
B12	12.3	12.6								295.5	3.5	1.4	
D1	12.7	13.0					31		16.35	153.8	4.6	4.4	2
B13	13.0	13.5					33	43.63		245.2	3.9	2	1

Data shown are: horizontal hydraulic conductivity (K_h); vertical hydraulic conductivity (K_v); anisotropy (K_h/K_v); porosity from mercury intrusion, $\varnothing_{MICP}$; porosity from SEM image analysis, $\varnothing_{SEM}$; modal pore-throat size, pt_{mod}; from mercury intrusion conducted in the vertical (V) and horizontal (H) direction; modal grain size, d_{mod}; standard deviation of grain-size distribution, sd; clay content from grain-size distribution, cl_{GS}; and clay content from SEM measurements, cl_{SEM}.

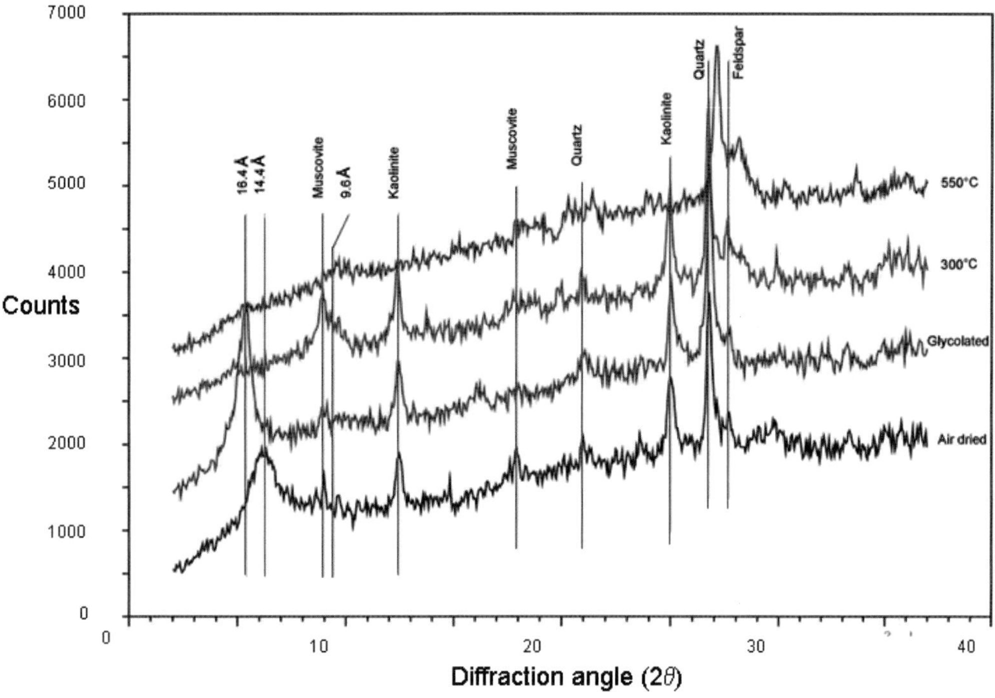

Fig. 4. Example of X-ray diffractogram results (sample A5 from Eggborough showing the presence of smectite, illite/muscovite, quartz and feldspar).

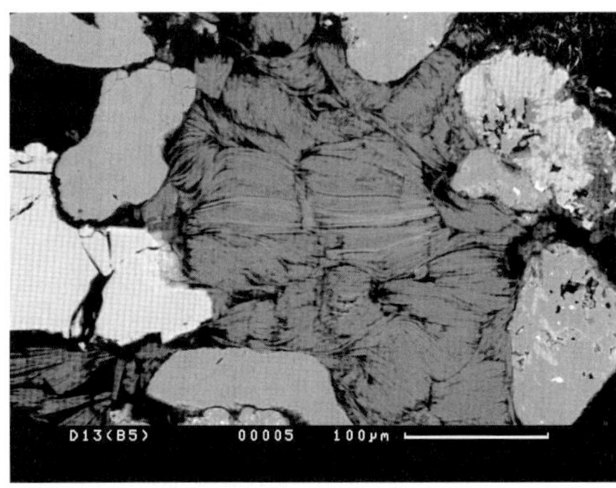

Fig. 5. SEM micrograph of kaolinite (sample B5 from the Eggborough field site).

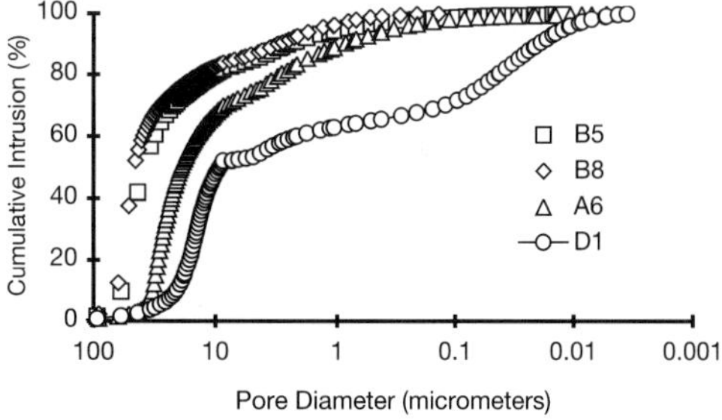

Fig. 6. Cumulative intrusion volumes v. pore-throat diameter from mercury intrusion porosimetry tests for samples from the Eggborough field site.

Analysis and interpretation of results

Comparison with published literature

The hydraulic conductivity data obtained from samples from Eggborough are compared below with some available literature values. In particular, the extensive database obtained by Koukis (1974) is used because it incorporates measurements of the Sherwood Sandstone from various locations in Yorkshire, although the localities tested are in the Vale of York area which is further north than the study area investigated here.

Hydraulic conductivity data from Koukis (1974), measured on cylindrical samples measuring 2.5 cm in diameter by 3.75 cm in length and using tap water as a permeant, ranged from 9.0×10^{-10} to 2.5×10^{-5} m s^{-1}, with a mean of 3.0×10^{-6} m s^{-1} or 0.26 m day^{-1} for K_v and from 7.1×10^{-9} to 3.6×10^{-5} m s^{-1}, with a mean of 4.2×10^{-6} m s^{-1} or 0.36 m day^{-1} for K_h. The anisotropy ratio (K_h/K_v) ranged from 1.2 to 2.5. Porosity measured using the water saturation technique, ϕ_{WS}, ranged from 11 to 34%, and high hydraulic conductivity values were associated with high ϕ_{WS}. Figure 7a & b shows both Koukis' hydraulic conductivity data v. ϕ_{WS} (crosses) and those from the Eggborough site (circles) v. ϕ_{MICP}. The hydraulic conductivity values from Eggborough samples fall in the middle of the range measured by Koukis. However, the porosities ϕ_{MICP} measured here are about 5% higher than the ϕ_{WS} values measured by Koukis (1974) for samples with

equivalent hydraulic conductivity. This difference may be because the water resaturation technique to measure porosity used by Koukis uses much lower pressures than high-pressure mercury intrusion, resulting in lower porosity values for equivalent samples. Alternatively, hydraulic conductivity measurements from the Eggborough samples may be lower due to the use of distilled water rather than tap water as the permeant, which may have resulted in clay swelling and/or particle migration processes. However, the high stability of measured K values in the repeat tests conducted here suggests that the latter explanation is unlikley.

Allen *et al.* (1997) summarized core-scale hydraulic conductivity data for the Sherwood Sandstone from NE England and found that they mainly lie in the range 0.01–1 m day^{-1}. Data measured here fall within this range, i.e. 0.02–0.17 m day^{-1} for K_h and 0.007–0.075 m day^{-1} for K_v for the Eggborough core, although the most permeable lithologies in the tests summarized by Allen *et al.* (1997) may not be represented. However, observations at quarries across the Selby–Doncaster region (Truss 2004) suggest that the sandstones cored here are representative of the dominant lithologies in the area, i.e. in-channel dune and bar deposits of the Triassic braided river system.

Statistical analyses

Principal component analysis (PCA) was used to identify the most important factors controlling hydraulic conductivity, and least-squares multiple regression techniques were used to obtain empirical relationships between hydraulic conductivity and these factors. The statistical software package Minitab (Minitab Inc. 2000) was used for the analyses. Only data from the Eggborough site were used as the hydraulic conductivity data from the Hatfield site may have been affected by the core being taken at 30° to vertical.

The variables included in the PCA were K_h, K_v, ϕ_{MICP}, ϕ_{SEM}, modal pore-throat diameter, modal grain size, standard deviation of grain size, kurtosis and skewness from grain-size distribution and clay content. The logarithmic counterparts of all the parameters were also included, on the grounds that some of the variables, such as hydraulic conductivity, are likely to show log-normal distributions. A correlation matrix was used for data reduction to identify where several very highly correlated variables can be represented by only one variable. For example, standard deviation of grain-size distribution was found to be strongly correlated to

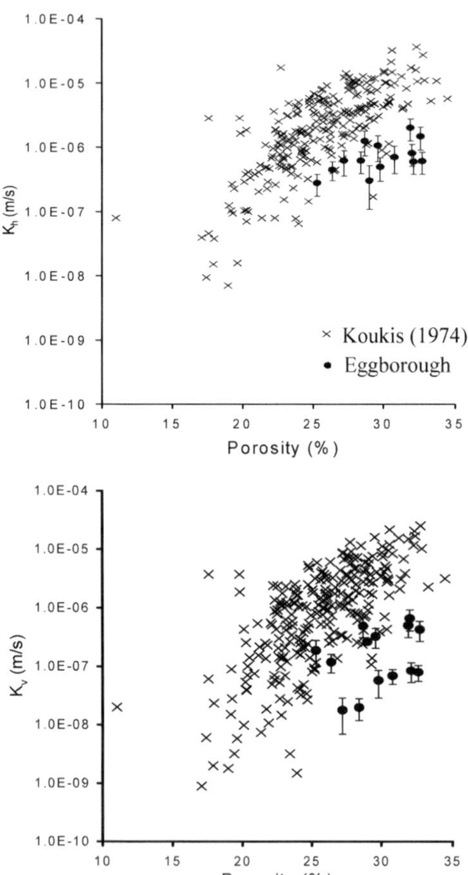

Fig. 7. (**a**) K_h v. ϕ_{WS} data from Koukis (1974) and K_h v. ϕ_{MICP} data from the Eggborough field site. Error bars represent one standard deviation. (**b**) K_v v. ϕ_{WS} data from Koukis (1974) and K_v v. ϕ_{MICP} data from the Eggborough field site. Error bars represent one standard deviation.

kurtosis and skewness, and so was used to represent the latter variables in PCA. This reduces the number of variables in the PCA analysis, and so reduces redundancy and repetition in the results.

PCA is sensitive to the absolute magnitudes of variables used in the analysis. Therefore the data are standardized, i.e. by normalizing to zero mean and unit standard deviation before performing the analysis. Normalization ensures that each variable has equal weight in the analysis. However, normalization may produce spurious correlations for variables with small variance (Davis 1986).

Table 3 shows the values (loadings) of the first three principal components for the analysis of eight variables from the Sherwood Sandstone petrophysical analysis. It shows that the first three principal components describe 81% of the variance; the first principal component (PC1) describes 48% of variance; PC2 describes a further 18% and PC3 a further 15%. It can be seen that most of the variables have high communality (80–90%), except for variables log K_h and log d_{mod} with a communality of only 60.4 and 60.7%, respectively.

In order to visualize this information, the loadings of the first three principle components for each variable, shown in Table 3, are plotted using direction cosine rules in Figure 8 (i.e. the three loadings are converted into an angular orientation and plotted using stereographic projection). If the variables plot further than 90° apart then they are statistically not related; variables plotting close together are strongly correlated.

The results indicate that the logarithm of vertical hydraulic conductivity, log K_v, is closely related to the standard deviation of the grain-size distribution (i.e. sorting), sd, and that log

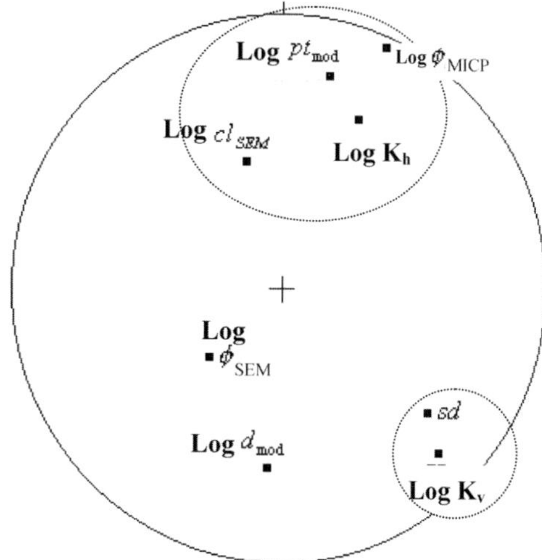

Fig. 8. Directional cosine plot of loadings from principle component analysis. The variables are: standard deviation from grain-size distribution, sd; the logarithms of hydraulic conductivities, K_h and K_v; porosity from SEM image analysis, ϕ_{SEM}; porosity from mercury intrusion, ϕ_{MICP}; modal pore-throat diameter, pt_{mod}; modal grain size, d_{mod}; and clay content from SEM image analysis, cl_{SEM}.

ϕ_{SEM} may be related to the logarithm of modal grain size, log d_{mod}. The logarithm of horizontal hydraulic conductivity, log K_h, plots close to the logarithm of porosity measured by mercury intrusion, log ϕ_{MICP}, modal pore-throat diameter, log pt_{mod}, and clay content estimated from SEM micrographs, log cl_{SEM}. However, K_h has a relatively limited range of values compared with K_v, so this association may not be physically meaningful.

Least-squares regression analysis was used to

Table 3. *Results of principal component analysis for eight variables: standard deviation from grain-size distribution, sd; the logarithm of hydraulic conductivity in the horizontal and vertical direction, K_h and K_v; porosity from SEM image analysis, ϕ_{SEM}; porosity from mercury intrusion, ϕ_{MICP}; modal pore-throat size, pt_{mod}; modal grain size, d_{mod}; and clay content from SEM image analysis, cl_{SEM}*

Variable	PC1	PC2	PC3	Communality (%)
log K_h	0.65	0.31	0.29	60.4
log K_v	0.64	−0.64	−0.15	83.5
log ϕ_{MICP}	0.87	0.39	0.04	90.8
log ϕ_{SEM}	0.42	0.45	−0.72	89.2
log pt_{mod}	0.87	0.20	0.20	84.3
log d_{mod}	0.71	0.06	−0.31	60.7
sd	−0.56	0.71	0.31	91.5
log cl_{SEM}	−0.69	0.20	−0.57	83.6
% Variance	48	18	15	81

obtain empirical relationships between some of the variables shown by PCA to be related. For example, PCA showed log K_v to be closely related to standard deviation of the grain-size distribution. A least-squares regression analysis between log K_v and *sd* (with K_v in m s⁻¹ and *sd* in μm) gave the following relationship with an R^2 coefficient of 0.56

$$\log K_v = -(1.18 \pm 0.27)sd - (2.01 \pm 1.08). \quad (2)$$

Figure 9 shows a plot of log K_v v. standard deviation of the grain-size distribution, and the trend confirms the positive correlation between these variables. A plot of the measured values of K_v against the values predicted using equation (2) is given in Figure 10. This plot shows that equation (2) is potentially useful for predicting vertical hydraulic conductivity from

grain-size data for the locality from which the samples were taken. Least-squares regression analysis is not appropriate for K_h due to the small range of values measured. However, the low variability in K_h suggests that parameter selection for modelling should not be problematic.

The hydraulic conductivity values quoted above were obtained from measurements made on core plugs, i.e. centimetre scale. It is also important to obtain data that are representative of field-scale conditions, i.e. outcrop scale and regional scale. Hence, field-scale hydraulic conductivity values $\overline{K_h}$ and $\overline{K_v}$ for the sandstone sampled by the core taken at the Eggborough site were found using thickness-weighted arithmetic and harmonic means of the core hydraulic conductivity values, respectively. The basic assumptions made in applying these scaling relations to core data are that the units seen in the core represent continuous horizontal layers, and that the measured values are representative of these layers. Analyses of gamma logs from boreholes reported by Giustiniani *et al.* (2001) suggest that the assumption of horizontal layering is valid for the Eggborough field site. The calculated $\overline{K_h}$ value for the Sherwood Sandstone from Eggborough is 1.0×10^{-6} m s⁻¹ (0.1 m day⁻¹) and $\overline{K_v}$ is 1.1×10^{-7} m s⁻¹ (0.01 m day⁻¹), i.e. field-scale hydraulic conductivity in the vertical direction $\overline{K_v}$ is about 10 times lower than field-scale hydraulic conductivity in the horizontal direction, $\overline{K_h}$. Thus, a $\overline{K_h}/\overline{K_v}$ ratio of 10 is considered to be appropriate for use in outcrop-scale flow modelling of the Sherwood Sandstone in the Eggborough locality.

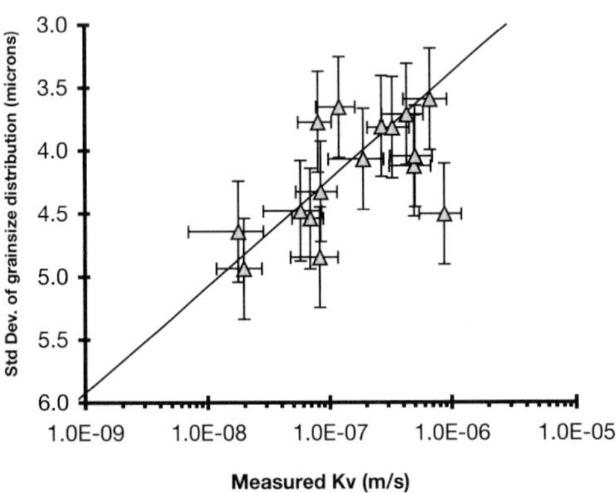

Fig. 9. Scatter plot of standard deviation from grain-size distribution, *sd*, v. measured log K_v. Error bars represent one standard deviation. Line is linear regression result (equation 1).

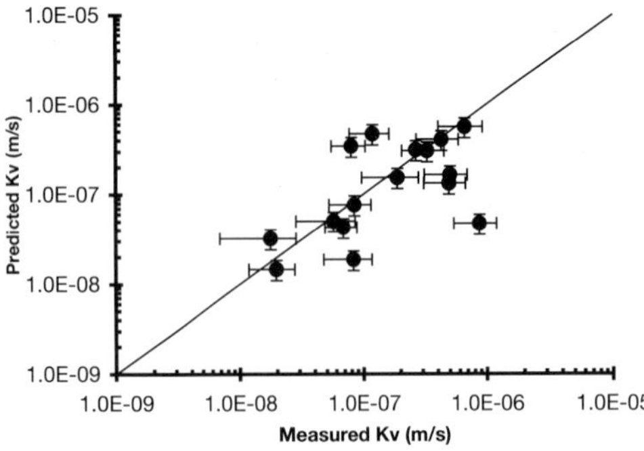

Fig. 10. Measured K_v v. predicted K_v. Error bars for the predicted K_v data were found from errors associated with *sd*.

Discussion and conclusions

Laboratory measured physical properties of the Sherwood Sandstone from two field sites in Yorkshire have been determined, and the relationships between them have been investigated in order to facilitate predictive modelling of fluid flow and contaminant transport. The visual appearance and properties of the sandstones from the two sites (which are 17 km apart) were essentially similar, and the results are probably broadly representative of the Permo-Triassic sandstones between Selby and Doncaster. Core-scale horizontal hydraulic conductivity values ranged from 0.01 to 0.17 m day⁻¹, with vertical hydraulic conductivities from 0.004 to 0.12 m day⁻¹. Finer grained specimens were relatively poorly sorted and showed significant permeability anisotropy arising from millimetre-scale bedding-parallel laminations of fine-sand/silt-sized quartz and feldspar clasts and/or clay. Layer thickness-weighted arithmetic and harmonic mean horizontal and vertical hydraulic conductivity values for the sequence at one of the sites were 0.1 and 0.01 m day^{-1}, respectively, which indicates that there is also an anisotropic effect due to macroscopic layering. Other workers have also found significant anisotropy in Permo-Triassic sandstones. For example, core permeametry studies of the Bromsgrove Sandstone Formation in the Midlands reported by Ramingwong (1974) in Allen *et al.* (1997) showed that the hydraulic conductivity of fine-grained beds was 0.02 (K_h) and 0.01 m day^{-1} (K_v). These values were 500 times lower than cleaner sandstone, where horizontal and vertical hydraulic conductivities were 8 and 6 m day^{-1}, respectively. The Sherwood Sandstone at Edwinstowe near Mansfield had a core-scale K_h/K_v ratio of 5 (Allen *et al.* 1997). This anisotropy was attributed to thin low hydraulic conductivity laminations by Allen *et al.* (1997). Bloomfield *et al.* (2001) also reported that Permo-Triassic sandstones have anisotropic permeability at field scale due to both interbedding of fine and coarse layers (i.e. layered heterogeneous conductivity at the bed scale) and hydraulic anisotropy within individual beds.

Permeability anisotropy is important to pollutant migration in both saturated and unsaturated zones. Low hydraulic conductivity layers may result in lateral flow in the unsaturated zone with possible funnelling through breaches leading to rapid bypass flow to the water table. This will significantly affect groundwater pollution vulnerability. In the saturated zone, permeability anisotropy may reduce vertical migration of solutes, restricting agrochemical and industrial contamination to the shallower parts of the aquifer.

Statistical analysis on the Eggborough data set found that log K_v correlated with standard deviation from grain-size distribution, *sd*, which is a parameter that measures the amount of sorting; in contrast, no strong relationship with modal pore-throat size was indicated. As the vertical hydraulic conductivity is likely to be highly influenced by the millimetre-scale laminations found in some of the sandstone beds, correlation between log K_v and *sd* is not surprising. Laminated samples tend to have poor sorting due to the presence of grain-size contrasts between coarser and finer grained laminations, which become mixed together when samples are disaggregated and tested. However, SEM images show that the finer grained laminations tend to be relatively thin, so their pore-throats sizes are not the most frequently occurring pore-throat sizes. This means that modal pore-throat size is not a useful predictor of their presence, or of the core-scale vertical hydraulic conductivity.

An empirical equation for the prediction of K_v based on grain-size data was established for the Eggborough field site. It was not possible to establish a predictive equation for K_h because this parameter did not vary much between samples. This was probably because the K_h values for laminated specimens are controlled by the coarser grained component between fine laminations, and are therefore similar to K_h values for non-laminated specimens.

The authors would like to acknowledge financial support for this work from NERC grant GR3/11500 and a PhD studentship from the British High Commission Chevening Awards. We would also like to acknowledge Rock Deformation Research for the use of mercury porosimetry testing and permeability testing facilities. We would like to thank J. Cann, K. Handley, M. Kay, Q. Fisher, A. Bolton, D. Condliffe, E. Condliffe, L. Neve, R. A. Clark, C. Guerrero and R. Marshall for their contribution to the results in this paper.

References

ALLEN, D.J., BREWERTON, L.J. *ET AL.* 1997. *The Physical Properties of Major Aquifers in England and Wales*. British Geological Survey, Technical Report, **WD/97/34**, 157–287.

BINLEY, A.M., WINSHIP, P., MIDDLETON, R., POKAR, M. & WEST, L.J. 2001. High resolution characterisation of vadose zone dynamics in the Sherwood Sandstone using cross-borehole radar. *Water Resources Research*, **37**, 2639–2652.

BINLEY, A., WINSHIP, P., WEST, L.J., POKAR, M. & MIDDLETON, R. 2002. Seasonal variation of moisture content in unsaturated sandstone inferred from borehole radar and resistivity profiles, *Journal of Hydrology*, **267**, 160–172.

BLOOMFIELD, J.P., GOODDY, D.C., BRIGHT, M.I. & WILLIAMS, P.J. 2001. Pore-throat size distributions in Permo-Triassic sandstones from the United Kingdom and some implications for contaminant hydrogeology. *Hydrogeology Journal*, **9**, 219–230.

BRINDLEY, G.W. & BROWN, G. 1984. *Crystal Structures of Clay Minerals and Their X-ray Identification*. Mineralogical Society Monograph, **5**.

BURLEY, S.D. 1984. Patterns of diagenesis in the Sherwood Sandstone group (Triassic) United Kingdom. *Clay Minerals*, **19**, 403–440.

DAVIS, J.C. 1986. *Statistics and Data Analysis in Geology*, 2nd edn. Wiley, New York.

GIUSTINIANI, M., CASSIANI, G. & BINLEY, A. 2001. Use of borehole geophysical data for stochastic characterisation of a sandstone aquifer. *In: Proceedings of the 7th Meeting of the Environmental and Engineering Geophysics Society (European Section), Birmingham, UK, September 2001*, 218–219.

KOUKIS, G. 1974. *Physical mechanical and chemical properties of the Triassic sandstone aquifer of the Vale of York*. PhD thesis, University of Leeds.

LINDHOLM, R. 1987. *A Practical Approach to Sedimentology*. Allen & Unwin, Winchester, MA.

MINITAB INC. 2000. *User's Guide 2: Data Analysis and Quality Tools*. Minitab Inc., State College PA, USA.

NETTO, A.S.T. 1993. Pore-size distribution in sandstones. *AAPG Bulletin*, **77**, 1101–1104.

POKAR, M. 2002. *Investigation of the unsaturated zone in the Sherwood Sandstone using petrophysical and geophysical monitoring methods*. PhD thesis, University of Leeds.

RAMINGWONG, T. 1974. Hydrogeology of the Keuper sandstone in the Droitwich syncline area. PhD thesis, Birmingham University, UK.

TRUSS, S.W. 2004. *Characterisation of sedimentary structure and hydraulic behaviour within the unsaturated zone of the Triassic Sherwood Sandstone aquifer in North East England*. Unpublished PhD thesis, University of Leeds.

WEBB, P.A & ORR, C. 1997. *Analytical Methods in Fine Particle Technology*. Micromeritics Instruments Corps, GA, USA.

Calcrete as a source of heterogeneity in Triassic fluvial sandstone aquifers (Otter Sandstone Formation, SW England)

ANDREW J. NEWELL

British Geological Survey, Maclean Building, Wallingford, Oxfordshire OX10 8BB, UK
(e-mail: a.newell@bgs.ac.uk)

Abstract: Carbonate is not generally considered as a potential source of large-scale heterogeneity in Permo-Triassic sandstone aquifers. This study shows that carbonate, in the form of early diagenetic calcrete, forms an abundant component of the Triassic Otter Sandstone Formation in south Devon. Three types of calcrete are described from the outcrop of this fluvial sandstone dominated aquifer: rhizocretions; calcrete sheets; and calcrete conglomerates. Data obtained from core plugs show that calcrete reduces sandstone permeability to less than 10^{-15} m^2 and porosity to less than 12%. Calcrete can be sufficiently abundant to produce a significant (up to 30%) reduction in the total effective porosity of the aquifer. Moreover, calcrete conglomerates can form laterally extensive (up to 1 km) low-permeability sheets that will represent major baffles to vertical flow.

Carbonate is not generally considered as a potential source of large-scale heterogeneity in Permo-Triassic sandstone aquifers. Indeed, many studies have shown that the process of aquifer development, which generally involves uplift and an influx of oxidizing surface-derived water, generally results in the extensive removal of carbonate cement from sandstone formations (Burley 1984; Strong 1993; Parnell 2002). This study aims to show that carbonate, in the form of early diagenetic calcrete, can survive uplift and dissolution to form an abundant component of Permo-Triassic sandstone aquifers. Outcrop of the Triassic Otter Sandstone Formation aquifer in south Devon is used to examine the form and facies distribution of calcrete. Calcrete is shown to have an important effect in reducing the porosity and permeability of sandstone within the aquifer. Moreover, when reworked into intraclast conglomerates, calcrete can form laterally extensive (up to 1 km wide) low-permeability sheets that may represent major baffles to flow within the aquifer.

Calcrete

Calcrete is an accumulation of carbonate in a near-surface terrestrial setting that results from the cementation and/or replacement of host sediment by calcium carbonate precipitated from soil water or groundwater (Wright & Tucker 1991). Calcretes are widespread in many semi-arid and arid areas at the present time (Wright 1990). Carbonate precipitation is driven by a number of factors including high rates of evaporation, degassing of carbon dioxide and various biogenic processes (Wright 1990; Goudie 1996). Calcretes are most frequently associated with soil profiles, where they are termed pedogenic calcretes, but they can also form around the water table as groundwater calcretes that can cement large volumes of sediment. In Australia, groundwater calcretes are commonly kilometres wide, tens of kilometres long and have an average thickness of 10 m (Mann & Horwitz 1979). Calcretes are typically composed of micritic, densely crystalline calcium carbonate. On a worldwide basis, most calcretes contain little magnesium carbonate, probably because of the relatively higher solubility of magnesium carbonate in the presence of CO_2 (Reeves 1976). The dense micritic fabric, massive development and relatively lower solubility of low-magnesium carbonate may account for the high resistance of calcrete to dissolution and its widespread preservation in aquifers including at outcrop. Calcretes are common in many ancient 'red-bed' successions, such as the Permo-Triassic of NW Europe when the prevailing semi-arid and arid climate provided suitable conditions for their formation.

Otter Sandstone Formation

The Otter Sandstone Formation is a 100–180 m thick, Lower–Middle Triassic, sandstone succession that crops out along the western margin of the Wessex Basin (Fig. 1). The formation is a major aquifer in East Devon (Allen *et al.* 1997) and an important oil reservoir at Wytch Farm in East Dorset (McKie *et al.* 1998). The Otter

From: BARKER, R. D. & TELLAM, J. H. (eds) 2006. *Fluid Flow and Solute Movement in Sandstones: The Onshore UK Permo-Triassic Red Bed Sequence*. Geological Society, London, Special Publications, **263**, 119–127.
0305-8719/06/$15 © The Geological Society of London 2006.

Sandstone Formation overlies the conglomeratic Budleigh Salterton Pebble Beds, and together they comprise the Sherwood Sandstone Group in this area. Both units are exposed in coastal outcrop between Budleigh Salterton and Sidmouth (Fig. 1). The conglomerates and sandstones were deposited largely by river systems that flowed northward through a system of fault-bounded basins. The Sherwood Sandstone Group is sandwiched between the thick mudstone successions of the Aylesbeare and Mercia Mudstone groups (Holloway *et al.* 1989).

Superficially, the Otter Sandstone Formation appears to be a monotonous sandstone succession and it has not been subdivided into formal lithostratigraphic members. However, close examination of the sedimentology shows that the formation can be subdivided into a number of discrete facies associations (Fig. 2) that have an important control on the distribution and form of calcrete (Purvis & Wright 1991; Lorsong & Atkinson 1995; Hounslow & McIntosh 2003). Four main facies associations are recognized.

Aeolian sandstone association

At the base of the Otter Sandstone Formation is a 10 m succession of cross-bedded and low-angle laminated, fine- to medium-grained sandstone. The sandstones rest sharply on a layer of wind-facetted pebbles (ventifacts) and

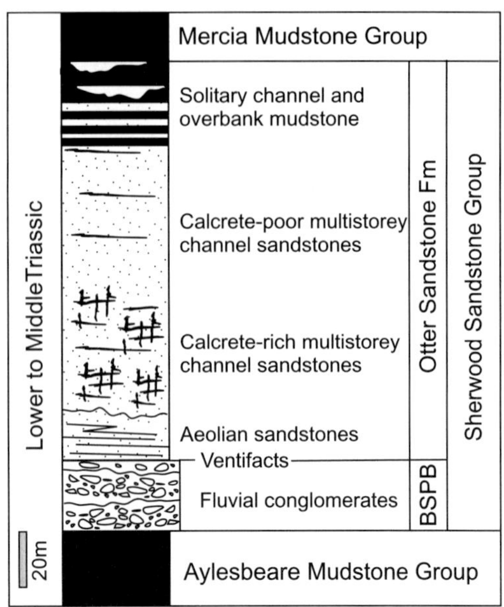

Fig. 2. Stratigraphy of the Otter Sandstone Formation in East Devon coastal exposures. BSPB, Budleigh Salterton Pebble Beds.

associated reg palaeosol developed on the top of the Budleigh Salterton Pebble Beds (Wright *et al.* 1991). These sandstones, which are well sorted, friable and do not contain calcretes or channel forms, are interpreted as aeolian.

Calcrete-rich, multistorey channel association

Overlying the basal aeolian sandstones is a fluvial unit that is dominated by sandstone arranged into vertically and laterally amalgamated sandstone sheets. Calcrete is abundant within this unit, while mudstone is absent. Sandstone sheets are typically in the order of 1–4 m thick and are generally separated by a conglomerate, which consists largely of reworked calcrete intraclasts (Fig. 3a). The primary sedimentary structure of the sandstones is generally faint and has locally been completely obliterated by the growth of calcrete. However, where visible the primary structure is dominated by trough cross-bedding, which may show an abrupt upwards transition into horizontal lamination.

The abundance of cross-cutting erosion surfaces, reworked intraclast conglomerates and trough cross-bedding suggests a fluvial depositional environment. The abundance of calcrete suggests relatively arid conditions with intense

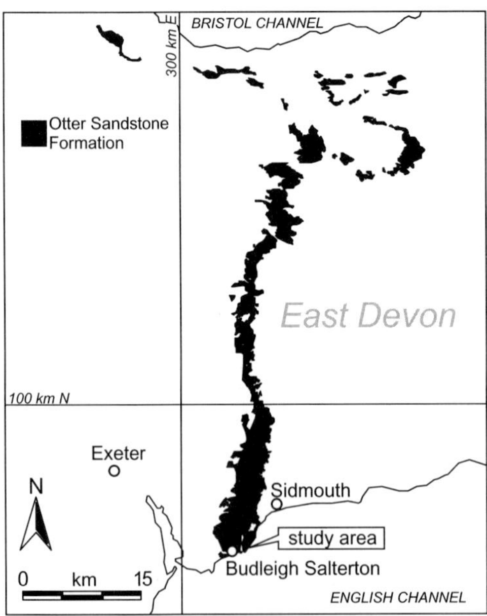

Fig. 1. Map showing outcrop of the Triassic Otter Sandstone Formation and location of the study area.

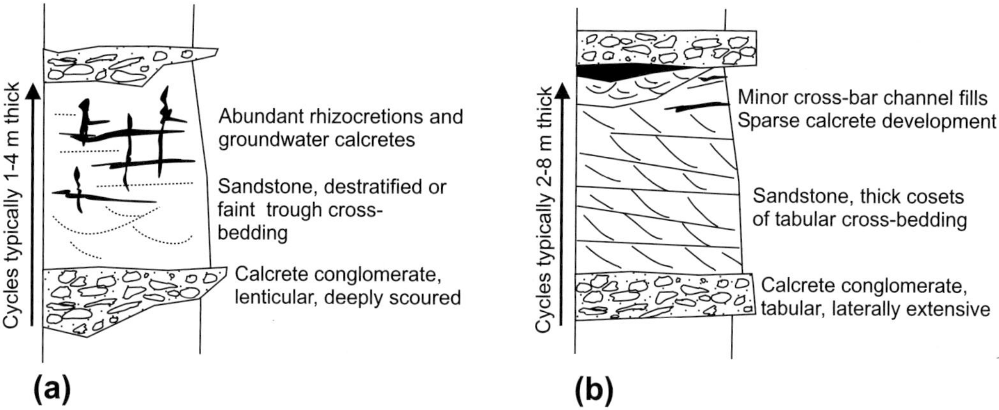

Fig. 3. Schematic logs showing the typical vertical facies successions through: (**a**) calcrete-rich, multistorey facies association; and (**b**) calcrete-poor, multistorey channel association.

evaporation and long dry seasons (Wang *et al.* 1994). It is probable that stream flows were of short duration and high intensity. This is not discounted by the predominance of large-scale trough cross-bedding and horizontal lamination in channel fills. These structures have been shown to dominate the fill of modern sand-bed ephemeral streams in arid central Australia (Williams 1970).

Calcrete-poor, multistorey channel association

The architecture of this facies association is broadly similar to that of the underlying unit in being composed of sandstone sheets that inter-lock vertically and laterally. The thickness of individual sheets increases, however, and can reach 8 m. Each sandstone sheet is typically floored by a laterally extensive intraclast conglomerate composed mainly of reworked calcrete fragments. Above the calcrete conglom-erate, the sheets are commonly dominated by thick cosets of tabular cross-bedding that can extend in a flow-parallel direction for several hundreds of metres (Fig. 3b). Set boundaries commonly incline in a down- or cross-current direction. Cross-sections perpendicular to palaeoflow direction generally show the development of large-scale inclined stratifi-cation (lateral accretion surfaces). The tops of tabular cross-bedded cosets are sometimes cut by minor channels infilled with trough cross-bedded sand and mudstone plugs, which can contain plant and fish debris. Calcretes do occur towards the top of sandstone sheets, but these

are much less abundant than in the underlying calcrete-rich association.

An increase in the thickness of sandstone sheets and a change in stratification type suggests a change in fluvial style. Tabular cross-bedding is produced by the migration of straight-crested dunes. Thick cosets of tabular cross-bedding with inclined set boundaries are generally interpreted as bars formed within major, possibly braided, channels that may have experienced perennial flow. Channel fills cut into bars may represent minor cross-bar channels. The dramatic reduction in the proportion of calcrete within this unit and preservation of plant material and fish remains within channel plugs further supports perennial flow and may indicate a change towards a subhumid climate.

Solitary channel sandstone and mudstone association

The top of the Otter Sandstone Formation is composed of interbedded, micaeous sandstones and mudstones. The sandstones range from sharp-based, sheet-like bodies to channel form with erosional bases. Sandstones are massive or trough cross-bedded and are encased within generally massive reddish brown mudstones. Calcretes are extremely rare within this unit, which probably represents a channel–floodplain complex (McKie *et al.* 1998). This unit is tran-sitional to the Mercia Mudstone Group by a progressive decrease in the thickness of sand-stone bodies and an increase in the proportion of mudstone.

Calcrete types in the Otter Sandstone Formation

Three main types of calcrete are found in the Otter Sandstone Formation: (1) rhizocretions; (2) calcrete sheets; and (3) reworked calcrete conglomerates. Purvis & Wright (1991) discuss the microstructure and stable isotope characteristics of the primary calcretes. Strong & Milodowski (1987) and Holloway *et al.* (1989) discuss calcrete petrography in the Otter Sandstone Formation at subcrop.

Rhizocretions

Rhizocretions are elongate, cylindrical concretions that reach 1.5 m in length and 0.20 m in diameter (Fig. 4a). The rhizocretions are orientated perpendicular to bedding surfaces, but may show offshoots parallel to bedding. Rhizocretions occur mainly within the calcrete-rich multistorey channel association in the lower part of the Otter Sandstone Formation, where they are generally found concentrated in certain horizons (see Fig. 5). The calcretes are composed of non-ferroan calcite and floating and corroded silicate grains (Purvis & Wright 1991). These cylindrical carbonates are interpreted as rhizocretions based on the general root-like morphology and a characteristic concentric zoning of a dense tubular micritic coating passing gradationally into less well-cemented inner zones (Purvis & Wright 1991). The rhizocretions closely resemble those from the Plio-Pleistocene Koobi Fora Formation of Kenya where micrite precipitated around roots and as later partial infills to root moulds (Mount & Cohen 1984). Evapotranspiration is a possible driving mechanism for precipitation. The length of the rhizocretions suggests that they formed around the tap roots of phreatophytic plants, which descend into the subsurface to reach low water tables (Purvis & Wright 1991).

Calcrete sheets

Calcrete sheets comprise laterally continuous carbonate accumulations typically up to 0.2 m thick and up to 10 m in length (Purvis & Wright 1991). The deposits generally follow bedding planes and form subhorizontal sheets, but the calcretes can also preferentially cement

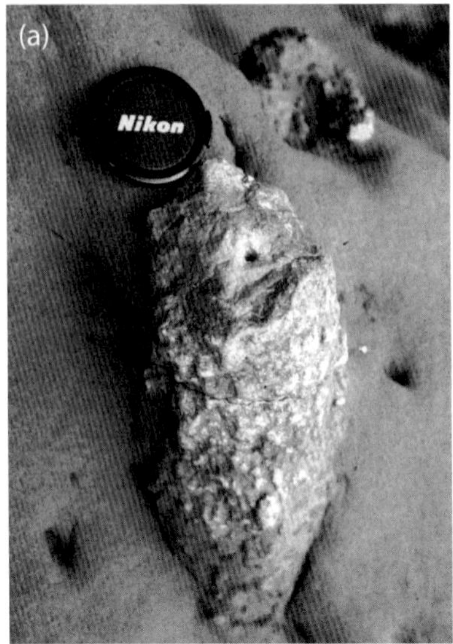

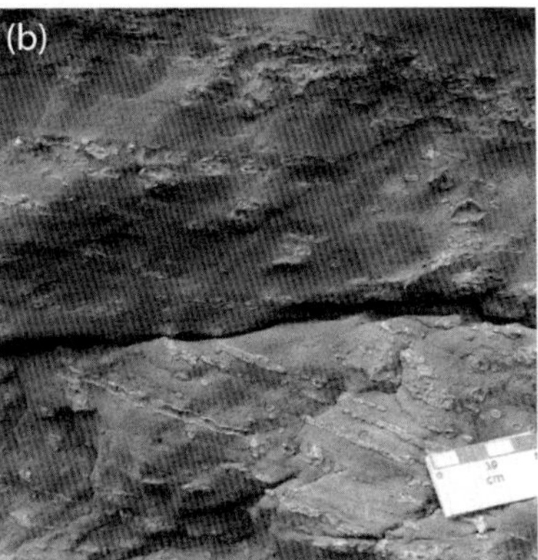

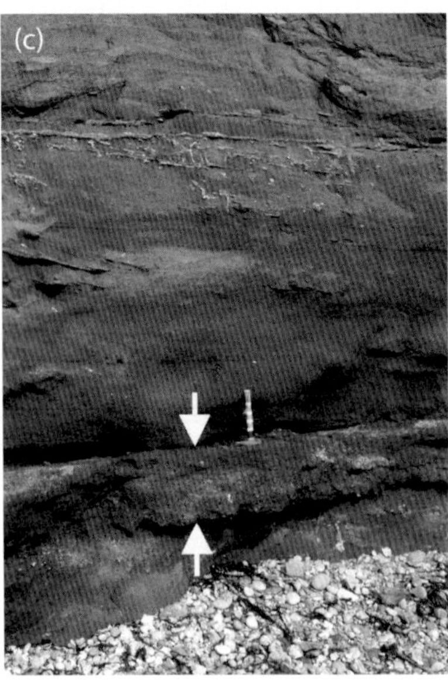

Fig. 4. Calcrete types in the Otter Sandstone Formation at outcrop in SE Devon: (**a**) rhizocretions; (**b**) calcrete sheets; and (**c**) calcrete conglomerate.

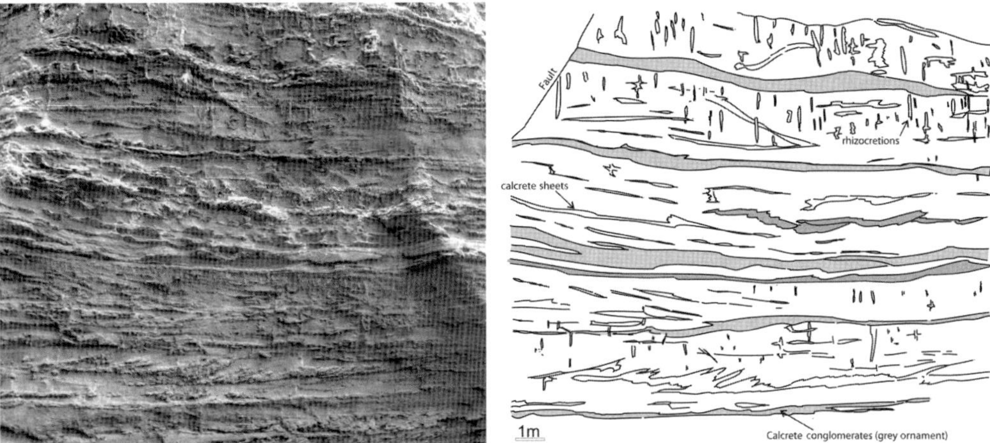

Fig. 5. Outcrop map of coastal cliff near Otterton Point [SY 082 822] showing the abundance and distribution of calcrete in the calcrete-rich multistorey channel association.

cross-stratification (Fig. 4b) and mimic the concave-up geometry of channel-fill deposits (Fig. 5). Calcrete sheets are composed of silicate grains with a fine- to medium-grained crystalline sparry calcite cement (Purvis & Wright 1991). The close association of calcrete sheets with rhizocretions suggests a similar near surface site of formation (Purvis & Wright 1991). The lack of distinctive calcrete profiles or associated palaeosols suggests a non-pedogenic origin and a groundwater origin has been suggested (Purvis & Wright 1991). The presence of carbonate nodules and sheets along stratification planes suggests that the carbonate may have formed inorganically from water travelling preferentially along such planes (Khadkikar *et al.* 1998). It is possible that the sheets developed around the water table in the capillary fringe zone with precipitation driven either by evaporation or by carbon dioxide degassing. Carbonate cementation by groundwaters has been widely described from sandy fluvial deposits in many semi-arid–arid environments (Mann & Horwitz 1979). Closely comparable bedding-concordant sheets have been described from the Quaternary of India (Khadkikar *et al.* 1998).

Calcrete conglomerates

Calcrete conglomerates occur at vertical intervals of 1–10 m throughout the multistorey channel associations. Relative to the sandstones, the conglomerates are extremely well cemented and form projecting ledges on cliff faces and broad wavecut platforms at the base of the

outcrop. The conglomerates generally form laterally continuous, horizontal beds with a basal erosion surface and a sharp top (Fig. 4c). Conglomerates are clast supported with a sandy matrix and are composed predominantly of reworked calcrete fragments with subordinate clasts of sandstone, mudstone, vertebrate bone, quartz and slate. The calcrete fragments are mostly well rounded, but often retain their branching, cylindrical or sheet-like form indicating their origin as rhizocretions or groundwater calcretes. Maximum clast size is generally around 0.15 m. The conglomerates generally display crude discontinuous horizontal stratification or trough cross-stratification. The conglomerates are generally well cemented by sparry calcite, the early diagenetic calcrete clasts probably acting as nucleation sites for later burial cements (Burley 1984; Lorsong & Atkinson 1995). The dimensions and form of calcrete conglomerates is variable and can be related to sedimentary facies. In the lower calcrete-rich, multistorey channel association the conglomerates are generally several tens of metres in width and are lenticular in form, reaching up to 1 m thick in the centre and tapering toward the edges. The erosional base is often highly irregular with evidence of deep scouring. In the upper, calcrete-poor multistorey channel association the conglomerates generally form tabular sheet-like units that are less than 0.5 m thick. They generally range up to 150 m wide, but Lorsong & Atkinson (1995) have traced one calcrete conglomerate for 1000 m.

Comparable calcrete conglomerates have

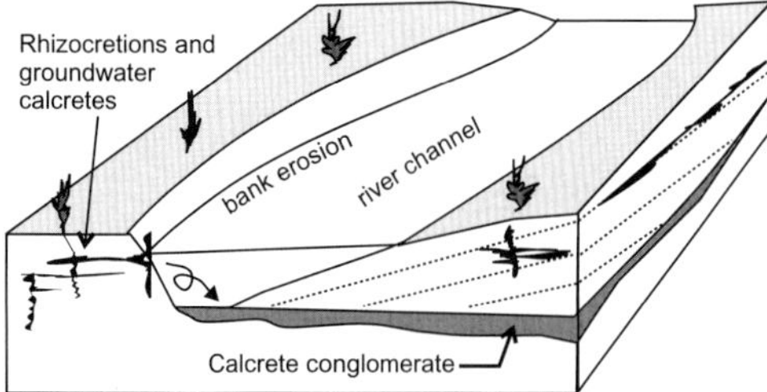

Fig. 6. Block diagram showing the process of lateral-channel migration and bank erosion, which generates calcrete conglomerates.

been described from the Permo-Triassic of Minorca (Gomez-Gras & Alonso-Zarza 2003) and from Late Quaternary deposits in western India (Khadkikar *et al.* 1998). The conglomerates, which form the base of channel fills and major channel belts, are interpreted as channel lags. The cutting and migration of a channel generally causes underscouring and bank collapse, which introduce large sediment loads into the channel (Fig. 6). The compositional homogeneity of the calcrete conglomerates may be due to the sorting of bank debris on account of different entrainment velocities of sand and calcrete fragments (Khadkikar *et al.* 1998). In the Otter Sandstone Formation, the spacing, form and dimensions of calcrete conglomerates can be broadly related to evolving fluvial style. The calcrete-rich multistorey channel association is characterized by lenticular calcrete conglomerates that are closely spaced vertically and, in general, do not extend laterally for more than a few tens of metres. This may reflect an episodic, ephemeral flow regime where channels were cut and filled rapidly with relatively little time available for extensive lateral migration. The large perennial channels of the calcrete-poor multistorey channel association produced calcrete conglomerates up to 1000 m wide, possibly reflecting lateral and downstream channel migration over large distances. River channels on sandy braidplains are rarely stable and are prone both to avulsion and lateral migration.

Porosity and permeability of calcretic sandstones

Porosity and permeability measurements of the Otter Sandstone Formation are held within the British Geological Survey's Aquifer Properties Database (Allen *et al.* 1997). These data are based on core plugs that were laboratory tested for gas permeability, porosity, bulk density and grain density. At each sampling interval, wherever possible, both a horizontal and vertical core plug was prepared. Horizontal samples were cut subparallel to bedding and vertical samples were cut subperpendicular to bedding. The core plugs were cut as cylindrical plugs of about 24 mm in diameter and 27.5 mm in length. All samples were dried prior to physical properties tests. A standard liquid saturation method was used to determine effective porosity, while gas permeability was determined using nitrogen under steady-state conditions. Core-plug data are taken to represent point estimates of permeability and bulk porosity within an aquifer. As the size of core samples is relatively small, laboratory results provide information on the aquifer properties at a matrix scale.

Porosity and permeability data from 36 samples of the Otter Sandstone Formation were examined (Fig. 7). Approximately half of the samples were taken from outcrop at Ladram Bay [SY 0975 8515] and the other half were from core of the Harpford Borehole [SY 0930 9040]. Core plugs were classified into one of three facies types: (1) sandstone showing no or minor calcrete development; (2) sandstone showing extensive calcrete development; and (3) calcrete conglomerates. Porosity values for the 36 samples range from around 7 to 34%. These values are consistent with those previously reported for UK Triassic sandstone aquifers, which vary in the range 2–35% with a median value of 26% (Allen *et al.* 1997). Gas permeability varies over five orders of magnitude from around 4×10^{-17} to 7×10^{-12} m². This wide range is consistent with reported permeabilities for UK Triassic sandstone aquifers (Allen *et al.* 1997). Allen *et al.* (1997) did not attempt to analyse the relationship between sedimentary facies and pororosity/permeability. However, this subset of Otter Sandstone Formation core plugs shows that samples with a porosity of less than 12% and a permeability of 10^{-15} m² are either calcretes or calcrete conglomerates. There is a clear porosity and permeability contrast between calcretes and non-calcretized sandstone (Fig. 7). Calcretes and calcrete conglomerates have a sufficiently low permeability to constitute baffles to flow within the aquifer.

Calcretes and aquifer heterogeneity

Spatial variability of porosity and permeability has been recognized as a dominant control on fluid flow through natural porous media. Many

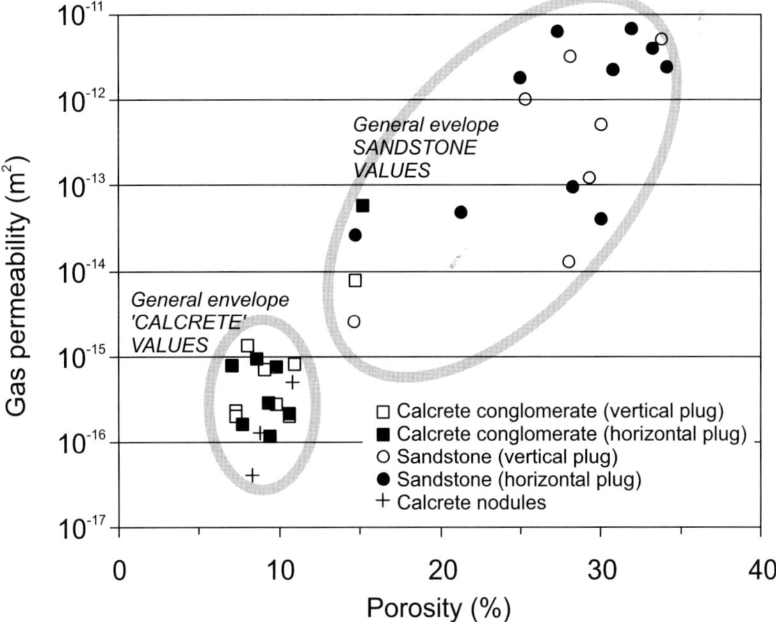

Fig. 7. Cross-plot of gas permeability against porosity for core plugs from the Otter Sandstone Formation.

studies have shown that the porosity and transmissive properties of a fluvial aquifer can be directly tied to depositional lithofacies and to the complex infill and stacking patterns of channel deposits and overbank mudstones (Aigner *et al.* 1996; Anderson *et al.* 1999). Heterogeneity can generally be considered at a range of scales: from the *megascopic*, which relates to the gross interbedding of coarse-grained channel belt and fine-grained floodplain deposits; the *macroscopic*, which describes large-scale inclined strata within bars and channel fills; the *mesoscopic* or medium- and small-scale cross-bedding; to *microscopic*, which considers pores and pore throats in individual strata. Mudstones commonly constitute the most important barriers to fluid flow with fluvial sandstone aquifers, and these can vary from floodplain deposits tens of metres in thickness, through to metre-scale channel plugs, to millimetre-thick drapes on foresets. In the Otter Sandstone Formation mudstones are rare, except in the uppermost part of the formation, and it is probable that calcretes and carbonate-cemented calcrete conglomerates are the dominant control on aquifer heterogeneity (Fig. 8) acting in the following possible ways.

- Calcrete conglomerates can form low-permeability sheets up to 1000 m wide, forming baffles to vertical flow at a megascopic scale.

- At a macro- to mesoscopic scale, groundwater calcretes form horizontal or inclined sheets that range up to several tens of metres in width. These would form significant flow baffles and increase the tortuosity of fluid flow paths. The preferential cementation of channel forms, bedding planes and lamination by groundwater calcrete will amplify the anisotropy that these stratal features normally introduce into an aquifer.

- Rhizocretions, as vertical cylindrical structures, are unlikely to have such an important role as baffles to flow, except to introduce more tortuousity. In combination with other calcrete types, however, they will have a significant effect in occluding a significant proportion of the porosity, possibly up to 20–30% in parts of the Otter Sandstone aquifer. Rhizocretions are also important as a source of clasts for calcrete conglomerates.

Allen *et al.* (1997, p. 273) report that the Otter Sandstone Formation in the Otter Valley (i.e. within the lower calcrete-rich part of the fluvial succession) shows typical bulk horizontal hydraulic conductivity for the Sherwood Sandstone Group but a lower overall vertical

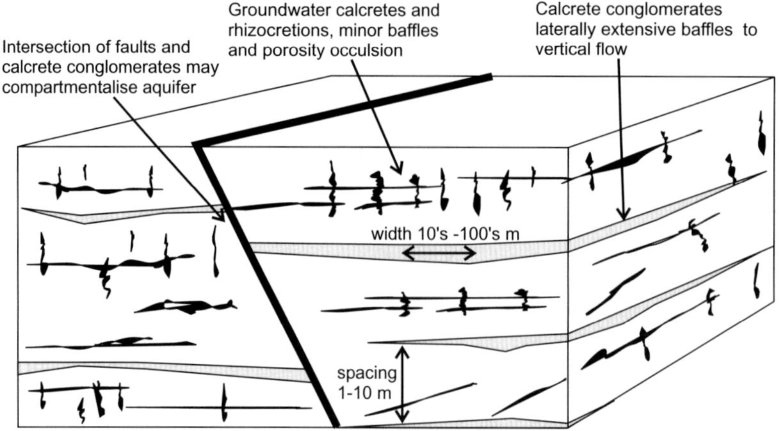

Fig. 8. Conceptual block diagram showing heterogeneity produced by calcrete in a sandstone aquifer.

hydraulic conductivity consistent with low-permeability layers acting as baffles. In addition, initially confined storage responses to pumping tests in the Otter Sandstone Formation are characteristic of a layered sandstone (Allen *et al.* 1997).

Conclusions

Calcrete is an early diagenetic carbonate cement that can form an abundant component of Triassic sandstone aquifers. In the English Triassic, calcrete occurs mainly within the Otter Sandstone Formation of SW England and the Bromsgrove Sandstone Formation of the West Midlands (Old *et al.* 1991). Outcrop study assists in understanding the form and distribution of calcrete, and its likely impact on the physical properties of the aquifer. In the Otter Sandstone calcrete is shown to occur as rhizocretions, as calcrete sheets and as reworked calcrete conglomerates. In parts of the aquifer, calcrete can become sufficiently abundant to produce a significant (up to 30%) reduction in the total effective porosity of the aquifer. Moreover, calcrete conglomerates can form laterally extensive (up to 1 km) low-permeability sheets that will represent major baffles to vertical flow.

An understanding of the aquifer sedimentology assists in considering the potential importance of calcrete as a source of heterogeneity. The distribution of calcrete is facies-specific. Calcrete is rare in aeolian sandstones and mudstone-rich channel–floodplain deposits. In the multistorey fluvial channel sandstones, calcrete is most abundant in the lower unit characterized by relatively small ephemeral channel

fills. An increase in channel size, possibly related to increased humidity, causes a reduction in the total abundance of calcrete. Conversely, an increase in channel size produces a corresponding increase in the width of calcrete conglomerates.

Calcrete may be more difficult to detect in limited subsurface data sets compared with other common types of flow barrier such as mudstone, highlighting the need for an awareness of its potential importance. For example, in many boreholes the gamma-ray log may be the only geophysical log available and, although this will indicate flow barriers formed by mudstone, it will not detect calcretes or calcrete conglomerates, unless these have a high proportion of mudstone clasts. A monotonous, low gamma-ray response from stacked channel sandstone deposits such as the Otter Sandstone Formation may suggest a degree of aquifer homogeneity that is not realistic.

J. Bloomfield, N. Jones, P. Turner and J. Tellam are thanked for their constructive reviews of the manuscript. This paper is published with the permission of the Director of the British Geological Survey (NERC).

References

AIGNER, T., ASPRION, U., HORNUNG, J., JUNGHANS, W.D. & KOSTREWA, R. 1996. Integrated outcrop analogue studies for Triassic alluvial reservoirs: Examples from Southern Germany. *Journal of Petroleum Geology,* **19,** 393–406.

ALLEN, D.J., BREWERTON, L.J. ET AL. 1997. *The Physical Properties of Major Aquifers in England and Wales.* British Geological Survey Technical

Report, **WD/97/34**. Environment Agency R&D Publication, **8**.

ANDERSON, M.P., AIKEN, J.S., WEBB, E.K. & MICKELSON, D.M. 1999. Sedimentology and hydrogeology of two braided stream deposits. *Sedimentary Geology*, **129**, 187–199.

BURLEY, S.D. 1984. Patterns of Diagenesis in the Sherwood Sandstone Group (Triassic), United-Kingdom. *Clay Minerals*, **19**, 403–440.

GOMEZ-GRAS, D. & ALONSO-ZARZA, A.M. 2003. Reworked calcretes: their significance in the reconstruction of alluvial sequences (Permian and Triassic, Minorca, Balearic Islands, Spain). *Sedimentary Geology*, **158**, 299–319.

GOUDIE, A.S. 1996. Organic agency in calcrete development. *Journal of Arid Environments*, **32**, 103–110.

HOLLOWAY, S., MILODOWSKI, A.E., STRONG, G.E. & WARRINGTON, G. 1989. The Sherwood Sandstone Group (Triassic) of the Wessex Basin, southern England. *Proceedings of the Geologist's Association*, **100**, 383–394.

HOUNSLOW, M.W. & MCINTOSH, G. 2003. Magnetostratigraphy of the Sherwood Sandstone Group (Lower and Middle Triassic), south Devon, UK: detailed correlation of the marine and non-marine Anisian. *Palaeogeography, Palaeoclimatology, Palaeoecology*, **193**, 325–348.

KHADKIKAR, A.S., MERH, S.S., MALIK, J.N. & CHAMYAL, L.S. 1998. Calcretes in semi-arid alluvial systems: formative pathways and sinks. *Sedimentary Geology*, **116**, 251–260.

LORSONG, J.A. & ATKINSON, C.D. 1995. *Sedimentology and Stratigraphy of Lower Triassic Alluvial Deposits, East Devon Coast*. Petroleum Group Excursion Guide. Geological Society, London.

MANN, A.W. & HORWITZ, R.C. 1979. Groundwater calcrete deposits in Australia: some observations from Western Australia. *Journal of the Geological Society of Australia*, **26**, 293–303.

MCKIE, T., AGGETT, J. & HOGG, J.C. 1998. Reservoir architecture of the upper Sherwood Sandstone Wytch Farm field, southern England. *In*: UNDERHILL, J.R. (ed.) *Development, Evolution and Petroleum Geology of the Wessex Basin*. Geological Society, London, Special Publications, **133**, 399–406.

MOUNT, J.F. & COHEN, A.S. 1984. Petrology and geochemistry of rhizoliths from Plio-Pleistocene fluvial and marginal lacustrine deposits, East Lake Turkana, Kenja. *Journal of Sedimentary Petrology*, **54**, 263–275.

OLD, R.A., HAMBLIN, R.J.O., AMBROSE, K. & WARRINGTON, G. 1991. *Geology of the Country around Redditch*. Memoir of the British Geological Survey, Sheet 183 (England and Wales).

PARNELL, J. 2002. Diagenesis and fluid flow in response to uplift and exhumation. *In*: DORE, A.G., CARTWRIGHT, J.A., STOKER, M.S., TURNER, J.P. & WHITE, N. (eds) *Exhumation of the North Atlantic Margin: Timing, Mechanisms, and Implications for Petroleum Exploration*. Geological Society, London, Special Publications, **196**, 433–446.

PURVIS, K. & WRIGHT, V.P. 1991. Calcretes related to phreatophytic vegetation from the Middle Triassic Otter Sandstone of South West England. *Sedimentology*, **38**, 539–551.

REEVES, C.C. 1976. *Caliche, Origin, Classification, Morphology and Uses*. Estacado Books, Lubbock, TX.

STRONG, G.E. 1993. Diagenesis of Triassic Sherwood Sandstone Group rocks, Preston, Lancashire, UK: a possible evaporitic cement precursor to secondary porosity? *In*: NORTH, C.P. & PROSSER, D.J. (eds) *Characterization of Fluvial and Aeolian Reservoirs*. Geological Society, London, Special Publications, **73**, 279–289.

STRONG, G.E. & MILODOWSKI, A.E. 1987. Aspects of the diagenesis of the Sherwood Sandstones of the Wessex Basin and their influence on reservoir characteristics. *In*: MARSHALL, J.D. (ed.) *Diagenesis of Sedimentary Sequences*. Geological Society, London, Special Publications, **36**, 325–337.

WANG, Y.F., NAHON, D. & MERINO, E. 1994. Dynamic-model of the genesis of calcretes replacing silicate rocks in semiarid regions. *Geochimica et Cosmochimica Acta*, **58**, 5131–5145.

WILLIAMS, G.E. 1970. Flood deposits of sand-bed ephemeral streams of Central Australia. *Sedimentology*, **17**, 1–40.

WRIGHT, V.P. 1990. Estimating rates of calcrete formation and sediment accretion in ancient alluvial deposits. *Geological Magazine*, **127**, 273–276.

WRIGHT, V.P. & TUCKER, M.E. 1991. Calcretes: an introduction. *In*: WRIGHT, V.P. & TUCKER, M.E. (eds) *Calcretes*. Reprint Series of the International Association of Sedimentologists, **2**, 1–24.

WRIGHT, V.P., MARRIOTT, S.B. & VANSTONE, S.D. 1991. A reg paleosol from the Lower Triassic of south Devon – Stratigraphic and paleoclimatic implications. *Geological Magazine*, **128**, 517–523.

Sedimentary and diagenetic environments of the Wildmoor Sandstone Formation (UK): implications for groundwater and contaminant transport, and sand production

JON E. BOUCH[1], ED HOUGH[1], SIMON J. KEMP[1], JOHN A. McKERVEY[1], GEOFFREY M. WILLIAMS[1] & RICHARD B. GRESWELL[2]

[1]*British Geological Survey, Kingsley Dunham Centre, Keyworth, Nottingham NG12 5GG, UK (e-mail: jbouch@bgs.ac.uk)*

[2]*Hydrogeology Research Group, Earth Sciences, School of Geography, Earth and Environmental Sciences, University of Birmingham, Birmingham B15 2TT, UK*

Abstract: The Wildmoor Sandstone Formation, proved in three boreholes drilled at Birmingham University, is dominated by fine- to medium-grained sandstones deposited in a braided river environment, within which channel lag, channel fill and abandoned channel facies are recognized. Minor proportions of aeolian sandsheet are present, as are dolocretes, not previously reported in the formation.

The sandstones are feldspathic and lithic arenites, and typically are clay-poor. Early dolomite dominates the diagenetic overprint, and is preferentially developed in channel-lag deposits. Burial diagenetic effects are minor. Late calcite occurs as a pore-filling phase and within fractures.

Minor fractures and granulation seams are oriented parallel to the NE–SW Birmingham Fault. 'Conventional' granulation seams, with comminution of detrital material, and more complex seams containing comminuted dolomite cement with a millimetre-wide halo of dolomite cement are present, the latter implying that the sandstone was dolomite-cemented at the time of fracturing.

Several scales of heterogeneity will affect groundwater solute transport. The palaeosols and abandoned channel mudstones may act as barriers to vertical flow at the decimetre scale. Dolomite-cemented channel-lag deposits may act similarly at smaller scales. Granulation seams have permeabilities of two–three orders of magnitude lower than their host sandstones, but their limited occurrence may limit their impact on larger scale flow. Matrix permeability is controlled by grain size and dolomite cement.

The fines in the fine-grained, ripple cross-laminated sandstones were extensively washed out during coring, and this lithology may be a source of sand yields in some sandstone boreholes. Although no enhancement of particle yields was seen during packer testing, the possibility remains that more comprehensive failure may occur at higher pumping rates.

Owing to extensive drift cover, urbanization and its soft, easily eroded nature, outcrops of the Wildmoor Sandstone Formation (Sherwood Sandstone Group) are relatively rare in the Permo-Triassic basins of central England. Consequently, detailed sedimentological or petrographical information is sparse. An opportunity to study the formation arose when three closely spaced boreholes were drilled on the Birmingham University campus as part of a project investigating the hydrogeological properties of sediment-filled fractures in Triassic sandstones (Pearce *et al.* 2001; Wealthall *et al.* 2001). This paper provides an account of the sedimentological, structural and petrographical features of the formation as proved in the boreholes. In particular we describe material considered to represent *in situ* palaeosols, which have not previously been reported in the Permo-Triassic of this part of the UK, but are well known in the stratigraphically equivalent Otter Sandstone of Devon (Purvis & Wright 1991; Newell 2006). Furthermore, we discuss the likely influence of sedimentological, diagenetic and structural heterogeneity on fluid migration and contaminant transport in the subsurface. Finally, we present details of ripple cross-laminated, poorly-cemented facies that liberate fines during drilling. We suggest this facies may account for some instances of sand production observed in some water-abstraction boreholes in Triassic aquifers.

From: BARKER, R. D. & TELLAM, J. H. (eds) 2006. *Fluid Flow and Solute Movement in Sandstones: The Onshore UK Permo-Triassic Red Bed Sequence*. Geological Society, London, Special Publications, **263**, 129–153.
0305-8719/06/$15 © The Geological Society of London 2006.

Background

Regional setting

The Wildmoor Sandstone Formation (Warrington *et al.* 1980) (Table 1) forms part of the Triassic Sherwood Sandstone Group, which crops out on the western side of Birmingham, on the Triassic structural high of the South Staffordshire Horst (Fig. 1). This horst was one of a series of broadly N–S-trending highs separating basins developed in response to extension during the early Triassic. The horst is bounded to the east by the Birmingham Fault, which downthrows 60–200 m to the east, and juxtaposes the sediments of the Sherwood Sandstone Group against the Mercia Mudstone Group. To the NW of the Birmingham Fault, the Wildmoor Sandstone is between 30 and 86 m thick (Eastwood *et al.* 1925; Butler & Lee 1942); to the SE of the fault the formation is less well developed and/or preserved and of uncertain thickness (Powell *et al.* 2000). The precise timing of movement on the Birmingham Fault has not been ascertained and, whilst the fault is considered to have been active during deposition of the Sherwood Sandstone Group, most movement is inferred to be post-Triassic (Powell *et al.* 2000). The contact between the underlying Kidderminster Formation and the Wildmoor Sandstone is conformable and gradational; the Wildmoor Sandstone is unconformably overlain by the Bromsgrove Sandstone Formation (Wills 1970, 1976).

Table 1. *Stratigraphical nomenclature for the Sherwood Sandstone Group in the Birmingham area, including comparison with previous stratigraphic nomenclature for the area (after Powell* et al. *2000)*

System	Series	Stage	Group	Lithostratigraphy	
				This account (Warrington *et al.* 1980)	Previous Nomenclature (e.g. Hull 1869; Eastwood *et al.* 1925).
				Formation	
Triassic	Middle	Anisian	Sherwood Sandstone	Mercia Mudstone	Keuper Marl
	Scythian	Induan-Olenekian		Bromsgrove Sandstone	Lower Keuper Sandstone
				Wildmoor Sandstone	Upper Mottled Sandstone
				Kidderminster	Bunter Pebble Beds
				Hopwas Brecciaa	Hopwas/ Quartite/ Sutton Breccias and Barr Beacon Beds

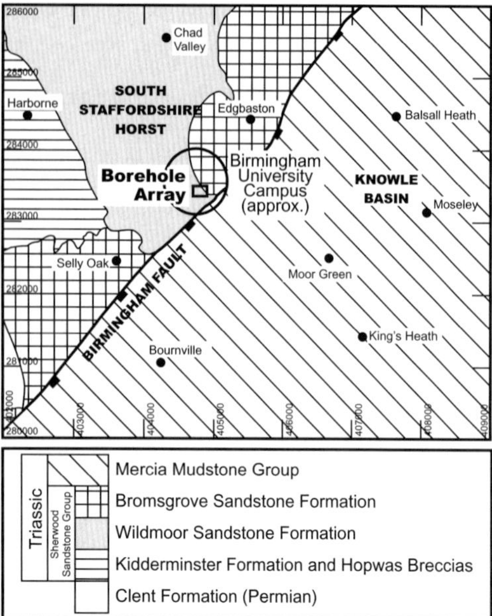

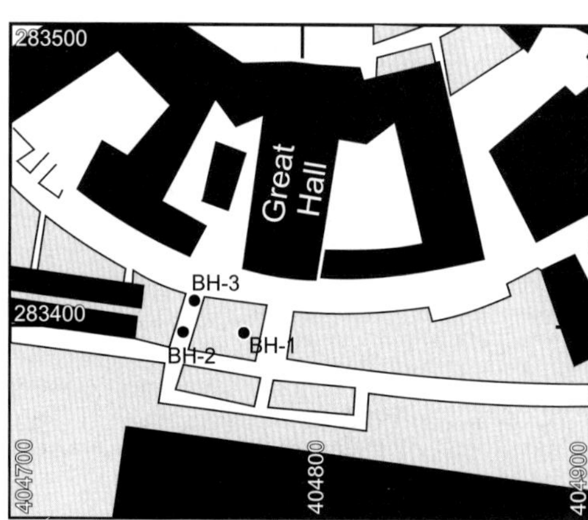

Fig. 1. Map showing the location of the Birmingham University Great Hall Borehole array. (**a**) Generalized geological map (based on British Geological Survey 1996). (**b**) Detailed locations of the boreholes.

Sediments of the Wildmoor Sandstone Formation are characterized as comprising orange-red, fine-grained, soft, silty, micaceous sandstones, containing only rare exotic or extraformational clasts. This is in contrast to the Bromsgrove Sandstone and Kidderminster Sandstone formations which are typified by abundant extraformational material, and commonly preserved mudstone beds in the case of the Bromsgrove Sandstone Formation (Bridge & Hough 2002). The sandstones are poorly cemented, and, when disaggregated, are easily moulded by hand and, consequently, were formerly worked for foundry sand. At outcrop, bedforms typically comprise tabular or low-angle cosets, with planar cross-bedding, or trough cross-bedding, with scour surfaces overlain by granules and mudstone rip-up clasts. The sandstones are interbedded with sub-ordinate red-brown and green-grey mudstones. The formation is interpreted to have been deposited in a fluvial environment, possibly in a relatively distal braidplain setting, with sediment sourced from the reworking of the upper part of the underlying Kidderminster Formation (Powell *et al.* 2000). The burial history for the formation in the study area is poorly constrained, and the maximum depth of burial is not clear as the thickness of any Jurassic or younger strata, that have since been eroded, is unknown.

Boreholes

During autumn 2001 three vertical boreholes were continuously cored (nominal 4 inch diameter) to a depth of 50 m on the University of Birmingham campus adjacent to the University Great Hall (Fig. 1 and Table 2). BH-2 and BH-3 are approximately 7 m apart, with BH-1 offset by approximately 20 m. In order to minimize any impact of the drilling on the formation, the boreholes were drilled using water flush at just sufficient pressure to achieve good-quality coring. Measurement of water levels in BH-1 during drilling of BH-2 and BH-3 indicated no significant variations in water levels that could be attributed to the drilling process, suggesting that there is no rapid inter-connection between them (a feature also confirmed by subsequent packer testing).

Although near-surface core recovery in made ground and weathered rockhead was relatively poor, overall core recovery was good (typically 80%; Table 2). Following drilling, each borehole was completed by the installation of a grouted casing to just below unstable or unconsolidated rockhead, with the remainder of the borehole left uncased. The boreholes were logged using resistivity and gamma tools, and image logs were acquired using a Robertson's Group Optical TeleViewer, producing a continuous high-resolution 360° 'unwrapped' true-colour image of borehole walls. The TeleViewer logs were processed to extract fracture and bedding plane orientation data. All depths quoted are in metres below ground level (m bgl).

Techniques

The cores were sealed in plastic liners and sleeving to prevent drying out and collapse of any clay minerals. They were sedimentologically (1:50 scale) and discontinuity (1:20 scale) logged, and on this basis a suite of samples were removed for petrographical analysis.

A Temco Inc. MP-402 mini-permeameter was used to measure permeabilities throughout BH-3, and to measure permeabilities of petro-graphical samples from all three boreholes. These analyses targeted both granulated and non-granulated sandstones (see the subsection on 'Fracture types and distribution' later), and derived repeatable data for the higher perme-ability sandstones. In mud-prone, cemented or granulated lithologies, permeability could not be determined as impracticably long analysis times were required for the permeameter to reach steady-state conditions.

Standard stained and cover-slipped thin

Table 2. *Location and depth information for the Birmingham University Great Hall Borehole array. Depths are in metres below ground level (mbgl). Elevation at the site of all three boreholes is approximately 125 m above Ordnance Datum*

Borehole name (including BGS borehole identification number)	Grid reference (BNG)	Depth (TD, m)	Depth to casing (mbgl)	Cored interval (mbgl)	Core recovery (m)	Optical TeleViewer run interval (mbgl)
Birmingham University 1 (Eastern Borehole; SP08SW 525)	[SP ^{4}04780 ^{2}83397]	50.15	12.35	7.40–50.15	40.94	12.35–48.40
Birmingham University 2 (Southwestern Borehole; SP08SW 526)	[SP ^{4}04760 ^{2}83397]	50.00	15.65	6.25–50.00	39.29	15.65–49.80
Birmingham University 3 (Northwestern Borehole; SP08SW 527)	[SP ^{4}04762 ^{2}83408]	50.00	12.30	6.40–49.94	39.74	12.30–48.90

sections were point counted. To enable calculation of intergranular volumes (IGVs), authigenic components were differentiated according to whether they occurred in primary pores, identified as 'cement', or secondary pores, identified as 'replacive' (Houseknecht 1987; Ehrenberg 1995). Small rock chips and polished thin sections were analysed using a LEO 435VP scanning electron microscope, fitted with a Oxford Instruments ISIS 300 energy dispersive X-ray analysis system (EDXA) and a KE Developments four-element solid-state backscattered electron detector, in secondary electron (SEM) and backscattered electron (BSEM) modes. Cathodoluminescence (CL) analysis was undertaken using Technosyn Mark II apparatus.

Particle size analysis (PSA; sedigraph), was undertaken on samples following dissolution of carbonate species. X-ray diffraction (XRD) analysis was undertaken on the <2 μm fraction of seven samples. The <2 μm material was re-suspended in 3 ml of deionized water and pipetted onto a ceramic tile in a vacuum apparatus to produce an oriented mount. The mounts were then Ca-saturated using 2 ml 0.1 M $CaCl_2 \cdot 6H_2O$ solution and washed twice to remove excess reagent. XRD analysis was carried out using a Philips PW1700 series diffractometer fitted with a cobalt-target tube and operated at 45 kV and 40 mA. The mounts were scanned from 2° to 32° 2θ at 0.55° 2θ min^{-1} as air-dried mounts, after glycol solvation and after heating to 550 °C for 2 h. Diffraction data were analysed using Philips X'Pert software coupled to an International Centre for Diffraction Data (ICDD) database.

Sedimentology

The sandstones proved by the boreholes have characteristics typical of the Wildmoor Sandstone Formation. In particular, pebbles are predominantly intra-formational (reworked mudstone, sandstone and palaeosol), with few extra-formational clasts observed (quartz, chert and limestone). Furthermore, the sandstones are typically silty, micaceous, poorly cemented and easily moulded by hand when disaggregated. The following subsections describe the sedimentological characteristics of the recovered cores in more detail.

Facies associations

Sedimentological logging indicates that the formation is made up of massive sandstone, planar laminated sandstone, ripple cross-laminated silty sandstone, pebbly sandstone, wind-ripple laminated sandstone, siltstone/claystone and mudstone with dolocrete nodules (summary log for BH-3 in Fig. 2, representative photographs are in Figs 3 & 4). These lithotypes are interpreted to be the result of sedimentation in a moderate- to low-sinuosity braided, fluvial environment within which fluvial and subaerial facies associations were recognized. Although minor aeolian sandstones are observed in the Birmingham boreholes, the bulk of the sandstone, mudstones and dolocrete nodule-rich mudstones appear to have been deposited under conditions sufficiently wet to maintain a permanent fluvial system. Optical TeleViewer data indicate that bedding predominantly dips SW, with a minor component to the SE (Fig. 5), but is unable to resolve the dips of cross-bedding foresets. Trigonometry on correlatable palaeosol and mudstone units indicates a dip of approximately 14° to the SW.

Fluvial facies association

This facies is dominated by erosionally based sandstone units. Siltstone and claystone lithologies towards the tops of units may represent channel waning and abandonment. Owing to the absence of surface expression of the Wildmoor Sandstone Formation from similar horizons, it has not been possible to differentiate confidently between channel and sheetflood facies, and consequently some beds identified as channels may in fact represent ephemeral sheetflood deposits. In more detail the following facies are recognized.

- *Channel-lag facies*: this represents approximately 5% of the recovered core, and forms erosively based, coarse-grained pebbly sandstones, that range from a few centimetres to 10 cm thick (Fig. 3a). Pebbly sandstones are not characteristic of the Wildmoor Sandstone Formation in general (Bridge & Hough 2002). However, pebble lags at channel bases would be expected within the moderate- to low-sinuosity braided, fluvial depositional environment usually ascribed to the formation, although they would also be expected at the bases of meandering channels. This facies commonly, but not invariably, contains well-developed to pervasive dolomite cement, which is described further below.
- *Channel-fill facies*: is the dominant facies encountered in the boreholes (approximately 60%), and comprises very-fine- to coarse-grained, massive, planar-laminated

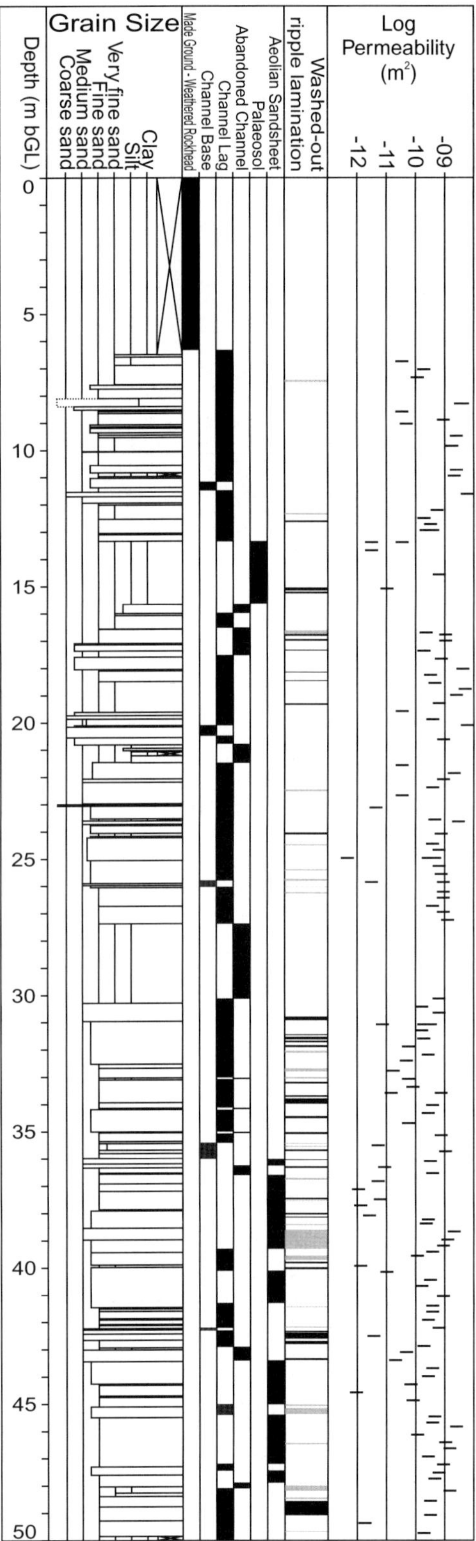

or low-angle cross-bedded sandstones, which occur in beds at the decimetre to metre scale, commonly stacked into bodies of sandstone up to approximately 5 m thick (Fig. 3b, c). The sandstones are typically moderately to well sorted, locally contain frosted, rounded grains indicative of reworking of aeolian material, and are poorly cemented and porous.

- *Abandoned channel facies*: this facies (approximately 15%) represents deposition of finer grained material during channel abandonment and infilling of the abandoned channel, including some lacustrine deposits. This facies is gradational with the channel-fill facies described above, and contains very-fine-grained, commonly ripple cross-laminated sandstones and siltstones (Fig. 3d). However, this facies is dominated by mudstones that are typically up to approximately 1 m thick, but locally up to 2.5 m thick. The mudstones are laminated or blocky and contain rare sand-grains. Desiccation features, which would indicate emergence and drying of mudstone units, were not identified. Ripple cross-laminated fine-grained sandstone beds up to 2 cm thick within the mudstones indicate that sporadic sediment charged flows fed into the standing body of water within which silt and clay particles settled. This facies has relatively low preservation potential and may be eroded by successive channels.

Throughout the formation, ripple cross-laminated sandstones commonly display features consistent with the preferential removal of fines during coring, leaving core that readily breaks apart into disks a few millimetres to a few centimetres thick (Fig. 3d, e). This 'washing out' is noted in approximately 10% of the core recovered (Fig. 2 and Table 3), and is best developed in the finer grained sandstones, but also occurs in some coarser grained sandstones.

Subaerial facies association

Two distinct subaerial facies are recognized.

- *Aeolian sandsheet facies*: this facies (15%) is represented by fine- to medium-grained sandstones containing a good degree of grain sorting, with individual grains having

Fig. 2. Summary log for BH-3, showing a simplified grain-size profile, interpreted facies, distribution of washed-out ripple cross-laminae and mini-permeameter data.

Fig. 3. Representative photographs of sandstone facies. (**a**) Coarse-grained channel lag with sharp contact over finer grained channel-fill sandstone. The channel lag contains abundant mudclasts that float in a matrix of pervasive early dolomite cement (BH-2: 20.80–21.02 mbgl). (**b**) Pebbly medium- to coarse-grained channel fill (BH-3: 18.40–18.63 mbgl). (**c**) Faintly laminated channel-fill sandstone. The lower part of this image contains several high-angle granulation seams, and minor washing out of finer grained laminae due to coring and core-handling processes (BH-2: 34.90–35.14 mbgl). (**d**) Fine-grained ripple cross-laminated sandstone. The clay-rich drapes over ripple laminae have been washed out during coring (BH-1: 39.23–39.43 mbgl). (**e**) Coarse-grained sandstone with washed out clays, probably on set boundaries (BH-2: 40.70–40.88 mbgl).

a high degree of rounding and sphericity, bimodal ('pin-stripe') lamination, and an absence of mica. Sandsheet deposits occur in units 0.2–2.5 m thick, and are considered to represent subaerial reworking and deposition on interfluves.

- *Palaeosol (dolocrete) facies*: dolocrete horizons (5%) have been identified in all three boreholes. In BH-1, two discrete horizons are present (13.45–14.18 and 19.20–20.30 mbgl). BH-2 contains a single horizon (13.15–16.61 mbgl) and BH-3 contains two horizons (13.30–14.40 and 15.0–15.50 mbgl; Fig. 4a) separated by a thin channel sandstone, which itself contains evidence for minor pedogenic modification.

Fig. 4. Representative optical TeleViewer, core and thin-section images of the palaeosols encountered in the Birmingham University Great Hall Borehole array. (**a**) Extract from TeleViewer log showing the large-scale characteristics of a palaeosol-dominated horizon. The top of the palaeosol, at 13.3 mbgl, has a relatively sharp contact with the overlying mudclast-bearing sandstone. The upper portion of the palaeosol contains abundant, well-defined, large (up to approximately 10 cm diameter) dolomite concretions that appear pale on the image log, with interstitial fine-grained/silty material (appears darker). These concretions become less well developed further down the profile, and by approximately 14.0 mbgl they are only poorly developed. Below 14.4 mbgl, carbonate cement is seen to occur along distinct high-angle to vertical fracture planes, and spread laterally along bedding planes within a more sandy unit. This fabric dies away by approximately 14.90 mbgl. A second palaeosol horizon is developed between approximately 15.0 and 15.5 mbgl, with a fabric dominated by irregular fractures with associated carbonate cement that brecciate the precursor sediment into 10–20 cm-diameter clasts (BH-3: 12.80–16.00 mbgl, horizontal scale = vertical scale). (**b**) Hand specimen of palaeosol with well-developed dolomite nodules in silty/sandy matrix. The nodules have yellow margins, and are red-stained in their cores (BH-2: 13.72–13.80 mbgl). (**c**) Thin-section photomicrograph showing zoned dolomite crystals replacing interstitial clay and silt-grade material between centimetre-scale dolomite nodules (arrowed, 1). The crystals have relatively euhedral cores and more irregular feathery or lobate margins (arrowed, 2; plane polarized light, BH-2: 13.72–13.80 mbgl). (**d**) Core photograph showing a detail of the lower, less well-developed palaeosol pictured in (a). The core contains dolomite nodules, millimetre-scale carbonate-cemented fractures and, in the centre of the image, a near-vertical feature that is filled with sediment of slightly different colour to the surrounding material (?rhyzolith; BH-3: 15.51–15.70 mbgl).

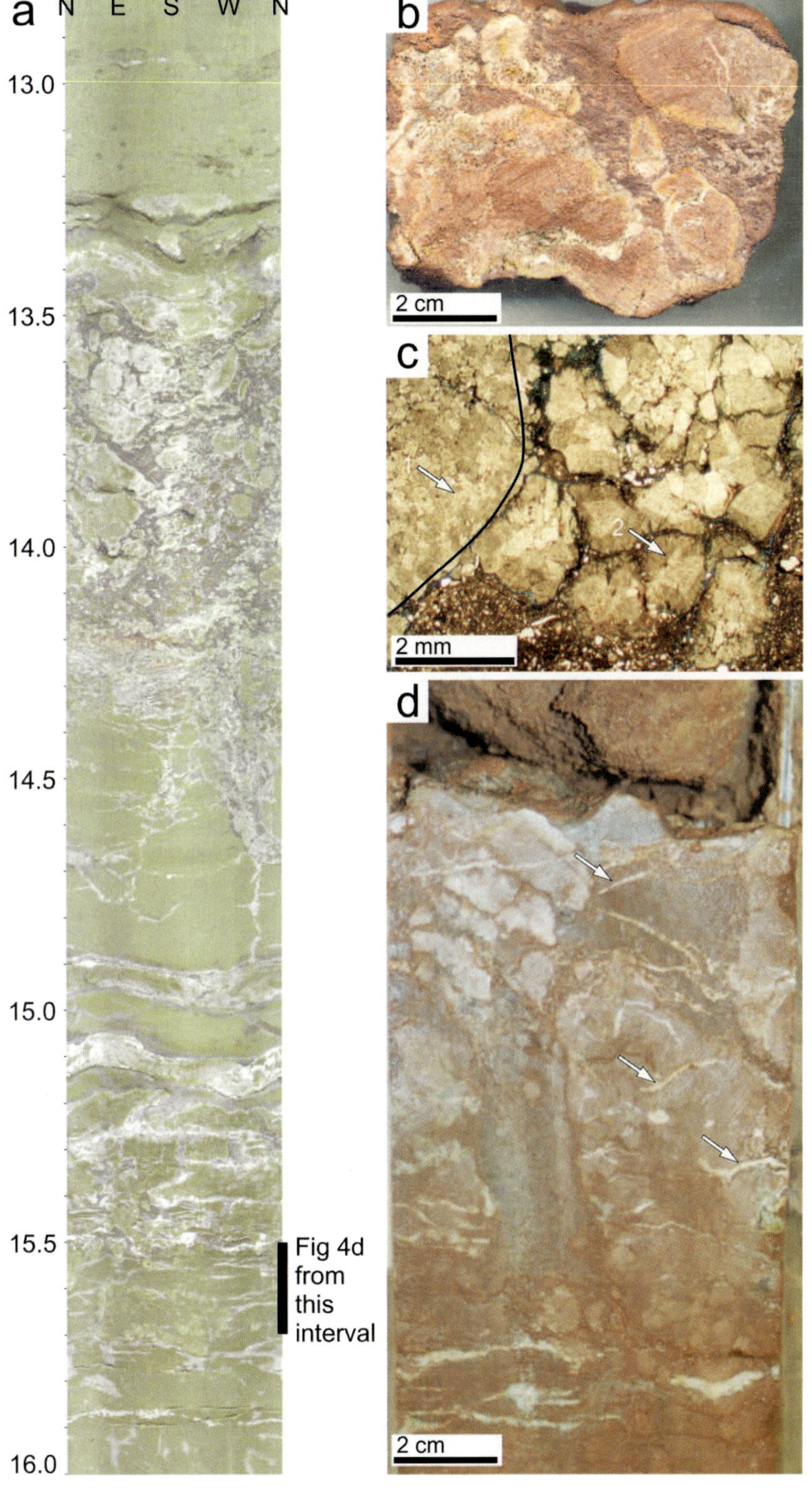

a
N E S W N
13.0
13.5
14.0
14.5
15.0
15.5
16.0
Fig 4d
from
this
interval
b
2 cm
c
2 mm
d
2 cm

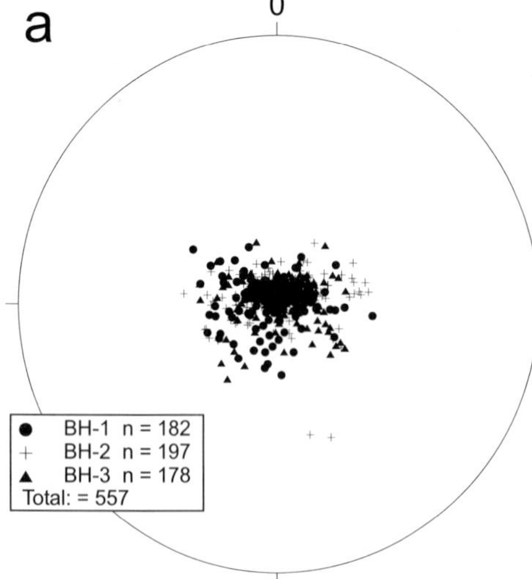

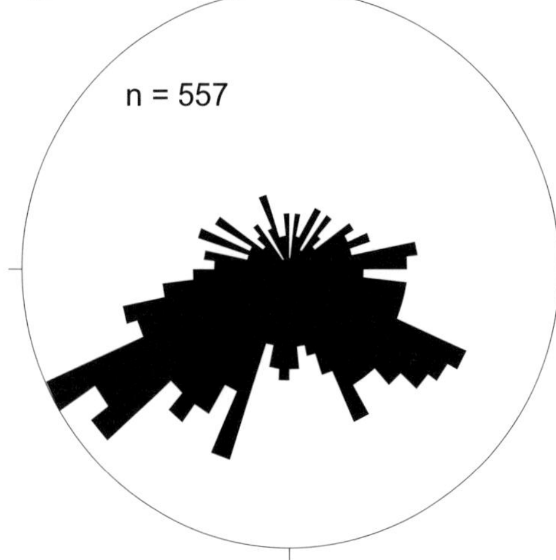

Fig. 5. Bedding-plane orientation data derived from optical TeleViewer logs for boreholes in the Birmingham University Great Hall Borehole array. (**a**) Stereoplot showing poles to planes (lower-hemisphere projection). (**b**) Rose diagram of dip directions.

In all cases the dolocretes have gradational bases and relatively sharp tops, with minor modification of sandstones beneath the dolocrete horizons limited to the development of dolomite extending down vertical fractures. The dolocretes are composed of variably well-developed dolomite nodules of up to approximately 10 cm diameter, which tend to be better developed towards the top of the horizons, in a matrix of green, red and grey silt and clay (Fig. 4b, d). The dolocretes are differentiated from dolomite-cemented fluvial-channel sandstones by their nodular/brecciated character, sharp tops and more gradational bases, their occurrence within typically fine-grained and mudstone-dominated facies (channel abandonment) and by their thicker character (1–2 m thick, compared with dolomite cemented intervals in fluvial sandstones which are typically 10–20 cm thick). These features are consistent with a pedogenic origin rather than a vadose or phreatic origin (which are more likely to be associated with coarser grained, more permeable units: Spötl & Wright 1992; Pimentel *et al.* 1996; Pimental 2002). However, a later generation of probable vadose dolomite can also be recognized (see below) in the dolocretes and the sandstones.

Petrographical analysis indicates that the central parts of the nodules comprise 100–200 µm diameter dolomite crystals in a subhedral–anhedral planar mosaic. Individual crystals commonly have inclusion-rich cores and clearer overgrowths, with abundant interstitial red-stained Fe-oxides or clays present between crystals. The nodule margins comprise clearer, more coarsely crystalline (up to 2 mm) dolomite, the earlier generations of which have clearly defined rhombic outlines. The outer portions of these crystals, however, have less well-defined crystal outlines, commonly with feathered or lobate forms where they engulf, displace and/or replace interstitial fines (Fig. 4c). The dolocretes are commonly cut by millimetre-scale fractures that are lined by euhedral dolomite and filled with later weakly-ferroan and non-ferroan calcite (Fig. 4d). Under CL, the dolomite within the nodules and that replacing interstitial material display similarly zoned yellow-orange and red-orange luminescence. There are no textures or calcite relicts that would suggest that the dolomite is a later diagenetic replacement of calcite, and the dolomite is therefore interpreted to be of primary origin. Schmid *et al.* (2003) suggest that in arid environments, the development of early dolomite rather than calcite cement may be controlled by progressive evaporation of river water and/or water in the shallow subsurface downstream. Calcite saturation and precipitation occurs further upstream, causing an increase in Mg/Ca ratio as the water moves basinward, until dolomite saturation is reached.

Further downstream, gypsum saturation may be reached.

The palaeosols form marker horizons and are important environmental indicators within the otherwise fairly monotonous, sandstone-dominated upper part of the proved succession. Fragments of dolocrete within channel-lag deposits below the lowest preserved palaeosol suggest that palaeosols exist, or previously existed, lower in the sequence. These have not been intersected by the boreholes, due to partial or complete fluvial reworking, or non-deposition in the vicinity of the boreholes.

Given the generally arid conditions that prevailed during the deposition of the Sherwood Sandstone Group and its equivalents, aeolian deflation would be expected to dominate over pedogenic processes on inter-fluves (Mader 1992). However, the development of carbonate concretions in mudstones has been alluded to in the Sneinton Formation in the Nottingham area (Mader 1992). In addition, Hough & Barnett (1998) have described calcrete nodules in the Permian Bridgenorth Formation in the Stafford Basin. With these exceptions, palaeosols have not previously been described in the Wildmoor Sandstone Formation, although calcretes are well documented in the stratigraphically equivalent Otter Sandstone of Devon (Purvis & Wright 1991; Newell 2006). However, the preservation of apparently *in situ* palaeosols here is a notable find.

Fracture types and distribution

Owing to the relatively soft nature of the recovered material, cores from all three boreholes contain very abundant drilling-induced or drilling-exaggerated discontinuities that tend to form horizontal to low-angle fractures that commonly exploit bedding planes (921 low-angle ($<15°$) discontinuities were identified during core logging; Table 3). These features are interpreted as being artificially induced as they occur at lithological boundaries, fail to be imaged on the TeleViewer logs, have tight fits across fractures and have no coatings on fracture surfaces. Bedding-plane (as opposed to bedding-plane fracture) orientations interpreted from the TeleViewer logs are presented in Figure 5.

In contrast, natural fractures are relatively

Table 3. *Summary of the numbers of discontinuities of different types logged in core and from Optical TeleViewer logs from the Birmingham University Great Hall Borehole array. 'Ambiguous' and 'drilling-enhanced fractures' predominantly represent fractures along natural planes of weakness within the core such as bedding planes, which may have been opened or artificially widened/enhanced during coring and/or core handling (e.g. Fig. 7c). 'Rubbled' intervals are intervals where the core has broken into rubble fragments possibly indicating less well-consolidated sediments*

Core data	BH-1	BH2	BH-3	Total
Granulation seams	16	30	22	68
Natural fractures	41	15	9	65
Cemented – healed	6	4	3	13
Cemented – open	1	–	–	1
Uncemented – closed	10	6	4	20
Uncemented – open	24	5	2	31
Ambiguous/drilling enhanced fractures	73	117	177	367
Bedding-plane fractures	2	2	11	15
Drilling-enhanced bedding-plane fractures	49	102	142	293
Ambiguous fractures	2	–	–	2
Drilling-enhanced fractures	20	13	24	57
Drilling-induced fractures	152	189	213	554
Drilling-induced fractures	125	162	181	468
Rubbled intervals	27	27	32	86
Total	282	351	421	1054
Intervals with washed out ripple cross-lamination (n)	59	112	79	250
Thickness of washed out washed out ripple cross-lamination (m)	4.89 m	5.13 m	5.19 m	15.21 m
Optical Televiewer data				
'Pale veins' and other cemented features	17	43	21	81
Open fractures	32	31	14	77

scarce but are readily identified in core and typically are imaged clearly on the TeleViewer logs (Fig. 6 and Table 3). Various fracture types are observed including granulation seams (also referred to as deformation bands, microfaults or shear bands), cemented fractures and uncemented fractures. The distribution and general appearance of these features are

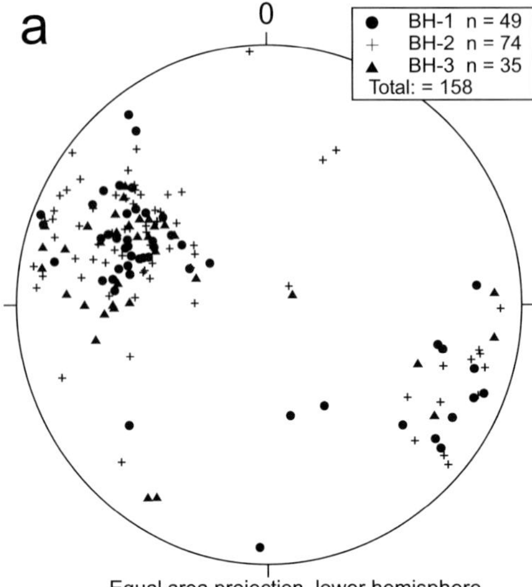

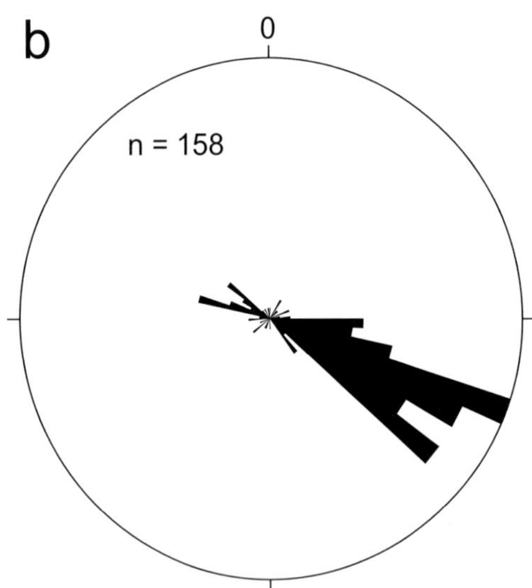

Fig. 6. Fracture orientation data for boreholes in the Birmingham University Great Hall Borehole array as measured using optical TeleViewer logging.
(**a**) Stereoplot showing poles to planes (lower-hemisphere projection). (**b**) Rose diagram of dip directions.

described below, with detailed information about their petrographical characteristics presented in a later section, in the context of the diagenetic evolution of the formation.

- *Granulation seams*: are resolved on the TeleViewer logs as white, high-angle, apparently cemented features that are not readily distinguished from other cemented fractures (see below). However, core observations indicate that, in all three boreholes, granulation seams only occur at depths of greater than approximately 25 mbgl, where they tend to occur in clusters spaced several metres apart. In detail, granulation seams form discrete, moderate- to high-angle (30°–60°) millimetre-scale features with single slip surfaces; however, some examples have multiple phases of displacement, with numerous subparallel slip surfaces (Fig. 7a–c). Determination of the magnitude of displacement across granulation seams is hindered by the limited diameter of the core; however, normal displacements of up to at least 10 cm are observed in some cases.

- *Cemented fractures*: cemented fractures are relatively rare ($n = 14$), and are characterized as high-angle to vertical, millimetre-scale calcite veins that occur typically, but not exclusively, within dolomite cemented intervals (Fig. 7d). These features are not readily resolved on the TeleViewer logs due to their occurrence within otherwise cemented intervals, or their similarity in appearance to granulation seams. In BH-1, a single calcite vein is associated with the development of vuggy, centimetre-scale, calcite-lined cavities along its length (Fig. 7e).

- *Uncemented fractures*: it is difficult to identify confidently natural uncemented fractures in core due to the abundance of drilling-induced and drilling-enhanced features (Fig. 7b, c); however, a number of moderate- to high-angle features ($n = 51$; typically 20°–90°) are observed (Fig. 7f). Where open, these features typically have rough uncemented surfaces, with a poor fit across the feature. Where closed, some small degree of displacement is usually evident, but there is no visible sign of the development of granulated material.

Orientation data from the TeleViewer logs indicate that the fractures, almost exclusively, have dips of 30°–90° to the SE (Fig. 6), with no difference in orientation between fracture types. These orientations are similar to that of the

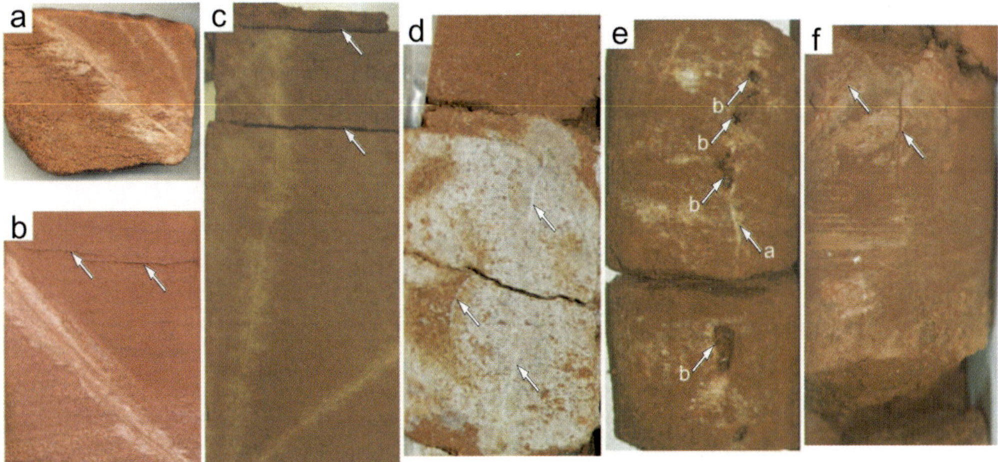

Fig. 7. Representative photographs of discontinuities in cores from the Birmingham University Great Hall Borehole array. (**a**) An example of a complex, high-angle granulation seam showing centimetre-scale normal displacement, multiple slip surfaces and adjacent preservation of dolomite cement (BH-2: 35.07–35.15 mbgl). (**b**) A similar granulation seam to that shown in (a), displaying at least 7 cm of normal displacement. A montage showing the textural variations that occur within this granulation seam is given in Figure 10. Also evident are thin laminae (arrowed) that have been washed out during sample handling, illustrating the relative ease with which fine-grained laminae can be preferentially removed from the sandstones (BH-3: 31.05–31.15 mbgl). (**c**) Two simple, high-angle granulation seams/fractures with negligible apparent displacement. Dolomite cement is preferentially developed/preserved in coarser laminae adjacent to the fracture. The image also shows two horizontal, drilling-induced fractures that exploit the primary lamination of the sandstone (arrowed, BH-3: 41.48–41.66 mbgl). (**d**) A dolomite-cemented coarse-grained sandstone, cut by a number of thin, near-vertical calcite veins (arrowed, BH-2: 22.02–22.25 mbgl). (**e**) Near-vertical calcite vein (arrowed, a) through lightly dolomite-cemented coarse-grained sandstone with centimetre-scale, calcite-lined vugs developed along the length of the fracture (arrowed, b: BH-1: 18.78–18.99 mbgl). (**f**) Near-vertical, uncemented fractures in sandstone (arrowed, BH-1: 11.47–11.61 mbgl).

Birmingham Fault, which is approximately 200 m to the SE of the study site (Fig. 1). This may indicate that the fracturing is related to the larger scale structure of the area, possibly representing part of a damage zone associated with movement along the Birmingham Fault.

Sandstone and fracture petrography

As described in more detail below, the selected sandstone samples (n = 29) are broadly typical of Sherwood Sandstone Group reported elsewhere (e.g. Waugh 1978; Ali & Turner 1982; Strong & Milodowski 1987; Bath *et al.* 1987; Strong 1993; Strong *et al.* 1994). The sandstones have average grain sizes from coarse silt to coarse sand, poor to good sorting, and commonly well-developed millimetre-scale grain-size lamination. Table 4 gives the mineralogical composition of the analysed sandstones as determined by modal analysis, and summary grain-size data (from PSA).

Detrital components

The sandstones classify predominantly as feldspathic and lithic arenites (Fig. 8a), with framework grain assemblages dominated by monocrystalline quartz, feldspar and rigid lithic material. Greater proportions of lithic material are present in the coarser grained sandstones (Fig. 8b, c). Feldspar grains commonly show evidence for corrosion and authigenesis with the development of overgrowths on K-feldspar grains. Plagioclase is notably less abundant than K-feldspar. Modal analysis may underestimate plagioclase due to the difficulty in differentiating fresh, untwinned plagioclase from quartz grains; however, where identified, plagioclase appears typically to be more extensively altered than K-feldspar, suggesting more extensive dissolution, as would be suggested by its lower overall stability (Waugh 1978). The impact of feldspar dissolution and precipitation on sandstone classification is uncertain. However, the bulk of the rigid lithic material comprises grains

Table 4. *Summary modal analysis data for sandstone samples from the Birmingham University Great Hall Borehole array. The mineralogical data are summarized such that total minerals = 100%, the porosity and summary total data are calculated as a proportion of total rock volume (i.e. minerals + porosity = 100%). '–' denotes that the phase was not observed within the sample, and 'tr' denotes a trace component that was observed during analysis, but was not formally counted during the modal analysis. Grain-size categories: f, fine; m, medium; c, coarse; s, sand; L, lower; U, upper*

Borehole	BH-1	BH-1	BH-1	BH-1	BH-1	BH-1	BH-1	BH-1	BH-2	BH-2	BH-2	BH-2	BH-2	BH-2	BH-2	BH-2	BH-2	BH-2	BH-3	BH-3	BH-3	BH-3	BH-3	BH-3	BH-3	BH-3	Min.	Max.	Mean
Sample identifier	H215	H217	H219	H220	H222	H223	H224	H225	H226	H228	H229	H230	H232 (frac.)	H232 (sst)	H233	H234	H235	H236	H248	H250	H251	H252	H253	H256	H257	H258			
Top depth (mbgl)	12.79	18.73	21.29	25.26	34.48	37.70	43.62	47.75	12.65	20.00	25.10	26.22	35.07	35.07	37.18	41.84	43.60	48.34	9.45	18.44	21.50	22.41	30.10	41.45	42.50	49.16			
Bottom depth (mbgl)	12.86	18.78	21.35	25.32	34.53	37.77	43.66	47.80	12.72	20.05	25.18	26.30	35.15	35.15	37.27	41.91	43.68	48.40	9.56	18.54	21.59	22.54	30.19	41.55	42.68	49.27			
Particle size data																													
Grain size (microns) (µm)	323	494	366	450	157	356	171	291	225	526	351	313	–	312	87	211	230	204	418	402	237	446	343	249	192	178	87	526	301
Grain size (category)	msL	msU	msU	msU	fsL	msU	fsL	msL	fsU	csL	msL	msL	–	msL	vfsL	fsU	fsU	fsU	msU	msU	fsU	msU	msL	fsU	fsU	fsU			
Sorting	poor	good	mod	good	good	mod	mod	good	very bimod, good	bimod, poor	mod	mod	–	mod	mod	good	good	mod	good	mod	good	good	mod	good	mod	mod			
Framework components (% of solid volume)																													
Monocrystalline quartz	20.5	19.0	20.5	28.0	48.5	40.0	40.5	39.5	33.0	19.5	24.0	23.0	28.0	35.5	29.0	37.5	32.0	40.5	28.3	19.5	34.5	23.0	36.5	34.0	26.5	29.5	19.0	48.5	30
Polycrystalline quartz	4.0	6.5	10.0	10.5	4.0	22.0	6.5	14.0	9.0	17.5	7.5	6.5	11.0	9.0	2.5	8.5	8.0	8.0	13.9	15.0	9.0	14.5	8.0	12.0	5.0	14.0	2.5	22.0	10
Feldspar	13.5	12.5	9.0	10.5	19.5	10.5	16.5	20.0	25.0	9.0	6.0	19.5	8.0	15.0	13.5	8.5	11.0	25.5	20.4	17.5	26.0	6.0	22.0	10.5	10.0	33.5	6.0	33.5	15
K-feldspar	12.0	10.5	5.5	8.0	11.0	7.5	12.0	15.5	20.0	7.0	5.0	17.0	4.0	11.0	8.5	6.0	7.0	23.0	16.1	15.0	17.5	4.5	17.0	6.5	5.0	26.5	4.0	26.5	11
Plagioclase	1.5	2.0	3.5	2.5	8.5	3.0	4.5	4.5	5.0	2.0	1.0	2.5	4.0	4.0	5.0	2.5	4.0	2.5	4.3	2.5	8.5	1.5	5.0	4.0	5.0	7.0	1.0	8.5	4
Heavy minerals	–	–	–	–	–	0.5	0.5	1.0	–	0.5	–	–	0.5	0.5	1.0	–	tr.	–	0.9	–	–	–	0.5	–	–	–	–	1.0	0.2
Dolomitized mudclasts/dolocrete clasts	13.5	7.0	4.5	0.5	–	–	–	–	–	23.0	1.0	1.5	–	–	–	–	–	–	–	10.0	–	0.5	–	–	–	–	–	23.0	2.4
Sandstone/metsedimentary clasts	3.0	2.0	1.5	4.0	2.5	4.5	1.0	5.0	5.0	1.5	3.5	0.0	2.5	4.0	0.0	3.0	3.0	4.5	4.8	0.5	4.5	0.5	2.0	5.5	1.5	2.0	0.0	5.5	2.8
Other rigid lithic clasts	2.0	1.0	1.5	0.5	2.0	3.0	1.0	2.5	3.5	2.5	1.5	1.0	0.5	3.0	1.0	3.5	2.5	2.5	3.5	3.0	2.5	2.5	1.5	2.0	2.5	1.5	0.5	3.5	2.1
Ductile components (% of solid volume)																													
Mudstone/siltstone clasts	10.5	1.0	2.5	1.0	1.5	tr.	1.0	1.5	3.0	8.5	–	1.0	0.5	4.0	–	1.5	1.5	2.5	16.9	4.5	3.5	1.0	3.5	3.0	1.0	0.5	–	16.9	2.9
Detrital clay	9.5	4.0	3.0	1.0	17.5	1.5	21.0	11.0	14.0	7.0	4.0	3.5	9.0	15.0	50.0	5.0	9.0	11.5	7.0	4.0	5.0	5.0	3.5	5.5	5.0	13.5	1.0	50.0	9
Mica	tr.	0.5	1.0	–	0.5	tr.	1.0	1.0	4.0	0.5	0.5	1.0	0.5	1.0	0.5	tr.	–	3.5	2.2	0.5	6.0	0.5	3.0	1.0	–	2.5	–	6.0	1.2
Authigenic Mineralogy (% of solid volume)																													
K-feldspar overgrowths (cement)	0.5	0.5	1.5	1.0	1.0	1.5	–	2.5	2.5	0.5	tr.	0.5	–	2.0	–	1.0	0.5	0.5	1.7	2.0	1.0	0.5	0.5	tr.	–	0.5	–	2.5	0.9
Quartz overgrowths (cement)	–	tr.	tr.	tr.	–	1.5	–	0.5	0.5	0.5	0.5	–	–	–	–	tr.	1.0	–	–	0.5	–	–	–	0.5	–	–	–	1.5	0.2
Kaolinite (predominantly replacive)	0.5	–	–	–	–	–	–	tr.	–	–	–	–	–	–	–	–	–	–	–	–	–	–	–	–	–	–	–	0.5	0.02
Illite	–	–	tr.	–	–	0.5	–	–	0.5	–	–	–	0.5	–	–	–	–	0.4	–	8.0	0.5	8.5	1.5	–	2.5	–	–	8.5	0.9
Cement	–	–	–	–	–	0.5	–	–	0.5	–	–	–	0.5	–	–	–	–	0.4	–	7.5	0.5	8.0	1.5	–	2.5	–	–	8.0	0.8
Replace	–	–	tr.	–	–	–	–	–	–	–	–	–	–	–	–	–	–	–	–	0.5	–	0.5	–	–	–	–	–	0.5	0.0
Dolomite	16.0	30.5	44.0	42.5	3.0	11.5	9.5	1.0	–	7.5	49.5	41.5	39.5	10.5	2.5	31.5	31.5	1.0	–	23.0	–	45.5	10.5	24.0	47.5	–	–	49.5	20
Non-ferroan cement	15.0	30.0	42.0	42.5	3.0	11.5	8.0	1.0	–	7.5	49.5	41.5	39.0	10.5	2.5	31.5	31.5	1.0	–	23.0	–	45.0	10.5	24.0	47.5	–	–	49.5	20
Non-ferroan replacive	1.0	0.5	2.0	–	–	–	1.5	–	–	–	–	–	0.5	–	–	–	–	–	–	–	–	0.5	–	–	–	–	–	2.0	0.2
Calcite	6.5	15.5	1.0	0.5	tr.	2.0	1.5	0.5	–	2.0	2.0	1.0	tr.	–	–	–	–	–	–	–	–	–	0.5	1.0	–	–	–	15.5	1.3
Non-ferroan cement	5.0	9.0	0.5	0.5	tr.	–	1.5	0.5	–	1.0	1.5	0.5	–	–	–	–	–	–	–	–	–	–	0.5	0.5	–	–	–	9.0	0.8
Non-ferroan replacive	–	–	–	–	0.5	–	–	–	–	–	–	–	–	–	–	–	–	–	–	–	–	–	–	–	–	–	–	0.5	0.02
Ferroan cement	1.5	6.5	0.5	–	tr.	1.0	–	–	1.0	0.5	0.5	–	tr.	–	–	–	–	–	–	–	–	–	–	0.5	–	–	–	6.5	0.5
Ferroan replacive	–	–	tr.	–	0.5	–	–	–	–	–	–	–	–	–	–	–	–	–	–	–	–	–	–	–	–	–	–	0.5	0.02
Haematite (cement)	–	–	–	–	1.0	–	–	–	–	–	–	–	–	–	–	–	–	–	–	–	–	–	–	–	–	–	–	1.0	0.04

continued

Table 4. *Continued*

	H215	H217	H219	H220	H222	H223	H224	H225	H226	H228	H229	H230	H232 (frac.)	H232 (sst)	H233	H234	H235	H236	H248	H250	H251	H252	H253	H256	H257	H258	Min.	Max.	Mean
Borehole	BH-1	BH-1	BH-1	BH-1	BH-1	BH-1	BH-1	BH-1	BH-2	BH-2	BH-2	BH-2	BH-2	BH-2	BH-2	BH-2	BH-2	BH-2	BH-3	BH-3	BH-3	BH-3	BH-3	BH-3	BH-3	BH-3			
Top depth (mbgl)	12.79	18.73	21.29	25.26	34.48	37.70	43.62	47.75	12.65	20.00	25.10	26.22	35.07	35.07	37.18	41.84	43.60	48.34	9.45	18.44	21.50	22.41	30.10	41.45	42.50	49.16			
Bottom depth (mbgl)	12.86	18.78	21.35	25.32	34.53	37.77	43.66	47.80	12.72	20.05	25.18	26.30	35.15	35.15	37.27	41.91	43.68	48.40	9.56	18.54	21.59	22.54	30.19	41.55	42.68	49.27			
Porosity (% of whole-rock volume)																													
Primary interparticle macroporosity																													
(PIM)	6.9	5.1	5.5	0.5	10.8	19.9	5.2	16.9	18.5	11.1	1.0	11.2	3.8	18.6	1.0	13.9	12.2	14.9	16.5	11.8	10.9	4.7	4.7	10.1	3.2	16.9	0.5	19.9	10
PIM	6.9	4.7	5.1	0.5	10.8	19.6	5.2	16.9	18.5	10.6	1.0	10.3	–	18.6	1.0	13.0	12.2	14.9	16.2	11.8	10.9	4.3	4.7	10.1	0.5	16.9	–	19.6	9
resurrected PIM	–	0.5	0.5	tr.	–	0.4	–	–	–	0.4	tr.	0.9	3.8	–	–	0.8	tr.	–	0.3	–	–	0.5	–	tr.	2.8	–	–	3.8	0.4
Secondary oversized macroporosity																													
(SOM)	–	0.9	2.3	3.3	1.3	4.1	0.5	1.6	2.3	0.4	tr.	1.3	0.5	1.6	–	1.3	0.4	–	4.1	3.4	1.7	0.5	–	1.8	4.2	–	–	4.2	1.4
SOM	–	0.9	0.9	2.4	1.3	4.1	0.5	1.6	2.3	0.4	tr.	1.3	0.5	1.6	–	1.3	0.4	–	4.1	2.9	1.7	0.5	–	1.8	4.1	–	–	4.1	1.2
vuggy porosity	–	–	1.4	1.0	–	–	–	–	–	–	–	–	–	–	–	–	–	–	–	0.4	tr.	–	–	–	3.2	–	–	3.2	0.2
Secondary intraparticle macroporosity																													
(SIM)	0.9	0.5	tr.	1.0	1.3	2.2	0.5	0.8	1.9	–	0.5	1.3	0.5	0.8	–	0.8	tr.	–	0.3	0.8	0.4	–	1.4	–	–	0.8	–	2.2	0.6
Whole rock proportions																													
(including porosity, excluding fractures)																													
Framework components	52.1	44.9	43.3	51.4	66.2	59.4	62.0	66.1	58.3	65.0	42.9	44.4	48.1	53.0	46.5	51.3	49.3	68.9	56.7	55.0	66.5	44.5	66.2	56.4	42.1	66.3	42.1	68.9	55
Ductile components	18.4	5.1	6.0	1.9	16.9	1.1	21.6	10.9	16.2	14.2	4.4	4.7	9.5	15.8	50.0	5.5	9.2	14.9	20.6	7.6	12.6	6.2	9.4	8.4	5.6	13.6	1.1	50.0	12
Diagenetic components	21.7	43.5	42.9	41.9	3.5	13.3	10.3	3.6	2.7	9.3	51.2	37.1	37.6	10.3	2.5	27.3	28.8	1.3	1.7	21.4	7.8	44.1	18.3	23.4	44.9	2.5	1.3	51.2	21
Total cements	20.3	43.0	41.0	41.9	3.5	12.5	8.9	3.6	2.7	9.3	51.2	37.1	37.1	10.3	2.5	27.3	28.8	1.3	1.7	21.4	7.4	43.6	17.8	23.4	44.9	2.5	1.3	51.2	21
Total carbonate cement	19.8	42.5	39.6	41.0	2.6	9.2	8.9	1.2	–	8.4	50.7	36.6	37.1	8.3	2.5	26.5	27.5	0.9	–	19.3	–	42.7	9.9	21.6	44.9	–	–	50.7	19
Total replacements	1.4	0.5	1.9	0.0	–	0.7	1.4	–	–	–	–	–	0.5	0.0	–	–	–	–	–	–	0.4	0.5	0.5	–	–	–	–	1.9	0.3
Porosity	7.8	6.5	7.8	4.8	13.4	26.2	6.1	19.4	22.8	11.5	1.5	13.8	4.8	20.9	1.0	16.0	12.7	14.9	21.0	16.0	13.0	5.2	6.1	11.9	7.4	17.7	1.0	26.2	12
Primary	6.9	5.1	5.5	0.5	10.8	19.9	5.2	16.9	18.5	11.1	1.0	11.2	3.8	18.6	1.0	13.9	12.2	14.9	16.5	11.8	10.9	4.7	4.7	10.1	3.2	16.9	0.5	19.9	10
Secondary	0.9	1.4	2.3	4.3	2.6	6.3	0.9	2.4	4.2	0.4	0.5	2.6	1.0	2.4	–	2.1	0.4	–	4.5	4.2	2.2	0.5	1.4	1.8	4.2	0.8	–	6.3	2.1
Intergranular volume	36	50	49	42	21	32	26	22	24	26	53	50	46	33		43	44	23	23	36	22	51	26	36	53	23	20.8	52.8	36
Detrital clay + illite	8.8	3.7	2.8	1.0	15.2	1.5	19.7	8.9	11.2	6.2	3.9	3.0	8.6	12.3	49.5	4.2	7.9	9.8	5.8	3.4	11.3	5.2	11.3	6.2	4.6	13.2	1.0	49.5	9
Mini-permeametry data																													
Permeability 1 (10^{-15} m^2)	30	334	1437			2027	1.3	342	695	5464		1484		2087		319	475	551	3944	2788	308	427	21	275	342	92			
Permeability 2 (10^{-15} m^2)		1996										2501		273		1148								363	362				
Permeability 3 (10^{-15} m^2)		3242																											

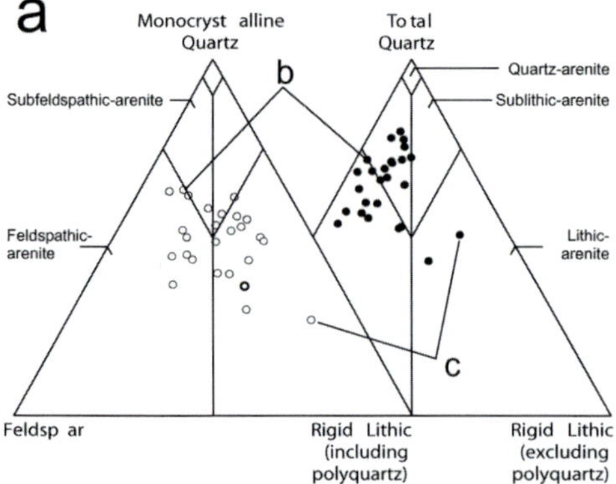

Fig. 8. (**a**) Sandstone classification maturity plots for sandstone samples from the Birmingham University Great Hall Borehole array. Photomicrographs (**b**) and (**c**) show the variation in detrital mineralogy observed between finer grained (more feldspathic/less lithic) and coarser grained sandstones (less feldspathic/more lithic) (b, BH-2: 20.00–20.05 mbgl; c, BH-1: 43.62–43.66 mbgl).

of polycrystalline quartz, and when this is considered with total quartz the sandstones classify a subfeldspathic and feldspathic arenites (chemical maturity plot; Fig. 8a). Other lithic material is dominated by intra-formational clasts of dolocrete and dolomitized mudstone fragments, reworked sandstone/meta-sedimentary material and other grains of igneous/metamorphic material (predominantly quartz–mica schist).

Detrital ductile material is variably abundant, and comprises reworked fragments of mudstone and siltstone, detrital clay and micas. Mudstone and siltstone clasts are typically more abundant within the coarser grained sediments. They are similar in character to *in situ* mudstones and siltstones seen in the formation, and are therefore interpreted as reworked, intraformational material (abandoned channel and palaeosol facies). Detrital clay is more abundant in the finer grained sandstones, where it fills intergranular porosity, and is locally concentrated with mica in specific ductile-rich laminae. Minor amounts of haematized pellicle clay and/or infiltrated clay forms grain coatings and meniscus-style pore bridges (Walker 1976) in some of the medium- and coarse-grained sandstones. SEM and thin-section analysis indicate that these clays are only lightly neomorphosed, with the development of incipient webby fabrics and rare wispy outgrowths (Fig. 9). XRD analysis

Fig. 9. Paragenetic sequence and representative photomicrographs showing the paragenetic relationships within the sandstones. (**a**) Relative timing of mineral precipitation and dissolution. (**b**) Photomicrograph of early dolomite cement, which preserves a high intergranular volume with floating detrital grains and rounded secondary pores (PPL; BH-2: 25.36–25.32 mbgl). (**c**) Early dolomite cement showing some corrosion, which is cut by a thin fracture (arrowed) containing later calcite (PPL; BH-2: 25.10–25.18 mbgl). (**d**) Corroded authigenic dolomite between loosely packed/floating detrital quartz grains with well-developed Fe-oxide pellicles (PPL; BH-2: 35.07–35.15 mbgl). (**e**) Incipiently developed quartz overgrowth partially inhibited by and locally engulfing webby, neomorphosed detrital clay (SEM; BH-1: 37.70–37.77 mbgl). (**f**) Late, weakly ferroan and manganoan calcite (purple (dark) stain) engulfs thin K-feldspar and quartz overgrowths, and is overlain by a later generation of pure (pink (light) stained) calcite (PPL; BH-3: 31.05–31.15 mbgl). (**g**) Fracture-lining, late calcite. This image shows early red-orange luminescent dolomite cement, with later orange luminescent overgrowths (arrowed; 1). Two generations of calcite are evident: a bright luminescent, strongly zoned generation (arrowed; 2), overlain by a non-luminescent generation (arrowed, 3; BH-1: 18.73–18.75 mbgl).

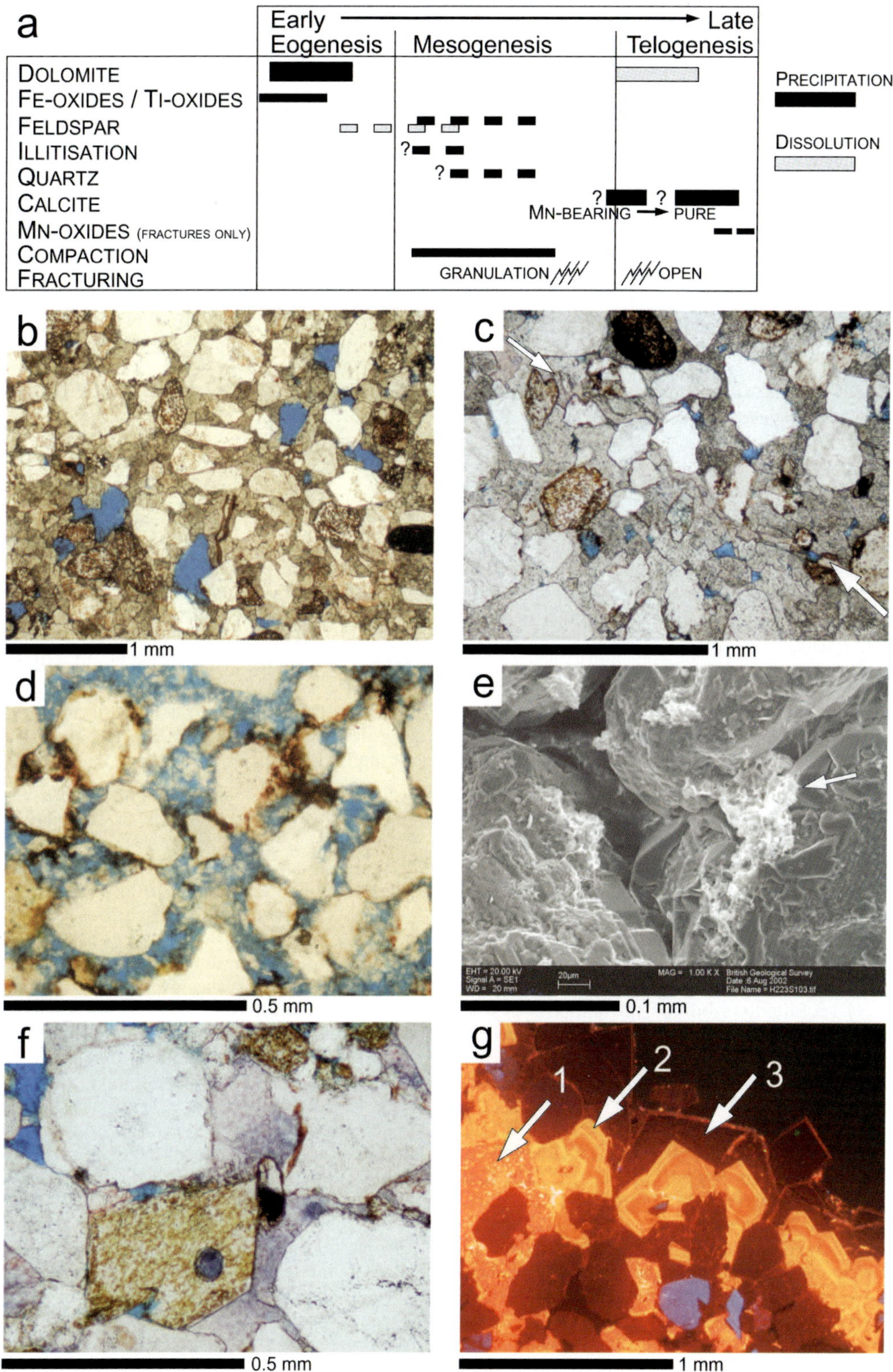
a
Early
Eogenesis
Mesogenesis
Late
Telogenesis
DOLOMITE
FE-OXIDES / TI-OXIDES
FELDSPAR
ILLITISATION
QUARTZ
CALCITE
MN-OXIDES (FRACTURES ONLY)
COMPACTION
FRACTURING
MN-BEARING → PURE
GRANULATION
OPEN
PRECIPITATION
DISSOLUTION
b
1 mm
c
1 mm
d
0.5 mm
e
EHT = 20.00 kV
Signal A = SE1
WD = 20 mm
20µm
MAG = 1.00 K X
Date :6 Aug 2002
British Geological Survey
File Name = H223S103.tif
0.1 mm
f
0.5 mm
g
1
2
3
1 mm

Table 5. *Clay-fraction (<2 μm) XRD data from selected samples from the Birmingham University Great Hall Borehole array*

Sample depth (top–base (mbgl):	Sample identifier	Illite	Chlorite	Kaolinite	Smectite	Quartz	Feldspar	Goethite
Borehole-1								
12.79-12.86	MPLH215	mj	mi	?tr	–	–	?tr	–
29.53-29.63	MPLH221	mj	mi	mi	–	tr	–	–
34.48-34.53	MPLH222	mj	mi	?tr	?mi	–	–	–
Borehole-2								
13.72-13.80	MPLH227	mj	mi	mi	–	tr	–	?tr
37.18-37.27	MPLH233	mj	mi	?tr	?mi	tr	–	–
43.60-43.68	MPLH235	mj	mi	–	?tr	–	–	–
Borehole-3								
21.50-21.59	MPLH251	mj	mi	mi	?mi	tr	–	–

mj, major component; mi, minor component; tr, trace component.

suggests that the majority of the clay within the <2 μm size fraction is illitic (Table 5). Mica preferentially occurs within the finer grained sandstones, where it tends to occur within ductile clast-rich laminae.

Diagenetic overprinting and fracture petrography

A summary of the paragenetic scheme interpreted from petrographical observations is given in Figure 9a. A range of early (eodiagenetic), burial (mesodiagenetic), tectonic (fracture-related) and, more recent, meteoric (telodiagenetic) effects are discernible, which are described below.

Syndepositional/early diagenesis (eodiagenesis)

Syndepositional and early diagenetic effects have profoundly modified the mineralogy and pore systems of the formation. As described above, the sandstones contain small volumes of detrital clay pellicles and evidence for early clay infiltration. Haematization of these imparts the characteristic orange-red colouration. Early, near-surface diagenetic oxidation (Walker 1976; Turner 1980) is supported by the fact that haematized clays are engulfed by the early dolomite cements (Fig. 9c, d). In addition to haematization, rare examples of euhedral Ti-oxide crystals are present within secondary pores, supporting early dissolution of unstable ferro-magnesian minerals (Ixer *et al.* 1979). In bleached or green-coloured sandstone intervals (typically less than 10 cm thick, but rarely up to 1 m thick), any haematite associated with pellicle clays appears to have been reduced and haematite is restricted to isolated patches (100 μm diameter) of pore-filling cement.

Dolomite is the most abundant authigenic component in the sandstones. It is typically more abundant in coarser grained material, but is present in all the samples analysed (abundance range 1.0–49.5%; mean 24%). Within the sandstones, three dolomite habits are observed (in addition to the dolomite in the palaeosols described above).

- Dolomite occurs predominantly in coarser grained horizons (typically in channel-lag facies) as bedding/lamination parallel bands, of 0.5–10 cm thickness. Some higher angle cemented bands are also observed. This type of dolomite occurs as coarse, interlocking, poikilotopic crystals, which completely occlude the primary pore system and commonly preserve high inter-granular volumes (IGVs) (up to 40–50%; Fig. 9b, c). Within certain parts of the dolomite-cemented bands, even higher IGVs are present, suggesting that dolomite may locally be displacive. These pervasively dolomite-cemented intervals are readily differentiated from dolocretes using the criteria described earlier.
- Outside of the pervasively cemented horizons dolomite also occurs as poikilotopic nodules (up to millimetre-scale) and as isolated microrhombic and rhombic crystals (typically 100–300 μm diameter), which occlude primary pore spaces. This dolomite has similar optical character to the dolomite described above and also

preserves higher IGVs than occur in the surrounding uncemented sandstones.

- Finally, dolomite also forms irregular mudclast-replacive crystals, with irregular outgrowths into adjacent porosity. These crystals typically have regular, rhombic cores, but more irregular feathered or lobate, less well-defined crystal margins.

CL analysis reveals three generations of dolomite, each of which occurs in all the habits described above. The earliest generation is yellow-orange luminescent with variably euhedral crystal outlines. This is overgrown, typically with a diffuse contact, by a later generation of orange-red luminescent dolomite (Fig. 9g); these generations have comparable luminescence characteristics to the dolomite seen within the palaeosols described above. A final generation of yellow-orange luminescent dolomite is developed as thin (10 μm) euhedral overgrowths that tend to be present overlying corroded earlier dolomite. This indicates that this latest dolomite generation was either more resistant to corrosion, or developed following corrosion of earlier dolomite.

Formation of dolomite prior to significant grain packing/compaction is indicated by the high IGVs. Further support for an early diagenetic origin is the occurrence of dolomite of similar character to dolocretes elsewhere in the formation. SEM-EDXA, coupled with staining and CL observations, indicate that the dolomite is non-ferroan and weakly manganoan, suggesting precipitation from fresh, oxidizing groundwaters. These characteristics suggest therefore that the dolomite is of vadose or phreatic origin.

Burial diagenesis (mesodiagenesis)

Authigenic quartz and feldspar are poorly developed, and form discontinuous, incipient euhedral overgrowths (Fig. 9e, f). Quartz and feldspar overgrowths do not appear to be engulfed by dolomite cement, indicating that they developed post-dolomitization, but due to their low abundances it has proven impossible to time quartz and feldspar authigenesis relative to each other. In addition to feldspar authigenesis, a degree of feldspar corrosion has also occurred, and, as noted above, plagioclase appears to be more extensively corroded than K-feldspar. Corrosion ranges from partial to complete removal of grains (total secondary porosity 0.0–6.3%, mean 2%), and consequently the porosity generated by this corrosion is variably well connected. Where complete dissolution has occurred, oversized pores are generated that may allow enhanced flow; where dissolution is incomplete the generated secondary pores are less well connected to the primary pore system and, hence, are likely to contribute to a lesser degree to flow. The presence of grain-sized, isolated pores in dolomite indicates that this corrosion also continued following dolomitization (Fig. 9b).

Grain-rimming clays locally have developed minor webby and wispy outgrowths, which have compositions suggestive of illite. Illite more rarely replaces feldspar and other lithic grains, especially the mica components of quartz–mica schist fragments. Illite inhibits and is engulfed by quartz cement suggesting neomorphism prior to significant quartz cementation (Fig. 9e), possibly associated with corrosion of feldspar and other unstable detrital components, during relatively early burial. Authigenic kaolinite is absent typically, but is present rarely both replacing grains and filling primary intergranular pores.

Tectonization

The precise timing of the development of granulation seams and other fractures within the sandstones is poorly defined. However, granulated dolomite cement within some fractures indicates that granulation post-dates at least the dolomite. Petrographic analysis reveals two different types of granulation seam in the sandstones: those that are characterized by the presence of comminuted detrital material and clay; and those characterized by the presence of crushed and/or corroded dolomite cement.

- Granulation seams containing comminuted detrital material and clays are of similar petrographical character to granulations described elsewhere in the literature, with a reduction in grain size caused by fracturing of framework grains, and a concomitant reduction in pore volume, pore size and permeability (e.g. Pitman 1981; Underhill & Woodcock 1987; Edwards *et al.* 1993).
- Those associated with dolomite cement are more unusual. A montage of photomicrographs through such a feature is shown in Figure 10. The granulation seam is bounded by a variably continuous zone, up to 1 cm wide, within which dolomite cement is well developed to pervasive and preserves the high IGVs seen in unfractured, dolomite-cemented intervals. Towards the outer margins of this zone the dolomite is increasingly corroded, passing from lightly corroded dolomite, through relict skeletal

and irregular dolomite crystals, to lightly cemented sandstone. Within the dolomite-cemented band there are a number of discrete, <1 mm-thick zones, within which dolomite is crushed into approximately 10 μm-diameter fragments (Figs 10 & 11). The reduction in sandstone grain size and closer packing due to comminution of the framework grain assemblage is less well developed than in a conventional granulation seam. These features are all consistent with the development of the granulation seam within a dolomite-cemented interval, with dolomite leaching during or after the granulation developed. This leaching removed dolomite cement from the surrounding sandstone, but left a band of dolomite cement in the direct vicinity of the actual granulation seam, possibly due to the lower permeability of the granulations. The implication of this observation is that dolomite cement within the formation was at one time more widespread than is the case today. Whilst the former extent of dolomite remains unproved, similar leaching of earlier-formed cements has been reported in other Sherwood Sandstone Group sandstones (Bath *et al.* 1987; Strong 1993; Strong *et al.* 1994), although in those examples the phase dissolved was early anhydrite cement, which is not seen here. We did not observe any anhydrite or gypsum in our samples; however, the possibility that anhydrite cement was present in these sandstones, and has since been removed, cannot be discounted. Indeed, Jackson & Lloyd (1983) suggest that the cause of high sulphate in the groundwaters pumped from wells in the confined aquifer SE of the Birmingham Fault is the presence of small amounts of mineral sulphate in the Bromsgrove Sandstone Formation.

Uplift and meteoric diagenesis (telodiagenesis)

Corroded dolomite, quartz overgrowths and K-feldspar overgrowths are all engulfed by calcite, which occurs throughout as a minor (typically <5%) pore-filling, poikilotopic cement, typically nucleated on corroded dolomite (Fig. 9f). IGVs

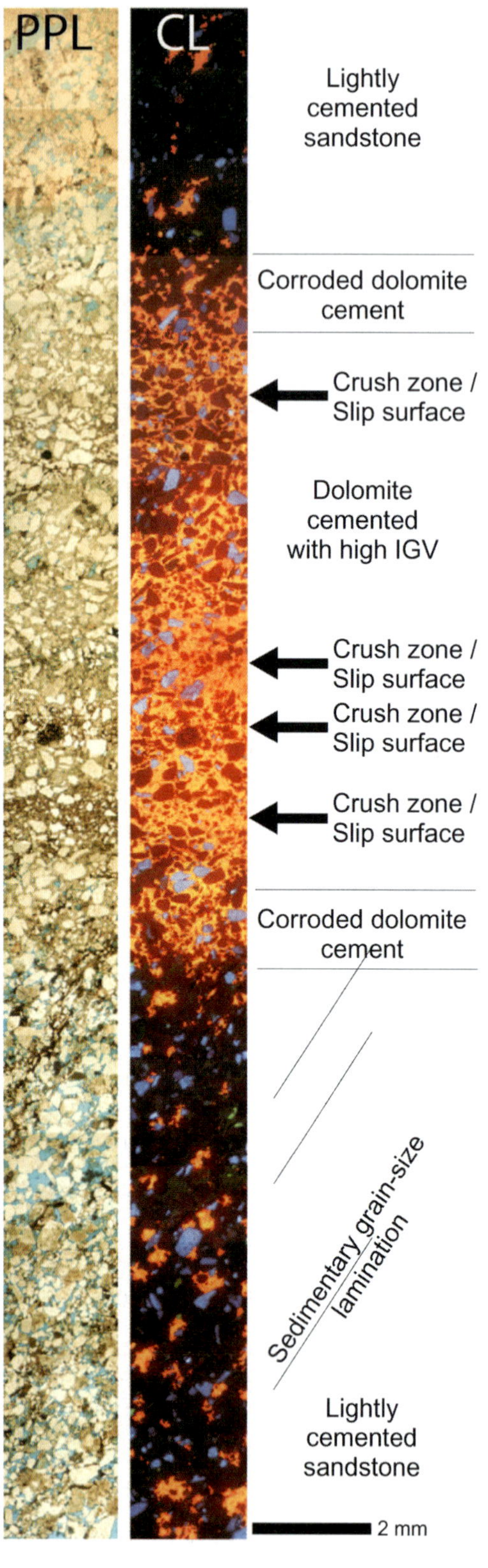

Fig. 10. Montage of photomicrographs showing the mineralogical and textural variability that occurs within a granulation seam through the dolomite-cemented sandstone shown in Figure 7b (BH-3: 31.05–31.15 mbgl).

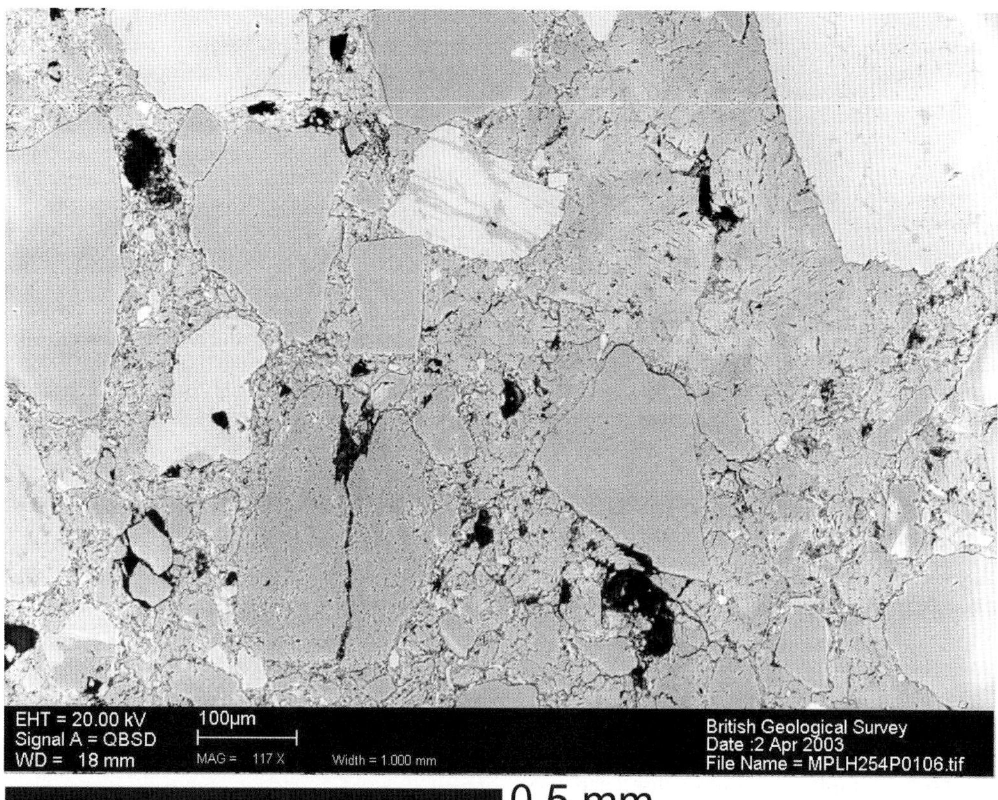

Fig. 11. BSEM image showing crushed dolomite cement from within one of the crush zones shown in Figure 10.

within calcite nodules appear to be lower than in the early dolomite cement nodules, indicating formation following grain packing/compaction. Calcite is also prominently developed within high-angle fractures, which are most common in intervals retaining significant or pervasive early dolomite cement. Dolomite at the margins of these fractures is commonly corroded, and vuggy pores up to 2 cm in diameter filled or coated by calcite may be developed adjacent to the fracture (Figs 7e & 9g).

Two calcite generations are present. The earlier is typically brightly luminescent under CL, with multiple internal growth zones of varying luminescence (Fig. 9g), which, coupled with staining and EDX observations, indicate manganoan and weakly ferroan compositions. This calcite is overlain by a generation of non-luminescent calcite, which is Mn- and Fe-free. There is currently a considerable interest in diagenetically 'late' calcite as a potential palaeo-hydrogeological tool, and calcite of similar character to that seen here has been reported

previously lining fractures at a number of sites including: Sellafield (Cumbria, UK: Milodowski *et al.* 1998; Bath *et al.* 2000), Tono (Japan: Iwatsuki *et al.* 2002), Äspo (Sweden: Pedersen *et al.* 1997) and in Derbyshire–Leicestershire (UK: Bouch *et al.* 2004). At these sites, relationships between calcite chemistry, calcite morphology and groundwater chemistry have been established, with changes in composition and morphology reflecting changes in the salinity and/or redox potential (E_h) of the groundwater. Minor amounts of fine, irregular crystals and aggregates of Mn-oxides coat the surfaces of calcite crystals lining the open fracture described above. Similar Mn-oxides have been described by Milodowski *et al.* (1998) in the Sellafield region, where they are considered to be the insoluble residues of oxidative, near-surface (telodiagenetic; possibly Tertiary–Recent) alteration and dissolution of ferro-manganoan carbonates. A similar origin may be envisaged for the Mn-oxides present in the Birmingham University boreholes.

Discussion

As well as providing a relatively rare opportunity to study the Wildmoor Sandstone Formation in detail, these boreholes also offer potential insights on the heterogeneity of this and similar formations and into sand-production problems encountered in some water abstraction boreholes.

Impact of heterogeneities on fluid flow and contaminant transport

These boreholes demonstrate a number of scales of heterogeneity within the Wildmoor Sandstone Formation that need to be considered in modelling of groundwater flow or contaminant transport. These are summarized in Figure 12, and discussed below in order of decreasing scale.

At the largest scale, the features that are most likely to influence both groundwater flow and contaminant transport are the abandoned channel and the palaeosol facies. These facies both have intrinsically low permeabilities due to their high clay content and/or their dolomite-dominated compositions. Data on the lateral extent of these facies in braided river systems are scarce (Miall 1977). However, on deposition, both facies would have formed subhorizontal zones of up to 1 m in thickness. The fact that they can be correlated between boreholes indicates lateral extents of tens of metres. However, even within the close spacing of the boreholes, correlation is not always obvious due to incision and erosion by later fluvial sandstones. These features are therefore likely to act as baffles to vertical fluid flow on the scale of tens of metres, and as such they may locally support perched water tables (Fig. 12a), such as those observed by Wills (1976) in exposures of the Wildmoor Sandstone Formation just to the south of Birmingham.

At a finer scale the distribution of early dolomite cement will also influence fluid flow. The sandstones are typically lightly cemented and overall they have good porosities, with modal analysis of uncemented samples indicating primary intergranular macropore volumes of up to approximately 20%. However, early dolomite and late calcite cements locally severely degrade primary interparticle macro-pore volume, with a good relationship observed between the amount of primary macroporosity and the amount of dolomite (±calcite) cement present (Fig. 12d). Core observations indicate that dolomite and calcite cements are significant within approximately 2–5% of the cored intervals, and that they are most prominently developed in the coarser grained sandstones of channel-lag deposits, although not all channel lags contain these cements. Consequently, whilst these cemented bands are relatively thin (typically 0.5–10 cm), their occurrence in channel-lag deposits suggests that they may be laterally continuous over distances comparable to the extent of the bedforms (Fig. 12b). Outcrop evidence from exposure of Sherwood Sandstone Group fluvial sandstones in the Sellafield area (Cumbria, UK: Jones & Ambrose 1994) indicates that channel bodies are typically tabular or lenticular bodies of up to 5 m thick and may be traced for tens or hundreds of metres. As such these cemented bands are likely to be readily bypassed by horizontal fluid flow, but they may act as baffles to vertical fluid percolation.

Minor numbers of granulation seams occur at the decimetre scale. Mini-permeametry of these was not always possible due to their low permeabilities. In the few examples where direct measurement was possible, permeabilities were one–three orders of magnitude lower than in the host sandstones (Table 6). Whilst these values are consistent with the results of studies by Pitman (1981), Ballard (2000), Bunch (2001) and Taylor & Pollard (2000), more detailed analysis of would be required to quantify their effect. However, these features are likely to act as localised baffles to lateral fluid flow (Fig. 12b).

In detail, variations in sandstone porosity and permeability in uncemented 'matrix' sandstones are caused by variations in primary texture and degree of compaction. Permeability is controlled by grain size, with coarser grained sandstones having larger, better connected pores and hence higher permeabilities than finer grained ones (Fig. 12c). Porosity and permeability reduction due to compaction is probably comparable in all the uncemented sandstones, although local variations may be controlled by differences in the abundance of deformable ductile material (clay, micas, mudclasts). IGVs

---

Fig. 12. Scales of heterogeneity (**a**) and (**b**), and detailed controls on permeability and porosity (**c**) and (**d**) that are likely to influence fluid flow and contaminant transport within the Wildmoor Sandstone Formation as proven in the Birmingham University Great Hall Borehole array. Features in black represent likely barriers or baffles to fluid flow.

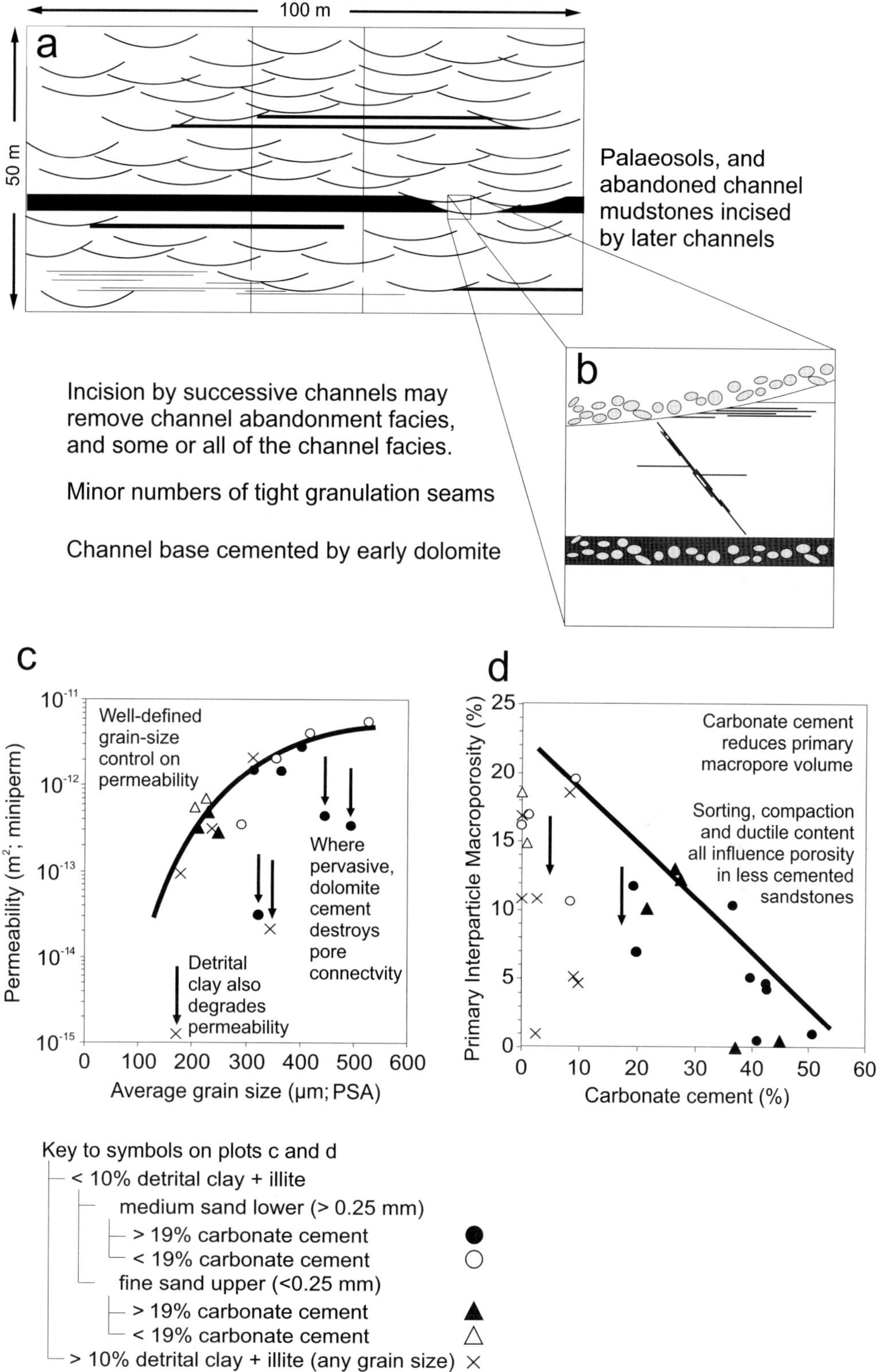
100 m
a
50 m
Palaeosols, and abandoned channel mudstones incised by later channels
Incision by successive channels may remove channel abandonment facies, and some or all of the channel facies.
Minor numbers of tight granulation seams
Channel base cemented by early dolomite
b
c
Well-defined grain-size control on permeability
10^-11
10^-12
10^-13
10^-14
10^-15
Permeability (m²; miniperm)
Where pervasive, dolomite cement destroys pore connectvity
Detrital clay also degrades permeability
0   100   200   300   400   500   600
Average grain size (µm; PSA)
d
25
20
15
10
5
0
Primary Interparticle Macroporosity (%)
Carbonate cement reduces primary macropore volume
Sorting, compaction and ductile content all influence porosity in less cemented sandstones
0   10   20   30   40   50   60
Carbonate cement (%)
Key to symbols on plots c and d
< 10% detrital clay + illite
medium sand lower (> 0.25 mm)
> 19% carbonate cement
< 19% carbonate cement
fine sand upper (<0.25 mm)
> 19% carbonate cement
< 19% carbonate cement
> 10% detrital clay + illite (any grain size)

Table 6. *Mini-permeametry data for granulation seams and adjacent sandstones. These measurements were made on slabbed sections of core that cut across granulation seams. The granulation seam permeability is measured on the actual granulation seam. The sandstone permeability(s) were measured on the unfractured sandstone directly adjacent to the granulation*

Borehole	Depth (mbgl)	Lithology	Granulation seam permeability (10^{-15} m^2)	Sandstone permeability (10^{-15} m^2)
BH-3	25.00	Sandstone	0.5	$437,254$ ($n = 2$)
BH-3	31.10	Ripple cross-laminated sandstone	1.9	$163,322,794$ ($n = 3$)
BH-3	31.15	Ripple cross-laminated sandstone	7.3	$328,205$ ($n = 2$)
BH-3	33.60	Sandstone	14.2	764 ($n = 1$)

in uncemented sandstones are typically 20–30%, indicating significant reduction of intergranular pore volume relative to depositional porosities that, based on the IGVs of dolomite cemented sandstones, may have been up to 40–50% (consistent with porosity values of uncompacted sandstones reported by Beard & Weyl 1973). In spite of this compaction, present-day porosities are still able to support moderate volumes of fluid flow with permeabilities of 10^{-13}–10^{-12} m^2. The pore system is locally augmented by minor amounts (up to 6.3%) of secondary porosity, which may slightly improve permeability. Minor cements such as quartz and feldspar have only limited impact on the primary pore system, as they are only developed as incipient overgrowths. Similarly, authigenic clays are not sufficiently well developed to significantly impact on permeability.

Sand production in boreholes

Some water abstraction boreholes in the Sherwood Sandstone Group are known to produce running sand, which has been known to be associated with metre-scale collapse features at the surface (Peacock & Seymour 1980; Campbell & Nelson 1988; Vines 1988; Fletcher 1994). However, much of the information related to these features is anecdotal, and the source of the pumped sand is not known. The main hypothesis is that it flows from unconsolidated sand, possibly including near-surface Quaternary sand and gravel deposits, along open fractures that are intersected by the abstraction boreholes.

An alternative source of sand production in boreholes is through borehole breakouts, which reflect borehole instability. These are relatively common features of water-extraction or oil-production boreholes in weakly consolidated and/or poorly cemented sandstones. Sand production is considered to be controlled by a number of factors, including the stress regime of the borehole (mechanical failure), hydrodynamics (hydrodynamic erosion) and the lithology's mechanical strength. Mechanical failure has been simulated in samples of high-porosity sandstone by Haimson & Kovacich (2003), who reported the development of open fracture-like breakouts, oriented perpendicular to the maximum principal stress direction. At the tips of breakouts, narrow zones of stress concentration were developed, within which disaggregation, grain fracturing and compaction occurred in the absence of any shear displacement (resembling 'compaction bands' described by Mollema & Antonellini 1996). These zones advanced ahead of the breakout, with disaggregated material flushed out by the drilling fluid leading to propagation of the features. In a study of perforation cavity stability Tronvoll & Fjær (1994), whose samples included one of Wildmoor Sandstone (from an unspecified outcrop), reported that, whilst fluid flow did not appear to significantly influence the stress at which initial rock failure occurred, fluid flow could influence the nature and direction of failure propagation, with plastification in a zone around the perforation cavity and the development of washed-out channels or fractures parallel to the direction of fluid flow. Furthermore, minor sand production occurred related to the initial failure of the samples, but, following this initial failure, no significant sand production was observed until the test specimen failed globally. Once rock failure has occurred, particles may be removed from the site of the failure by internal and surface erosion due to hydrodynamic instabilities (Skjærstein *et al.* 1997; Unander *et al.* 1997). In ultra weak rocks, the role of fluid flow and hydrodynamic erosion become more significant as erosion may occur prior to rock failure (Tronvoll *et al.* 1997).

Although Tronvoll & Fjær (1994) tested the effect of anisotropy due to lamination in one of their samples, the studies described above have largely concentrated on mechanical and hydrodynamic effects in relatively homogeneous fine- to medium-grained sandstone samples. The Birmingham University boreholes provide a location at which to investigate a possible facies dependence on production of particulate material from pumped boreholes. In all three boreholes approximately 10% of the cored interval was composed of ripple cross-laminated sandstone with mud drapes, which were extensively washed out during drilling/coring, leaving the core with a 'comb-like' profile (Fig. 3d and Table 3). To investigate the possibility that the ripple cross-laminated intervals may be more susceptible to particulate production during water abstraction, packer tests were conducted. Enhanced particle yields were not seen from these intervals, although the possibility remains that more comprehensive failure may occur at higher pumping rates than those used in the tests.

Conclusions

Three closely spaced boreholes on the Birmingham University campus provide a rare opportunity to characterize the Wildmoor Sandstone Formation, which is one of the major aquifers in the UK Midlands. The sedimentological characteristics of the formation are consistent with deposition in a fluvial, probable braidplain, environment. Of particular significance is the preservation of distinct dolocretes in all three boreholes. Typically, these facies have relatively low preservation potential, and have not previously been reported in the Wildmoor Sandstone Formation.

Diagenetic overprinting is dominated by early dolomite precipitation, which preserves high intergranular volumes, and is interpreted to be of vadose or phreatic origin. Dolomite-cemented intervals are readily differentiated from dolocretes on the basis of their coarse sedimentary grain size, preferential occurrence within channel-lag facies, non-nodular fabrics and a general absence of fissures. Burial diagenetic effects are relatively limited with only minor volumes of quartz and feldspar cement and some illitization of detrital clay.

Minor numbers of granulation seams and other cemented fractures are orientated approximately parallel to the Birmingham Fault. The preservation of crushed dolomite in the vicinity of low-permeability granulation seams that cut otherwise uncemented sandstones may indicate that dolomite cement was previously more abundant than at the present day, although the extent of corrosion is uncertain. Although further work would be required to confirm this, two generations of diagenetically late calcite that occur in fractures and as a disseminated cement may provide an indicator of changes in groundwater chemistry (in particular redox) at the site since uplift.

A number of scales of heterogeneity are observed within the formation. At the largest scale the palaeosol and mudstone facies both have intrinsically low permeabilities, are up to 1 m thick and may be laterally continuous over tens of metres. A key uncertainty in any model of groundwater and contaminant transport will be the preservation and, hence, the lateral extent of these facies. Where palaeosols/mudstones are poorly preserved, their dissected remnants may act as baffles to vertical fluid migration; however, should these facies be sufficiently well preserved to form continuous layers, then their influence on fluid migration would be more significant. At another borehole array on the Birmingham campus and in the same Formation, matrix permeability of mudstone has been measured at 10^{-6} m day^{-1}, but vertical tracer tests have shown that solute can traverse the mudstone beds, albeit at limited rates (J. H. Tellam pers. comm.). At a finer scale, dolomite cement preferentially occurs within channel-lag deposits and may form additional baffles to vertical flow. Granulation seams and fractures locally form thin barriers to fluid flow; however, their relatively low abundances means that their impact on lateral fluid migration is probably limited. At the finest scale, permeability is controlled by grain size and compaction in uncemented 'matrix' sandstones. Dolomite cement is locally responsible for occlusion of the primary pore system and, where abundant, may significantly reduce permeability.

Finally, the coring process caused significant damage to fine-grained, ripple cross-laminated sandstones, with the washing out of fines from the ripples. To investigate the possibility that this facies may be more susceptible to particulate production pumping tests were conducted. Although enhanced particle yields were not seen, the possibility remains that more comprehensive failure may occur at higher pumping rates than those used in the tests.

This work was conducted as part of the 'Sediment Filled Fractures in Triassic Sandstones – Pathways or Barriers to Contaminant Migration?' project funded by the EPSRC Waste and Pollution Management Programme, grant number GR/M89737 to BGS, Birmingham University and Sheffield University. P. Turner and P. L. Younger both provided constructive reviews which improved the quality of this paper.

This paper is published with the permission of the Executive Director of the British Geological Survey (Natural Environment Research Council).

References

ALI, A.D. & TURNER, P. 1982. Authigenic K-feldspar in the Bromsgrove Sandstone Formation (Triassic) of central England. *Journal of Sedimentary Petrology*, **52**, 187–197.

BALLARD, M. 2000. *The role of granulation seams in pollutant transport in the UK Triassic Sandstone.* Unpublished MSc Project Report, Earth Sciences, University of Birmingham.

BATH, A., MILODOWSKI, A.E., RUOTSALAINEN, P., TULLBORG, E.-L., CORTÉS RUIZ, A. & ARANYOSSY, J.F. 2000. *EUR 19613 – Evidence From Mineralogy and Geochemistry for the Evolution of Groundwater Systems During the Quaternary for Use in Radioactive Waste Repository Safety Assessment (EQUIP Project).* Project Report, Nuclear Science and Technology Series, **2000-XIV**. Luxembourg Office for Official Publications of the European Communities.

BATH, A., MILODOWSKI, A.E. & STRONG, G.E. 1987. Fluid flow and diagenesis in the East Midlands Triassic Sandstone aquifer. *In*: GOFF, J.C. & WILLIAMS, B.P.J. (eds) *Fluid Flow in Sedimentary Basins and Aquifers*. Geological Society, London, Special Publications, **34**, 127–140.

BEARD, D.C. & WEYL, P.K. 1973. Influence of texture on porosity and permeability of unconsolidated sand. *AAPG Bulletin*, **57**, 349–369.

BOUCH, J.E., MILODOWSKI, A.E. & AMBROSE, K. 2004. Dolomitisation, fracturing and pore-systems in Dinantian carbonates from Leicestershire, South Derbyshire and West Cumbria, UK. *In*: BRAITHWAITE, C.J.R., RIZZI, G. & DARKE, G. (eds) *The Geometry and Petrogenesis of Dolomite Hydrocarbon Reservoirs*. Geological Society, London, Special Publications, **235**, 325–348.

BRIDGE, D. McC. & HOUGH, E. 2002. *Geology of the Wolverhampton and Telford District.* Sheet description of the British Geological Survey 1:50 000 series sheet 153 (England & Wales). HMSO, London.

BRITISH GEOLOGICAL SURVEY. 1996. *England and Wales Sheet 168 (Birmingham), Solid and Drift Edition.* HMSO, London.

BUNCH, M. 2001. *Detecting granulation seams in the Triassic Sandstone using resistivity.* Unpublished MSc Project Report, Earth Sciences, University of Birmingham.

BUTLER, A. J. & LEE, J. 1942. Water supply from underground sources in the Birmingham–Gloucester district. Geological Survey Wartime Pamphlet, **32**.

CAMPBELL, J.E. & NELSON, E. 1988. *TV and Geophysical Survey, Manley Common No. 1 Borehole.* Groundwater Services Report, **CHS/88/18**, North West Water Authority.

EASTWOOD, T., WHITEHEAD, T.H. & ROBERTSON, T. 1925. *The Geology of the Country Around Birmingham.* Memoir of the Geological Survey of Great Britain, Sheet 168 (England and Wales).

EDWARDS, H.E., BECKER, A.D. & HOWELL, J.A. 1993. Compartmentalization of an aeolian sandstone by structural heterogeneities: Permo-Triassic Hopeman Sandstone, Moray Firth, Scotland. *In*: NORTH, C.P. & PROSSER, D.J. (eds) *Characterization of Fluvial and Aeolian Reservoirs*. Geological Society, London, Special Publications, **73**, 339–365.

EHRENBERG, S.N. 1995. Measuring sandstone compaction from modal analyses of thin-sections: how to do it and what the results mean. *Journal of Sedimentary Research*, **A65**, 369–379.

FLETCHER S.W. 1994. *Report on the Pumping tests conducted at Nurton in 1985 and 1988.* Internal report. NRA, Severn Trent Region.

HAIMSON, B. & KOVACICH, J. 2003. Borehole instability in high-porosity Berea sandstone and factors affecting dimensions and shape of fracture-like breakouts. *Engineering Geology*, **69**, 219–231.

HOUGH, E. & BARNETT, A.J. 1998. *Geology of the Beckbury and Worfield Area.* British Geological Survey Technical Report, **WA/97/82**.

HOUSEKNECHT, D.W. 1987. Assessing the relative importance of compaction processes and cementation to reduction of porosity in sandstones. *AAPG Bulletin*, **71**, 633–642.

HULL, E. 1869. *The Triassic and Permian Rocks of the Midland Counties of England.* Memoir of the Geological Survey of Great Britain.

IWATSUKI, T., SATAKE, H., METCALFE, R., YOSHIDA, H. & HAMA, K. 2002. Isotopic and morphological features of fracture calcite from granitic rocks of the Tono area, Japan: a promising palaeohydrogeological tool. *Applied Geochemistry*, **17**, 1241–1257.

IXER, R.A., TURNER, P. & WAUGH, B. 1979. Authigenic iron and titanium oxides in Triassic red beds: (St. Bees Sandstones), Cumbria, Northern England. *Journal of Geology*, **14**, 179–192.

JACKSON, D. & LLOYD, J.W. 1983. Groundwater chemistry of the Birmingham Triassic Sandstone aquifer and its relationship to structure. *Quarterly Journal of Engineering Geology*, **16**, 135–142.

JONES, N. & AMBROSE, K. 1994. Triassic sandy braidplain and aeolian sedimentation in the Sherwood Sandstone Group of the Sellafield area, west Cumbria. *Proceedings of the Yorkshire Geological Society*, **50**, 61–76.

MADER, D. 1992. *Evolution of Palaeoecology and Palaeoenvironment of Permian and Triassic Fluvial Basins in Europe, Volume 1. Western and Eastern Europe.* Gustav Fischer, Stuttgart.

MIALL, A.D. 1977. A review of the braided-river depositional environment. *Earth Science Reviews*, **13**, 1–62.

MILODOWSKI, A.E., GILLESPIE, M.R., NADEN, J., PEARCE, J.M,. SHEPHERD, T.J., FORTEY, N.J. & METCALFE, R. 1998. The petrology and paragenesis of fracture mineralisation in the Sellafield area, west Cumbria. *Proceedings of the Yorkshire Geological Society*, **52**, 215–242.

MOLLEMA, P.N. & ANTONELLINI, M.A. 1996. Compaction bands: a structural analog for antimode I cracks in aeolian sandstone. *Tectonophysics*, **267**, 209–228.

NEWELL, A. 2006. Calcite as a source of heterogeneity in Triassic fluvial sandstone aquifers (Otter Sandstone Formation, SW England). *In*: TELLAM, J.H. & BARKER, R.D. (eds) *Fluid Flow and Solute Movement in Sandstones: The Onshore UK Permo-Triassic Red Bed Sequence*. Geological Society, London, Special Publications, **263**, 119–127.

PEACOCK, A.J. & SEYMOUR, K.J. 1980. *Investigation into Excessive Sediment Loads Being Pumped from Lower House Borehole Mottram St Andrew*. Report, **66**. North West Water Authority.

PEARCE, J.M., WILLIAMS, G.W., WEALTHALL, G.P., HOUGH, E., TELLAM, J., HERBERT, A. & LERNER, D.N. 2001. Sediment-filled fractures in Triassic sandstones: pathways or barriers to contaminant migration? *In: Fractured Rock 2001, Toronto, Abstracts*, **19**.

PEDERSEN, K., EKENDAHL, S., TULLBORG, E.-L., FURNES, H., THORSETH, I. & TUMYR, O. 1997. Evidence of ancient life at 207 m depth in a granitic aquifer. *Geology*, **25**, 827–830.

PIMENTAL, N.L. 2002. Pedogenic and early diagenetic processes in Palaeogene alluvial fan and lacustrine deposits from the Sado Basin (S Portugal). *Sedimentary Geology*, **148**, 123–138.

PIMENTAL, N.L., WRIGHT, V.P. & AZEVEDO, T.M. 1996. Distinguishing early groundwater alteration effects from pedogenesis in ancient alluvial basins: examples from the Palaeogene of southern Portugal. *Sedimentary Geology*, **105**, 1–10.

PITMAN, E.D. 1981. Effect of fault-related granulation on porosity and permeability of quartz sandstones, Simpson Group (Ordovician), Oklahoma. *AAPG Bulletin*, **65**, 2381–2387.

POWELL, J.H., GLOVER, B.W. & WATERS, C.N. 2000. *Geology of the Birmingham Area*. Memoir of the Geological Survey of Great Britain, Sheet 168 (England and Wales).

PURVIS, K. & WRIGHT, V.P. 1991. Calcretes related to phreatophtic vegetation from the Middle Triassic Otter Sandstone of South West England. *Sedimentology*, **38**, 539–551.

SCHMID, S., WORDEN, R. & FISHER, Q.J. 2003. The origin and regional distribution of dolomite cement in sandstones from a Triassic dry river system, Corrib Field, offshore west of Ireland. *Journal of Geochemical Exploration*, **78–79**, 475–479.

SKJÆRSTEIN, A., STAVROPOULOU, M., VARDOULAKIS, I. & TRONVOLL, J. 1997. Hydrodynamic erosion: a potential mechanism of sand production in weak sandstones. *International Journal of Rock Mechanics and Mining Science*, **34**, 463.

SPÖTL, C. & WRIGHT, V.P. 1992. Groundwater dolocretes from the Upper Triassic of the Paris Basin, France: a case study of an arid, continental diagenetic facies. *Sedimentology*, **39**, 1119–1136.

STRONG, G.E. 1993. Diagenesis of Triassic Sherwood Sandstone Group rocks, Preston Lancashire: a possible evaporitic cement precursor to secondary porosity. *In*: NORTH, C.P. & PROSSER, D.J. (eds) *Characterization of Fluvial and Aeolian Reservoirs*. Geological Society, London, Special Publications, **73**, 279–289.

STRONG, G.E. & MILODOWSKI, A.E. 1987. Aspects of the diagenesis of the Sherwood Sandstones of the Wessex Basin and their influence on reservoir characteristics. *In*: MARSHALL, J.D. (ed.) *Diagenesis of Sedimentary Sequences*. Geological Society, London, Special Publications, **36**, 325–337.

STRONG, G.E., MILODOWSKI, A.E., PEARCE, J.M., KEMP, S.J., PRIOR, S.V. & MORTON, A.C. 1994. The petrology and diagenesis of Permo-Triassic rocks of the Sellafield area, Cumbria. *Proceedings of the Yorkshire Geological Society*, **50**, 77–89.

TAYLOR, W.L. & POLLARD, D.D. 2000. Estimation of in situ permeability of deformation bands in porous sandstone, Valley of Fire, Nevada. *Water Resources Research*, **36**, 2595–2606.

TRONVOLL, J. & FJÆR, E. 1994. Experimental study of sand production from perforation cavities. *International Journal of Rock Mechanics and Mining Science and Geomechanics Abstracts*, **31**, 393–410.

TRONVOLL, J., SKJÆRSTEIN, A. & PAPAMICHOS, E. 1997. Sand production: mechanical failure or Hydrodynamic erosion? *International Journal of Rock Mechanics and Mining Science*, **34**, 465.

TURNER, P. 1980. *Continental Red Beds*. Developments in Sedimentology, **29**, Elsevier, Asterdam.

UNANDER, T., PAPAMICHOS, E., TRONVOLL, J. & SKJÆRSTEIN, A. 1997. Flow geometry effects on sand production from an oil producing perforation cavity. *International Journal of Rock Mechanics and Mining Science*, **34**, 464.

UNDERHILL, J.R. & WOODCOCK, N.H. 1987. Faulting mechanisms in high-porosity sandstones; New Red Sandstone, Arran, Scotland. *In*: JONES, M.E. & PRESTON, R.M.F. (eds) *Deformation of Sediments and Sedimentary Rocks*. Geological Society, London, Special Publications, **29**, 91–105.

VINES, K.J. 1988 *Groundwater Services Report: Scales Demesne Abstraction Boreholes*. Report, **NLS/88/02**. North West Water Authority.

WALKER, T.R. 1976. Diagenetic origin of continental red beds. *In*: FAULKE, H. (ed.) *The Continental Permian in Central, West, and South Europe*. Reidel, Dordrecht, 240–282.

WARRINGTON, G., AUDLEY-CHARLES, M.G. *ET AL.* 1980. A correlation of Triassic rocks in the British Isles. Geological Society, London, Special Report, **13**.

WAUGH, B. 1978. Authigenic K-feldspar in British Permo-Triassic sandstones. *Journal of the Geological Society, London*, **135**, 41–49.

WEALTHALL, G.P., STEELE, A., BLOOMFIELD, J.P., MOSS, R.H. & LERNER, D.N. 2001. Sediment filled fractures in the Permo-Triassic sandstones of the Cheshire Basin: Observations and implications for pollutant transport. *Journal of Contaminant Hydrogeology*, **50**, 41–51.

WILLS, L.J. 1970. The Triassic succession in the central Midlands in its regional setting. *Journal of the Geological Society, London*, **126**, 387–397.

Wills, L.J. 1976. *The Triassic of Worcestershire and Warwickshire*. Institute of Geological Sciences Report, **76/2**.

Estimation of vertical diffusivity from seasonal fluctuations in groundwater pressures in deep boreholes near Sellafield, NW England

M. J. STREETLY[1,2], J. A. HEATHCOTE[1] & P. J. DEGNAN[3]

[1]*Entec UK Ltd, Canon Court, Abbey Lawn, Abbey Foregate, Shrewsbury SY2 5DE, UK*

[2]*Present address: ESI New Zealand House, 160 Abbey Foregate, Shrewsbury SY2 6FD, UK*
(e-mail: mikestreetly@esinternational.com)

[3]*United Kingdom Nirex Limited, Curie Avenue, Harwell, Didcot,
Oxfordshire OX11 0RH, UK*

Abstract: In 1998, as part of the process to assess the suitability of a site near Sellafield, Cumbria, for a deep geological repository for solid intermediate level and some solid low-level radioactive waste, United Kingdom Nirex Limited (Nirex) commissioned a study into the seasonal fluctuations in groundwater pressure observed in many of the monitoring boreholes in the area. Many of the monitoring zones in the deep boreholes at Sellafield show some response in groundwater pressure to annual variations in recharge. These seasonal fluctuations were quantified in terms of amplitude and lag over two full recharge cycles (1994–1996). The extremely detailed monitoring array installed by Nirex at the potential repository zone gives a unique opportunity to observe in detail the attenuation of the recharge signal, as it is propagated vertically downwards through the Sherwood Sandstone Group and into the underlying basement rocks. Use of an analytical approach to model the data provides constraints on values of the vertical diffusivity of the strata. The values of hydraulic formation parameters derived by this methodology are broadly consistent with results from borehole testing, though somewhat higher. This may in part be due to the large scale associated with the cyclic recharge signal. The attenuation of the seasonal fluctuations in three dimensions (3D) throughout the study area provides information at a scale suitable for use in constraining regional flow models.

In 1998, as part of the process to assess the suitability of a site near Sellafield, Cumbria, for a deep geological repository for solid intermediate level and some solid low-level radioactive waste, United Kingdom Nirex Limited (Nirex) commissioned a study into the seasonal fluctuations in groundwater pressure observed in many of the monitoring boreholes in the area. Nirex had completed 27 deep monitoring boreholes in the area by that time, and Figure 1 shows the location of the 20 boreholes relevant to this study. Each borehole had been completed with discrete sealed monitoring zones in order to monitor groundwater pressures in the deep basement and overlying Carboniferous, Permian and Triassic strata. A number of the boreholes are focused on the 'potential repository zone' (PRZ) (see Fig. 1). Many of the monitoring zones in the deep boreholes near Sellafield show some response in groundwater pressure to seasonal variations in recharge.

The seasonal variation in recharge to the groundwater system at Sellafield forms the source for the lowest frequency signal that can be observed on the timescale of the current monitoring operations. The extremely detailed monitoring array installed by Nirex at the PRZ gives a unique opportunity to observe in detail the attenuation of the recharge signal, as it is propagated vertically downwards through the Sherwood Sandstone Group and into the underlying basement rocks. The attenuation of this signal by the formations at Sellafield is therefore indicative of the hydrogeological properties of the strata at a larger scale than can be measured by borehole tests.

The aim of this paper is to describe a simple approach to quantifying the amplitude and lag of the seasonal fluctuations. The data derived by this approach were used to calibrate an analytical model of the vertical attenuation of the recharge signal and thus derive values of vertical diffusivity of the various strata. These values are compared with those derived from different approaches.

The work described here forms part of a larger study presented in Nirex (1997).

From: BARKER, R. D. & TELLAM, J. H. (eds) 2006. *Fluid Flow and Solute Movement in Sandstones: The Onshore UK Permo-Triassic Red Bed Sequence.* Geological Society, London, Special Publications, **263**, 155–167.
0305–8719/06/$15 © The Geological Society of London 2006.

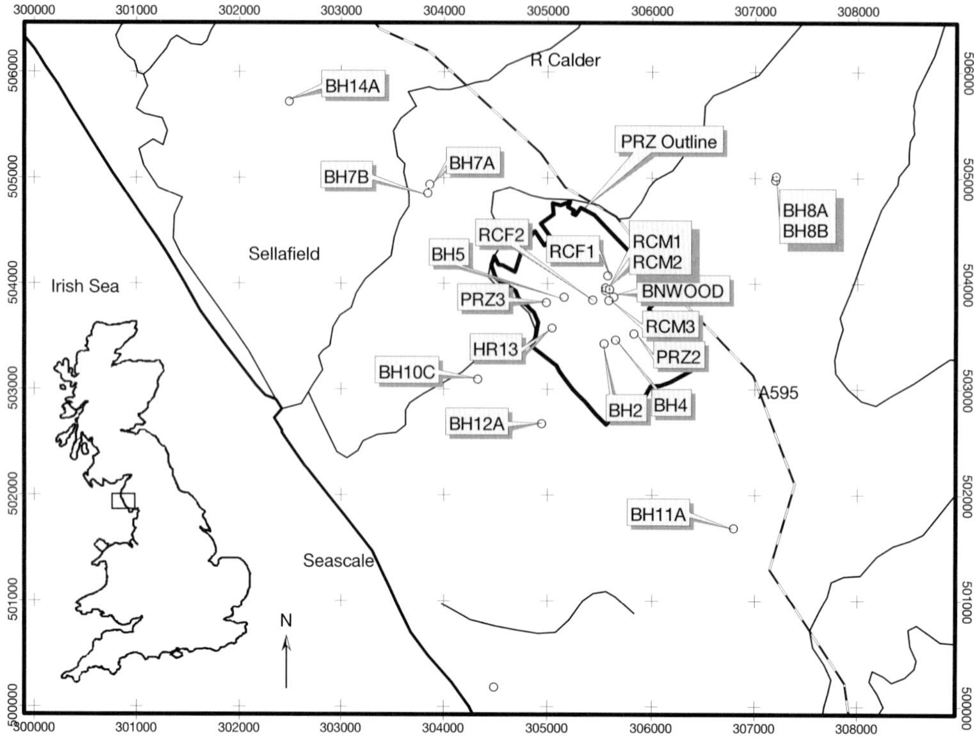

Fig. 1. Location map showing the locations of the deep boreholes used in the study and the location of the potential repository zone (PRZ).

Geology

The geological framework of the Sellafield area and its relationship to the local hydrogeology have been summarized by Michie (1996), and are described in detail in Akhurst *et al.* (1997). The stratigraphic units of relevance to this paper are as follows.

- *Borrowdale Volcanic Group (BVG)*: the Ordovician BVG is the oldest unit instrumented as part of the investigation. The upper part of the BVG sequence largely consists of tuffs and lapilli tuffs of the Longlands Farm Member.
- *Carboniferous Limestone*: the Carboniferous Limestone overlies the Borrowdale Volcanic Group in the western part of the area, but is absent in the PRZ.
- *Brockram*: in the central PRZ, the Permian Brockram overlies the BVG and consists of about 75 m of poorly sorted, poorly bedded breccia. The matrix is generally of low porosity.
- *St Bees Sandstone*: the lowermost formation of the Sherwood Sandstone Group (SSG),

the St Bees Sandstone, is a predominantly fluviatile, fine-grained, well-cemented sandstone that has been proved from near surface (about 100 m above Ordnance Datum (mAOD)) to around −370 mAOD in the central PRZ. Siltstone/claystone beds ascribed to sheet floods increase in frequency towards the base of the unit, and the lowermost 80 m are referred to as the North Head Member (NHM) (Nirex 1996*a*). About 15–20% of the NHM is comprised of argillaceous strata. The lowermost 24 m of the NHM have a very high proportion of siltstone/claystone beds and are referred to as the lower NHM here (also known as the St Bees Shale).

- *Calder Sandstone*: this formation, part of the Sherwood Sandstone Group, overlies the St Bees Sandstone Formation. It is a sandstone-dominated succession, generally coarser grained and less well sorted than the overlying Ormskirk Sandstone. Although predominantly aeolian in origin, there are also fluvial sandstones.
- *Ormskirk Sandstone*: this is the highest

formation of the Triassic SSG in West Cumbria. The sandstone comprises aeolian sandstones with common–abundant coarse–very coarse frosted quartz grains.

Hydrogeology

A conceptual model of the hydrogeology of the area is described by Black & Brightman (1996). The water table in the central PRZ lies within the Calder Sandstone at around 70–80 mAOD, decreasing to near to 0 mAOD at the coast. Groundwater flow in the area is therefore interpreted to be predominantly from NE to SW. Throughout the area, there is a change of water chemistry with depth, from fresh waters at shallow depth to saline waters or brines at greater depth. In the central PRZ, this change in hydrochemistry is observed in the vicinity of the Brockram–SSG boundary and is sharp (Bath *et al.* 1996).

Data/data processing

The Nirex deep boreholes were drilled to investigate the geology and hydrogeology of the PRZ. Nirex has completed 27 boreholes since 1989. The deepest of these penetrates to over 1500 m below ground level (mbgl). In all of these boreholes, inflatable packers were used to isolate individual monitoring zones. Groundwater pressures and temperatures in the zones were monitored at 10 or 15 min intervals. A total of over 150 of these monitoring zones were created to give a comprehensive picture of the distribution of groundwater pressures in three

dimensions (3D). Sutton (1996) describes the monitoring system in more detail.

In order to quantify the amplitude and time lag of pressure responses in each zone to the seasonal cycles in recharge, plots were prepared of all the head data for the period January 1995–December 1997, recalculated as environmental heads to account for the varying salinities (Lusczynski, 1961). During this period, groundwater levels in the Permo-Triassic Sandstone aquifers in the UK fell from high to low levels. Each zone was examined to see whether seasonal fluctuations were present, and, if so, the date and value of four turning points were measured from the plots (see Fig. 2 for an example hydrograph showing these points):

- Winter peak   1994–1995   WP1;
- Summer low   1995   SL1;
- Winter peak   1995–1996   WP2;
- Summer low   1996   SL2.

From these data, the amplitude and lag of the pressure response to seasonal recharge were calculated. To determine the amplitude of the seasonal fluctuations, the average difference in environmental head between each winter–summer pair was calculated using:

$$\text{Range} = \frac{(WP1 - SL1) + (WP2 - SL1) + (WP2 - SL2)}{3}. \quad (1)$$

The largest range observed was 5.4 m.

The lag of each turning point relative to the earliest occurrence of that turning point in any of the zones was calculated. The four values of lag calculated for each zone in this way were

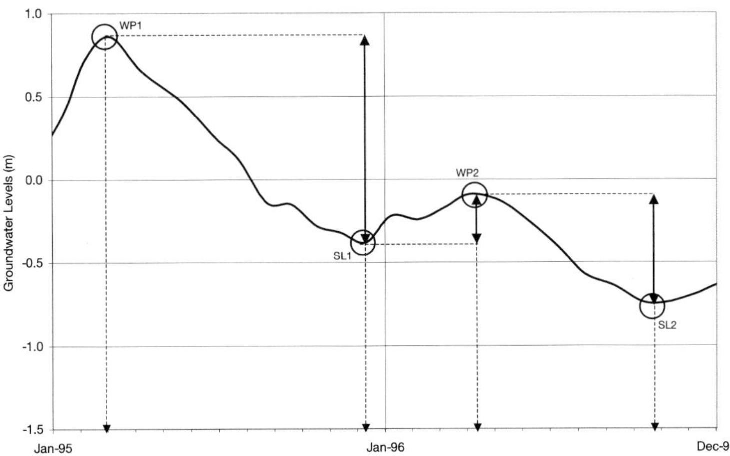

Fig. 2. Example hydrograph showing key turning points. WP, winter peak; SL, summer low.

then averaged. Because of the nature of the fluctuations, the lags were less well defined than the ranges. For zones in which seasonal fluctuations were discernible, the longest lag was around 200 days.

Monitoring data, including the seasonal range of fluctuation, are also available for shallow boreholes in the area (Nirex 1996*b*). The range in water level in these boreholes has been determined by visual assessment of the hydrographs. However, these values have been included in Figure 5 below for comparison as they provide a greater spatial coverage.

Results

The data

The size of lags and ranges in the zones vary in 3D as well as with lithology, as illustrated in Figures 3–8, which are discussed in more detail below. In Figure 4 lag has been plotted relative to the shortest lag observed in the vicinity of the PRZ (20 days).

Variation with depth/lithology

Figure 3 shows the variation in seasonal range with depth for zones of differing lithology in the central PRZ. A gradual decrease in the seasonal range with depth is apparent in the upper parts of the St Bees Sandstone, but a sharp decrease occurs in the NHM. Below this, the results are more scattered, but significant seasonal fluctuations are observed in the Brockram and BVG (1–1.5 m). The larger responses in the BVG were observed in zones that had already been shown to be in hydraulic contact with the St Bees Sandstone during the RCF3 pumping test (Streetly *et al.* 2000).

Figure 4 shows the variation in lag time with depth for zones in the central PRZ. There is an inverse pattern to that seen in the range plot, with a marked increase in lag at the base of the St Bees Sandstone.

Variation with distance from the coast

Figure 5 shows the variation of the seasonal range with perpendicular distance from the coast. This shows the largest fluctuations occurring inland with smaller ranges towards the coast. The pattern is complicated by two additional factors: the effect of depth/lithology (discussed above); and the differing positions of the boreholes relative to streams. In contrast, Figure 6 does not appear to show any systematic pattern of seasonal lag with distance from the coast.

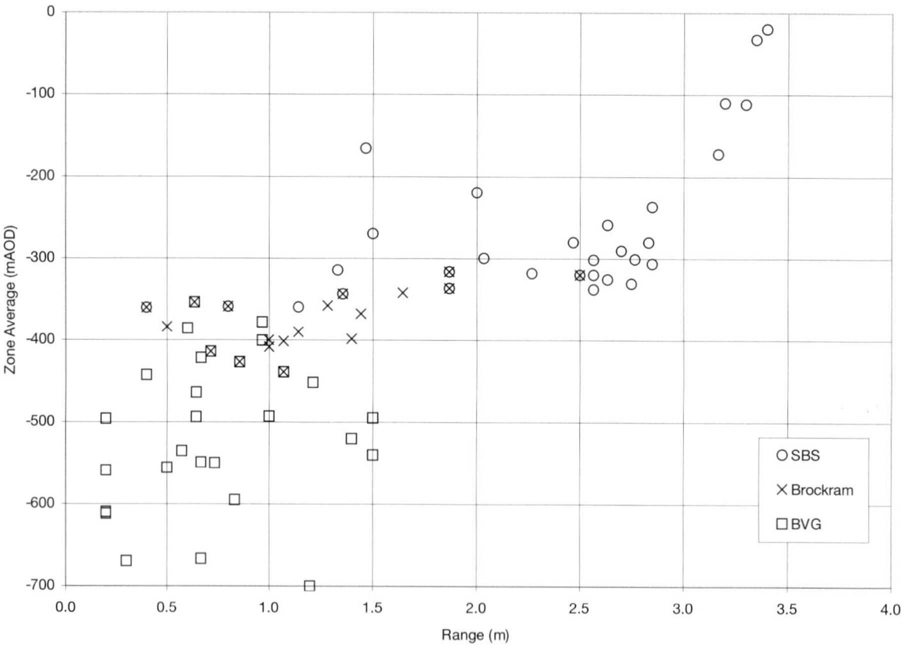

Fig. 3. The relationship between the range in seasonal head fluctuations and the elevation of the measurement zone in the central PRZ.

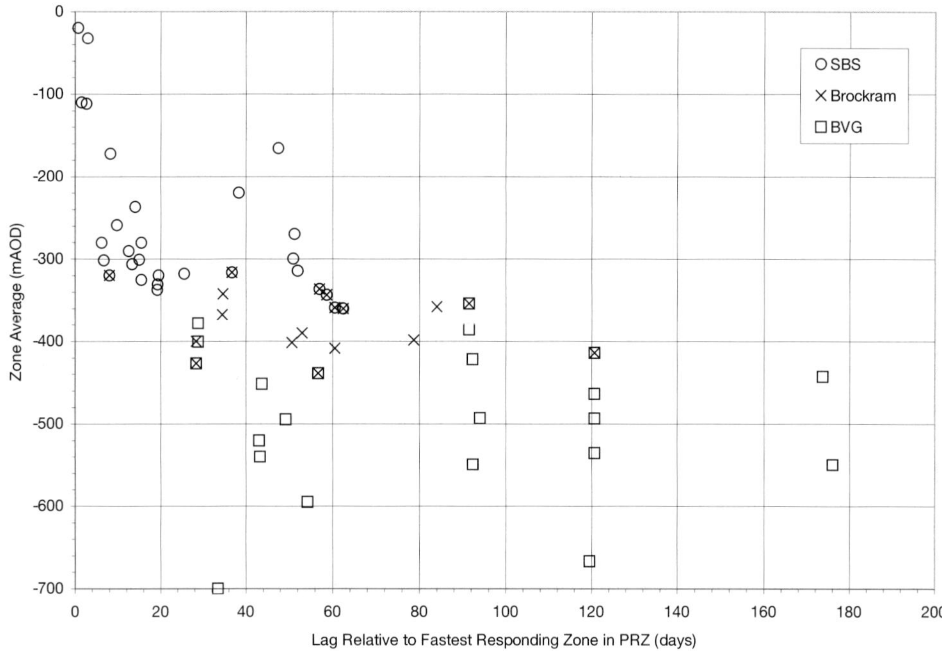

Fig. 4. The relationship between the head fluctuation lag time (measured relative to the fastest responding zone) and the elevation of the measurement zone.

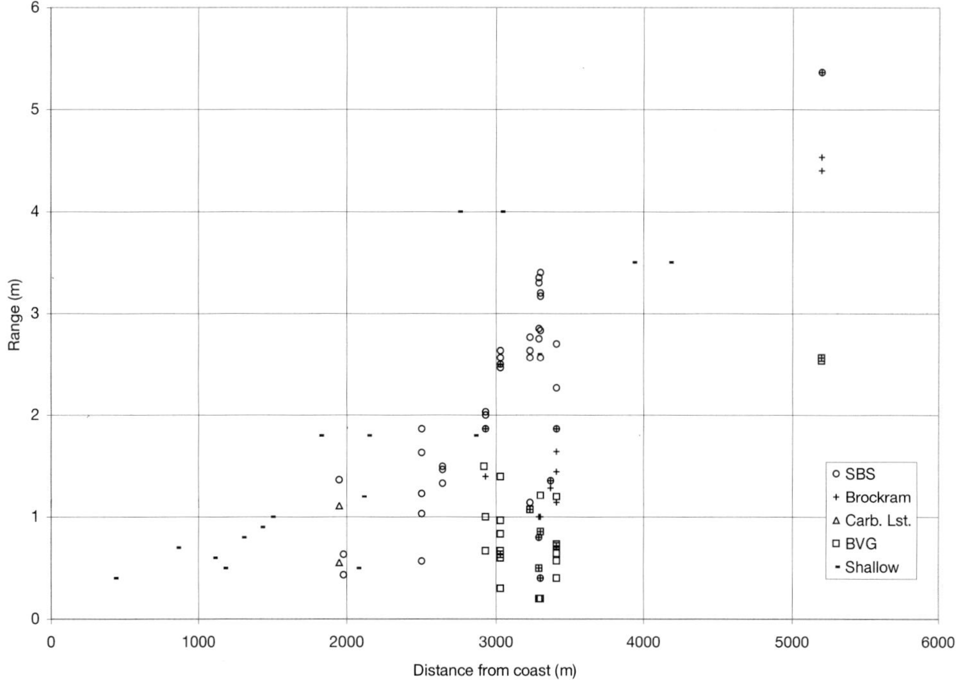

Fig. 5. The relationship between the range in seasonal head fluctuations and distance from the coast for the deep and shallow boreholes at Sellafield.

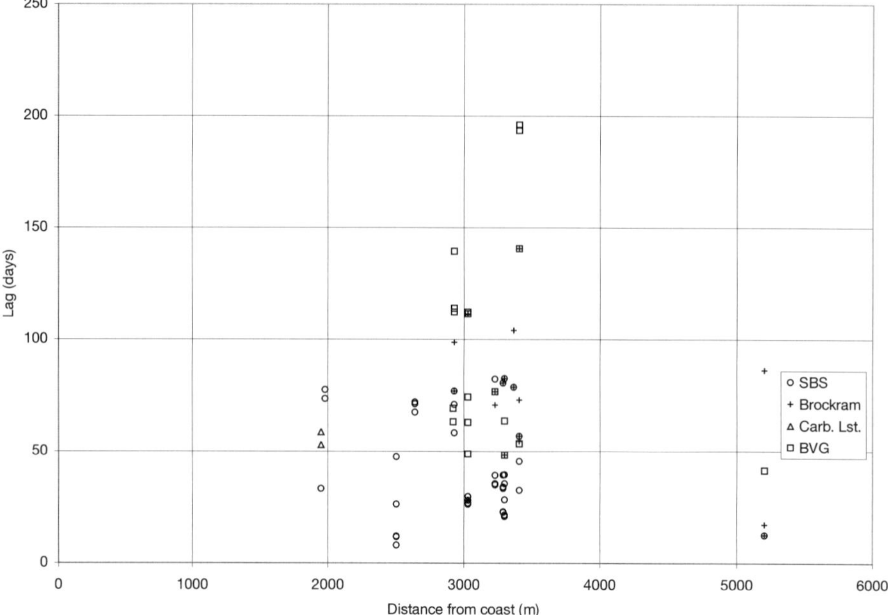

Fig. 6. The relationship between the lag in seasonal fluctuations and distance from the coast for the deep boreholes at Sellafield.

Variation in 2D and with lithology

The patterns noted above can be presented on a single, schematic cross-section through the PRZ perpendicular to the coast. This is presented in Figures 7 and 8, which show the range and lag in the various zones along the line of section. (Note that there is some transposition required to project all the data onto a single cross-section.)

Conceptual model

The patterns observed in Figures 3–8 can be interpreted in terms of a relatively simple conceptual model in which two significant processes are occurring.

- Lateral control of groundwater heads by the 'fixed head' at the coast (and, to a lesser degree, streams on the sandstone outcrop), the range being greatest furthest from the points of head control and decreasing to zero at them; this affects the range of fluctuation observed, but not the lag.
- Vertical attenuation and delay of the recharge signal, as it is transmitted downwards from the water table, by the vertical hydraulic diffusivity (vertical hydraulic conductivity divided by specific storativity) of the units present.

Figure 7 shows the effect of these two processes on the seasonal amplitudes. In contrast, Figure 8 shows that the lag mainly increases downwards. In both these plots, the results are more scattered in the BVG and Brockram than in the St Bees Sandstone. This may be anticipated as the SBS conforms more closely to a uniform, porous medium than the BVG and Brockram. Owing to their fractured nature and resultant heterogeneity, the behaviour of the latter two units may therefore be anticipated to be more variable than the SBS.

The variation of the range and lag of the seasonal fluctuations with depth and lateral position thus reflects fundamental hydrogeological parameters of the units in which these seasonal fluctuations occur. In the remaining part of this paper the attenuation of the recharge signal, as it is transmitted vertically downwards, is examined with the aid of a simple analytical model.

Analysis of vertical attenuation of recharge signal

In this assessment, a simple analytical solution was used to check the self-consistency of the field results and also to see whether they are compatible with the existing understanding of

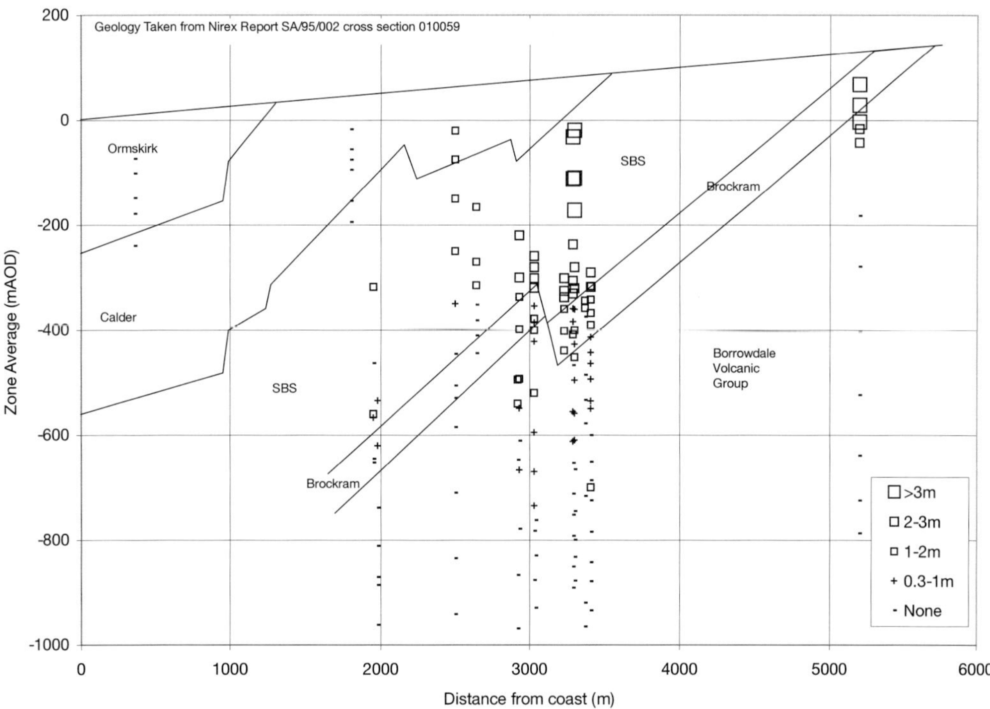

Fig. 7. The variation of the range in seasonal head fluctuations with depth and distance from the coast.

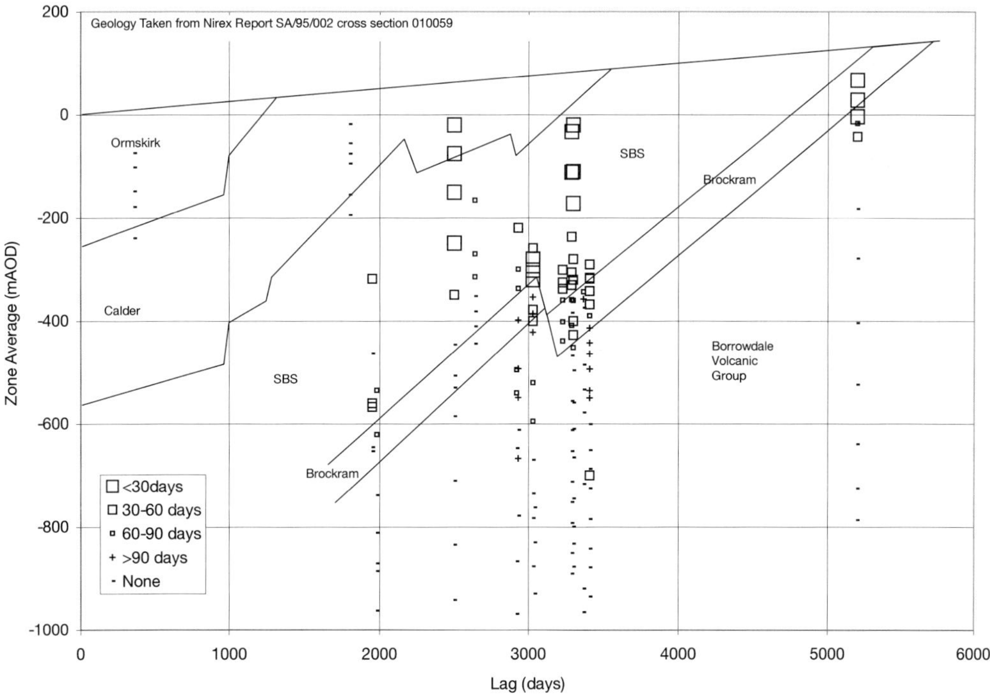

Fig. 8. The variation of head fluctuation lag time with depth and distance from the coast.

the hydrogeology of the area, particularly in the central PRZ area.

Todd (1959) presents the following analytical solution for the propagation of pressure waves through a a 1D semi-infinite confined aquifer with a sinusoidally fluctuating head imposed at one end:

$$h = h_0 e^{-x\sqrt{\frac{\pi}{t_0 D}}} \sin\left(\frac{2\pi t}{t_0} - x\sqrt{\frac{\pi}{t_0 D}}\right) \qquad (2)$$

where h is the net rise or fall [L] of the piezometric surface with reference to the mean level, at a distance x from the origin and at time t; h_0 is the amplitude of applied impulse [L] at $x = 0$; t_0 is the period of applied impulse [T] – i.e. 365 days in this case; D is the diffusivity [L²T⁻¹], which equals hydraulic conductivity/specific storage.

As the seasonal fluctuations in the study area are very approximately sinusoidal and appear to be propagated vertically downwards in the PRZ, this equation was used to interpret the seasonal head data. Issues related to delayed yield and unconfined behaviour at the water table can be eliminated by taking the highest responding zone as the origin rather than the water table.

Equation 2 yields two useful solutions for variation of amplitude and lag of the propagated wave with distance:

$$\frac{h_x}{h_0} = e^{-x\sqrt{\frac{\pi}{t_0 D}}} \qquad (3)$$

$$t_x = x\sqrt{\frac{t_0}{4\pi D}} \qquad (4)$$

where h_x is amplitude at distance x [L]; t_x is time lag at distance x [T].

It can be seen from equations 3 and 4 that short period fluctuations are attenuated more rapidly with distance than long period fluctuations, and that short period fluctuations are propagated faster than long period fluctuations.

Combination of equations 3 and 4 yields a relationship between lag and amplitude that depends only on the period (t_0) and is independent of aquifer properties or distance of travel:

$$\frac{h_x}{h_0} = e^{\frac{-2\pi t_x}{t_0}} . \qquad (5)$$

Figure 9 is a plot of log-normalized seasonal range (h_x/h_0) v. lag for data from zones in the central PRZ. The theoretical relationship between lag and amplitude for a sinusoidal wave with a 365-day period has been added.

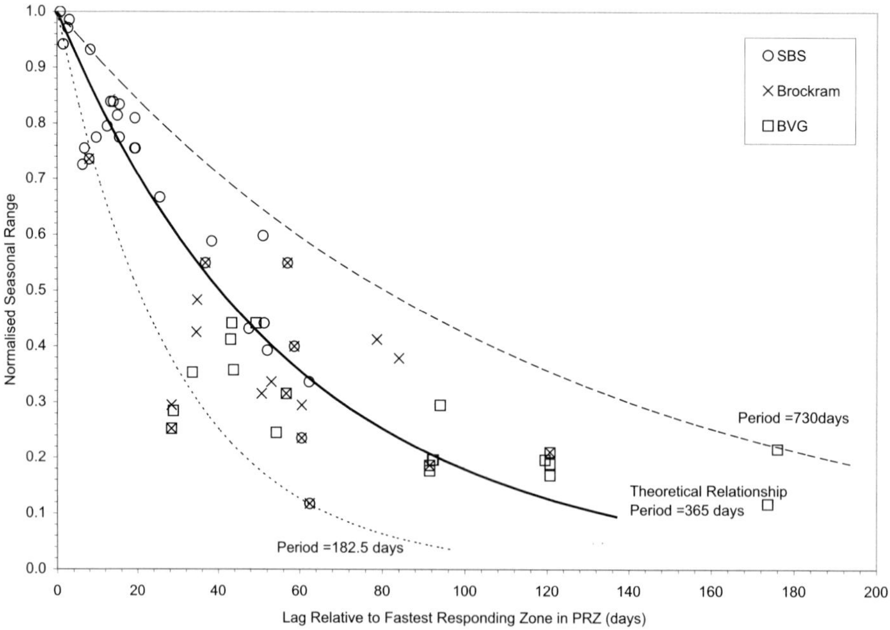

Fig. 9. The relationship between the head fluctuation lag time and the normalized seasonal head range for measurement zones in the central PRZ.

From the figure, it can be seen that most of the data are approximately consistent with the theoretical relationship, particularly for the Sherwood Sandstone Group zones. This suggests that the independent measurements of lag and range made during this study yield self-consistent results.

Equations 3 and 4 were then used to construct an analytical 1D stacked model of the central PRZ area. The aquifer was subdivided into a series of layers, which corresponded with the main, lithologically distinct units (SBS – from the water table downwards, NHM, lower NHM, Brockram and BVG). The amplitude of the input signal at the top of the stack was 3.5 m (the largest seasonal range observed in these zones) and a start lag of 20 days was used as the shortest lag observed in these zones. The attenuation of the fluctuation was calculated using equations 3 and 4 (for range and lag, respectively). The only variables to be applied to the equations are: distance (taken from the measured thickness of each unit); period (365 days); and diffusivity (which was adjusted to match the observed attenuation). The output from the transmission of the signal through each layer was then treated as the input to the underlying layer. The results from the 'best-fit' model are shown in Figures 10 and 11 for range and lag, respectively, together with data from the central PRZ area only. The values of vertical diffusivity used in the 'best-fit' model are shown in Table 1. A sensitivity analysis is also presented on Figures 10 and 11 (for the vertical diffusivity increased/decreased by a factor of 2). The good fit of the model to the data shows that the vertical attenuation of the recharge signal can be explained successfully by this approach and results of the sensitivity analysis show that the values of vertical diffusivity derived are reasonably constrained (i.e. well within an order of magnitude), particularly for the St Bees Sandstone zones.

Comparison with results from other approaches

The values of vertical diffusivity derived by this methodology for the central PRZ are compared in Table 2 with those obtained by careful partial penetration and two-layer radial flow modelling of the RCF3 Sherwood Sandstone Group pumping tests (Streetly *et al.* 2000). Values of vertical and horizontal hydraulic conductivity from other investigations in the Sellafield area are also shown for comparison.

For the St Bees Sandstone, the analysis in this paper gives a higher value of vertical diffusivity than those derived from the RCF3 test, by a

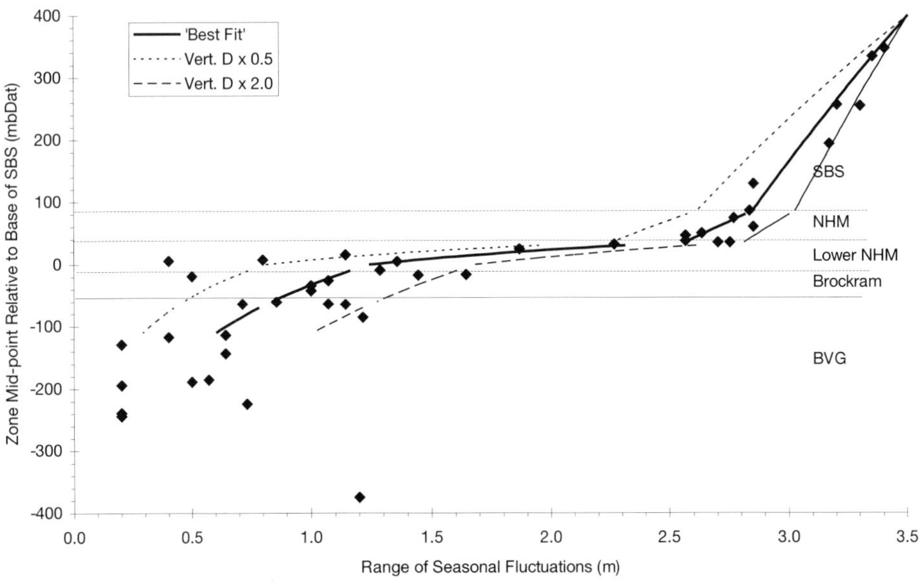

Fig. 10. The relationship between the elevation of the measurement zone relative to the base of the St Bees Sandstone Formation and the seasonal range in head. Lines show modelled values. The sensitivity of the model solution is indicated by the broken lines (Df × 0.5 for halving the 'best-fit' diffusivity value; Df × 2 for doubling the 'best-fit' diffusivity value).

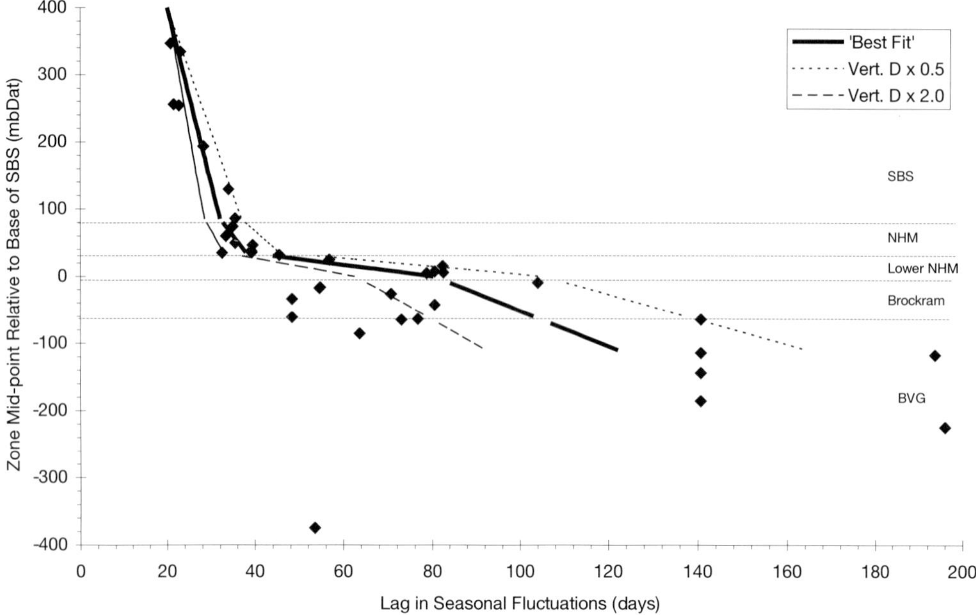

Fig. 11. The relationship between the elevation of the measurement zone relative to the base of the St Bees Sandstone Formation and the lag in seasonal head fluctuation relative to the fastest responding measurement zone in the central PRZ. Lines show modelled values. The sensitivity of the model solution is indicated by the broken lines (D $\times$ 0.5 for halving the 'best-fit' diffusivity value; D $\times$ 2 for doubling the 'best-fit' diffusivity value).

Table 1. *'Best-fit' parameters used in the multilayer model based on equations 3 and 4*

Geological unit	Vertical diffusivity $(m^2\ s^{-1})$
SBS	0.23
NHM	0.023
Lower NHM	0.00023
Brockram	0.0023
BVG	0.0023

factor of 5–10 (which would seems to rule out permeability anisotropy in this unit as a cause for the discrepancy). However, the RCF3 analysis includes part of the NHM in the section of aquifer used to derive values of vertical hydraulic conductivity for the St Bees Sandstone, and the vertical hydraulic conductivity given in Table 2 therefore may be too low. The values derived for the NHM itself are slightly closer (a factor of 5), although again are higher. This difference could be due to the differing scale of the two 'tests', or perhaps due to the steep, short-term hydraulic gradients imposed by the pumping test in comparison with more gradual transmission of the recharge signal.

Discussion and application

The very detailed data available in the vicinity of the PRZ have provided a unique opportunity to examine the attenuation and delay of seasonal variations in head, as this signal is propagated vertically downwards through the layered aquifer. The approach outlined above indicates that this type of data is amenable to analysis yielding vertical diffusivity. Diffusivity can be used to derive values of vertical hydraulic conductivity on a scale appropriate to the assessment of regional groundwater flows: to do this, specific storage values need to be estimated, but this can usually be done relatively easily to within an order of magnitude. Combining the resultant values of vertical hydraulic conductivity with the observed head gradients in the same monitoring arrays provides an opportunity to derive estimates of vertical flux. It is considered that this 'passive' approach offers significant advantages over more intrusive approaches (e.g. pumping tests), particularly in low-permeability horizons. These advantages principally relate to the larger scale at which the determination can be made and the fact that the system does not need to be disturbed

Table 2. *Comparison of the diffusivities determined using the multi-layer model based on equations 3 and 4 with those determined using other approaches*

Unit	Hydraulic conductivity (m s^{-1})			Vertical diffusivity (m^2 s^{-1})	
	Horizontal*	Horizontal†	Vertical†	(Previous analysis)‡	(This analysis)
St Bees Sandstone	2.3×10^{-8}–8×10^{-8}	4.3×10^{-8}–9.1×10^{-8}	1.7×10^{-8}–3.0×10^{-8}	2.3×10^{-2}–4.0×10^{-2}	2.3×10^{-1}
Upper North Head Member		6.0×10^{-8}–1.1×10^{-7}	3.5×10^{-9}	4.7×10^{-3}	2.3×10^{-2}
Lower North Head Member			Significantly lower	Significantly lower	2.3×10^{-4}
Brockram	2.0×10^{-10}–1×10^{-9}				2.3×10^{-3}
Borrowdale Volcanic Group	1.5×10^{-11}–3×10^{-9}				2.3×10^{-3}

* Bedding-parallel hydraulic conductivity ranges used by Heathcote *et al.* (1996).

† From partial penetration analysis of RCF3 Upper SBS and NHM pumping tests (Streetly *et al.* 2000).

‡ Derived using the values of vertical hydraulic conductivity in the previous column combined with a specific storage calculated by assuming Confined Storage Coefficient $= 3 \times 10^{-4}$ and aquifer thickness $= 400$ m (see Streetly *et al.* 2000).

significantly. However, a lengthy period of data acquisition (2 years or more) is required.

The approach is considered to be applicable in a number of different cases.

- Developing a clearer understanding of vertical flows in the Permo-Triassic sandstone. Groundwater hydrographs from boreholes completed in the Permo-Triassic sandstone often show preferential attenuation of the seasonal fluctuations in recharge in comparison with longer term cycles in recharge. The significance of the period of the fluctuation in the rate of attenuation outlined above can help to explain this.
- Developing a clearer understanding of the attenuation of the recharge signal in the unsaturated zone. In many cases, some of the attenuation of the recharge signal is likely to occur in a thick, layered unsaturated zone. Whilst the simple equations presented in this paper are not directly applicable, analogous processes are presumed to occur. In particular, as the propagation of the recharge signal in the unsaturated zone of the Permo-Triassic sandstones appears typically to occur at a rate of 0.1–0.25 m day^{-1} (Streetly & Shepley 2005), lags of 100 days or so typically may be anticipated, and would thus be large enough to be observed and hence analysed. Figure 9 suggests that, if the process of attenuation in the unsaturated zone is analogous to that in the saturated zone, then the amplitude of the recharge signal should be damped by a factor of around 0.2 for a lag of 100 days. Delay factors are commonly applied to the recharge signal in groundwater modelling projects, but the damping of amplitude that should be associated with any lag is rarely applied. As a consequence, the modelled variability in the recharge signal is likely to be too exaggerated and a larger value of specific yield may be required in the model in order to simulate the observed fluctuation of groundwater levels.
- Providing field-based evidence to help to quantify vertical fluxes in low-permeability environments (e.g. clay layers underlying contaminated land sites). The rate of vertical flux is often one of the key controlling parameters in contaminated land risk assessments, and yet determination of vertical hydraulic conductivities on an appropriate scale is particularly difficult. One practical difficulty in applying this approach is the requirement to estimate a

specific storage capacity in the relatively unconsolidated clayey drift that typically forms the underlying aquitard. This problem has been noted by other authors (e.g. Keller *et al.* 1989; Neuman & Gardner 1989).

The EU Water Framework Directive (2000/60/EC) includes a requirement for the monitoring of vertical variations in groundwater conditions, particularly in recharge zones. Compliance with this requirement could generate a potentially significant volume of data related to both groundwater heads and chemistry that would be amenable to this type of analysis, and thus provide a clearer understanding of the links between geo-hydrochemical processes and groundwater flow in 3D.

Conclusions

The extremely detailed monitoring array installed by Nirex at the potential repository zone gives a unique opportunity to observe in detail the attenuation of the signal generated by seasonal fluctuations in recharge as it is propagated vertically downwards through the Sherwood Sandstone Group and into the underlying basement rocks. This example may be useful in interpreting fluctuations due to seasonal cycles in recharge observed in other deep boreholes, where less comprehensive monitoring is available.

In this study, a simple measurement of the seasonal fluctuations has created a new data set that characterizes the response of the flow system to a large-scale impulse. Features that can be observed in the data are:

- attenuation with depth of the pressure response to seasonal fluctuations in recharge flux, particularly at the base of the Sherwood Sandstone Group in the PRZ;
- increasing delay in the seasonal fluctuations with depth, particularly at the base of the Sherwood Sandstone Group in the PRZ (of particular note in this respect is the fact that, in some of the deepest zones, the pressure response to recharge may be 180° out of phase with the recharge signal at the water table);
- attenuation of the amplitude of seasonal pressure fluctuations in response to recharge with proximity to the coast.

An analytical model was used successfully to show that the results obtained from the hydrograph analysis were self-consistent and, although giving greater values than the other methods, could be used to derive plausible values of vertical hydraulic diffusivity. The recharge signal is the largest scale signal likely to be imposed on local strata within an observable timescale, and the values derived from this analysis are likely to be most appropriate for input into regional models.

The Cyclic Fluctuations Report (Nirex 1997) on which this work was based was commissioned by Nirex. Their permission to publish this paper is gratefully acknowledged. All Nirex reports are available on request from Nirex. The project was managed by S. Sutton of Entec UK Ltd and valuable advice was provided by W. Lanyon of Geoscience Ltd – the authors are grateful for their input.

References

AKHURST, M.C., CHADWICK, R.A. *ET AL.* 1997. *The Geology of the West Cumbria District.* Memoir of the British Geological Survey, Sheets 28, 37 and 47 (England and Wales).

BLACK, J.H. & BRIGHTMAN, M.A. 1996. Conceptual model of the Sellafield area: development of the way forward. *Quarterly Journal of Engineering Geology*, **29**, S83–S94.

BATH, A.H., MCCARTNEY, R.A., RICHARDS, H.G., METCALFE, R. & CRAWFORD, M.B. 1996. Groundwater chemistry in the Sellafield area: a preliminary interpretation. *Quarterly Journal of Engineering Geology*, **29**, S39–S58.

HEATHCOTE, J.A., JONES, M.A. & HERBERT A.W. 1996. Modelling groundwater flow in the Sellafield area. *Quarterly Journal of Engineering Geology*, **29**, S59–S81.

KELLER, C.K., VANDERKAMP, G. *ET AL.* 1989. A multiscale study of the permeability of a thick clayey till. *Water Resources Research*, **25**, 2299–2317.

LUSCYNSKI, N.J. 1961. Head and flow of groundwater of variable density. *Journal of Geophysical Research*, **66**, 4247–4256.

MICHIE, U.McL. 1996. The geological framework of the Sellafield area and its relationship to hydrogeology. *Quarterly Journal of Engineering Geology*, **29**, 29, S13–S27.

NEUMAN, S.P. & GARDNER, D.A. 1989. Determination of aquitard aquiclude hydraulic-properties from arbitrary water-level fluctuations by deconvolution. *Ground Water*, **27**, 66–76.

NIREX. 1996a. *The 3D Geological Structure of the Potential Repository Zone – Summary Report.* Nirex Science Report, **S/95/005**.

NIREX. 1996b. *Rock Characterisation Facility, Longlands Farm West Cumbria: Report on Baseline Groundwater Pressures and Hydrochemistry.* Nirex Science Report, **S/96/006**.

NIREX. 1997. *Cyclic Fluctuations of Pressures Monitored in Deep Boreholes at Sellafield: Assessment, Methodology and Results.* Nirex Science Report, **SA/97/086**.

STREETLY, M.J. & SHEPLEY, M.G. 2005. *East Shropshire Permo-Triassic Sandstone Groundwater Modelling Project – Task 8 Final Report to ESI.*

STREETLY, M.J., CHAKRABARTY, C. & MCLEOD, R. 2000. Interpretation of pumping tests in the Sherwood Sandstone Group, Sellafield, Cumbria, UK. *Quarterly Journal of Engineering Geology*, **33**, 263–281.

SUTTON, J.S. 1996. Hydrogeological testing in the Sellafield area. *Quarterly Journal of Engineering Geology*, **29**, S29–S38.

TODD, D.K. 1959. *Groundwater Hydrology*. J. Wiley, Chichester.

Structural controls on groundwater flow in the Permo-Triassic sandstones of NW England

K. J. SEYMOUR, J. A. INGRAM & S. J. GEBBETT

Environment Agency, Richard Fairclough House, Knutsford Road, Warrington, Cheshire WA4 1HG, UK (e-mail: keith.seymour@environment-agency.gov.uk; john.ingram@environment-agency.gov.uk; simon.gebbett@environment-agency.gov.uk)

Abstract: There is increasing evidence that groundwater flow in many parts of the major Permo-Triassic sandstone aquifers of NW England is influenced strongly by predominantly N–S-trending faults. These structural controls on groundwater flow may only become apparent when the aquifers are subject to abstraction stress. A series of case examples are presented, from the Fylde Sandstone aquifer north of Preston, and from the sandstone aquifers of the Lower Mersey Basin, Manchester and Wirral areas. In these studies the 'compartmentalization' of the aquifers by faults has been recognized in field investigations and also in numerical modelling studies related to groundwater resources development on both local and aquifer-wide scales.

The Permo-Triassic sandstone aquifers of NW England have a long history of heavy abstraction for both industrial and public water supply. The major sandstone sequence consists of the Sherwood Sandstone Group, which is underlain in places by the Manchester Marl Formation or Bold Formation and the Collyhurst Sandstone Formation, and is overlain by the Mercia Mudstone Group. The rocks were deposited in basins formed in Permo-Triassic rift systems, which cut N–S across Britain and its continental shelf. In the NW of England there were two interlinked basins, the Cheshire Basin and the East Irish Sea Basin (Plant *et al.* 1999). The basins are heavily faulted and most of the faults are normal faults with throws decreasing upwards, indicating syndepositional movement. The large normal faults divide the basins into several smaller structural provinces, each comprising a set of tilted fault blocks (Plant *et al.* 1999).

There is increasing evidence that present groundwater flow within the sandstone aquifers of these basins is strongly influenced by some of the N–S-trending normal faults that subdivide the basins. Evidence of the effects on groundwater flow and groundwater chemistry is presented here in a series of case examples from across NW England (Fig. 1). The examples include both site-scale and regional-scale investigations.

Case example 1 – Fylde area

Background

The Fylde Sandstone aquifer is located in central Lancashire and extends from the River Ribble and Preston in the south, northwards to Morecambe Bay (Fig. 2). Structurally the Fylde area is part of the West Lancashire Basin, an onshore extension of the East Irish Sea Basin, which is connected en echelon to the Cheshire Basin to the SE (Allen *et al.* 1997). The aquifer comprises the Permo-Triassic Sherwood Sandstone Group, which is bounded in the east and underlain by Carboniferous Namurian (Millstone Grit Series) and Dinantian (Limestone Series) rocks. The Permo-Triassic sediments were deposited on a marked Carboniferous topography. The aquifer generally dips westwards and increases in thickness from its contact with the Carboniferous uplands in the east to over 500 m close to the Woodsfold Fault, where the Sherwood Sandstone is downthrown by up to 600 m and is overlain by the Mercia Mudstone Group (Seymour *et al.* 1998). The Sherwood Sandstone across the Fylde is almost entirely covered by Quaternary glacial deposits.

The aquifer is exploited for industrial and public water supply, providing an important component of the 'Lancashire Conjunctive Use Scheme' (LCUS). The scheme was developed in the early 1970s in order to optimize the water resources of the area; it involved the interlinking and conjunctive use of upland reservoirs, river abstraction and transfer, and a group of 26 abstraction boreholes tapping the Fylde aquifer. The groundwater sources, which are situated in a N–S line parallel to the M6 motorway (Fig. 2) and now operated by United Utilities (UU), are used intermittently to meet peak seasonal demand.

From: BARKER, R. D. & TELLAM, J. H. (eds) 2006. *Fluid Flow and Solute Movement in Sandstones: The Onshore UK Permo-Triassic Red Bed Sequence.* Geological Society, London, Special Publications, **263**, 169–185. 0305–8719/06/$15 © The Geological Society of London 2006.

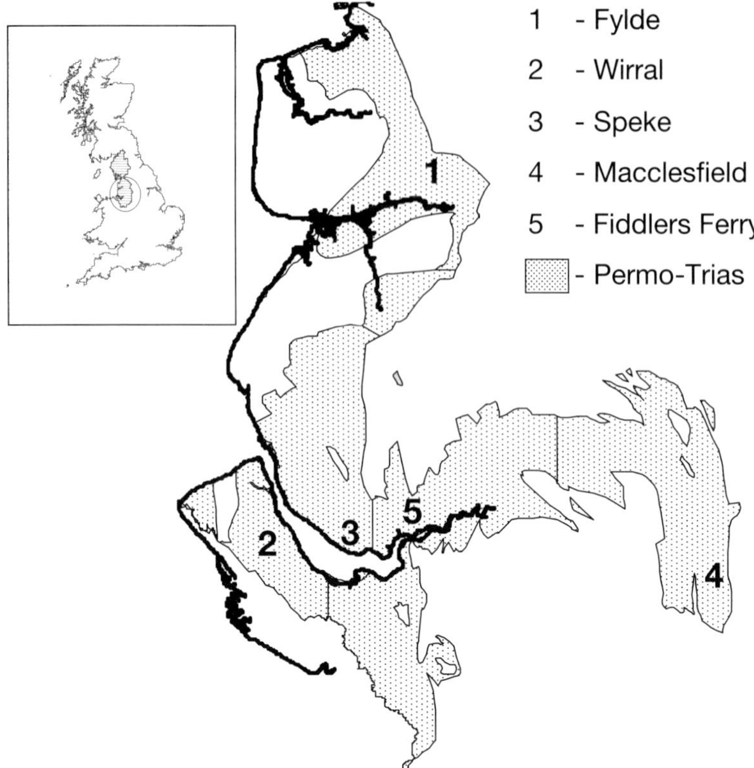

Fig. 1. Location of case examples.

As part of the original LCUS investigations, comprehensive testing of the aquifer was undertaken, including the construction of over 70 observation boreholes, short-term individual and extended group test pumping, and the development of a numerical model of the aquifer. The results of these investigations were used to formulate a set of complex abstraction licence conditions aimed at protecting existing abstractors and river flows. Since then the LCUS sources have been used operationally for about 30 years, with ongoing groundwater level monitoring.

In the mid 1990s the Environment Agency undertook a review of the impact of the Fylde abstractions, given concerns about the sustainability of the original abstraction licence and the environmental impact (Seymour *et al.* 1998). The study involved the development of an integrated surface water/groundwater numerical model of the whole of the Fylde aquifer, which was calibrated over the period from 1969 to 1994. This was used to test the validity of the assumptions underlying the original licence conditions and to test proposed future opera-

tion scenarios (Mott MacDonald & Environment Agency 1997).

Findings

In the central part of the aquifer, within the Wyre catchment where the UU abstractions are located, modelled groundwater levels accurately reflected the observed hydrographs, in terms of both elevation and magnitude of response to the strong seasonal stress applied by the public supply sources (Fig. 3a).

Similarly, in the north, towards the Wyre Estuary and Morecambe Bay, the model showed a good agreement with observed levels, with a more muted but still measurable response to UU's seasonal abstraction (Fig. 3b).

However, further south in the Ribble catchment around Preston, early simulations of the numerical model failed to match the observed response, as illustrated for the observation boreholes referred to as T74 and T68 (Fig. 3c, d). At T74, to the SW of the LCUS abstractions, the groundwater level has declined by 2.5 m and shows a limited response to UU's seasonal

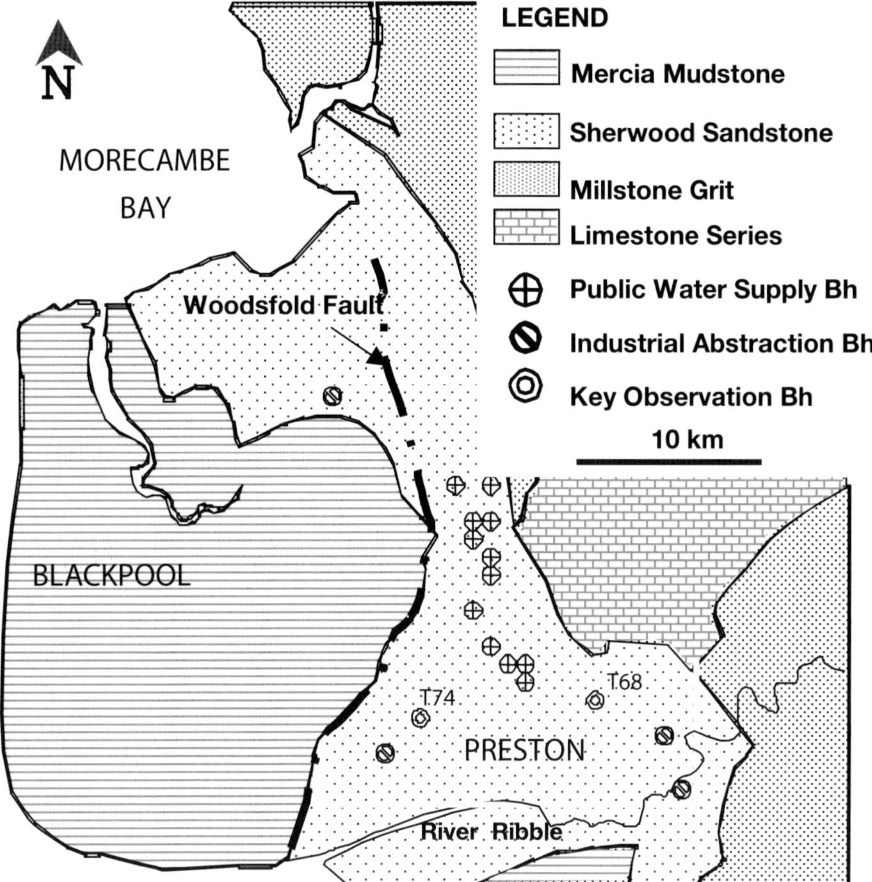

Fig. 2. Fylde aquifer location plan. bh, borehole.

abstraction. Similarly at T68, to the SE of the public supply abstractions, there has been a gradual decline of 2.5 m in groundwater level since the LCUS was commissioned, with no signature of the seasonal abstraction pattern. It can be seen that the model, assuming a uniform aquifer thickness and hydraulic properties, was not able to replicate the overall observed decline and dampened response to pumping, nor the observed groundwater elevations at T68. Therefore it was necessary to re-examine the initial conceptual model and consider the possibility that structural controls were affecting the hydraulic response to pumping.

The available geological and hydrogeological evidence was reviewed critically. The possibility of a series of N–S 'horst and graben' structures was postulated. The Agency commissioned the British Geological Survey (BGS) to produce structural maps of the study area using surface geophysical (seismic and gravity) data combined with a review of existing borehole lithological logs. The resulting maps (Evans *et al.* 1996*a*) showed depth contours on the base of the Permo-Triassic sequence and the position of both major and minor faults. The Environment Agency complemented this work by drilling a number of new observation boreholes to confirm the geological structure around Preston in an area where seismic coverage was poor or absent. This work resulted in a fundamental revision and enhancement of understanding of the geological structure of the Fylde aquifer. It confirmed the dominance of N–S faulting and that there are marked variations in aquifer thickness from east to west (Fig. 4a, b).

The revised geological structure was incorporated into the numerical model by adjusting the aquifer thickness and hydraulic properties. The horizontal hydraulic conductivity in the

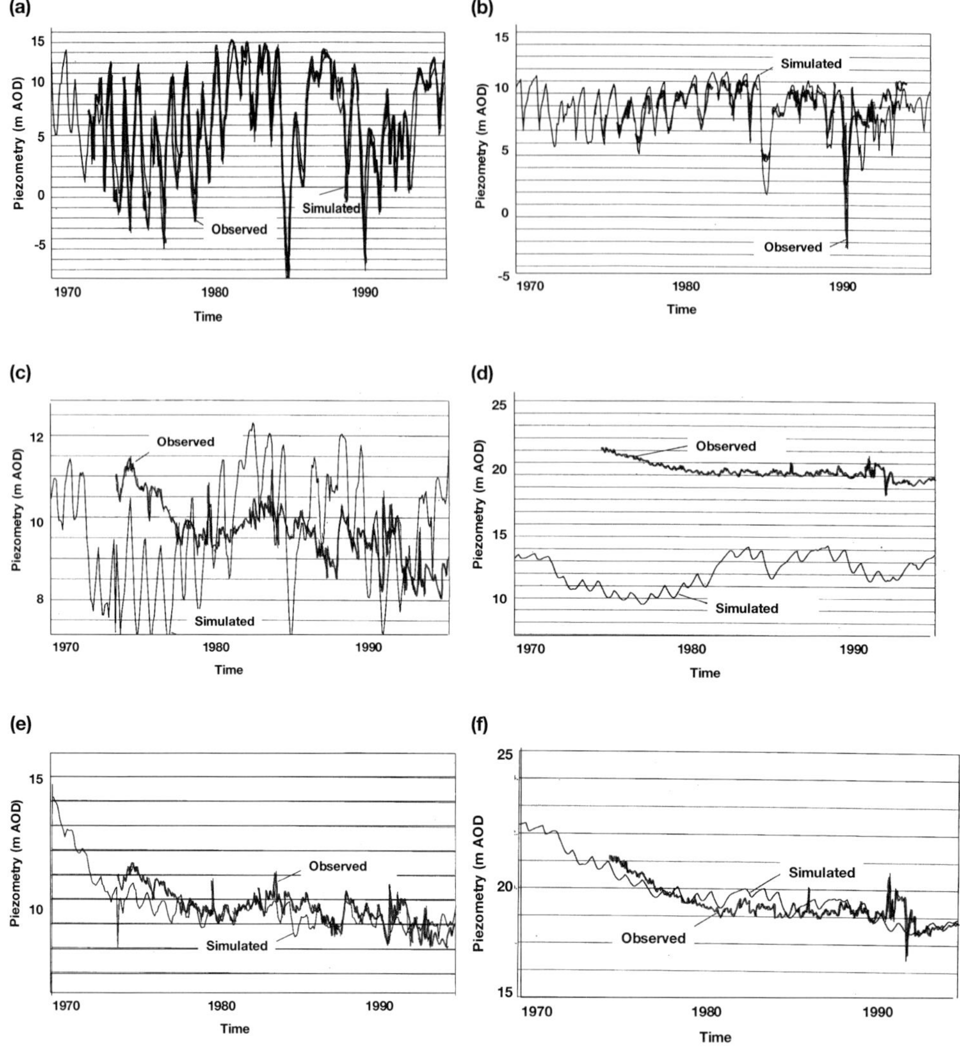

Fig. 3. Fylde groundwater flow model calibration results. (**a**) Calibration in the central area; (**b**) calibration in the northern area; (**c**) early simulation for observation borehole (Obh) T74; (**d**) early simulation for Obh T68; (**e**) final calibration for Obh T74; and (**f**) final calibration for Obh T68.

E–W direction across the faults adjacent to T74 was decreased to 0.2 m day^{-1} within the sandstone while the hydraulic conductivity in the N–S direction was set at 3 m day^{-1}, the same as the regional hydraulic conductivity. The same parameters were used for the faults separating the 'Red Scar Basin' (see Fig. 4b) from the block containing the UU abstractions. These refinements improved the model calibration significantly; however, it was also necessary to reduce the vertical hydraulic conductivity of the Quaternary deposits to achieve the overall decline in groundwater levels in T74. Combin-ing these modifications, the model was able to match the observed groundwater levels closely at T74 and T68 (Fig. 3e, f).

Conclusions

The marked anisotropy in aquifer permeability caused by the N–S faulting effectively hydraulically isolates the UU abstractions from the aquifer to the west where observation well T74 is located. Faulting also isolates the UU abstractions from other abstractions in the 'Red Scar Basin' to the east: in this case, the faults form

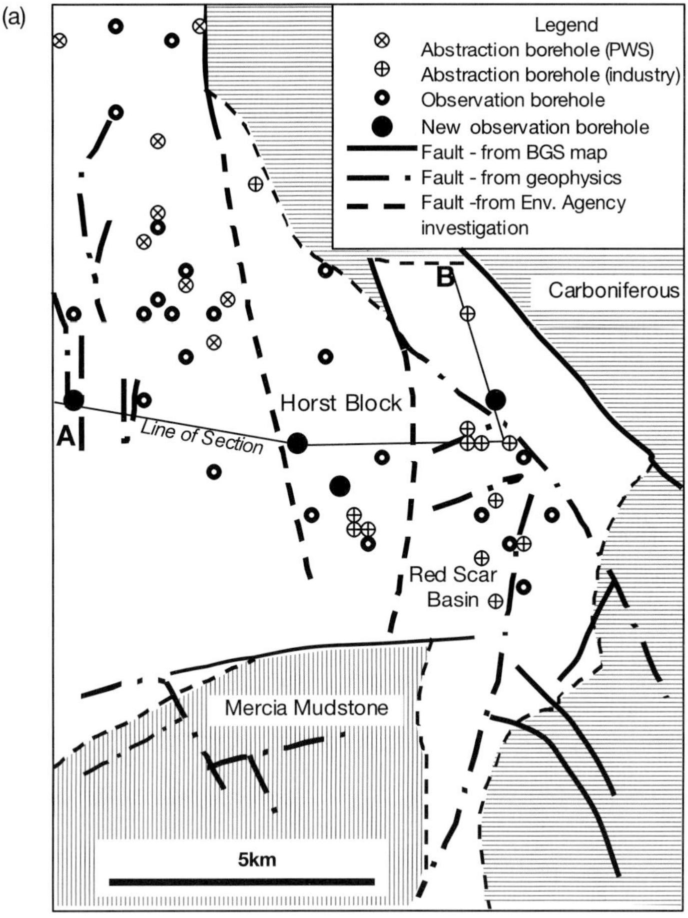

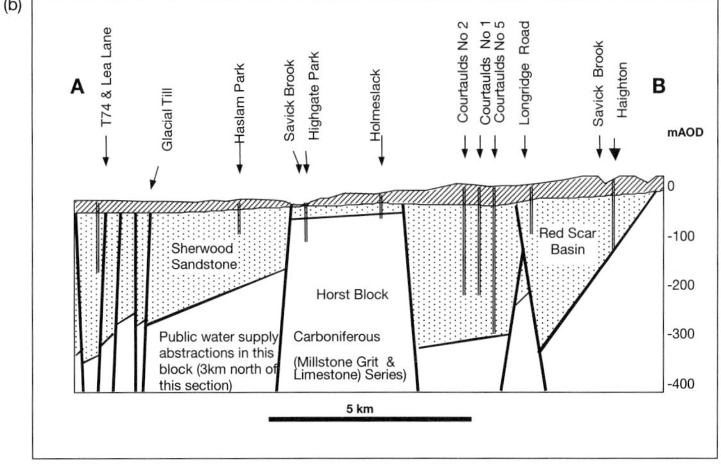

Fig. 4. Reinterpreted geological structure of the SE Fylde area. (**a**) Map; (**b**) cross-section A–B.

the boundary to a horst structure, the sandstone on top of which drains into the aquifer blocks to both the east and west.

Case example 2 – Wirral area

Background

The Wirral peninsula is bounded on three sides by the Dee Estuary, the Mersey Estuary and the Irish Sea (Fig. 1), and, except for a small area of Mercia Mudstone outcrop in the NW, consists of Sherwood Sandstone, overlain in places by glacial deposits (mainly tills). The peninsula lies on the western margin of the Cheshire Basin and the eastern margin of the East Irish Sea Basin. The 'Ellesmere Saddle' (Plant *et al.* 1999) separating the two basins crosses the peninsula. Groundwater has been exploited on the Wirral for at least 160 years, with the first public water supply wells sunk in 1843. Most of the exploitation has been in the north and the east of the area, with many of the industrial supplies located along the Mersey Estuary to the east.

Most of the industrial supplies have developed in the last 80 years, and have led to saline intrusion as groundwater levels in the area adjacent to the Mersey Estuary have fallen below sea level. Dewatering associated with road and rail tunnels beneath the Mersey has also depressed groundwater levels in Liverpool and Birkenhead (Fig. 5). As with many industrial areas in the UK, the amounts of water abstracted have reduced in the last 20 years, leading to some recovery of groundwater levels. Heads across the area have a steep gradient from the recharge area in the centre of the peninsula towards the Mersey Estuary (Fig. 5).

As part of a groundwater resources study looking at the whole of the Wirral and West

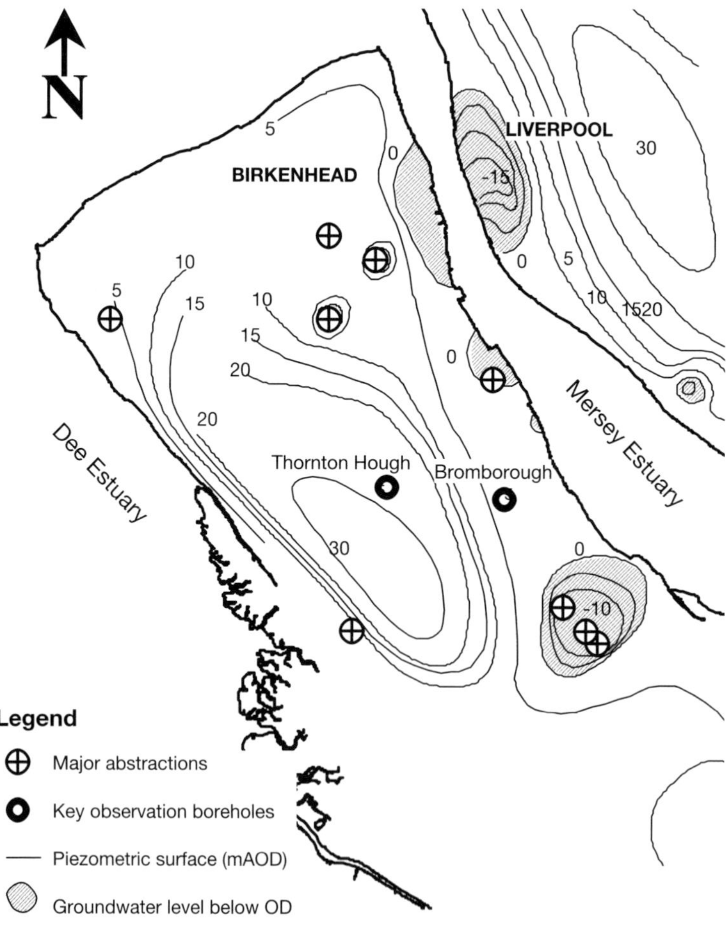

Fig. 5. Piezometric surface in metres above Ordnance Datum (September 2000) for the Wirral peninsula.

Cheshire Permo-Triassic aquifer, a groundwater flow model of the Wirral was constructed in order to investigate the major components of the flow regime and to produce groundwater balances (Gebbett & Johnson 2002).

Findings

The initial modelling work, which assumed a uniform distribution of horizontal hydraulic conductivity values across the area, failed to produce the steep head gradient seen in the aquifer between Environment Agency observation boreholes at Bromborough and Thornton Hough (Fig. 5). The initial results produced heads that were too high near to the coast at Bromborough (Fig. 6a) and too low in the recharge area at Thornton Hough (Fig. 6b). The model calculated a difference in heads between the two sites was in the region of 7.5 m, whereas the observed difference was 30 m.

Geophysical mapping by the BGS (Evans *et al.* 1996*b*) proved that there are a number of N–S-trending faults between the higher ground that is the recharge area and the industrial area nearer the coast. The model was re-run with a number of the major faults included as barriers to flow perpendicular to the fault. The horizontal hydraulic conductivity of the model cells through which the faults pass was reduced from the regional value of 1 to 0.05 m day^{-1}: for a 250 m-wide cell, this is equivalent to a conductivity of 2×10^{-4} m day^{-1} across a 1 m-wide fault. This change in the horizontal hydraulic conductivity in the fault zones produced a much better representation of groundwater heads at the Bromborough and Thornton Hough observation boreholes (Fig. 6a, b).

Conclusions

The observed steep groundwater head gradient between the recharge area in the centre of the Wirral peninsula and the area adjacent to the Mersey Estuary is due to restricted flow across certain N–S-trending faults. This is consistent with observations made elsewhere on the Wirral peninsula and in Liverpool, north of the Mersey, as far back as the 1890s (Hewitt 1898; Moore 1898; Tellam 2004).

Case example 3 – Speke area

Background

Speke is on the north side of the Mersey Estuary and lies on the NW margin of the Cheshire Basin (Fig. 1). At an industrial site in the Speke area near to Liverpool (John Lennon) Airport, the Environment Agency has been dealing with

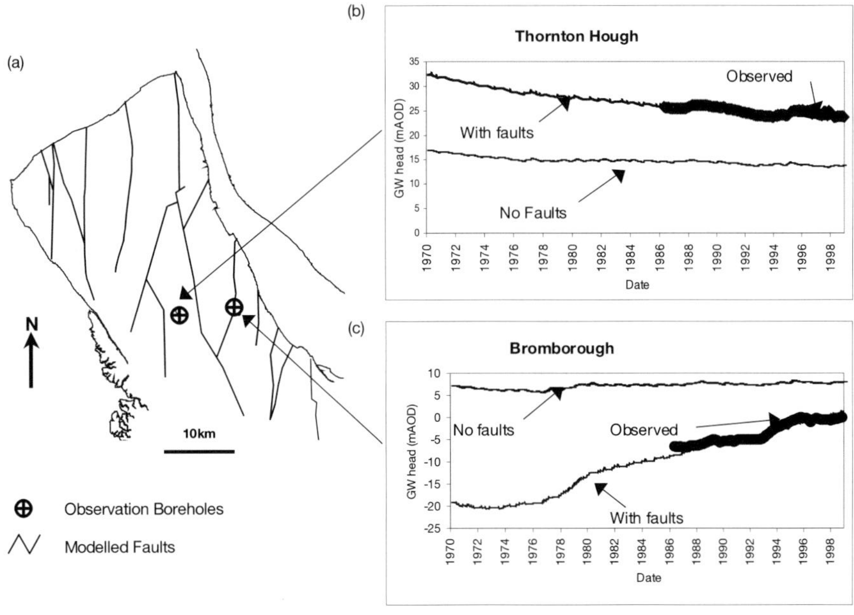

Fig. 6. The Wirral groundwater flow model. (**a**) Bromborough Observation Borehole hydrograph and model results; (**b**) Thornton Hough Observation Borehole hydrograph and model results; and (**c**) faults included in the final model, and location of Bromborough and Thornton Hough Observation Boreholes.

an application from an existing industrial abstractor for the construction of a new abstraction borehole and a significant increase in overall licensed quantities. A phased approach to the application has been adopted because of concerns about the sustainability of the proposed increase, given:

- the dominance of low-permeability drift in the area, restricting recharge;
- the historic problems of saline intrusion, which have been caused by the reversal of the natural groundwater gradient elsewhere along the Mersey (the Speke site is 3 km from the estuary);
- the indications from previous investigations in the 1980s at this and an adjacent industrial site of marked anisotropy in aquifer properties in the locality, in terms of hydraulic response to test pumping and groundwater chemistry (Seymour 1987).

According to the published geology maps, the site in question is located on the boundary between the Chester Pebble Beds Formation and the overlying Wilmslow Sandstone Formation (both in the Sherwood Sandstone Group). Immediately to the east of the adjacent Ford Motor Company site is the major NNW-trending Croxteth Fault, which displaces the Wilmslow Sandstone against an inlier of Carboniferous strata (Fig. 7).

The interpretation of seismic data by Chadwick *et al.* (1997) indicates a more complex situation, with a series of faults aligned sub-parallel with the main Croxteth Fault. Significantly, the industrial site is situated between two faults which join to the north of the site. The more easterly fault runs between the site and the adjacent Ford site (Fig. 8). This is consistent with previous down-hole logging results, from which the presence of a fault between the two sites had been inferred (Seymour 1987).

Preliminary water-balance calculations and investigative numerical modelling suggested that, on an aquifer-wide scale, there should be sufficient recharge to balance abstraction without causing saline intrusion, assuming isotropic conditions (Kawecki 1984; Southern Science Ltd 1997). However, analysis of a short-term test pumping of the new borehole (W4) inferred the presence of boundary conditions, reinforcing previous concerns over local anisotropy and, hence, the difficulty of predicting impact and sustainability (Stanger Science & Environment 1998). Therefore an extended operational pumping test is being carried out on boreholes W2, W3 and W4 under a time-limited abstraction licence, with groundwater levels on site and in other nearby observation boreholes being monitored by data loggers, as shown in Figure 8 and summarized in Table 1.

The abstractor was required to construct a purpose-drilled observation borehole midway between the site and the estuary (referred to as Obh 2). As a protection against saline intrusion, the licence conditions included a 'hands-off level' condition set on the observation borehole to prevent a reversal of the hydraulic gradient

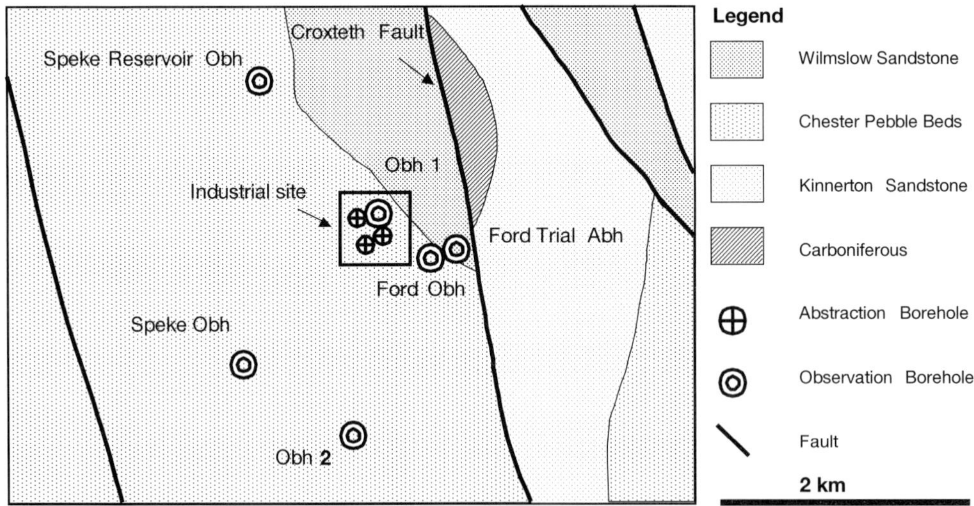

Fig. 7. Bedrock geology of the Speke area (based on published BGS maps).

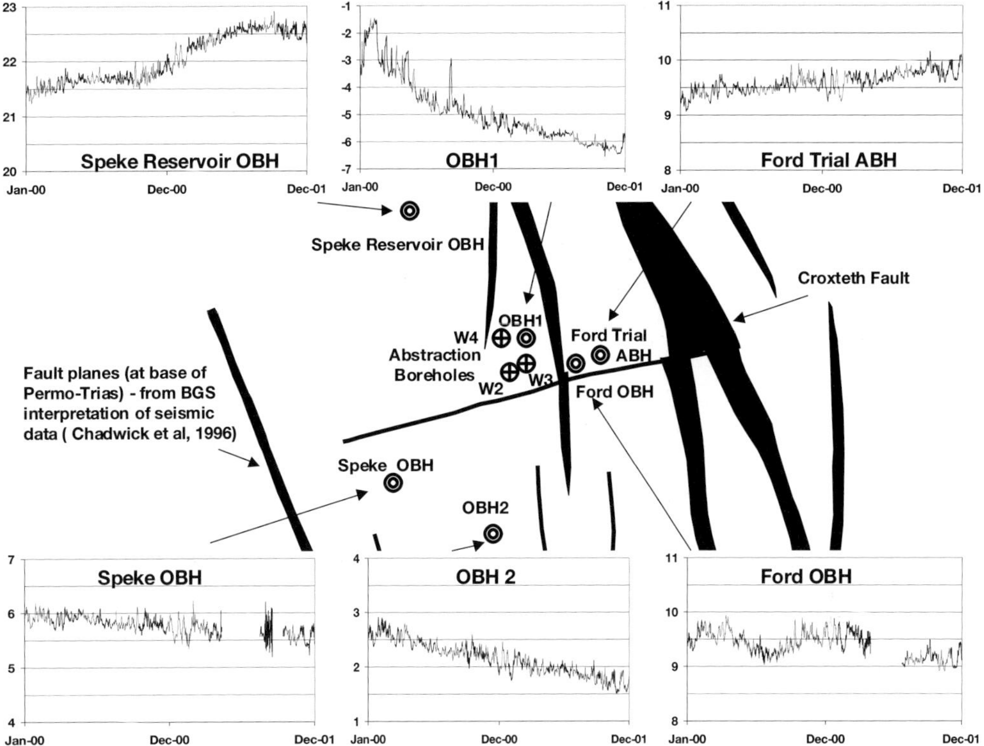

Fig. 8. Response to the 2-year operational pumping test in the Speke area.

 Speke – responses to operational pumping after 2 years (negative drawdown indicates a rise in elevation of the water table)

Observation borehole	Radial distance (m)	Drawdown (m)
Obh1	150	–5.0
Ford Obh	700	–0.5
Speke Reservoir Obh	1600	+1.0
Speke Obh	1800	–0.5
Ford Trial Borehole	1000	+0.8
Obh 2	1900	–1.0

to below sea level. Pumping of the new abstraction borehole must cease if the water level in the observation borehole falls below sea level.

Findings

Observation borehole 1 (Obh 1) (close proximity) and Obh 2 show a continuing decline in response to the increased pumping rate from the site, whilst the Speke Obh shows a much more muted response, despite being closer to the abstraction borehole than Obh2. The observed decline in this borehole was twice the magnitude predicted using analytical solutions, assuming homogeneous conditions. In contrast, at Speke Reservoir Obh and Ford Trial Borehole, groundwater levels have been recovering.

Conclusions

It is concluded that there is a marked hydraulic anisotropy at the Speke site, with preferential flow in the N–S direction and restricted flow in the E–W direction. The dominant N–S-faulting has produced a series of hydraulically poorly connected compartments, as was seen in the Fylde and the Wirral cases discussed above.

Case example 4 – Macclesfield area

Background

Macclesfield is situated on the eastern margin of the Cheshire Basin, within the 'Manchester and East Cheshire' aquifer unit (Fig. 1). The sandstone aquifer is bounded in the east by the Carboniferous Namurian (Millstone Grit

Series) and Westphalian (Coal Measures Series) strata, and in the west and south by the overlying Mercia Mudstone Group (Fig. 9). Again, there is dominant N–S faulting, with the eastern aquifer boundary marked by the Red Rock Fault. The Red Rock Fault is part of the Wem–Bridgemere–Red Rock Fault System, which forms the eastern boundary of the Cheshire Basin (Plant *et al.* 1999). Displacements on these large normal faults were partly syndepositional and exceed 2500 m further south in the basin (Evans *et al.* 1968; Plant *et al.* 1999).

The sandstone aquifer is exploited for public water supply and industrial use. The regional groundwater flow is from south to north, with discharge controlled by the major rivers, especially the River Mersey in the north. Groundwater level records from abstraction and observation boreholes indicated that there was a marked groundwater depression within a thin faulted wedge to the east of the Red Rock Fault,

just north of Macclesfield (Fig. 9a). This interpretation has now been confirmed following the recent completion of an array of observation boreholes constructed on behalf of a major industrial abstractor located within this faulted wedge. These boreholes were constructed specifically to assess the groundwater head relationships either side of the Red Rock Fault and to provide monitoring to assess the long-term sustainability of that abstraction (Fig. 9b).

From the groundwater level data in Table 2 it is evident that there is a marked groundwater head contrast across the Red Rock Fault.

Conclusions

The Red Rock Fault is acting as a significant barrier to groundwater flow. As in the case of the Speke abstraction, long-term monitoring will be necessary to establish the effects of the abstraction regime at the Macclesfield industrial site. It

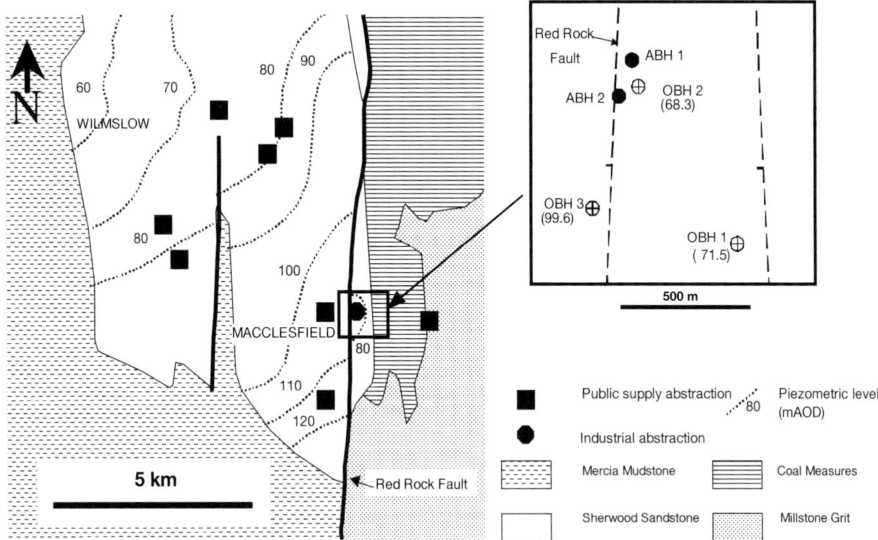

Fig. 9. The Macclesfield area. (**a**) Piezometric levels in 2000; and (**b**) groundwater levels on either side of the Red Rock Fault (March 2003).

Table 2. *Red Rock Fault water levels (for locations see Fig. 9b)*

Observation borehole	GWL mAOD* (March 2003)	Radial distance from abstraction boreholes (m)	Location
Obh 1	71.5	750	Within faulted block
Obh 2	68.3	100	Within faulted block
Obh 3	99.6	450	100 m west of Red Rock Fault

* Groundwater level in metres above Ordnance Datum (mAOD).

is possible that the abstraction is being sustained partially by cross-boundary flow from the Carboniferous to the east of the site and possibly also from a reversal of flow across the Red Rock Fault to the west of the site. Again, this case illustrates the importance of having adequate monitoring in place for large abstractions.

Case example 5 – Fiddlers Ferry area

Background

The Fiddlers Ferry area is in the 'Lower Mersey Basin' aquifer unit and lies on the northern side of the Cheshire Basin (Fig. 1). In 1982 and 1983 five trial abstraction boreholes, between 112 and 230 m deep, and five observation boreholes,

between 30 and 35 m deep, were drilled for a power station site at Fiddlers Ferry, 1 km north of the River Mersey. In 1984, a 2-month combined constant rate pumping test of the abstraction boreholes was carried out.

Findings

Prior to the start of the 1984 test, the rest groundwater levels in the abstraction boreholes and observation boreholes were measured. This revealed that there was a significant difference in groundwater levels across the site, from 17 m below Ordnance Datum (OD) at borehole W5 in the west to 2 m above OD at W4 in the east (Fig. 10a). The distance between W5 and W4 is 600 m.

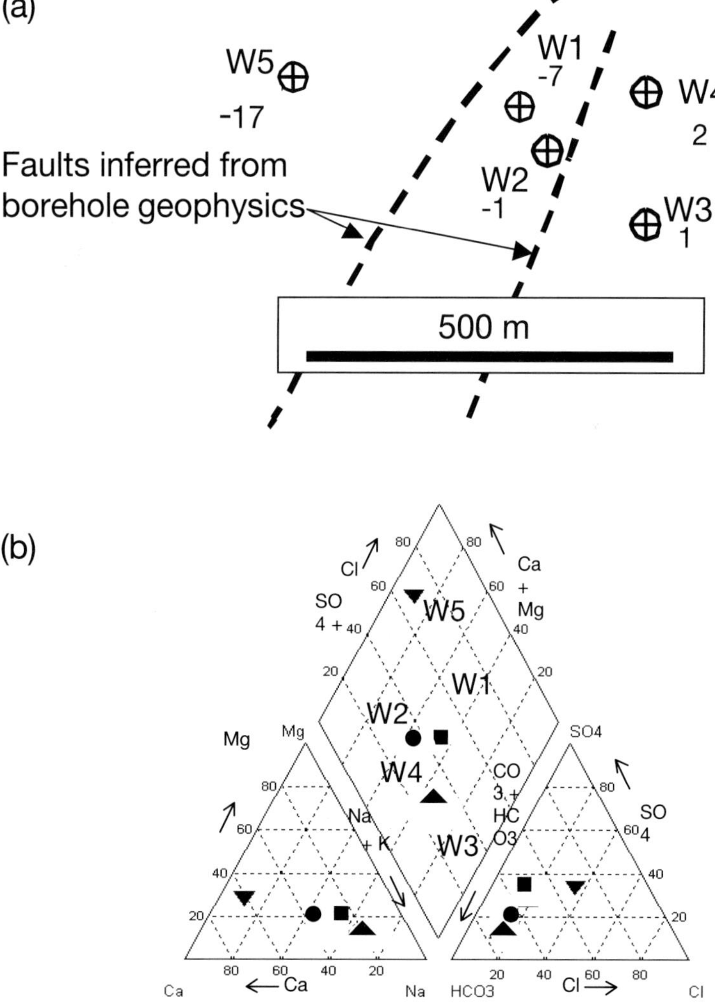

Fig. 10. The Fiddlers Ferry area. (**a**) Borehole locations and water levels in metres above Ordnance Datum in 1984; and (**b**) major ion chemistry.

In 1987 the abstraction boreholes were geophysically logged and an interpretation of the geological structure at the site was made using existing geological information and the new logs. It was concluded that two NE-trending faults cross the site, the western fault having the larger throw (Fig. 10a). This western fault is interpreted as being the southern extension of a large fault, known as the Roaring Meg Fault, which extends northwards into the outcrop of the Carboniferous. This interpretation has been confirmed by the BGS geophysical re-mapping of the area (Chadwick *et al.* 1999), which shows that the Roaring Meg Fault crosses the Fiddlers Ferry site. In previous work (University of Birmingham 1981; Howard 1988) the groundwater levels either side of the Roaring Meg Fault could not be represented adequately during numerical groundwater flow modelling unless the horizontal hydraulic conductivity across the fault was significantly reduced. The fault was assumed to be a 1 m-wide zone with a horizontal conductivity of 10^{-4} m day^{-1}. This had the effect in the model of reducing the gross permeability across a 1 km-zone represented by the model mesh to 0.095 m day^{-1} or about 5% of its initial value. More recent modelling work by Furlong (2002) has confirmed that the Roaring Meg Fault and other faults in the Lower Mersey Basin form significant barriers to groundwater flow.

The University of Birmingham (1981) (see also Tellam 1994) also showed that the apparent barrier to groundwater flow represented by the Roaring Meg Fault was well supported by hydrochemical evidence, which showed significant differences in the groundwater chemistry on either side of the fault.

The investigation at the power station site included the sampling and analysis of groundwater samples taken from all the abstraction boreholes during the 1984 test. Figure 10b is a Piper diagram showing the results of analyses of water samples taken from all the abstraction boreholes on the same day during the test. The samples taken from W1 to W4 have similar chemistry, very different from that of the sample from W5 across the western (Roaring Meg) fault. The compositions of all the samples have features that are unique in the region. The main differences between the samples are that the W5 waters contain much greater concentrations of Cl, SO_4 and NO_3, and have very much lower Na/Cl ratios. The latter (molar) ratios are particularly striking – 3–6 for W1–W4 and <0.3 for W5. Both sets of waters are significantly calcite- and dolomite-undersaturated. Without further investigation it is not possible to explain the chemistry in detail. However, W5 waters appear to be influenced by a high Cl, high SO_4, recent (NO_3, *c.* 30 mg l^{-1}) source, which could be the overlying estuarine deposits: the low Na/Cl ratio may reflect exchange of Na for Ca as the higher Cl water enters the aquifer. The composition of the waters from W1 to W4 is much more similar to that of the regional sandstone groundwater, suggesting less interaction with estuarine deposits. However, their unusually high Na/Cl ratios may imply flushing out of more saline, possibly estuary, water. Thus, one possibility is that intrusion of low-salinity estuary water has occurred in the past resulting in the chemistry seen in W5: increase in heads to the east of the Roaring Meg Fault have resulted in flushing out of this intruded water, the only chemical evidence now remaining being the high Na/Cl ratio as the sandstone exchange sites continue to re-attain equilibrium with the regional sandstone groundwaters. This difference in flow direction implied by the Na/Cl ratios is consistent with the very low rest water levels in W5 and much higher rest levels in W2–W4.

Conclusion

The conclusion from work at the power station site at Fiddlers Ferry is that there is evidence from both the water level and chemical data that NE-trending faults crossing the site form a significant barrier to groundwater flow. One of the faults is interpreted as being the southern extension of the Roaring Meg Fault, which in other work carried out on a regional scale has been shown to form a barrier to groundwater flow. Within the western fault block at the site, in which W5 is situated, there has been historic over-abstraction by other large industrial abstractors in the Widnes area, 2 km to the west of Fiddlers Ferry. This over-abstraction resulted in a lowering of the groundwater levels to below sea level, allowing both saline intrusion from the Mersey Estuary and increased recharge through the drift.

Discussion

Previous work on fault sealing in the region and elsewhere

As mentioned above, an earlier regional study by the University of Birmingham and the North West Water Authority recognized that certain faults, which displace the sandstone aquifers in the Lower Mersey Basin, appear to act as barriers to groundwater flow (University of Birmingham 1981). They are considered to be the reason for old saline groundwater still being present in the aquifer east of Widnes, despite the low sea-level stands in the Devensian and other glacial intervals (Tellam *et al.* 1986; Tellam 1995). Other examples of the barrier effects produced by faults in the Permo-Triassic sandstones of NW England have been previously recorded. Campbell (1987) described the barrier effect of a N–S-trending fault at Sandon Dock in Liverpool by examining the tidal response of observation boreholes on both sides of the fault. Barker *et al.* (1998) noted abrupt chemical changes across the Castle Street Fault in Liverpool. Following the installation of a dewatering scheme to protect the underground Liverpool Loop Line from the effects of rising groundwater, these changes, as well as marked differences in hydraulic responses across the same fault system, have become more apparent. Knott (1994) investigated the relationship between fault-zone width and displacement in the Cheshire Basin, and discussed the possibilities of using displacement as an indicator of fault permeability. Wealthall *et al.* (2001) note both sandy and clayey fracture fill material in Runcorn, adjacent to the Mersey Estuary. Looking even further back, the effects of faulting on groundwater flow and chemistry within and around Merseyside was debated at length in early papers of the Liverpool Geological Society (1847–1900) (Tellam 2004).

Further south, in the Permo-Triassic sandstone basins of the Midlands, faults have been inferred as acting as barriers to groundwater flow, for example in Shropshire (Hunter Williams 1995), Staffordshire (Wrathmell 1996; Chompusri 1997) and in Birmingham (Jackson & Lloyd 1983). Pokar *et al.* (2006) note clay-filled fractures in Yorkshire. To the north, Fowles & Burley (1994) describe examples of permeability reduction in fault zones in Cumbria (NW England) and SW Scotland.

The effects of fault seals in sandstones have been discussed extensively in the context of hydrocarbon development in a range of locations worldwide; for example, see Davies & Handschy (2003) and the references therein.

However, it is only recently that the significance and extent of the compartmentalization of groundwater flow in the Permo-Triassic sandstone aquifers of the NW of England has been fully appreciated in terms of catchment-scale groundwater resource management.

Mechanisms for reduced permeability across faults

This paper focuses on observed hydraulic responses to inferred low permeability across certain faults, rather than on an analysis of the nature of fault planes and the causes of the anisotropy. However, some of the case examples are clearly related to geometry, where fault displacement has resulted in the juxtaposition of geological units with different lithological and sedimentary characteristics on either side of the fault plane (Figs 4b & 11a). A number of the sandstone units within the Sherwood Sandstone Group include thin mudstone beds, particularly within the Chester Pebble Beds Formation and the Helsby Sandstone Formation. There are also numerous thin beds of very fine silty sandstone throughout the group. Immediately adjacent to the fault planes of major normal faults the bedding is likely to be affected by fault drag created by the dip-slip movement along the faults. This may lead to a zone of steeply dipping and contorted bedding that could result in a zone of low-horizontal permeability. The fault plane itself may be smeared by mudstone, may include low-permeability fault gouge and may be sealed by minerals deposited from migrating fluids. In the Cheshire Basin migrating ore fluids deposited calcite–barite–sulphide mineralization adjacent to major fault structures (e.g. Plant *et al.* 1999). In places within the Cheshire Basin there are fault-related 'granulation seams'. These are narrow seams, a few millimetres thick, formed by cataclasis and subsequently sealed by post-deformational quartz cement; they occur in swarms, and on the Wirral over 1000 seams have been observed in one swarm (Jeffcoat 2002). The permeability of these seams has been measured in the laboratory as being 10^{-5}–10^{-4} m day^{-1}, some four–five orders of magnitude lower than the host rock (Ballard 2000). They may be more significant than the fault plane itself in reducing lateral transmissivity. Occurrences of granulations seams in northern England and southern Scotland, and their potential effects on fault-zone permeability, have been have been described in detail by Fowles & Burley (1994).

It is known that at least some of the faults where these effects are observed have been subject to syndepositional movement, e.g. the Red Rock Fault in Macclesfield (Case example 4) and the Roaring Meg Fault (Case example 5). It is suggested that this may be a common factor in explaining why only certain faults act as barriers to groundwater flow in the Cheshire Basin and northwards.

In some cases where there is low permeability perpendicular to the fault plane, there may be enhanced permeability parallel to the fault (e.g. Fowles & Burley 1994; Knott 1994; Gibson 1994; Gutmanis *et al.* 1998). L. J. West (pers. comm. 2003) has noted open fractures parallel to fault planes in the Permo-Triassic aquifers of York-shire, and Chompusri (1997) has investigated a clay-filled fault zone with associated open anti-thetic fractures in a quarry in Stoke on Trent.

Aquifer management implications

The Environment Agency has been reviewing its conceptual understanding of the behaviour of the sandstone aquifers of NW England under abstraction stress and updating its groundwater contour maps in the light of new information on structural controls (Fig. 11). The structural information has been obtained primarily from the interpretation of seismic data carried out by the BGS. The interpretation of the data and the production of structural maps were specifically commissioned by the Environment Agency.

The revised conceptual understanding has been used to define more appropriate and tech-nically justifiable groundwater management units (GWMUs), as required under the Agency's Catchment Abstraction Management Strategies (CAMS), which set out to achieve sustainable, integrated groundwater and surface water management on a river catchment scale. An example is illustrated in Figure 12 (Environ-ment Agency 2003). CAMS work on a 6-year review cycle, and in conjunction with this the Agency is moving towards the use of time-limited abstraction licences. This is consistent with the case examples presented here, which have illustrated the importance of collecting sufficient groundwater level monitoring data to enable the effects of compartmentalization to be identified and the significance to be assessed in terms of sustainability.

Implementation of the EU Water Framework Directive will also require an understanding of the compartmentalization of the Permo-Triassic sandstone aquifers in terms of the designation of 'Groundwater Bodies', and the requirement to monitor and report on their qualitative and quantitative status.

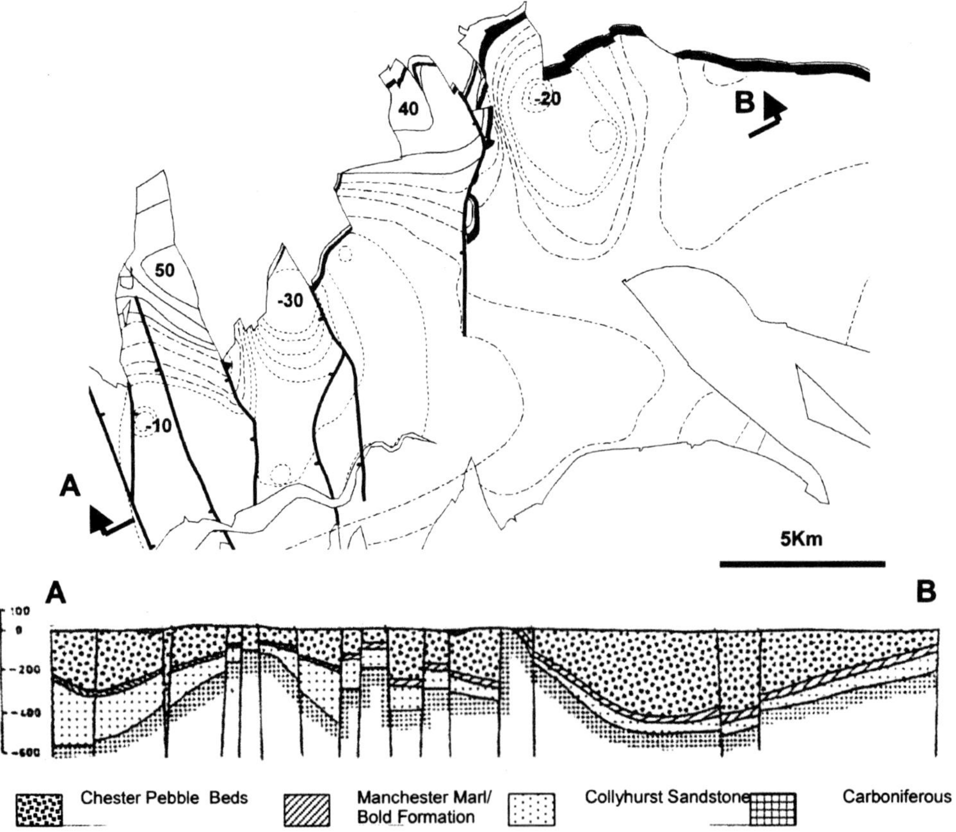

Fig. 11. Mersey Basin piezometric surface in 2000, and geological cross-section A–B (University of Birmingham 1981).

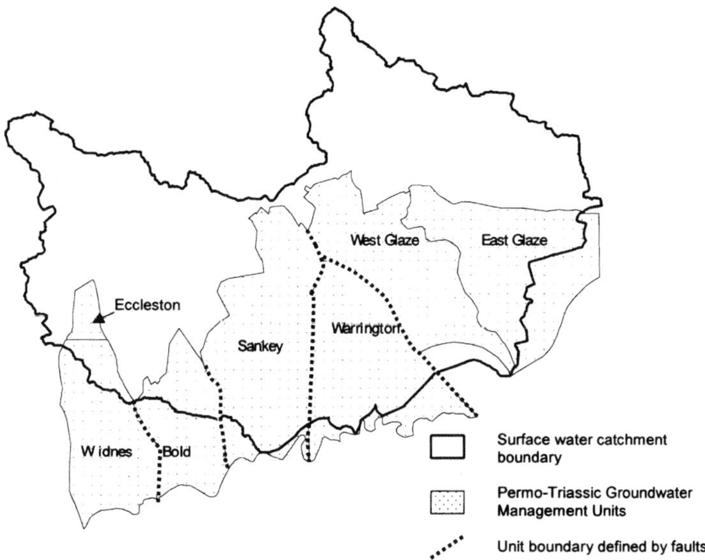

Fig. 12. Sankey/Glaze CAMS area in the Mersey Basin (cf. Fig. 11), and the Groundwater Management Units (after Environment Agency 2003).

Conclusions

The case examples presented have demonstrated that there is a significant degree of structural control on groundwater flow within the Permo-Triassic sandstone aquifers of NW England. In these examples, dominant N–S faulting appears to have divided the aquifers into a series of interconnected blocks, with restricted cross-boundary groundwater flow between the blocks. The effects of faulting have only become apparent when the aquifer has been subject to abstraction stress. Similar effects have been observed in the Cheshire Basin and further south in the Midlands.

Mechanisms for producing low permeability across faults include the juxtaposition of differing lithologies, granulation seams and fault-plane infill. It is suggested that syndepositional movement may be a common factor of the low-permeability faults described in these case studies.

For new groundwater resource developments in high storage aquifers, where fault controls or other anisotropy are likely, adequate spatial and temporal groundwater level (and quality) monitoring and extended operational test pumping is recommended to identify structural controls and limits on groundwater resource availability.

In terms of research, further investigation of the properties of low-permeability faults should include laboratory testing of samples obtained from quarries, tunnels or other excavations that intersect such fault planes. In critical situations, the drilling of inclined boreholes to intersect and sample fault planes may be justified, but this will be more for academic rather than practical value to abstractors.

Some of the case examples discussed in this paper include information provided to the Environment Agency (and its predecessor authorities) by consultants, in particular work carried out for the Agency by Mott MacDonald and by the British Geological Survey. Individual thanks are also due to J. Dodds, R. Sage and S. Wood for providing some of the data included in this paper. The views expressed above are those of the authors.

References

ALLEN, D.J., BREWERTON. L.J. ET AL. 1997. *The Physical Properties of Major Aquifers in England and Wales.* BGS Technical Report, **WD/97/34**. Environment Agency R&D Publication, **8**.

BALLARD, M. 2000. *The role of granulation seams in pollutant transport in the U.K. Triassic Sandstone.* MSc Project Report, Earth Sciences, University of Birmingham.

BARKER, A., NEWTON, R., BOTTRELL, S.H. & TELLAM, J.H. 1998. Processes affecting groundwater chemistry in a zone of saline intrusion in an urban aquifer. *Applied Geochemistry*, **6**, 735–750.

CAMPBELL, J.E. 1987. *Sandon Dock Outfall Shafts. Hydrogeological Investigations.* Hydrogeological

Report, **182**. Rivers Division, North West Water Authority, Warrington.

CHADWICK, R.A. & ROWLEY, W.J. 1996. *Structure Contour maps of the West Cheshire, Lower Mersey and Manchester & East Cheshire Aquifers*. BGS Technical Report, **WA/96/93C**. Report prepared for the Environment Agency.

CHADWICK, R.A., EVANS, D.J., BAILEY, H.E. & ROWLEY, W.J. 1997. *Geophysical Re-mapping of the Liverpool/Ormskirk Aquifer*. BGS Technical Report, **WA/96/95C**. Report prepared for the Environment Agency.

CHADWICK, R.A., ROWLEY, W.J., SMITH, N.J.P. & SHAW, K.L. 1999. *Geophysical Re-mapping of the Aquifers of the Western and Northern Cheshire Basin*. BGS Technical Report, **WH/99/39C**. Report prepared for the Environment Agency.

CHOMPUSRI, S. 1997. *Effects of faults on Triassic sandstones, UK*. MSc Project Report, Earth Sciences, University of Birmingham.

DAVIES, R.K., & HANDSCHY, J.W. (eds). 2003. *Fault Seal. AAPG Bulletin*, **87**, (3), 377–524 (theme issue).

ENVIRONMENT AGENCY. 2003. *The Sankey and Glaze Catchment Abstraction Management Strategy*. Environment Agency, North West Region, Warrington.

EVANS, D.J., BAILY, H.E., HULBERT, A.G. & KIRBY, G.A. 1996a. *Geophysical Re-mapping of the Fylde Aquifer*. BGS Technical Report, **WA/96/24C**. Report prepared for the National Rivers Authority.

EVANS, D.J., CHADWICK, R.A., ROWLEY, W.J. & SHAW, K.L. 1996b. *Geophysical Re-mapping of the Wirral Aquifer*. BGS Technical Report, **WA/96/94C**. Report prepared for the Environment Agency.

EVANS, W.B., WILSON, A.A., TAYLOR, B.J. & PRICE, D. 1968. *Geology of the Country around Macclesfield, Congleton, Crewe and Middlewich*. Memoir of the Geological Survey of Great Britain, Sheet 110. HMSO, London.

FOWLES, J. & BURLEY, S. 1994. Textural and permeability characteristics of faulted, high porosity sandstones. *Marine and Petroleum Geology*, **11**, 608–623.

FURLONG, B.V. 2002. *Regional scale solute transport in the Permo-Triassic sandstone aquifer of the Lower Mersey Basin, North West England*. PhD thesis, University of Birmingham.

GEBBETT, S. & JOHNSON, D. 2002. *Wirral: Quantitative Conceptual Model and Investigative Numerical Modelling*. Hydrogeological Report, **470**. Environment Agency, North West Region, Warrington.

GIBSON, R.G. 1998. Fault-zone seals in siliciclastic strata of the Columbus Basin, offshore Trinidad. *AAPG Bulletin*, **78**, 1372–1385.

GUTMANIS, J.C., LANYON, G.W., WYNN, T.J. & WATSON, C.R. 1998. Fluid flow in faults: a study of fault hydrogeology in Triassic sandstone and Ordovician volcaniclastic rocks at Sellafield, north-west England. *Proceedings of the Yorkshire Geological Society*, **52**, 159–175.

HEWITT, W. 1898. Notes of some sections exposed by excavations on the site of the new technical schools, Byrom Street, Liverpool. *Proceedings of the Liverpool Geological Society, Session Thirty-ninth*, **8**, 268–273.

HOWARD, K.W.F. 1988. Beneficial aspects of sea-water intrusion. *Ground Water*, **25**, 398–406.

HUNTER WILLIAMS, N. 1995. *Investigation of the response of the Triassic Sandstone aquifer, Nurton, to extended abstraction and of the controls on groundwater flow*. MSc Project Report, Earth Sciences, University of Birmingham.

JACKSON, D. & LLOYD, J.W. 1983. Groundwater chemistry of the Birmingham Triassic sandstone aquifer and its relationship to structure. *Quarterly Journal of Engineering Geology*, **16**, 135–142.

JEFFCOAT, A.M. 2002. *Exploring the hydraulic properties of discontinuity geometry in the UK Triassic sandstones*. PhD thesis, University of Birmingham.

KAWECKI, M.W. 1984. *Speke Groundwater Model Study*. Birmingham University for North West Water Authority.

KNOTT, S.D. 1994. Fault zone thickness versus displacement in the Permo-Triassic sandstones of NW England. *Journal of the Geological Society, London*, **151**, 17–25.

MOORE, C.C. 1898. The chemical examination of sandstones from Prenton Hill and Bidston Hill. *Proceedings of the Liverpool Geological Society, Session Thirty-ninth*, **8**, 241–267.

MOTT MACDONALD & ENVIRONMENT AGENCY. 1997. *Fylde Aquifer/Wyre Catchment Water Resources Study*. Final report. Environment Agency, North West Region, Warrington.

PLANT, J.A., JONES, D.G. & HASLAM, H.W. (eds). 1999. *The Cheshire Basin. Basin Evolution, Fluid Movement and Mineral Resources in a Permo-Triassic Rift Setting*. British Geological Survey, Keyworth, Nottingham.

POKAR, M., WEST, L.J. & ODLING, N.E. 2006. Petrophysical characterization of the Sherwood Sandstone from East Yorkshire, UK. This Volume.

SEYMOUR, K.J. 1987. *CW 13455 – Dista Products & CW 2531 – Ford Motor Co.* Hydrogeological Report, **186**, North West Water Authority, Warrington.

SEYMOUR, K.J., WYNESS, A.J. & RUSHTON, K.R. 1998. The Fylde aquifer – A case study in assessing sustainable use of groundwater resources. *In:* WHEATER, H. & KIRKBY, C. (eds) *Hydrology in a Changing Environment*, Volume 2. Wiley, Chichester.

SOUTHERN SCIENCE LTD. 1997. *Groundwater Modelling Contract Speke Summary Report*. Southern Science Ltd Report, **97/1/1729**, Manchester.

STANGER SCIENCE & ENVIRONMENT. 1998. *Final Report on the Environmental Impact of the Proposed Increase in Groundwater Abstraction at Eli Lilly Speke Operations*. Stanger Science & Environment, Manchester.

TELLAM, J.H. 1994. The groundwater chemistry of the Lower Mersey Basin Permo-Triassic Sandstone Aquifer System in 1980 and pre-industrialisation urbanisation. *Journal of Hydrology*, **161**, 287–325.

TELLAM, J.H. 1995. Hydrochemistry of the saline groundwaters of the lower Mersey Basin Permo-Triassic sandstone aquifer, UK. *Journal of Hydrology*, **165**, 45–84.

TELLAM, J.H. 2004. 19th century studies of the hydrogeology of the Permo-Triassic Sandstones of northern Cheshire Basin, England. *In*: MATHER, J.D. (ed.) *200 Years of British Hydrogeology*. Geological Society, London, Special Publications, **225**, 89–105.

TELLAM, J.H., LLOYD, J.W. & WALTERS, M. 1986. The morphology of a saline groundwater body; its investigation, description and possible explanation. *Journal of Hydrology*, **83**, 1–21.

UNIVERSITY OF BIRMINGHAM. 1981. *Saline Groundwater Investigation. Phase 1 – Lower Mersey Basin*. Final Report to the North West Water Authority.

WEALTHALL, G.P., STEELE, A., BLOOMFIELD, J.P., MOSS, R.H. & LERNER, D.N. 2001. Sediment filled fractures in the Permo-Triassic sandstones of the Cheshire Basin: observations and implications for pollutant transport. *Journal of Contaminant Hydrology*, **50**, 41–51.

WRATHMELL, E. 1996. *A hydrogeological investigation of the region around Croxden Quarry near Cheadle, Staffordshire*. MSc Project Report, Earth Sciences, University of Birmingham.

Towards understanding the Dumfries Basin aquifer, SW Scotland

M. C. AKHURST[1], D. F. BALL[1], L. BRADY[2], D. K. BUCKLEY[3], J. BURNS[4],
W. G. DARLING[3], A. M. MACDONALD[1], A. A. McMILLAN[1],
B. É. Ó DOCHARTAIGH[1], D. W. PEACH[3], N. S. ROBINS[3] & G. P. WEALTHALL[5]

[1]*British Geological Survey, Murchison House, West Mains Road,
Edinburgh EH9 3LA, UK*

[2]*Scottish Water, 419 Balmore Road, Glasgow G22 6NU, UK*

[3]*British Geological Survey, Maclean Building, Wallingford, Oxfordshire OX10 8BB, UK
(e-mail: nsro@bgs.ac.uk)*

[4]*Scottish Environment Protection Agency, Rivers House, Irongray Road,
Dumfries DG2 0JE, UK*

[5]*British Geological Survey, Kingsley Dunham Centre, Keyworth,
Nottingham NG12 5GG, UK*

Abstract: The Dumfries Basin aquifer supports groundwater abstraction for public supply, agriculture and industry. Abstraction is concentrated in the western part of the basin, where falling groundwater levels and deteriorating water quality both reflect the effects of intense pumping. There are two bedrock units: a predominantly breccia–coarse sandstone sequence in the west, interfingering with a predominantly sandstone sequence in the NE and east. The basin is bounded by weakly permeable Lower Palaeozoic rocks, and is largely concealed by variable superficial deposits. Surface water flows onto the basin from the surrounding catchment via the Nith and the Lochar Water and their respective tributaries. Direct rainfall recharge occurs via superficial sands and gravels, especially in the north, and discharge is predominantly to the rivers in the central area rather than the sea. A picture is developing of two main aquifer types within the basin: the high-transmissivity western sector underlain by a fracture-flow system with younger water and active recharge and a high nitrate content, compared with the east where groundwater residence times are longer and the storage capacity is higher.

Many UK Permian aquifers have both intergranular (primary) and fracture (secondary) permeability. Secondary permeability tends to be controlled to some extent by local structural trends, and flow patterns within these aquifers can be complex. This is the case in the Dumfries Permian Basin aquifer, located in the lower part of the catchment of the River Nith in SW Scotland (Fig. 1). Here, the aquifer is further complicated by superficial cover of various lithologies ranging from gravel to clay, and by a complex interaction between river and aquifer as the River Nith crosses the basin. The basin aquifer receives runoff from surrounding uplands underlain by Lower Palaeozoic rocks. It discharges to the river upstream of Dumfries, and possibly also within the tidal reaches south of the town as well as to the sea at the lower end of the basin, through marine and alluvial silts and clays. The basin aquifer is some 10 km wide by 20 km long (Fig. 1), whereas the total catchment area of the rivers Nith and Lochar Water is 1380 km².

A 2-year programme of study has been completed, which draws on over 30 years of earlier piecemeal work. The study identifies and addresses knowledge gaps and develops a new and robust conceptual flow model for the basin aquifer. This paper describes the current understanding of the geology, and the conceptual groundwater flow model. Although uncertainties still remain, it is anticipated that current understanding will provide a good basis for the numerical simulation of groundwater flow in the basin.

Geological setting

The bedrock aquifer sequence of the Dumfries Basin comprises the Doweel Breccia and

From: BARKER, R. D. & TELLAM, J. H. (eds) 2006. *Fluid Flow and Solute Movement in Sandstones: The Onshore UK Permo-Triassic Red Bed Sequence.* Geological Society, London, Special Publications, **263**, 187–198.

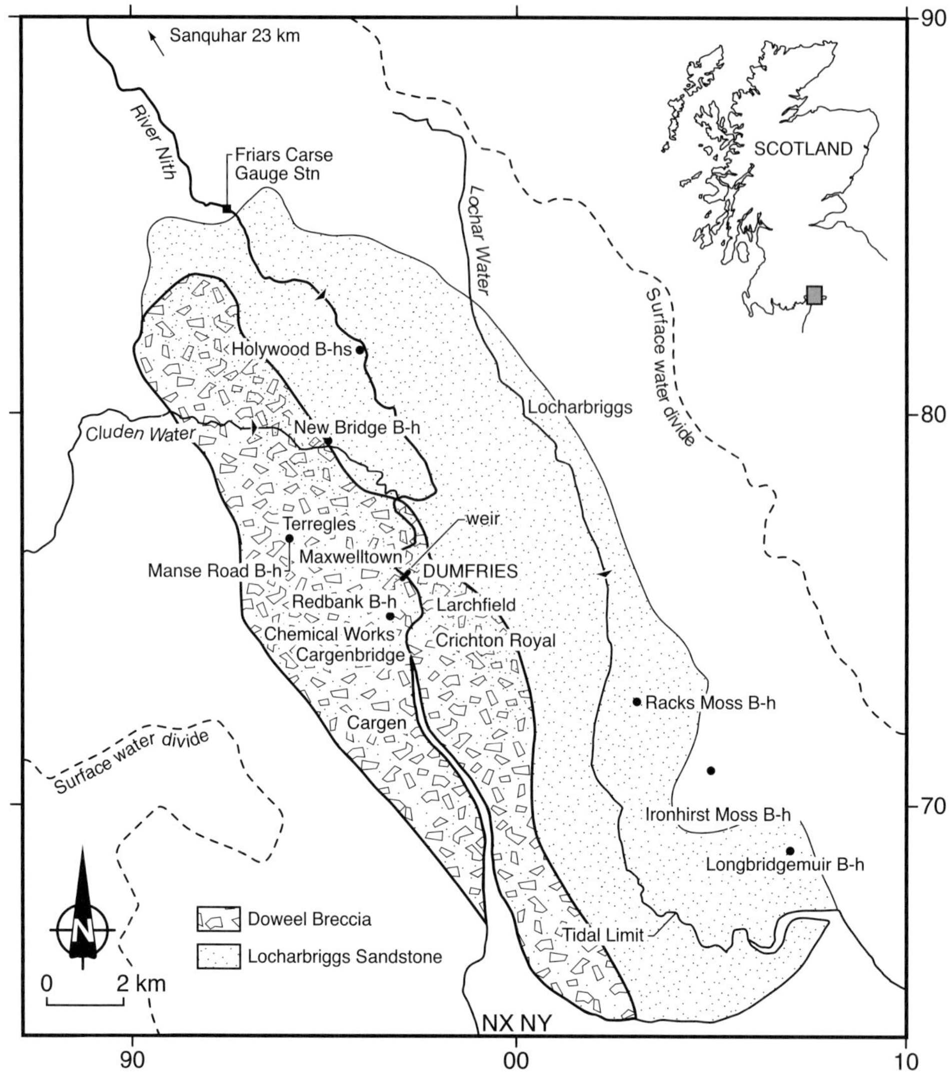

Fig. 1. The Dumfries Basin aquifer showing the location of sites mentioned in the text.

Locharbriggs Sandstone formations that are of Permian age (Fig. 1). The Doweel Breccia comprises predominantly sedimentary breccia interbedded with thin sandstones and underlies the western part of the basin (McMillan 2002). It extends eastwards towards the centre of the basin where it interfingers with the Locharbriggs Sandstone that underlies the eastern and northern parts of the basin. The Locharbriggs Sandstone itself comprises two facies: distinctive orange-red, cross-bedded sandstones, that are interpreted to have accumulated as a migrating dune field in arid desert conditions; and thin-bedded and laminated, orange-red,

silty sandstones containing pebbles of local derivation, generated by fluvial reworking of breccia and sandstone (Akhurst & Monro 1996).

The Permian basin-fill sequence unconformably overlies a steeply dipping succession of grey, fine-grained, wacke sandstone and mudstone of Silurian age that is intruded to the SW of the basin by the Criffel–Dalbeattie granodiorite (McMillan 2002). Carboniferous strata underlie the SE part of the Dumfries Basin, and continue east into the adjacent Annan Basin.

Regional gravity data suggest that the Permian basin-fill sequence has a maximum thickness of between 1.1 and 1.4 km, and the deepest part of the basin lies immediately to the north of Dumfries (Bott & Masson-Smith 1960). Bouger gravity data indicate that the basin is fault-bounded by a series of en echelon faults along its western and NE margins, although the precise locations are as yet unknown.

The superficial geology of the Dumfries Basin is dominated by an extensive development of glacigenic deposits, both granular and cohesive, deposited during the Dimlington (Late Devensian) glaciation. In the SE of the basin, rock ridges are aligned along a SE trend duc to the passage of ice from the NW. During deglaciation, retreat of the Nith glacier took place to the NW towards bedrock highs at Cargenbridge, Maxwelltown and Locharbriggs (Fig. 1). To the NW of Dumfries, the basal glacial deposit resting on the Permian strata is an overconsolidated sandy diamicton with wacke and sandstone clasts, and is regarded as the lodgement till of the Dimlington ice sheet. On lower lying ground the till is overlain by extensive discontinuous spreads of cobble gravel that form a distinctive morphology of 15 m high mounds (kames), kame terraces and ridges (eskers), the crests of which trend NW–SE. These deposits exhibit the normal faulting characteristic of an ice-contact origin. They are commonly overlain by a discontinuous, thin (usually less than 1 m), gravelly flow till. To the SE of Locharbriggs (in a more distal position in relation to the retreating ice front) fine sand, silt and clay, with dropstones, were deposited in ephemeral glacial lakes. These glaciolacustrine deposits are overlain by cross-bedded sand and pebbly gravel. A final re-advance of the Nith glacier deposited a moraine characterized by moundy topography, that extends in an arc between Locharbriggs and Cargenbridge. The deposits of the moraine comprise folded and sheared glaciolacustrine sand and silt which locally exceed 30 m in thickness.

Following final deglaciation, the rise in relative sea level resulted in the deposition of extensive Late-glacial estuarine deposits of bedded sand, clay and silt to the south of Dumfries (McMillan 2002). Isostatic rebound has left dissected remnants of these estuarine terraces at an elevation of 10–15 m above Ordnance Datum (AOD). Marine clays, once worked for brick-making, are overlain by bedded sands of littoral origin. In the last 1000 years, renewed estuarine and tidal-flat sedimentation associated with the main Post-glacial transgression has laid down fine-grained sediments that now form flat-lying ground up to 10 m AOD backing the coast. Extensive peat basins have developed locally on top of such Holocene alluvial, estuarine and tidal-flat deposits. The most recent alluvial sediments of the Dumfries Basin occupy the valley-floor floodplain and lowest terraces of the River Nith and its tributaries. These comprise gravel, sand and silt reworked from the glacigenic sequences.

Hydrogeological investigations

Introduction

Rainfall varies from 1000 mm in coastal areas to over 2000 mm on the high ground near the western watershed of the Nith catchment. Measurements at Crichton Royal, to the south of Dumfries town, show a 20% increase in long-term rainfall since records began in 1857, largely due to increased winter rainfall. This has resulted in a remarkable increase in winter flow in the River Nith of nearly 50% over the last 40 years (Scottish Environment Protection Agency unpublished data).

Until the late 1970s, the aquifer was exploited only on a modest scale to sustain mineral-water bottling, and groundwater supplies to a hospital, a milk-processing plant and large industrial premises (most notably rubber and chemical works). In 1978 the first public supply borehole, at Manse Road near the village of Terregles (Fig. 1), was commissioned. The long-term tested yield of this borehole is in excess of 4300 m^3 day^{-1}. The source originally had an artesian flow of 900 m^3 day^{-1} but the piezometric head has since declined below ground surface. In more recent years production boreholes have been drilled for both public supply and to sustain the demands of fish farming, so that the aquifer currently yields up to 30 000 m^3 day^{-1} (Gaus & Ó Dochartaigh 2000). However, when taken over an annual cycle, total abstraction may be less, perhaps only 15 000 m^3 day^{-1}. There is no licensing requirement under Scottish Law and there has been no regulatory control over this development. Today, areas of the western part of the aquifer may be considered overexploited, with excessive drawdown and declining water quality.

Nitrate concentrations are rising in parts of the aquifer and the area has recently been designated a 'Nitrate Vulnerable Zone'. The aquifer has also suffered pollution around a large industrial site due to historical production activities.

The Dumfries Basin aquifer has been studied periodically since the Manse Road borehole was commissioned. A series of piecemeal studies has provided an increasing understanding of how the aquifer works, but none has addressed its role within the catchment as a whole and none has been able to close the water balance successfully.

Early investigation – the 1980s

The aquifer was first described by Robins & Buckley (1988). Using borehole geophysics to identify inflow horizons, and comparing test pumping data with intergranular permeability measured in the laboratory on core specimens, it was estimated that up to 90% of total water input to boreholes in the Dowel Breccia derives from fracture flow. The porosity of the sandstones, however, ranges from 13 to over 20%. Generally, inflow horizons were observed up to 40 m below ground level, but occasional strong inflow horizons were observed to 100 m. The importance of secondary permeability is demonstrated in the Terregles area by strong hydraulic connection between two boreholes separated by 2200 m. These boreholes appear to be linked by fractures producing preferred flow directions at depth along a trend parallel to the western boundary fault (Robins 1990). No hydraulic contact is evident between one of these boreholes and an adjacent, but relatively shallow, observation borehole located 200 m away and perpendicular to this trend.

During the 1980s little attention was given to the sustainable potential either in terms of quantity or quality of available water. Neither recharge nor river–groundwater interaction were of concern in Scotland in what was then considered to be a 'land of plenty' (Robins *et al.* 2004). The occurrence of nitrate in groundwater and the risk from industrial pollution in the basin was similarly of limited concern. It was only in the latter part of the decade that intensive groundwater exploitation in the western part of the basin was recognized as an adverse influence on both borehole yields and discharge water quality. However, it was generally believed that groundwater flow was dominantly towards the coast and that groundwater discharge was to the sea (Ball *et al.* 1987); it was only later realized that the discharge was largely to the River Nith.

Investigations in the 1990s

Attention only turned to issues of resource sustainability, and in particular of recharge, during the 1990s. In places upstream from Dumfries, bedrock is separated from the River Nith by only a few metres of sand and gravel. Cheney & MacDonald (1993*a*) demonstrated that the hydraulic gradient perpendicular to and

towards the river in gravels in this area was about 0.01. Test pumping at three boreholes in sandstone with the shallow sand and gravel cased off, at Holywood (Fig. 1), sited at right angles to and 50, 100 and 150 m from the river bank, indicated that the dynamic hydraulic gradient was reversed in direction away from the river during the test. However, chemical sampling during the testing indicated that mixing with river water did not occur, and the specific electrical conductance (SEC) in the borehole discharge remained steady at 220 µS cm^{-1}, whereas that in the river was only 125 µS cm^{-1}.

Elsewhere new data on the piezometry of the aquifer indicated that the main lines of surface water drainage acted as the principal discharge areas for groundwater in both the Permian and superficial aquifers (Cheney & MacDonald 1993*b*). It was also now apparent that the low-permeability fluvio-marine silts and clays in the south of the basin, both onshore and offshore, allowed little groundwater flow directly to the sea, whilst also acting as a barrier to sea-water intrusion. Boreholes that penetrate the base of the silts and clays are confined.

Although there are a number of boreholes with high yields (Table 1), new information revealed that there is significant variation across the basin aquifer with some locations where yields are poor due to low transmissivity (Fig. 2). Individual borehole yields reflect the degree of bedding-parallel fracturing encountered in each borehole. In the SW of the basin three trial boreholes showed low yields. However, in the Crichton area east of the River Nith, boreholes penetrate a highly fractured zone of soft sandstone and their specific capacites are amongst the highest in Scotland, at over 10 l s^{-1} m^{-1} (Gaus & Ó Dochartaigh 2000).

New hydrochemical data confirmed that, although the groundwater is of weakly to moderately mineralized calcium–magnesium-bicarbonate type, it tends to be stable in the long term, with the major ions showing little variation with time. NO$_3^-$–N concentrations are, however, generally in the range 20–30 mg l^{-1}, with a distinct upwards trend discernible in some boreholes. For example, Robins & Ball (1998) reported an increase in the discharge from the Manse Road public supply borehole from 1 mg NO$_3^-$–N l^{-1} in 1978 to 6 mg NO$_3^-$–N l^{-1} in more recent years.

Localized pollution at a chemical works west of Dumfries was identified in the late 1980s. In the early 1990s the operators drilled several investigatory boreholes confirming that sulphate, nitrate, formaldehyde and chlorinated hydrocarbons (including trichloroethane) had

Table 1. *Estimates of current groundwater abstraction in the Dumfries Basin Aquifer*

Source name	Location (NGR*)	Borehole depth (m)	Average yield (m³ day⁻¹)
Manse Road Public Supply	NX 9400 7680	112	4300
Cargen Public Supply (two boreholes)	NX 9630 7210	115	2000
Larchfield Public Supply	NX 9810 7500	95	1100
Holywood Fish Farm (eight boreholes)	NX 9760 7780	*c.* 70	Total: 13 000
Terregles Fish Farm (six boreholes)	NX 9290 7730	30–130	Total: 8000
Nestlé (three boreholes)	NX 9690 7730	138–183	Total: 1380
Dupont	NX 9430 7450	71	800
Gates Rubber	NX 9890 7910	103	800
Crichton Royal Hospital	NX 9780 7330	150	1000
Golf Club	NX 9580 7570	54	300
Galloway Mineral Water	NX 9780 7330	138	150
Agriculture	Distributed widely		1000

*NGR, national grid reference.

entered the aquifer. Some residual source area soil remediation was undertaken and a number of scavenger wells were drilled to halt plume development and recover contaminated groundwater from the aquifer prior to remedial treatment and discharge into a sea outfall pipeline. Groundwater concentrations of all contaminants have reduced considerably during the past 10 years of corrective action.

Investigations since 2000

The principal uncertainties highlighted in the earlier work are listed below. These uncertainties preclude a comprehensive and effective management strategy for the aquifer being developed and operated:

- the catchment and aquifer water balance;
- the river–aquifer relationship along the River Nith;
- the location of areas of relatively high and low transmissivity;
- the renewable, exploitable resource both in the west and east of the basin.

A variety of different investigations have now been carried out in an attempt to address some of these issues. A holistic approach to understanding the basin aquifer has been taken that reviewed all previous work and in addition included:

- borehole drilling and site investigation;
- groundwater and surface-water chemistry investigation;
- fracture analysis of Permian sandstones at outcrop.

Three new boreholes have been drilled in the SE part of the basin in a previously unexplored area of the aquifer (Fig. 1: Racks Moss, Ironhirst Moss and Longbridgemuir). Two of these penetrated Locharbriggs Sandstone Formation with fine- to occasionally coarse-grained sandstones, one with breccia horizons, each with uncharacteristically poor hydraulic properties and low test yields (Ó Dochartaigh 2002). Twelve-hour specific capacity values of $0.3\,l\,s^{-1}\,m^{-1}$ compare unfavourably with up to $4\,l\,s^{-1}\,m^{-1}$ for sandstones in the north of the basin and $11\,l\,s^{-1}\,m^{-1}$ for the Doweel Breccia Formation in the western part of the basin (Fig. 2). Geophysical logging in the new boreholes indicates discrete flow horizons in the sandstone down to 100 m depth. The borehole at Ironhirst Moss, which was expected to penetrate the sandstone, revealed a weakly permeable purple siltstone of possible Devonian or Carboniferous age, but no Permian strata. The borehole logs (Table 2) also show the Permian basin sediments to thin towards the south and towards the eastern margin of the basin. The new boreholes reveal that this area appears to comprise weakly permeable material with occasional fractures offering limited movement of groundwater.

Analysis of pumped groundwater samples (Table 3) confirm that groundwater in the Doweel Breccia is slightly more mineralized, with HCO_3^- concentrations up to $200\,mg\,l^{-1}$, than water in the Locharbriggs Sandstone in the eastern part of the basin where HCO_3^- concentrations rarely exceed $100\,mg\,l^{-1}$ (MacDonald *et al.* 2003). CFC and SF_6 determinations indicate that the groundwater in the eastern part of the basin contains a greater proportion of old water than that in the western part of the basin. The older water has notably lower NO_3^- concentrations. Deteriorating water quality in the western area, for example at the Manse Road borehole,

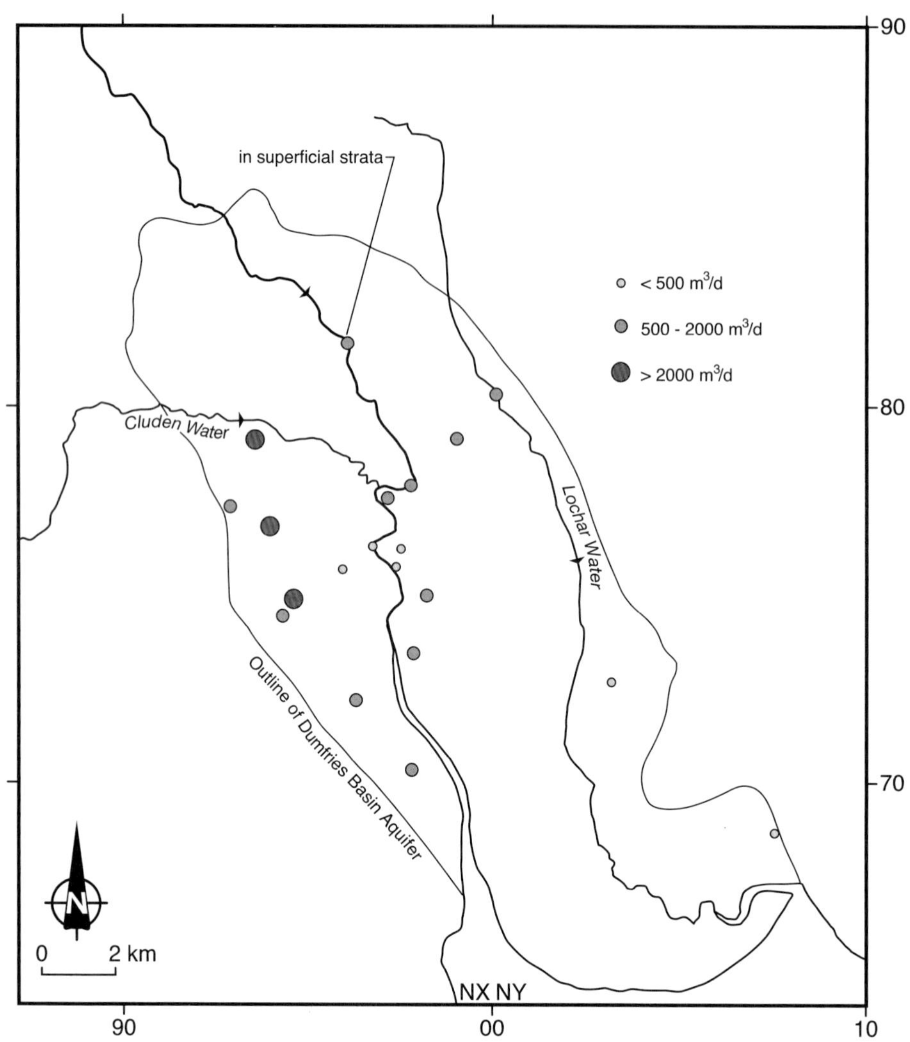

Fig. 2. Distribution of borehole yields in the Dumfries Basin aquifer (including historical data).

Table 2. *Results from three test boreholes in the eastern zone of the Dumfries Basin aquifer*

	Location (NGR*)	Elevation (m AOD)	SWL[†] (m AOD)	Test yield ($m^3\,day^{-1}$)	Specific capacity ($l\,s^{-1}\,m^{-1}$)	Stratigraphy
Racks Moss	NY 0297 7273	11	9.4	500	0.25	Rockhead –5 m AOD; Permian to –89 m AOD
Ironhirst Moss	NY 0490 7071	11	10.7	120	0.08	Rockhead +1 m AOD; Carboniferous–Devonian to –3 m AOD; Silurian proved to –89 m AOD
Longbridgemuir	NY 0699 6891	20	13.6	454	0.35	Rockhead +15 m AOD; Permian to –57 m AOD; Carboniferous to –60 m AOD

* NGR, national grid reference.
[†] SWL, static water level.

Table 3. *Groundwater chemistry (mg l^{-1}) (after MacDonald et al. 2003)*

Name	NGR*	Borehole depth (m)	pH	SEC* (μS cm^{-1})	Mg	Ca	Na	K	Cl	HCO$_3$	NO$_3$–N	SO$_4$	Si
Townfoot Farm[1]	NY 0625 7200	60	7.2	420	20.6	41.2	14.1	3.4	24.2	147	10.0	12.8	6.6
Holmhead Farm[1]	NY 0590 7133	30	7.1	680	19.2	86.1	18.4	3.3	33.1	162	28.5	40.1	5.06
South Bowerhouse Farm[1]	NY 0719 7034	62	7.4	410	15.5	40.2	11.3	1.7	18.9	108	10.7	26.6	3.67
Nestlé Borehole[2]	NX 9691 7733	183	7.3	390	19.7	42	12.9	1.3	20.1	159	3.3	20.3	5.5
Dundas Chemical[1]	NY 0002 7756	90	7.4	190	7.8	19.4	7.3	1.2	11.5	73.1	1.5	3.4	5.09
Gates Rubber, New Borehole[1]	NX 9893 7906	103	7.3	260	12.7	26.3	11.6	1.5	17	112	2.0	14.8	4.83
Galloway Mineral Water[2]	NX 9782 7321	150	7.1	360	18.1	38	11.1	1.6	17.8	127	6.3	21	5.13
Manse Road Public Supply[2]	NX 9402 7677	112	7.1	410	21.7	45.2	8.9	1.5	15.4	172	5.8	16.8	4.27
Larchfield Public Supply[2]	NX 9809 7506	95	7	350	17.6	35.5	10.8	1.4	17	117	6.1	25	5.33
Locharmoss[1]	NY 0118 7682	43	7.4	360	15.7	33.8	23.3	3.9	14.3	195	0.2	2.1	4.71
Shortridge Laundry[2]	NX 9679 7646	48	7.1	330	17.3	37	9.4	1.2	16.4	133	4.2	17.8	4.78
Hardthorne Road (30 m)[2]	NX 9356 7810	128	7	470	29.2	58.5	7.6	1.5	13.7	232	9.9	7.7	3.47
Hardthorne Road (125 m)[2]	NX 9356 7810	128	7.4	390	21.4	41.7	9.3	1.2	14.8	168	6.2	12.4	4.86
Manse Road observation[2]	NX 9392 7659	28	7.2	560	29.9	70.9	9.6	2.7	14.6	274	10.0	13.9	4.65
Terregles Fish Farm, bottom left[2]	NX 9295 7717	122	7.2	400	19.8	46.6	9.6	1.8	14.3	188	5.5	10.2	4.67
Terregles Fish Farm, dog kennel[2]	NX 9289 7737	122	7.3	480	23.1	58.7	10.1	1.7	16.1	226	7.1	13.4	4.07
Cargen Production[2]	NX 9638 7203	112	7.6	400	19.8	41.5	12.9	1.9	22.6	150	6.2	21.2	5.81
Greenmerse[2]	NX 9776 7048	75	7.1	370	12.9	32.4	25.9	2.3	23.4	122	2.0	29.4	5.37
Galloberry Farm[1]	NX 9680 8270	90	7.2	140	3.9	10.4	8.9	1.4	9.3	29.3	3.5	9.3	6.3
Hollywood Fish Farm[1]	NX 9754 7817	122	7.5	320	16.8	36.8	9.6	2	14.2	157	1.6	14.6	5.22
Dupont DGI[2]	NX 9450 7450	30	7	770	41	83.3	23.4	2.3	96	267	6.7	26.6	6.01
25th percentile			7.1	350	15.7	35.5	9.4	1.4	14.3	122	3.3	12.4	4.67
Median			7.2	390	19.2	41.2	10.8	1.7	16.4	157	6.1	14.8	5.06
75th percentile			7.4	420	21.4	46.6	12.9	2.3	20.1	188	7.9	21.2	5.37

* NGR, national grid reference; SEC, specific electrical conductance.
[1] Penetrating dominantly Locharbriggs Sandstone; [2] penetrating dominantly Doweel Breccia.

appears to be due to an increasing component of poorer quality, nitrate-rich younger water coming in from nearby recharge areas. Variations in groundwater chemistry in the basin do not appear to correlate with changes in drift lithology.

Out of 13 recently sampled sources in bedrock, six contain NO_3^-–N in the range 4 to 7 mg l^{-1}, with only three below 2 mg l^{-1}. Additional sampling of 29 private sources in the aquifer revealed that 60% of them had concentrations in excess of 5 mg NO_3^-–N l^{-1} (Ball 2002). Water quality in the superficial deposits remains largely unknown. To the north of Dumfries, elevated concentrations of iron and manganese in groundwater sampled from alluvial gravels adjacent to the river Nith have been noted. The surface water quality in the Nith north of the Dumfries Basin towards Sanquhar reflects discharges from former coal mines in the Sanquhar coalfield (Jameson 2001).

Given that the last 20 years have been the wettest since the 1850s, it is notable that borehole hydrographs indicate a decline in level at Redbank, in the west of the Dumfries Basin, since 1981 (Fig. 3), and that some other sites in the west of the basin also reflect this trend. However, at Newbridge, to the NW of the centre of the basin, the groundwater level is relatively stable. The implication is that the decline in the west is a result of intensive abstraction in this area, which commenced in the late 1970s and gathered pace in the early 1980s. Careful examination of the Redbank hydrograph indicates a stepped, rather than continuous, decline, with steps corresponding to times when new groups of abstraction boreholes were commissioned.

Annual variations in piezometric level in both the western and eastern parts of the Permian aquifer are less than 2 m. Piezometric heads tend to peak in February, with a steady natural decline through spring. Water levels in the Permian are above the rock–superficial strata interface across much of the basin (Cheney & MacDonald 1993*b*). Higher ridges of bedrock occur in some places, such as around Larchfield south of Dumfries, and there may be up to 15 m of unsaturated rock beneath some of these. In general, surface topography is reflected by the elevation of the piezometric surface. Confined conditions occur under low-lying areas south of Dumfries across the floodplain of the River Nith. Artesian flow occurs wherever the surface level is at or below 10 m AOD and laminated silty clay overlies the main aquifer. Elsewhere, individual fractures, and sandstone horizons separated by breccia, may be at different heads, as observed in a number of geophysical flow logs that have been measured in boreholes in the western part of the basin (Buckley 2000).

The relationship between surface water and groundwater, as indicated by river gauging and recession analysis carried out by the Scottish Environment Protection Agency, suggests that the Cluden Water does not gain from groundwater whereas the Lochar Water does show some small gain. Comparison of relative river stage and groundwater level in the gravels and underlying sandstone at Holywood indicates that groundwater discharges to the River Nith at all stages of river flow, save for brief periods of flood (Fig. 4). Using the data from test pumping and river-stage measurement conducted at this site and reported by Cheney & MacDonald (1993*a*), simple Darcian calculation suggests that under normal flow conditions some 5000 m^3 day^{-1} could discharge from the

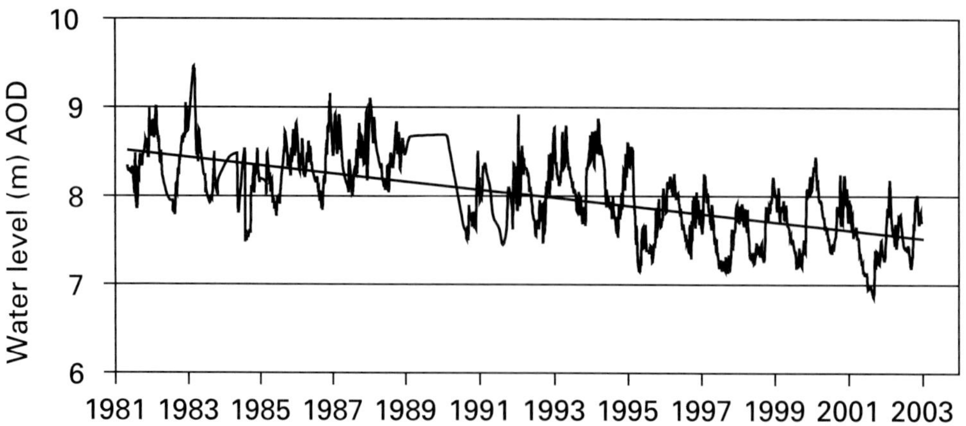

Fig. 3. Redbank borehole hydrograph showing a linear regression line.

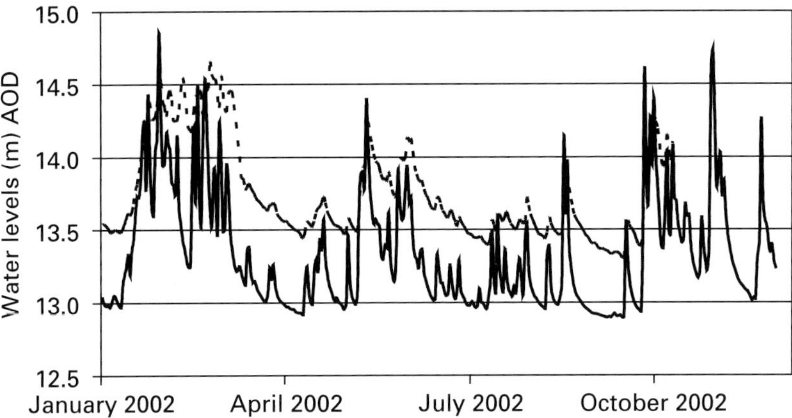

Fig. 4. Holywood borehole hydrograph (dotted line) and stage in the River Nith at Holywood, the latter is estimated (±0.1 m) from river gauging at Friars Carse.

gravel aquifer into the river along a 1 km-length of river (given a transmissivity of 250 m^3 day^{-1} m^{-1}; hydraulic gradient of 0.01 and width of flow of two 1000 m long river banks).

Statistical analysis of available measurements in the Doweel Breccia indicates that the intergranular hydraulic conductivity is 7 × 10^{-5} m day^{-1} (Wealthall 2002) and the bulk hydraulic conductivity is 10^{-3} m day^{-1}, increasing to 10^3 m day^{-1} where fracturing is present. Fracture flow concentrates on lines of weakness which in most of the Dumfries Basin are dominantly subhorizontal, including bedding-plane joints between sandstone and breccia, with subordinate subvertical fractures. As a consequence, horizontal permeability is commonly greater than vertical permeability. In the SW of the basin there is some subvertical fracturing that trends parallel to the western boundary, i.e. NW–SE. The boundary fault to the NE of the basin also imparts preferential orientation in the east of the basin. In some boreholes and at some outcrops shallow seepages occur from bedrock fractures above the water table after periods of rain (Wealthall 2002).

A conceptual model of groundwater flow

There appear to be two basic aquifer types in the basin. In the west, thin sandstone units are interbedded with breccias. Fracturing associated with sandstone–breccia boundaries appears to form the principal pathway for groundwater movement. It is these near-horizontal fractures that provide a broad interconnection between many of the main abstraction boreholes in the west. Coupled with the less well-developed near-vertical fracture system, the horizontal

fractures allow the development of a dynamic groundwater system, with young water moving relatively rapidly through the shallow fractures in the western part of the aquifer. The system is recharged via near-vertical fractures principally in the NW and other areas of higher ground where granular superficial deposits prevail. This contrasts with the eastern part of the Permian basin aquifer where breccia is largely absent and sandstone predominates. Here, a combination of higher specific yield, low groundwater abstraction and a covering of silty marine clay over part of the area has resulted in the presence of older groundwater that has a significantly lower nitrate concentration (the data in Table 3 are from pumped borehole samples that represent degrees of mixing between old and recently recharged water, so that all have elevated NO$_3$ concentrations to some extent).

Lack of extension of the main fracture system to the eastern part of the aquifer tends to isolate groundwater in the area east of the River Nith. Groundwater flow does not cross the line of the river, which, for the most part, acts as a groundwater sink wherever there is hydraulic connection with the river and the aquifer.

Groundwater occurs in both the Permian bedrock and the granular superficial deposits. The transmissivity of the Permian aquifer ranges up to 10^3 m^2 day^{-1} in the west, although the aquifer is not uniformly permeable with depth. Storage of groundwater, particularly west of the River Nith, is largely restricted to sandstone horizons, which account for only 20% of the aquifer thickness in the upper 100 m of the aquifer. In general, the hydraulic response is expected to be confined except where the aquifer is in contact with permeable superficial deposits beneath which an unconfined response will occur.

The Permian basin receives direct rainfall recharge through alluvial and glaciofluvial sands and gravels, and also receives some indirect recharge from losing rivers and streams in the upper part of the basin. In the main, however, the rivers are gaining from groundwater. The basin is traversed by the River Nith, but the eastern part of the basin is hydrologically separate and is part of the Lochar Water catchment. The effective surface catchment of the basin is much greater than the outcrop area of the aquifer, and encompasses the high hills surrounding the basin as far north as Sanquhar.

Direct recharge from rainfall occurs across the higher ground within the basin to the north and west of Dumfries. In these areas sand and gravel deposits promote infiltration to the water table. Discharge from the Permian aquifer is principally via superficial gravels to the River Nith north of Dumfries, with some also to the Lochar Water and possibly some to the sea.

The current two-dimensional horizontal conceptual model of groundwater flow in the Dumfries Basin aquifer is shown schematically in Figure 5. It is based on the work carried out during the 1980s and 1990s, and the additional investigation during the 2-year study that was targeted at infilling gaps in knowledge.

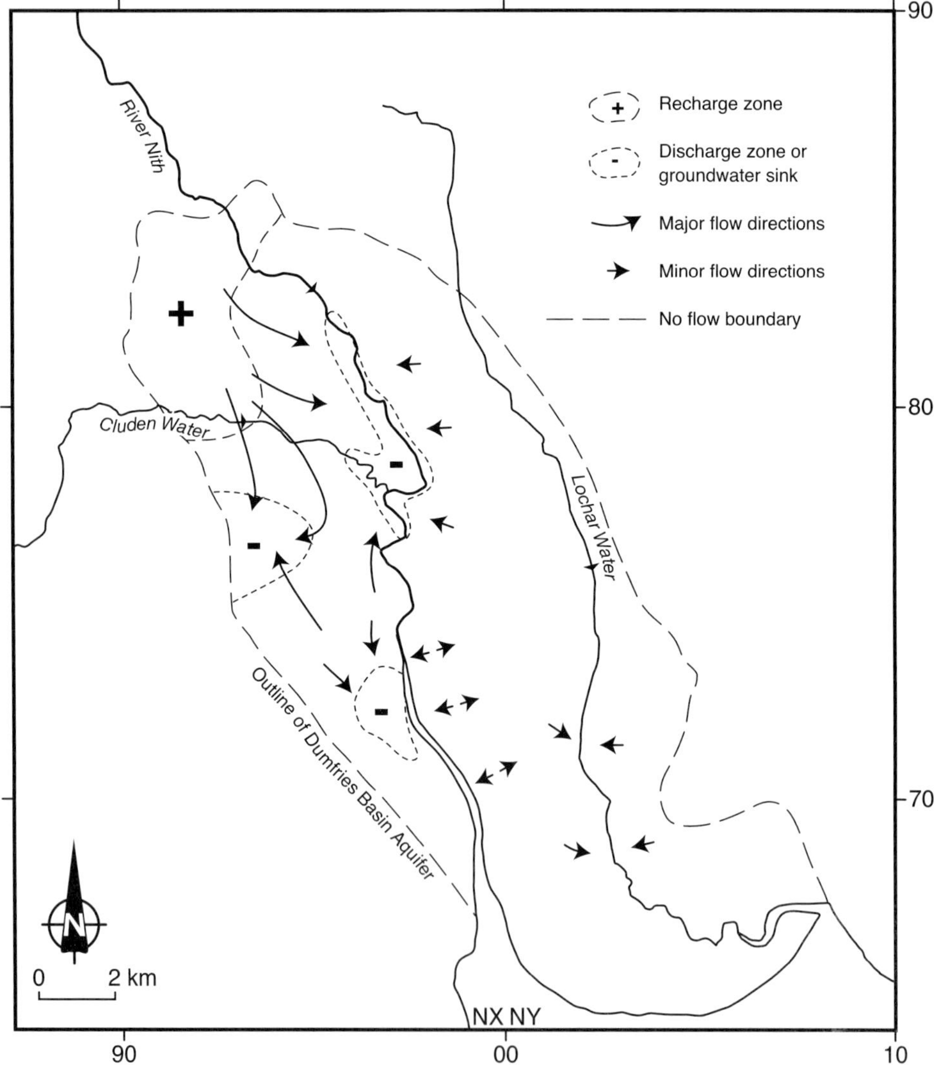

Fig. 5. Schematic conceptual flow model of the Dumfries Basin aquifer.

Although the conceptual groundwater flow model for the aquifer has grown in complexity as investigation has progressed, confidence in the model has remained weak. The current model has the following features.

- The basin edge is effectively a no-flow boundary given the comparatively limited hydraulic conductivity of the surrounding Palaeozoic rocks.
- Rainfall recharge occurs to the bedrock aquifer via superficial sands and gravels that principally occur in the NW and central part of the basin. Rainfall recharge is greatly inhibited in areas underlain by clay or silt-grade superficial material and peat.
- The drift and bedrock aquifers are not always in hydraulic contact.
- Some surface water indirectly recharges the aquifer, probably in the upper or northern-most part of the basin.
- Piezometry indicates both lateral flow towards the River Nith and groundwater sinks in the western central parts of the basin that are intensively pumped.
- Marine and alluvial silts inhibit discharge from the basin directly to the sea.

The water balance for the basin will need to be developed from numerical simulation yet to be carried out. Present estimates on flows include the following.

- Effective rainfall – estimated at 436 mm year^{-1} (Gaus & Ó Dochartaigh 2000); recharge is constrained by diversion to field drains and runoff induced over clayey areas, and a total average annual rainfall recharge volume of only 25 Mm^3 year^{-1} is estimated, equivalent to 68 000 m^3 day^{-1}.
- Surface water ingress to the aquifer – unknown but likely to be small.
- Loss from sewers and water mains, irrigation returns and other discharges to the ground – likely overall to be small (much of the Dumfries sewerage system is beneath the water table and is gaining from groundwater rather than losing to it).
- Groundwater discharge to the River Nith – potential for 5000 m^3 day^{-1} per km of river in the Holywood area, some also to the Lochar Water.
- Groundwater discharge directly to sea – unknown, but believed to be small.
- Groundwater abstraction – between 15 000 and 30 000 m^3 day^{-1}.

The implied balance, albeit only coarse, suggests abstraction may represent a significant component of the available renewable resource.

Conclusions

Investigation of the Dumfries Basin aquifer has been developing over the last 25 years. A considerable body of information has now been collected and a range of hypotheses and conceptual models describing the hydrogeology of the basin have been created over the years, and subsequently revised in the light of new information. As more has been learned about how the basin reacts under certain conditions, new questions have had to be asked and further investigation undertaken. The latest hypothesis on the hydraulics of the basin identifies a number of recharge and discharge mechanisms. The conceptualization has yet to be tested with a numerical simulation, so it may be amended and replaced by investigation into new areas and hitherto unrecognized controls or constraints.

A picture is developing of two main aquifer types within the basin: the high-transmissivity western sector underlain by a fracture-flow system with younger water and active recharge and a high nitrate content, compared with the east where groundwater residence times are longer and the storage capacity is higher.

The views expressed in this paper do not necessarily reflect the corporate views of either Scottish Water or the Scottish Environmental Protection Agency. The paper is published by permission of the Director, British Geological Survey (NERC).

References

AKHURST, M.C. & MONRO, S.K. 1996. Excursion 9. Dumfries: a Permian desert. *In*: STONE, P. (ed.) *Geology in South-west Scotland: An Excursion Guide.* British Geological Survey, Keyworth, Nottingham, 80–87.

BALL, D.F. 2002. *Additional Measurement of Nitrate Concentrations in Groundwater in the Nith Catchment.* Http://www.scotland.gov.uk/library5/environment/bgsnith.pdf BGS Commissioned Report, **CR/02/262N**.

BALL, D.F., BUCKLEY, D.K., PERKINS, M.A. & ROBINS, N.S. 1987. *The New Red Sandstone Aquifers of Scotland.* Scottish Development Department Report, **ARD 17**.

BOTT, M.H.P. & MASSON-SMITH, D. 1960. A gravity survey of the Criffel Granodiorite and the New Red Sandstone deposits near Dumfries. *Proceedings of the Yorkshire Geological Society*, **32**, 317–332.

BUCKLEY, D.K. 2000. Some case histories of geophysical downhole logging to examine borehole site and regional groundwater movement in Celtic regions. *In*: ROBINS, N.S. & MISSTEAR, B.D.R. (eds) *Groundwater in the Celtic Regions: Studies in Hard Rock Hydrogeology and Quaternary Hydrogeology.* Geological Society, London, Special Publications, **182**, 219–237.

CHENEY, C.S. & MACDONALD, A.M. 1993*a*. *Exploratory Drilling and Aquifer Testing in the Stranraer, Dumfries and Moffat Areas, 1992*. BGS Technical Report, **WD/93/6**.

CHENEY, C.S. & MACDONALD, A.M. 1993*b*. *The Hydrogeology of the Dumfries Basin*. BGS Technical Report, **WD/93/46**.

GAUS, I. & Ó DOCHARTAIGH, B.É. 2000. Conceptual modelling of data-scarce aquifers in Scotland: the sandstone aquifers of Fife and Dumfries. *In*: ROBINS, N.S. & MISSTEAR, B.D.R. (eds) *Groundwater in the Celtic Regions: Studies in Hard Rock and Quaternary Hydrogeology*. Geological Society, London, Special Publications, **182**, 157–168.

JAMESON, C.F. 2001. *An evaluation of the water quality patterns of a major river–aquifer system, Dumfries, Scotland*. MSc Dissertation, Water Resources Engineering Group, University of Newcastle-upon-Tyne.

MACDONALD, A.M., DARLING, W.G., BALL, D.F. & OSTER, H. 2003. Identifying trends in groundwater quality using residence time indicators: an example from the Permian aquifer of Dumfries, Scotland. *Hydrogeology Journal*, **11**, 504–517.

MCMILLAN, A.A. 2002. *Geology of the New Galloway and Thornhill District*. Memoir of the British Geological Survey, Sheets 9W and 9E (Scotland).

Ó DOCHARTAIGH, B.É. 2002. *Initial Report on Borehole Drilling and Testing in the Dumfries Aquifer*. BGS Technical Report, **IR/02/153**.

ROBINS, N.S. 1990. *Hydrogeology of Scotland*. HMSO, London.

ROBINS, N.S. & BALL, D.F. 1998. Groundwater exploitation and development: some current issues in Scotland. *Journal of the Institution of Water and Environmental Management*, **12**, 440–444.

ROBINS, N.S. & BUCKLEY, D.K. 1988. Characteristics of the Permian and Triassic aquifers of south-west Scotland. *Quarterly Journal of Engineering Geology*, **21**, 329–335.

ROBINS, N.S., BENNETT, J.R.P. & KULLEN, K.T. 2004. Groundwater versus surface water in Scotland and Ireland – the formative years. *In*: MATHER, J.D. (ed.) *200 Years of British Hydrogeology*. Geological Society, London, Special Publications, **225**, 183–191.

WEALTHALL, G.P. 2002. *Conceptual aspects of DNAPL penetration in fractured rocks*. PhD thesis, University of Sheffield.

DC electrical properties of Permo-Triassic sandstone

STEVE TAYLOR & RON BARKER

School of Geography, Earth and Environmental Sciences, University of Birmingham,
Edgbaston, Birmingham, UK (e-mail: s.b.taylor@bham.ac.uk)

Abstract: The DC electrical properties of fully and partially saturated, poorly cemented
Permo-Triassic sandstone samples have been measured in the laboratory and the results
analysed using three popular models. The results of this work, undertaken on samples from
the Wildmoor Formation, indicate that the Permo-Triassic sandstone of the UK is a typical
shaly sandstone, which cannot be satisfactorily modelled using the simple conventional
relationships proposed by Archie for application in the oil industry to non-shaly formations. Application of the more sophisticated models of Waxman–Smits and Hanai–
Bruggeman more faithfully model the electrical response of the sandstone. In addition, the
derived parameter estimates are better able to characterize the electrical properties of the
rock and correlate better with other independently determined hydraulic properties.
Application of these models in groundwater investigations will therefore lead to better and
more useful estimates of hydraulic parameters. More importantly, this knowledge will allow
more accurate quantitative interpretation of electrical monitoring of the vadose zone.

DC electrical resistivity measurements have
long been used to estimate the hydraulic and
petrophysical properties of reservoir and
aquifer rocks (e.g. Archie 1942; Keller 1953;
Barker & Worthington 1973). Such properties
are key to the understanding of fluid-flow
processes in rocks for the purposes of determining, for example, hydrocarbon reserves, aquifer
vulnerability and contaminant migration. Both
change in saturation level and change in the
conductivity of the saturating fluid affect the
bulk electrical response of a rock. Hence, resistivity measurements may be used to monitor
and understand changes in the vadose zone,
enabling better prediction of the distribution
and flow of both water and contaminants in the
near subsurface for the purposes of aquifer
vulnerability assessment and the understanding
of recharge processes.

Much work has been undertaken in attempt-
ing to relate the electrical properties of important sandstone oil and water reservoir rocks to
their hydraulic properties, such as porosity and
permeability. The first generally accepted
empirically determined relationship was that
presented by Archie (1942), who demonstrated
that the bulk resistivity of a sandstone was
directly proportional to the resistivity of its
saturating pore water. Other studies (Winsauer
et al. 1952; Mendelson & Cohen 1982) confirmed
the basic Archie relationship, although different
constants of proportionality were determined.
In most of this work measurements were made
on fully saturated samples, but a few studies
have been undertaken on partially saturated

rocks. For example, some authors have
examined the variation of electrical resistivity
with saturation for sandstone samples from
North America (Archie 1942; Vinegar &
Waxman 1984; Knight 1991), and from oilfields
from various locations around the world (Keller
1953; Waxman & Smits 1968), whilst similar
work has been conducted on other rock formations such as rhyolitic tuffs composed primarily
of quartz and alkali feldspar (Roberts & Lin
1997).

It was recognized at an early stage in the
study of sandstone electrical properties (e.g.
Patnode & Wyllie 1950; Waxman & Smits 1968)
that the electrical response of sandstone was not
straightforward and that the presence of
electrically conductive clay minerals caused
considerable complexity. The simple Archie
relationships appeared to be only valid for clean
(clay-free) formations under the high-salinity
conditions that prevail in oil reservoirs, and that
these relationships are inappropriate for
describing shaly sandstones or for sandstones
saturated with low-conductivity formation
waters. It is precisely these conditions that are
encountered in the Permo-Triassic sandstone in
the UK and in similar aquifers elsewhere in the
world.

More recently, considerable work on shaly
sandstone formations has led to the formulation
of numerous empirical relationships to describe
the observed data. Much early work concentrated around the need to better interpret oil
industry wireline logs (Patnode & Wyllie 1950;
Keller 1953; de Witte 1957; Waxman & Smits

From: Barker, R. D. & Tellam, J. H. (eds) 2006. *Fluid Flow and Solute Movement in Sandstones: The Onshore
UK Permo-Triassic Red Bed Sequence.* Geological Society, London, Special Publications, **263**, 199–217.
0305-8719/06/\$15 © The Geological Society of London 2006.

1968; Worthington 1982), whilst other authors developed relationships for the interpretation of measurements for hydrogeological studies (e.g. Worthington & Barker 1972; Worthington 1973, 1977). A more theoretical approach to the question of conduction mechanisms in shaly sands has been undertaken by Clavier *et al.* (1977), Berg (1995) and de Lima (1995); a review of some of these models is provided by Bussian (1983) and Worthington (1985).

Detailed studies on the variation of resistivity with partial saturation have been conducted by a number of authors, concentrating initially on samples from oilfields. For example, Keller (1953) presented resistivity v. saturation data for initially brine-saturated, oilfield, sandstone samples. The data indicated that resistivity increases with decreasing saturation and that this increase was most pronounced at lower saturation. More recent studies have investigated conditions more applicable to the vadose zone. Knight (1991) performed measurements to observe the difference in electrical response whilst saturating and draining three sandstones from the Alberta Basin. Her data showed that hysteresis effects, although not always repeatable, were observable at mid saturation levels and further complicated the electrical response of partially saturated shaly-sandstones. Roberts & Lin (1997) examined the variation in electrical resistivity with saturation of tuff samples from the Nevada test site. The data displayed the typical drainage response, and modelling of the electrical spectra was undertaken to investigate the relationship between the electrical and hydraulic properties using the equation of Waxman & Smits (1968).

The Permo-Triassic sandstone is the second most important aquifer in the UK, and much work has been undertaken to determine relationships between its electrical and hydraulic properties (e.g. Barker & Worthington 1973). This early work showed that the basic Archie relationships were not applicable in the case of UK Permo-Triassic sandstone, the properties of which were typical of a shaly sandstone. Although various attempts to model the shaliness have been undertaken (e.g. Worthington & Barker 1972; Barker & Worthington 1973), no fully satisfactory approach has been reported. The work presented here forms part of an investigation into the electrical properties of the Triassic sandstone in the English Midlands with the aim of better characterizing the electrical response of the sandstone. This paper focuses on the effects of partial saturation on the electrical response and on determining appropriate relationships to model the electri-

cal response. Such relationships are important in non-invasive hydrogeological studies, and have important implications in terms of monitoring and modelling the variation in saturation of the unsaturated zone of Triassic sandstone aquifers. In this paper we briefly describe the experimental procedure and the data collected for both partially and fully saturated sandstone samples. The data are then modelled using three models, those of Archie (1942), Waxman & Smits (1968) and the Hanai–Bruggeman mixing law equations (Bussian 1983). Finally, a comparison is made between the hydraulic parameters determined from the modelling and independently determined hydrogeological parameters.

Experimental procedure

Description of samples

The samples used in this study were obtained from Sandy Lane Quarry [grid reference SO955760] located near Bromsgrove in the English Midlands. The quarry is dug into the Wildmoor Formation of the Triassic sandstone, which forms part of a major aquifer for the region. The quarry provides good exposure, with a thick unsaturated zone and enables representative samples of the formation to be obtained. Part of the quarry is used as a landfill site operated by Cleanaway Ltd and has a permanent electrode array installed beneath the landfill liner (White & Barker 1997). This array can be used to monitor the electrical resistivity of the sandstone beneath the liner, and results of experiments from samples obtained from the site may be used to interpret these changes in terms of saturation.

At the quarry approximately 30 m of uniform, fine-grained Wildmoor strata are visible. Indistinct upwards-fining cycles and large-scale cross-bedding can be detected in the weathered faces of the quarry, along with a few mudstone bands (approximately 1 cm thick) and partings with large black and white mica flakes. Along the quarry wall a number of marl bands are present. Directly above these exist dark, damp, sand-rock layers from which vegetation is observed to grow, evidence of perched water tables. These layers are also found to contain a considerable amount of mica. The sandstone shows distinct bedding planes defining layers from 0.15 to 0.75 m and dipping at about 10° to the south.

Beneath the quarry floor, the Wildmoor Sandstone continues for a depth of approximately 75 m. Full sequences of the Wildmoor

Sandstone are present in boreholes at Washingstocks (a public supply borehole less than 2 km south of Sandy Lane) and Webheath, located 10 km to the SE, where the formation attains its maximum thickness of 134 m. The formation includes upwards-fining rhythms, which commence with a medium- to coarse-grained or even pebbly sandstone passing upwards through cross-bedded, fine-grained sandstone into planar-bedded, fine-grained sandstones and mudstones (Wills 1976).

The main rock is a red-brown, fine- to medium-grained, well-rounded and well-sorted sandstone. Its mineralogy is predominantly quartz with a few white feldspar grains. There is little to no calcite cement, hence its previous use as a moulding sand. The distinctive deep red-brown colour of the sandstone is probably due to the presence of iron oxides that coat the sand grains. Old *et al.* (1991) describe a typical sample taken from the quarry as containing the following particle sizes: 12% clay, 81% silt and sand finer than 0.2 mm, and 7% sand of 0.5 mm size.

A total of 26 core-plug samples were drilled from blocks of rock chiselled from the freshly quarried rock face at three locations around the quarry (samples from each location are prefixed SL1, SL2 and SL3) and also from a borehole drilled to just below the water table some 10 m below the quarry floor (samples prefixed SL4). The very nature of the sampling means that there is an inherent bias towards more cemented samples from the formation, as the less-cemented rock samples are not preserved in drilling. The core plugs had a nominal diameter of 33 mm, and lengths between 29 and 73 mm.

Visual analysis of the sandstone samples indicates that they differ in grain size, friability and colour. Samples from blocks SL1 and SL2 are typically bleached, medium- to coarse-grained, rounded, poorly sorted sandstone. Samples from block SL3 are orange-red in colour, fine grained, rounded, moderately to well sorted, and typically more friable than samples from blocks SL1 and SL2. Samples drilled from the borehole (labelled SL4) are typically a pale orange-red in colour, medium to fine grained, rounded and moderately to well sorted. These samples are similar to those obtained from block SL3 in appearance, but are slightly paler and typically less friable. X-Ray diffraction analysis indicates that all samples contain illite and chlorite clays with some mixed layer smectite–illite clay species, although in varying amounts. These clays appear in all samples and are evenly distributed throughout each of the samples. The clays are restricted to the surface of grains with the pore throats generally clay free, suggesting that the clays do not have a controlling influence on the permeability. These clays were less prominent in samples from blocks SL1 and SL2.

Measurement technique

In order to undertake low-frequency electrical measurements on poorly cemented sandstone core samples, while avoiding problems associated with electrode contact resistance and electrode polarization, an experimental procedure was developed that used a standard four-electrode measurement cell with reversible silver–silver-chloride electrodes and agar gel between the current electrodes and the sample (Fig. 1) (Taylor 2000; Taylor & Barker 2002). A Wavetek voltage-controlled waveform generator was used to generate a 1 Hz sinusoidal signal across the current electrodes with a peak-to-peak voltage of approximately 0.25 V. The voltage across a standard variable resistor and the sample were passed through a comparator circuit consisting of a pair of FET input amplifiers and then into a Le Croy 9310 digital oscilloscope. The digital oscilloscope enabled the quality of the signals across both the resistor and sample to be monitored and allowed for the direct subtraction of the signals to obtain a null. The cell resistance was obtained from the standard variable resistor recorded at resistance balance and, when corrected for the geometry of the sample and sample holder, yielded the bulk resistivity of the sample. Throughout the experiment the cell and core samples were maintained at a near-constant temperature by storing them inside a Gallenkamp laboratory incubator, which also behaved as a Faraday Cage.

Experimental procedure

Samples were initially saturated under vacuum using a de-aerated synthetic groundwater solution with a resistivity of 6.25 Ωm (a conductivity of 1600 μS cm^{-1}). The synthetic groundwater contained ions in the following weight ratios: 30 mg l^{-1} Ca^{2+}, 15 mg l^{-1} Mg^{2+}, 17 mg l^{-1} Na$^+$, 71 mg l^{-1} Cl$^-$, 60 mg l^{-1} SO$_4^{2-}$ and 15 mg l^{-1} HCO$_3^-$, a chemistry that closely matched the cation concentrations in the Triassic sandstone aquifer measured at the University of Birmingham borehole (Mitchener 2003). The core plugs were returned to atmospheric pressure after approximately 1 h and then left to soak in the solution for 1 week before commencing any measurements. Measurements were also made on samples initially saturated with groundwater

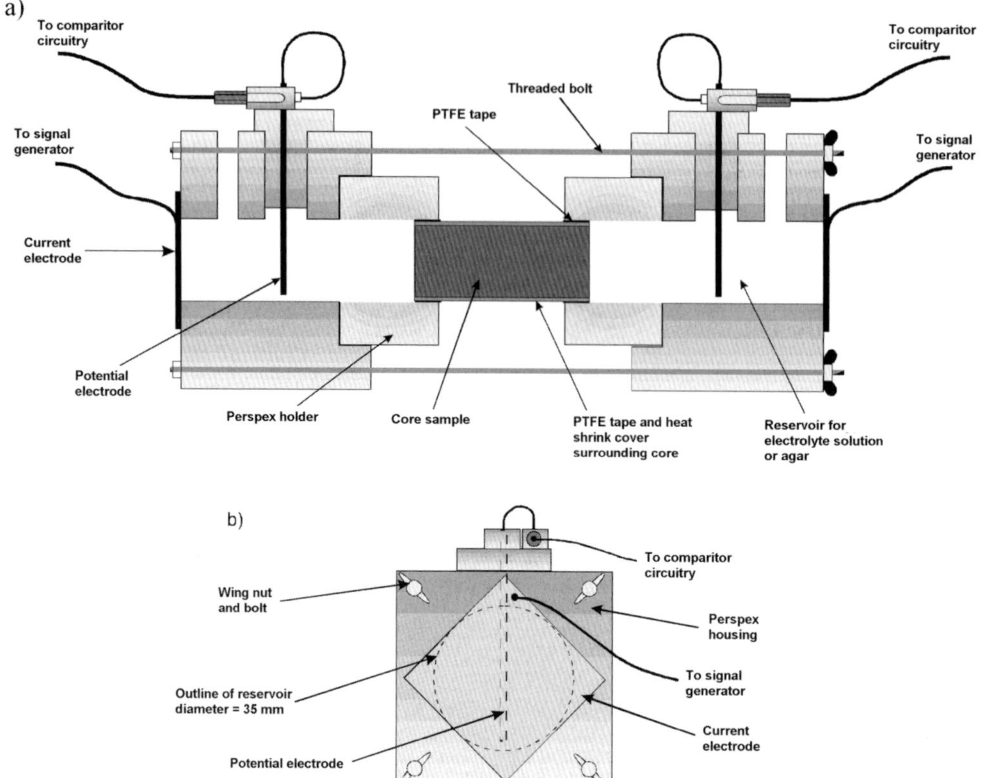

Fig. 1. Measurement cell design. (**a**) Side view and (**b**) end view.

solutions with resistivities of 25 (400 µS cm^{-1}) and 12.5 Ωm (800 µS cm^{-1}).

Measurements were made both on fully and partially saturated samples. In order to observe the change in resistivity of fully saturated samples with changing fluid resistivity, measurements were made on samples saturated with salinities ranging from 0.125 to 32 g l^{-1} (approximately 0.002–0.5 M), with electrolyte concentration increasing by a factor of 2 at each stage (Worthington & Barker 1972).

The partially saturated measurements were made only on samples that were initially fully saturated and then allowed to drain. This was to avoid the potentially problematic effects of hysteresis (Knight 1991; Roberts & Lin 1997), which is due to different fluid geometries being set up during drainage and imbibition. During drainage it is observed that resistivity generally increases with decreasing saturation and that the curves are repeatable. During imbibition, the resistivity v. saturation curve departs significantly from that observed during drainage and the measured resistivity is consistently less at low and mid-saturation levels. At higher saturation levels the measured resistivity rapidly returns to values similar to those seen during drainage.

The samples were dried through evaporation by placing them upright on filter papers in the open air and allowing the pore water to drain freely under gravity and to evaporate from the surface of the core sample. Saturation levels of approximately $S_w = 0.35$ were obtained in this way. Lower saturation levels were obtained using a desiccator where levels as low as $S_w = 0.02$ were obtained for some samples. Electrical measurements were made at saturation intervals of approximately 0.05–0.10. The time required for drainage meant that measurements were made approximately daily with a complete drainage curve taking approximately 3 weeks to achieve.

Saturation, defined as the volume fraction of the pore space filled with saturant, was determined by weighing the samples both before and after each electrical measurement. This was necessary to determine the level, if any, of imbibition into the sample from the agar or filters, or loss of mass where the samples were in contact with the filter papers. As a consequence of these processes a correction to the saturation level had to be made at the end of the experiment when the saturated and dry masses of the samples were once again determined. This equated to an additional error on each saturation value of less than $S_w = 0.02$.

On reaching the required level of saturation, each core plug was placed in the electrical measurement cell and its resistivity determined.

To demonstrate repeatability of measurement, a number of longer plugs were halved in length and the variation in resistivity v. saturation determined for each adjacent half. Ignoring any small-scale heterogeneity, the results were observed to be reproducible.

Measurements of effective (i.e. interconnected) porosity, saturated hydraulic conductivity and cation-exchange capacity (CEC) were determined for each sample using techniques described by Taylor (2000) and Taylor & Barker (2002).

Experimental results and observations

Resistivity v. electrolyte conductivity – fully saturated experiment

The variation of bulk resistivity for samples fully saturated with different fluid salinities is presented in Figure 2, where representative samples from each of the four sandstone blocks are displayed, plotted as bulk resistivity, ρ_o, v. electrolyte resistivity, ρ_w. The data for samples from each block follow the same basic trend of increasing bulk sandstone resistivity with increasing saturating electrolyte resistivity. The data follow a curve that varies greatly from a linear trend for values of ρ_w greater than about 5 Ωm (conductivity below 2000 μS cm^{-1}) with unit increase in electrolyte resistivity producing smaller and smaller increases in bulk resistivity. The fact that this is not a linear relationship indicates that there are conduction mechanisms, additional to electrolytic conduction, taking place within the sandstone samples and these are most probably related to the presence of a conductive matrix (Waxman & Smits 1968).

The level of scatter on data collected from samples drilled from the same block is small compared to the overall differences between samples from different blocks. This indicates that there are measurable electrical differences between the sandstone blocks (except for SL1 and SL2, which appear to be similar) and only small differences between samples drilled from the same block.

Resistivity v. partial saturation

The results of measurements made on partially saturated samples are displayed in Figure 3 as plots of resistivity index ($R.I.$ – the ratio of resistivity measured at partial saturation with that at full saturation) v. saturation (S_w). All data show the drainage cycle only.

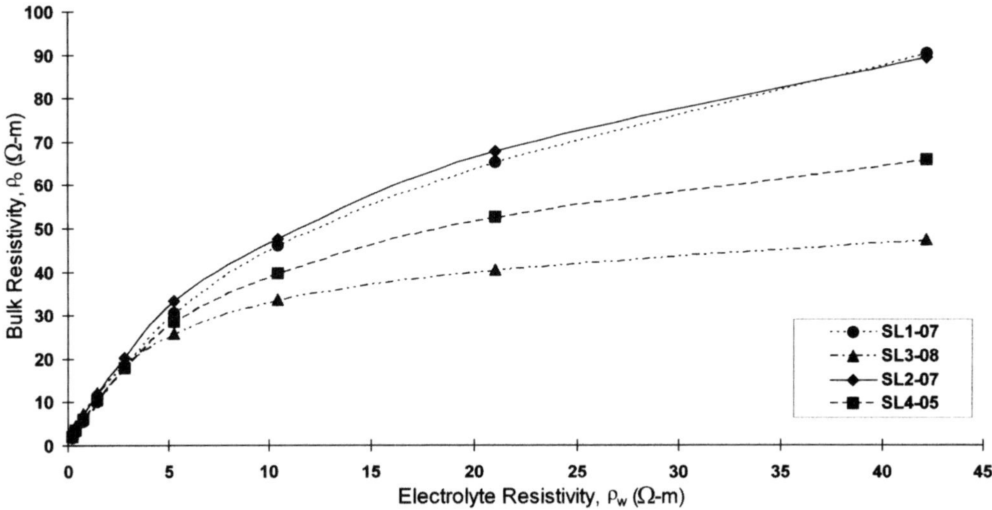

Fig. 2. Variation of bulk resistivity with saturating electrolyte resistivity for representative samples of Permo-Triassic sandstone from Bromsgrove, UK.

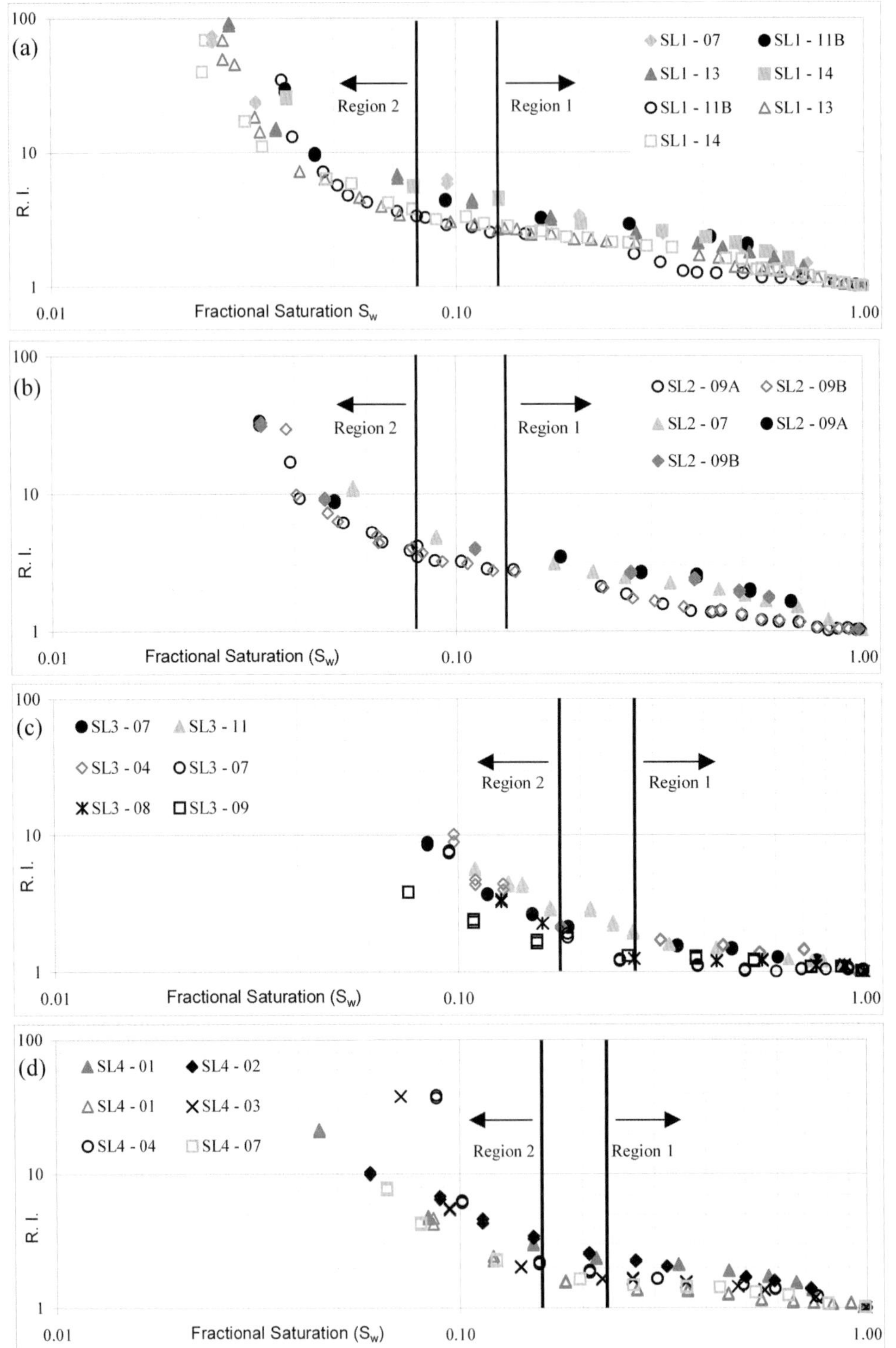
(a)
SL1 - 07
SL1 - 11B
SL1 - 13
SL1 - 14
SL1 - 11B
SL1 - 13
SL1 - 14
Region 2
Region 1
R.I.
100
10
1
0.01
0.10
1.00
Fractional Saturation Sw

(b)
SL2 - 09A
SL2 - 09B
SL2 - 07
SL2 - 09A
SL2 - 09B
Region 2
Region 1
R.I.
100
10
1
0.01
0.10
1.00
Fractional Saturation (Sw)

(c)
SL3 - 07
SL3 - 11
SL3 - 04
SL3 - 07
SL3 - 08
SL3 - 09
Region 2
Region 1
R.I.
100
10
1
0.01
0.10
1.00
Fractional Saturation (Sw)

(d)
SL4 - 01
SL4 - 02
SL4 - 01
SL4 - 03
SL4 - 04
SL4 - 07
Region 2
Region 1
R.I.
100
10
1
0.01
0.10
1.00
Fractional Saturation (Sw)

Figure 3a & b show data for representative samples obtained from blocks SL1 and SL2, respectively. Data are presented for sample cores initially saturated with synthetic groundwater solutions of 800 (open symbols) and 1600 μS cm^{-1} (closed symbols). Similarly, Figure 3c & d show data for representative samples drilled from blocks SL3 and SL4, respectively. In these figures data are presented for sample cores initially saturated with synthetic groundwater solutions of 400 (open symbols) and 1600 μS cm^{-1} (closed symbols).

Data from all samples show the same basic response of increasing resistivity, and hence resistivity index, with decreasing level of saturation. The basic curve shape can be split into two relatively well-defined regions (as illustrated in Fig. 3). In Region 1, at higher levels of saturation, the rate of increase in resistivity with decreasing saturation is relatively small and constant. This changes at the transition to Region 2, where the resistivity increases at an increasing rate with decreasing saturation. This occurs somewhere between $0.08 < S_w < 0.25$ for all samples.

The primary difference between samples drilled from different sandstone blocks is the location of the transition from Region 1 to Region 2. This occurs at a *critical* water saturation, S_w^0 (Knight & Nur 1987; Endres & Knight 1991), where there is a change in the pore-water geometry from a bulk-water phase to a purely adsorbed, surface-water phase. In the data for samples drilled from blocks SL1 and SL2 (Fig. 3a, b) this transition occurs between $0.08 < S_w < 0.13$. Once again, the lack of variation in the results between these samples indicates that sandstone blocks SL1 and SL2 are electrically very similar. In the case of samples obtained from block SL3 (Fig. 3c) the transition between the two regions occurs at a much higher saturation level, between $0.18 < S_w < 0.28$. The data for samples drilled from block SL4 (Fig. 3d) also show a transition at higher saturation levels ($0.17 < S_w < 0.21$). For a given pore-fluid salinity and at any given saturation in Region 1, the values of *R.I.* are generally highest for samples from sandstone blocks SL1 and SL2, lower for those from SL4 and lowest in the samples drilled from block SL3.

Samples saturated with a higher salinity pore fluid produce consistently higher values of *R.I.* for a given saturation than those saturated with lower salinity fluid. These samples also show a clearer, more pronounced increase in resistivity with decreasing saturation in Region 1. This can be seen most clearly in the data in Figure 3a & b, which display curves for the same core samples saturated with 6.25 (1600 μS cm^{-1}) and 12.5 Ωm (800 μS cm^{-1}) electrolyte solutions. Similar variation is also seen in Figure 3c & d for samples from SL3 and SL4 saturated with solutions of resistivity 25 (400 μS cm^{-1}) and 6.25 Ωm (1600 S cm^{-1}).

Data modelling

The Archie model

Archie (1942) originally proposed a simple linear empirical relationship between bulk resistivity and the pore-fluid resistivity of the form

$$F = \frac{\rho_o}{\rho_w} = \frac{a}{\phi^m} \tag{1}$$

where F is the constant of proportionality, referred to as the *formation factor*, ϕ is the fractional porosity, and a and m are formation constants. Archie empirically determined values of $a = 1$ and $m = 2$ for clean sandstone samples saturated with high-salinity brines (i.e. where $\rho_o \gg \rho_w$) – the conditions often encountered in oil fields. Other workers (Winsauer *et al.* 1952; Wyllie & Gregory 1953; Keller 1966; Carothers 1968) have reported values of a of between 0.5 and 2.0, and values of m of between 1.5 and 2.5. As m was sometimes observed to be related to the extent of cementation (Wyllie & Gregory 1953), it has come to be known as the cementation factor, although as we shall see it is more correctly associated with the pore-scale morphology. To avoid confusion, henceforth we shall refer to m as the pore-morphology factor or, simply, pore factor. For a porous medium where the matrix does not contribute to the conduction of electricity, the linear relationship of equation 1 is valid at all electrolyte salinities. However, if the rock matrix comprises clay minerals and is therefore conductive, as is generally the case in most Permo-Triassic sandstone, this equation will only be valid at very high electrolyte salinities. At the low levels of salinity found in most near-surface groundwater, there is a marked departure from equation 1 as F is observed to vary with

Fig. 3. Desaturation curves for samples from Sandstone blocks: (**a**) SL1 and (**b**) SL4. Closed symbols indicate samples initially saturated with a solution of conductivity 1600 μS cm^{-1}. Open symbols indicate samples initially saturated with a solution of conductivity: (a) 800 μS cm^{-1} and (b) 400 μS cm^{-1}.

electrolyte resistivity. This is clear from the way in which the fully saturated data presented in Figure 2 deviate significantly from a linear response. As F is the gradient of the curve, this is seen to vary considerably at high values of ρ_w. Therefore, it is only appropriate to fit Archie's equation at the lowest electrolyte resistivities (ρ_w <10 Ωm), as shown in Figure 4. The resulting values of F and m for individual samples calculated using Archie's equation are displayed in Table 1. These values suggest that the electrical properties of the samples from the blocks SL1, SL2 and SL4 are similar, whilst those of SL3 differ only slightly from the other blocks.

For a partially saturated rock, a decrease in saturation will be accompanied by a marked increase in resistivity due to replacement of the electrolyte with air, oil or gas. Archie expressed this using the linear relationship

$$\frac{\rho_t}{\rho_o} = S_w^{-n} \tag{2}$$

where ρ_t is the resistivity of the rock with fractional saturation, S_w. A value of $n = 2$ was proposed for clean sandstones.

The partial saturation data of Figure 3 clearly show that the linear trend predicted by the Archie equation is not followed and that the rock is more conductive for a given saturation than predicted. Applying equation 2 to representative sample data (Fig. 5) indicates that the adoption of $n = 2$ is clearly not appropriate for modelling partially saturated samples from this Permo-Triassic sandstone aquifer. Allowing a

best-fit value for n to be determined produces significantly lower values (Table 2) at the saturating electrolyte conductivities used in this study. However, the best-fit value of n increases as the conductivity of the saturating electrolyte increases, suggesting that at the highest electrolyte conductivities the use of Archie equation with a value of n nearer 2 may be appropriate.

Waxman & Smits shaly sandstone model

As has been noted, at low electrolyte conductivities, the data for fully saturated samples displayed in Figure 2 show a significant departure from the simple linear relationship proposed by Archie (equation 1). This is due to conduction mechanisms in the rock that are additional to the flow of ions through the electrolyte; this additional component of charge transfer has been referred to as *excess conductivity*. An important origin of excess conductivity is the presence of counterions near the surface of all mineral grains, but particularly significant at the surfaces of clay minerals. The measure of the number of counterions present in the rock is referred to as the *cation-exchange capacity*, or CEC, and is expressed here as Q_v, the CEC per unit pore volume measured in milli-equivalents per millilitre (meq ml^{-1}).

Many authors have recognized the need to apply correction factors to account for excess conductivity and the resulting curvature of the bulk resistivity v. electrolyte resistivity graph. Many have adopted empirical factors to enable information about the formation to be

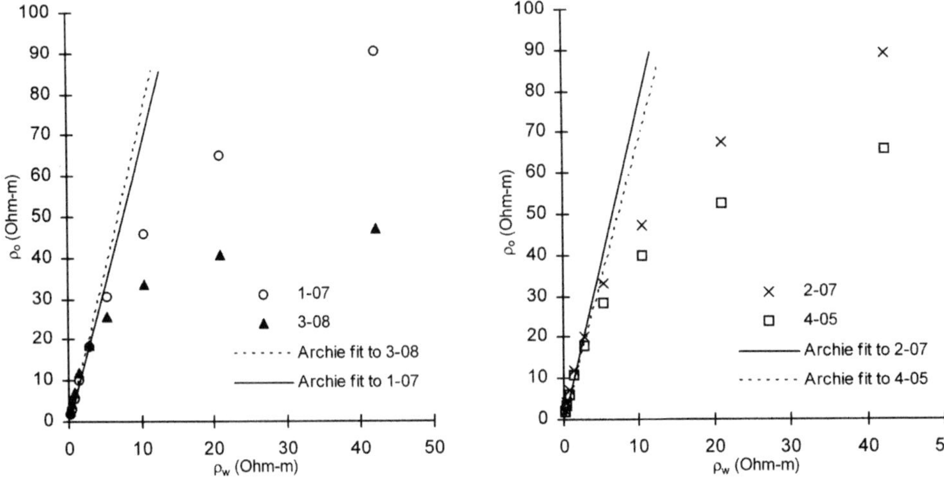

Fig. 4. Comparison of the modelled response using Archie's equation (equation 1) fit to data from a number of fully saturated cores.

Table 1. *Laboratory determined properties and parameters derived from modelling electrical measurements of fully saturated sandstone samples*

Block Sample	ϕ	K (m day^{-1})	CEC (meq 100 g^{-1})	Q_V (meq ml^{-1})	F	m	Q_V (meq ml^{-1})	F^*	m^*	ρ_r (Ωm)	m_{H-B}
SL1											
1-07	0.28	1.11	0.84	0.06	6.7	1.49	0.11	8.81	1.71	126.5	1.64
1-08	0.27	1.75	1.43	0.10	7.0	1.51	0.09	9.07	1.70	161.0	1.64
1-10A	0.28	1.18									
1-10B	0.28	1.18									
1-11A	0.28	1.25	2.67	0.18	6.7	1.49	0.13	9.02	1.72	115.0	1.64
1-11B	0.28	1.25	2.67	0.18	6.4	1.45	0.12	8.61	1.68	115.0	1.62
1-13	0.28	1.03	1.54	0.11	6.8	1.49	0.14	9.67	1.76	113.0	1.66
1-14	0.26	0.70			7.4	1.49	0.13	10.20	1.73	129.0	1.66
SL2											
2-05	0.26	1.25	1.43	0.11	7.3	1.46	0.13	10.18	1.71	123.0	1.63
2-07	0.26	0.27			7.6	1.52	0.16	11.43	1.82	118.0	1.73
2-09A	0.27	1.01	2.54	0.18	6.9	1.48	0.12	9.32	1.70	134.0	1.63
2-09B	0.28	1.01	2.54	0.18	6.8	1.49	0.12	9.30	1.73	125.0	1.65
SL3											
3-02		0.06	6.11								
3-04	0.27	0.02	5.61	0.40	5.3	1.29	0.32	12.74	1.96	57.8	1.86
3-05	0.27	0.02	5.19	0.37	5.4	1.30	0.50	16.97	2.17	51.1	2.05
3-07	0.26	0.02	4.21	0.33	6.6	1.38	0.44	14.99	1.98	49.3	1.77
3-08	0.26	0.01	6.74	0.51	7.3	1.47	0.55	18.34	2.16	51.5	1.93
3-09	0.26	0.01	5.45	0.41	7.1	1.45	0.50	17.31	2.11	54.5	1.89
3-10	0.25	0.01					0.41	15.33	1.99	54.6	2.00
3-11	0.25	0.03	4.13	0.33	7.3	1.44	0.54	18.01	2.09	50.5	1.87
SL4											
4-01	0.29	0.32			7.9	1.68	0.23	11.88	2.01	85.4	1.88
4-02	0.29	0.34	2.97	0.19	6.4	1.49	0.23	10.32	1.88	67.7	1.77
4-03	0.29	0.47	3.47	0.22	6.4	1.50	0.19	9.85	1.85	79.0	1.74
4-04	0.29	0.34	3.47	0.23	7.0	1.56	0.17	10.17	1.85	97.0	1.76
4-05	0.28	0.42	2.12	0.15	6.8	1.50	0.21	10.70	1.86	81.0	1.75
4-07	0.29	0.41	2.31	0.15	6.4	1.48	0.21	10.23	1.86	73.0	1.75

The column headers above group as: "Property measurements" (ϕ, K, CEC, Q_V); "Archie Fully saturated" (F, m); "W & S Fully saturated" (Q_V, F^*, m^*); "H–B Fully saturated" (ρ_r, m_{H-B}).

determined more accurately. For example, Worthington & Barker (1972) adopted an apparent formation factor equal to ρ_o/ρ_w and introduced a second term to provide a better estimate of the true formation factor, F, and hence the porosity of the Bunter Sandstone samples being investigated.

To account for the presence of the excess conductivity in their rock samples, Waxman & Smits (1968) suggested an empirical model (the W–S model) for shaly sands. Their model represents the conductive elements of the rock as two parallel resistors, one representing the resistance due to the free electrolyte in the pore space of the rock and the other a resistance associated with clay minerals within the rock, the conductance of which is brought about by cation exchange as described above. The basic relationship is

$$\frac{1}{\rho_o} = \frac{1}{F^*}\left(\frac{1}{\rho_w} + BQ_v\right) \qquad (3)$$

where F^* is the formation factor for shaly sands. Q_v is the CEC per unit pore volume (measured in meq ml^{-1}) and B is the equivalent ionic conductance of clay exchange ions, which is a function of ρ_w (calculated according to equation 12 of Vinegar & Waxman 1984). In the limit of a clean non-conductive rock matrix, $BQ_v = 0$ and F^* reduces to F as defined in equation 1.

The W–S equation is able to model our data of Figure 2 very closely (Fig. 6) producing estimates for Q_v and F that suggest that,

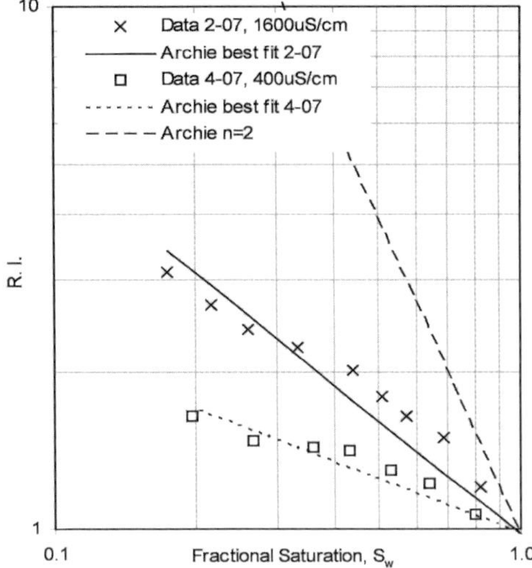

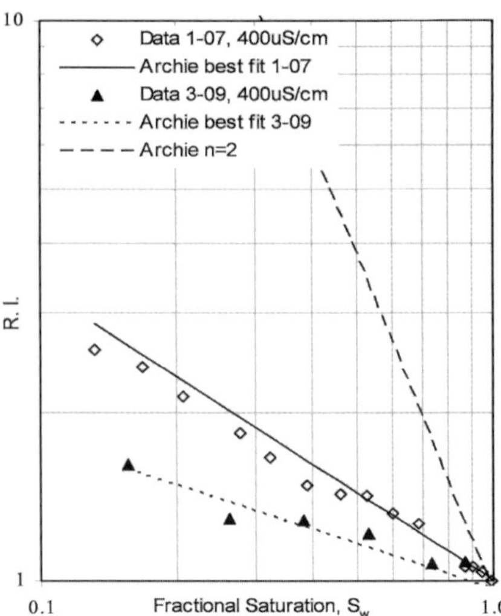

Fig. 5. Comparison of the modelled response using Archie's equation (equation 2) fit to data from a number of partially saturated cores.

contrary to the modelling using Archie's equations, there is substantial variation in the electrical properties of the sandstone blocks. It is also possible to calculate values for the Waxman–Smits derived pore factor (here labelled m^*) from F^* using equation 1. The parameter values are presented for each sample in Table 1. It is clear that the W–S equation reproduces the fully saturated response over all electrolyte resistivities and that the additional term accounts for the curvature in the observed data.

For partially water-saturated oil-bearing shaly sands, Waxman & Smits proposed the equation

$$\frac{1}{\rho_o} = \frac{S_w^{n^*}}{F^*}\left(\frac{1}{\rho_w} + \frac{BQ_v}{S_w}\right). \tag{4}$$

This equation assumes that the exchange ion population remains the same as water is replaced by the non-wetting fluid (in their case oil) and thus the effective concentration of exchange ions increases as Q_v/S_w. As this assumption has been shown to be valid in oil-bearing shaly sands (Waxman & Thomas 1974), we will test to see if the assumption is valid in our case as water is replaced by air. The parameters that result from applying this model to our data are displayed in Table 2 and indicate that the saturation exponent n^* varies only slightly from sample to sample. Similarly, varying saturating electrolyte conductivity has little effect on the value of n^*. The values are

also very different from those determined using equation 2 suggesting that, as with m and m^*, the values of n and n^* differ significantly the greater the departure from conditions of clean sand matrix and high-salinity pore fluid. Calculated values of Q_v and F are similar to those determined for the fully saturated samples using equation 3, and here also suggest that there is significant variation in the electrical properties of the samples.

Hanai–Bruggeman effective medium theory

A theoretical model based on the *effective medium theory* is the Hanai–Bruggeman (H–B) mixing law equation. This equation, extended by Hanai (1960) from the work by Bruggeman (1935), relates the electrical properties of a heterogeneous mixture comprising a component dispersed in a continuous medium, to the electrical properties of the individual components. Bussian (1983) adapted this model for lattice-like structures and at low frequencies, applicable to this study, the equation he presented can be expressed as

$$\rho_o = \rho_w \phi^{-m_{H\text{-}B}}\left(\frac{1 - \rho_w/\rho_r}{1 - \rho_o/\rho_r}\right)^{-m_{H\text{-}B}} \tag{5}$$

where the parameter $m_{H\text{-}B}$ is a pore factor, similar to Archie's m, and ρ_r is the matrix

Table 2. *Parameters derived from electrical measurements on partially saturated samples of Permo-Triassic sandstone*

Core	Archie partially saturated			W&S Partially saturated parameters constrained to 10% of fully saturated values												H–B Partially saturated parameters constrained to 10% of fully saturated values								
	400 µS cm⁻¹	800 µS cm⁻¹	1600 µS cm⁻¹	400 µS cm⁻¹				800 µS cm⁻¹				1600 µS cm⁻¹				400 µS cm⁻¹			800 µS cm⁻¹			1600 µS cm⁻¹		
Label	n	n	n	Q_v (meq ml⁻¹)	F^*	n^*	RMS (%)	Q_v (meq ml⁻¹)	F^*	n^*	RMS (%)	Q_v (meq ml⁻¹)	F^*	n^*	RMS (%)	ρ_r (Ωm)	m_{H-B}	RMS (%)	ρ_r (Ωm)	m_{H-B}	RMS (%)	ρ_r (Ωm)	m_{H-B}	RMS (%)
SL1																								
1-07	0.52		0.83	0.12	7.92	1.31	1.8					0.10	9.68	1.30	2.2	113.8	1.48	4.4				139.2	1.66	2.6
1-08			1.02									0.08	9.98	1.46	1.9							177.1	1.80	1.6
1-10A†	0.46	0.47		0.13	8.85	1.26	1.4	0.13	8.87	1.24	3.5					115.0	1.64	1.1	103.5	1.48	3.8			
1-10B†		0.45						0.13	8.90	1.27	4.0								103.5	1.48	4.0			
1-11A	0.39		0.80					0.14	8.66	1.25	4.4	0.12	9.92	1.43	4.2				106.0	1.60	5.7	126.5	1.80	4.4
1-11B	0.42		0.76					0.13	8.77	1.22	3.1	0.10	9.48	1.34	5.3				107.0	1.59	4.8	126.5	1.78	3.8
1-13	0.43	0.50	0.74	0.15	8.70	1.20	2.8	0.15	9.94	1.30	1.2	0.13	10.63	1.28	1.3	101.7	1.49	6.5	115.0	1.59	1.8	124.3	1.66	2.5
1-14		0.53	0.81					0.14	10.93	1.29	1.2	0.12	11.22	1.35	3.9				141.9	1.51	2.0	141.8	1.83	1.9
SL2																								
2-05																								
2-07			0.71									0.15	12.57	1.29	2.1							129.8	1.84	2.1
2-09A		0.45	0.85					0.12	8.80	1.24	4.0	0.10	10.25	1.40	4.7				120.6	1.48	4.8	147.4	1.79	3.3
2-09B		0.47	0.75					0.14	9.18	1.28	3.1	0.11	10.23	1.37	5.0				118.0	1.60	4.8	137.5	1.82	3.9
SL3																								
3-02†	0.33			0.37	15.00	1.17	4.3									60.0	1.85	4.1						
3-04	0.43			0.29	12.35	1.32	2.7									63.8	1.67	6.0						
3-05	0.41			0.45	17.10	1.25	3.9									56.1	1.85	4.9						
3-07	0.25		0.47	0.49	16.11	1.21	4.0					0.41	16.49	1.27	1.5	45.0	1.80	3.7				54.2	1.90	3.7
3-08	0.23			0.49	18.43	1.08	1.6									52.0	1.94	2.6						
3-09	0.26			0.51	19.04	1.19	1.6									55.0	1.95	1.6						
3-10	0.32			0.44	16.86	1.22	2.1									59.5	1.95	2.5						
3-11	0.16		0.52	0.59	18.29	1.17	2.6					0.58	19.81	1.45	3.7	45.0	1.90	2.6				53.0	1.95	4.3
SL4																								
4-01	0.26		0.67	0.24	13.07	1.14	1.3					0.21	13.07	1.31	3.4	80.5	1.90	3.0				94.0	1.90	3.5
4-02	0.24		0.64	0.25	10.96	1.19	2.5					0.21	11.35	1.31	2.2	67.0	1.77	2.0				74.4	1.90	2.0
4-03	0.38			0.17	9.80	1.21	1.5									84.0	1.74	0.9						
4-04	0.42			0.15	10.16	1.24	1.3									105.0	1.75	1.5						
4-05	0.42			0.19	10.50	1.25	2.4									85.0	1.80	1.5						
4-07	0.34			0.19	10.15	1.18	1.2									79.0	1.80	1.2						

† For Waxman and Smits model only, partially saturated parameters not constrained to ±10%.

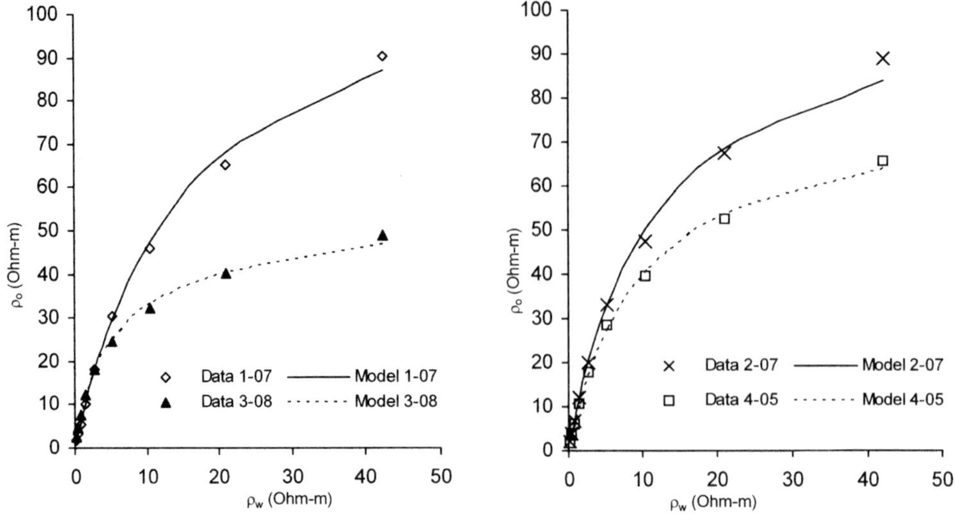

Fig. 6. Comparison of the modelled response using the Waxman–Smits empirical formulation (equation 3) fit to data from a number of fully saturated cores.

resistivity. Whereas Archie equation is only valid for high-salinity electrolytes, the H–B equation is valid at all salinities. Consequently, the pore factor, m, in the Archie equation 1 varies with salinity and only approaches the pore factor of the H–B equation, $m_{H–B}$, at the limit of very conductive electrolytes. At very high matrix resistivities equation 5 approaches the empirical equation of Archie for $a = 1$. A useful output from the H–B equation is the rock matrix resistivity, ρ_r, which represents the electrical resistivity of the rock grains surrounded by the associated electrical double layer (Revil 1999). The rock matrix resistivity in fact includes anything that is not the continuous phase (pore water) and will include clay particles and water that is molecularly bound to clay particles. As it includes the electrical double layer, it also includes adsorbed water. The matrix resistivity provides a further useful parameter with which to describe and characterize rock samples.

As with the W–S equation, the H–B equation is able to faithfully reproduce the fully saturated laboratory data over the full range of electrolyte resistivities (Fig. 7), yielding parameter estimates for the H–B pore factor, $m_{H–B}$, and the matrix resistivity, ρ_r. Once again, these parameters (Table 1) show a wide variation between sample blocks suggesting there are notable differences in the properties of the different sample blocks.

Partial saturation relationships based on theoretical models have been formulated and used by a number of authors. Knight & Endres (1990) use a complex three-component mixing model comprising a water-wetted rock, the bulk pore water and the air. Other models have been formulated for water displacement by an oil phase. In de Lima (1995), a simple transformation such that $\rho_w \rightarrow \rho_w S_w^{-n}$ is adopted in equation 5. This method is different from that adopted by Knight & Endres (1990) as it involves treating the pore fluids as a self-similar oil–water emulsion that satisfies the Hanai–Bruggeman equation.

In the formulation by Berg (1995), a general equation is derived for a partially saturated system from the two-component effective medium equation given in equation 5. The dispersive component comprises both the rock matrix and the non-wetting hydrocarbon, and the resistances of these are combined using the approximation of resistors-in-parallel. The water remains the effective medium in this case. We suggest that this formulation by Berg, although derived for a rock partially saturated with a hydrocarbon–water mix, may be equally applicable for an air–water mix appropriate here. This is because at low frequencies the resistivity of the hydrocarbon, like air, is infinite, and the hydrocarbon and water are not treated as an emulsion but instead the hydrocarbon is treated as part of the rock. The hydrocarbon in the model is the non-wetting fluid, which is also true for the air in the study presented here. By

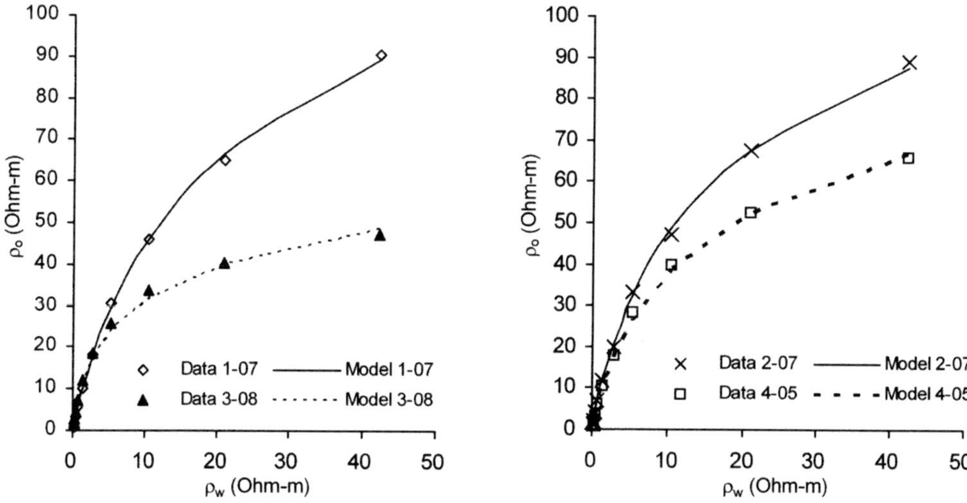

Fig. 7. Comparison of the Hanai–Bruggeman model (equation 5) fit to data from a number of fully saturated cores.

associating the non-wetting air phase with the matrix, equation 5 can be rewritten for the case of a partially saturated rock as

$$\rho_t = \rho_w S_w^{-m_{H\text{-}B}} \phi^{-m_{H\text{-}B}} \left(\frac{1 - \rho_w / \rho_d}{1 - \rho_t / \rho_d} \right)^{-m_{H\text{-}B}} \quad (6)$$

where ρ_d is the dispersed phase resistivity which is calculated for a parallel resistor combination of the rock and air resistivities. As the air resistivity is infinite, the disperse phase resistivity is given simply by

$$\rho_d = \rho_r \frac{(1 - S_w \phi)}{(1 - \phi)}. \quad (7)$$

Substituting this into equation 6 produces a quadratic equation, which can be easily solved for S_w (Berg 1995; Taylor & Barker 2006).

As the H–B equation is based on the electrical response of dispersed particles in a continuous medium, the model is only valid for a rock that is water-wet. As a consequence, the applicability of the model breaks down at saturation levels below the critical water saturation, S_w^0. At this point it is believed that there is a transition in the behaviour of the water in the rock pores from a bulk-water phase to a surface-adsorbed-water phase (Knight & Dvorkin 1992). This causes the resistivity of the rock to increase rapidly with further decreasing saturation. Hence, it is only appropriate to model the data in Region 1, as shown in Figure 8. The resulting parameter values from applying equation 6 to

the measured data are presented in Table 2 and reflect similar values to those determined from applying the H–B equation to the fully saturated data.

Discussion

The data indicate that the samples of Permo-Triassic sandstone from the Wildmoor Formation adopted for this study show a wide variation in both the independently determined hydraulic properties and those predicted from modelling of the measured electrical response. The values for the parameters calculated using the W–S and H–B equations suggest a greater variation in sample properties than those predicted from Archie's equations. This is not surprising given that Archie's equations are only appropriate for modelling the resistivity response at high electrolyte conductivities where the data from all samples show the least variation (Fig. 4). Therefore, in order to determine representative formation properties from electrical measurements, it is important to understand the limitations of Archie equation and adopt other models where appropriate.

From the results of the H–B and W–S models, it is clear that the medium- to coarse-grained, bleached sandstone core samples from blocks SL1 and SL2 are characterized by large values of matrix resistivity, low pore factors, m, and low Q_v values (Table 1). This suggests that these sandstone samples have limited clay content

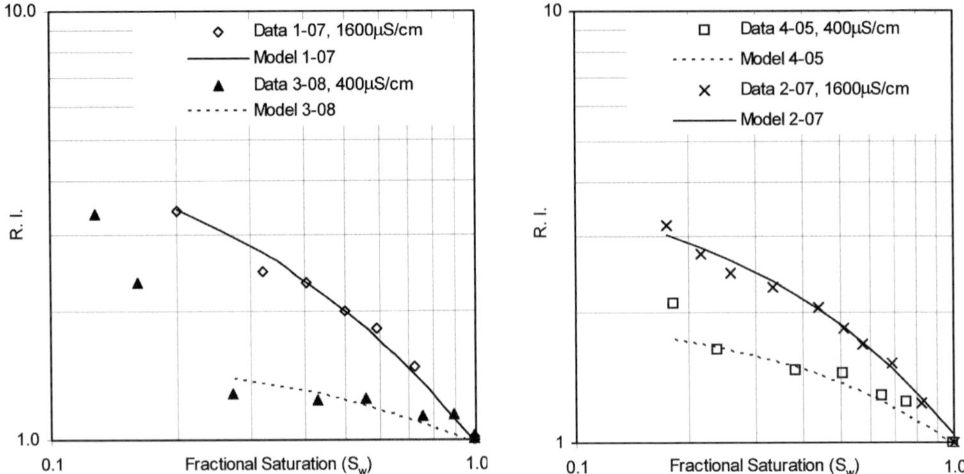

Fig. 8. Example fits of Berg's equation (equation 6) to partially saturated data above the transition zone.

and that the clay present is not a significant contributor to the measured CEC. The matrix resistivities for core samples from block SL3, characterized by its orange-red colouring and finer grain size, are much lower with much higher pore factors and higher Q_v values. The higher values of Q_v and conductive matrix suggest either clay or other conductive surface materials are present in this sample, although there may also be a contributing factor of smaller grain size (this would increase the overall surface area within the rock that was available for surface conduction effects). The redder colouring of these samples suggests the presence of an iron-related coating on the grains. This is confirmed by Mitchener (2002) who undertook a chemical stripping of core samples from the Wildmoor Formation, and identified significant iron and manganese oxides and oxyhydroxide groups to be present in samples with orange-red colouring. It is likely that these groups will contribute significantly to both CEC and surface conduction due to their high surface area per unit volume and charge characteristics (Langmuir 1997). The cores from the borehole sample, SL4, have a slightly paler red colour to those of SL3, and are characterized by a medium to fine grain size. The modelled parameters for these core samples fall between those for samples from blocks SL1 (and SL2) and SL3. These values suggest a relatively clean sandstone, perhaps better cemented than SL1 and SL2, and with a significantly more conductive matrix. Once again, the colour of these samples suggests that oxides

and oxyhydroxides are likely to be present coating the rock matrix.

We now compare the modelled parameter values for the cation-exchange capacity determined using the W–S equation, $Q_{v(W-S)}$, with those determined independently using laboratory methods, $Q_{v(meas)}$. These data are plotted in Figure 9 with a best-fit line that yields a gradient of 1.11 ± 0.07 meq ml^{-1}, which is within two standard deviations of unity, and a correlation coefficient of $r = 0.89$. Although the scatter around the straight-line fit is greater than that experienced by Vinegar & Waxman (1984) in their analysis of North American shaly sandstones, the level of agreement between the measured and modelled values of CEC suggests that adoption of the W–S equation to model the electrical response of these Permo-Triassic sandstone samples is appropriate over the moderate range of CECs represented here.

The laboratory results and subsequent modelling have indicated that the rock matrix is conductive and that this is believed to be due, at least in part, to exchange cations on the mineral surfaces of the rock. If the rock matrix conduction arises from cation exchange alone then it might be assumed that

$$1/\rho_r \approx BQ_v \qquad (8)$$

where B is the equivalent cation conductance (as in equation 3). In Figure 10 a plot of the measured CEC ($Q_{v(meas)}$) against the inverse of matrix resistivity calculated using the H–B model shows a strong correlation suggesting a

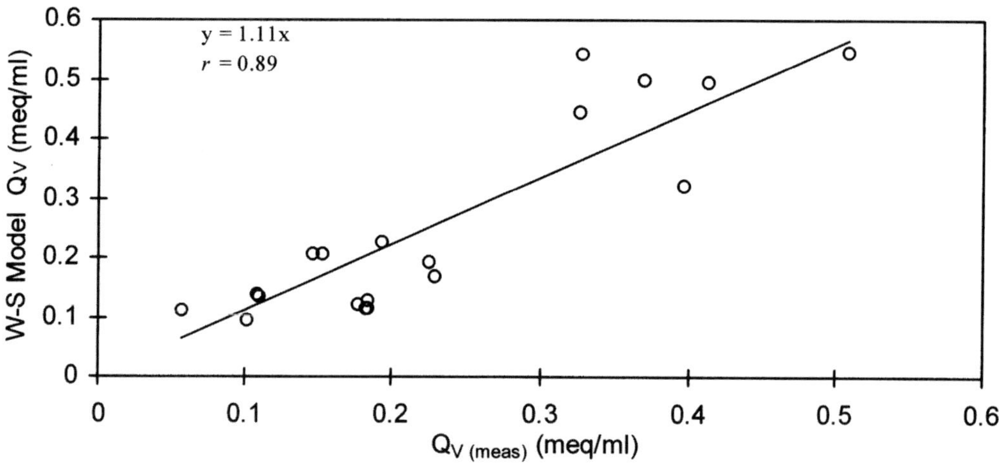

Fig. 9. Comparison between measured values of cation-exchange capacity, $Q_{v(meas)}$, and the values of Q_v predicted by the W–S model.

significant contribution from the exchange cations to matrix conduction. A best-fit linear-trending line produces values for B of 0.34 ± 0.05 mS-cm^2 meq^{-1} and an intercept of 0.05 ± 0.01 mS cm^{-1} with a correlation coefficient of r = 0.85. If the intercept is set to zero, then the gradient becomes 0.51 ± 0.03 mS-cm^2 meq^{-1} with r = 0.72. Assuming that matrix conduction results purely from volume cation exchange, this gradient represents the equivalent cation conductance. This is significantly smaller than the maximum ionic conductance for Na$^+$ cations at infinite dilution ($B \approx 50$ mS-cm^2 meq^{-1}) and clay surface ionic mobilities, which are typically a factor of 10 less (Revil 1999). As suggested by Bussian (1983), this reduced value for the equivalent conductance may be a result of charge on the surfaces of the clays reducing the mobility of exchange cations. However, the higher CEC

measurements associated with the redder coloured samples suggest that there may be contributors other than clay to the measured CEC. The red colouring observed in Triassic and other red-bed deposits is mainly attributed to oxides and oxyhydroxides of iron and with which manganese oxides are often associated. These groups can make a significant contribution to the exchange properties of the rock, and manganese oxyhydroxides, in particular, have CEC values similar to that of smectite clays (Buss 2000). Thus, the higher CEC values measured in the redder samples may be attributable to the presence of these surface-coating oxyhydroxide groups, and it may be these that are the primary source of the matrix conduction.

Note that if the non-zero intercept in Figure 10 is assumed real, and not just a result of scatter in the data, then this suggests residual matrix

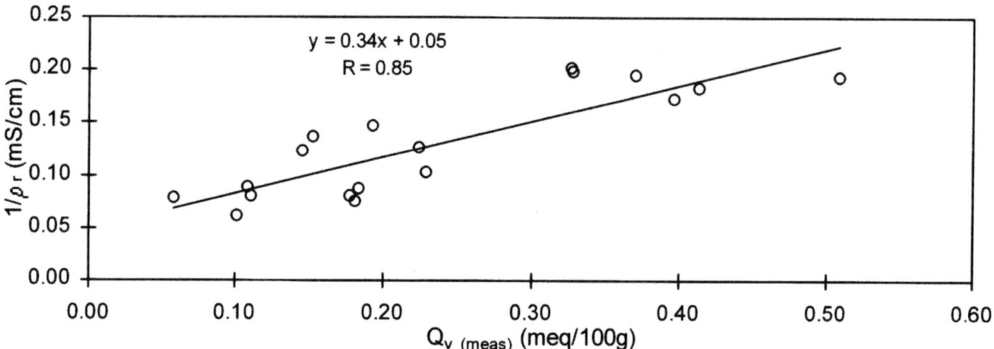

Fig. 10. Plot of the measured CEC ($Q_{v(meas)}$) against the inverse of matrix resistivity.

conductivity once cation exchange is accounted for. This residual conductivity might be attributable to other processes, perhaps redox exchange, but investigation of such is beyond the scope of this study.

Porosity is observed to vary little from sample to sample, and yet values of the formation factor, F, and pore factor calculated from the W–S and H–B models are seen to vary significantly (and although values of $m_{H–B}$ and m^* for the same sample differ slightly, they vary in a similar way). Thus, it may be appropriate in this setting to think of the variation in pore factor (and F) as relating to changes in pore morphology and not just as a predictor of porosity. Considering the range of values for m (and F) calculated from the W–S or H–B models and comparing these with the values of ρ_r determined using the H–B equation, it is clear that samples that exhibit a lower matrix conductivity also have lower formation and pore factors (and, as we have seen, lower Q_v). The implication here is that there is an overriding control on all of these parameters and that this is likely to relate to the pore morphology. If CEC is primarily associated with oxyhydroxide groups and not clays, then the pore factor is most likely to relate to pore-space geometry and thus electrical tortuosity. If we adopt Dullien (1992) then a smaller pore factor suggests preferential loss of dead-end pore space that might ordinarily be bypassed by both electrical and fluid flow. Thus, we might potentially expect to see a correlation between hydraulic conductivity, K, and calculated pore factor. In Figure 11 a plot of these two parameters is displayed for all samples and produces a relationship of the form $m_{H–B} = aK^{-b}$ with a high correlation coefficient $r = 0.88$. Relationships between hydraulic conductivity and formation factor have been presented for hydrocarbon reservoirs (Archie 1942), with

similar relationships identified for freshwater reservoirs (Worthington 1982, 1985). Identification of a strong correlation for the Permo-Triassic Sandstone suggests that the controls on electrical and fluid flow and on electrical and hydraulic properties may be similar in origin. The physical origin of such a relationship is not clear and more work would be required to justify using any such relationship beyond this dataset.

A final factor to consider is the variation in the critical saturation, S_w^0. This value is significant as it is related to the pore-water distribution and is believed to mark the change from bulk to surface phase water, as discussed previously. The critical saturation varies over the range $0.08 < S_w < 0.25$. For the cleaner cores of SL1 and SL2 this change occurs at a much lower saturation than for the iron-rich or clay containing cores of SL3 and SL4. A discussion into the pore-scale fluid distributions and their effect on the bulk resistivity measurements are beyond the scope of this paper. However, it is clear that the critical saturation level is affected by the pore-scale geometry, and that this assertion appears to be corroborated by the values of pore factor and CEC.

Implications for monitoring the variation in saturation in the vadose zone

In terms of the implications of these observations for field-scale monitoring of the unsaturated zone, a number of noteworthy comments may be made.

It is clear that at the site scale, there is significant heterogeneity in the Triassic sandstone. Although the samples were obtained from only one site in the West Midlands, there is observed variation in characteristics such as porosity,

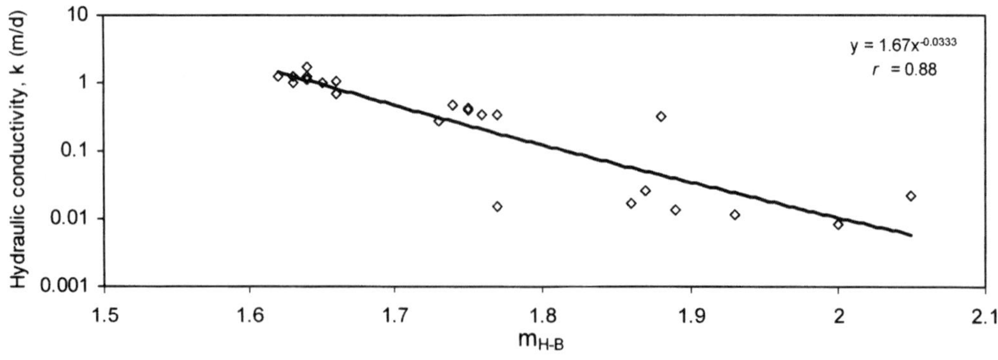

Fig. 11. Variation of permeability, K, with pore factor, $m_{H–B}$ calculated from the H–B equation.

Table 3. *Variation of measured and derived parameters for the Permo-Triassic of the Wildmoor Formation, Bromsgrove, UK*

		Derived parameters		
Model	F or F^*	m, m^* or m_{H-B}	Q_v (meq ml^{-1})	ρ_r (Ωm)
Archie	6.5–8.0	1.30–1.55		
Waxman & Smits	9.0–18.0	1.70–2.15	0.09–0.55	
Hanai–Bruggeman		1.65–2.05		50–135

		Measured properties		
	ϕ	K (m day^{-1})	CEC (meq 100 g^{-1})	Q_v (meq ml^{-1})
All samples	0.25–0.29	0.01–1.75	0.84–6.74	0.06–0.51

permeability, CEC and the parameters derived from modelling the electrical measurements of fully saturated sandstone. A summary of the ranges of the various properties of the fully saturated Permo-Triassic sandstone is presented in Table 3.

Resistivities of partially saturated sandstone extend far beyond the range of that measured with full saturation, with the largest observed changes in resistivity occurring at the lowest saturation levels (Region 2 of the desaturation curve). However, it should be noted that under typical field moisture conditions the adsorbed phase water, water in dead-end pore spaces and capillary-bound water (together referred to as the irreducible water content) is not free to drain and therefore it would not be expected that readings in Region 2 would normally be observed at the field scale in the UK. Conversely, however, the higher saturation levels of Region 1 will be commonly observed in the field. In this region the largest variation in resistivity is observed in samples characterized by low CEC, high hydraulic conductivity and where the saturant has a high conductivity. The smallest variation is observed in samples characterized by high CEC, low (saturated) hydraulic conductivity and where the saturant has a low conductivity.

The implication for monitoring the variation in saturation of the unsaturated zone, at this site at least, is that the largest variation in resistivity with saturation is likely to occur in relatively clean or heavily weathered sandstone areas. In these cases the resistivity of the partially saturated sandstone may vary by as much as 250% (i.e. between fully saturated and S_w^0). This allows for the assumption that in field resistivity monitoring surveys, where care is taken to ensure identical location of electrodes for each measurement, a change in resistivity of 5% may be observed without difficulty. Consequently, small changes in saturation of between 5 and 8% are probably observable. At the other extreme, in a very shaly or clay-rich sandstone, variation over Region 1 may only amount to a 20% change in resistivity, suggesting that the minimum measurable change in saturation may be nearer 15–30%.

Conclusions

DC resistivity measurements and modelling of data has been undertaken on a set of fully and partially saturated Permo-Triassic sandstone cores. A successful methodology for the collection of partially saturated resistivity data at low frequency has been developed. The partially saturated resistivity data for these sandstone samples show that there are two relatively well-defined regions, representing distinct pore-water geometries, the transition between which occurs at different saturation levels depending on the pore morphologies of the sample. Results indicate that under typical UK field conditions only saturation levels in Region 1 are likely to be encountered and that the largest variation in resistivity with saturation is likely to occur in relatively clean or heavily weathered sandstone.

Modelling of the data using three different models has indicated that the Permo-Triassic sandstone is a typical shaly sandstone, which cannot be satisfactorily modelled using the simple relationships proposed by Archie. The often quoted assumptions that $m = 2$ and $n = 2$ in Archie's equations are inappropriate for describing formations saturated (fully or partially) with low-salinity electrolyte or where the rock matrix is conductive. Furthermore, our experiments have shown that these basic relationships provide poor insight into the properties of the rock as the derived parameters exhibit little variation even though independently determined measurements of CEC indicate significant variation exists between our samples. Application of the more sophisticated models of Waxman & Smits and the Hanai–Bruggeman to our data more faithfully represents the electrical response of the sandstone under both fully and partially saturated conditions. In addition, the resulting estimates of derived parameters better characterize the electrical properties of the rock, (i.e. providing matrix resistivity) and provide improved correlation with other hydraulic and lithological properties. Comparison of laboratory determined values of CEC and those calculated for the same cores using the W–S model show a high degree of correlation. Modelling using the H–B equation has resulted in estimates for the matrix resistivity and pore factor that also indicate a significant variation between the samples and better correlation with hydraulic properties.

Application of these models in groundwater investigations will therefore lead to improved and more useful estimates of hydraulic parameters and more accurate predictions of the variability of the properties of aquifers.

The authors would like to thank Cleanaway Ltd for their co-operation in this project, and R. Mitchener for undertaking the hydraulic conductivity and CEC measurements on the samples used in this study.

References

ARCHIE, G.E. 1942. The electrical resistivity log as an aid in determining some reservoir characteristics. *Transactions of the American Institute of Mining, Metallurgy and Petroleum Engineers*, **146**, 54–67.

BERG, C.R. 1995. A simple, effective-medium model for water saturation in porous rocks. *Geophysics*, **60**, 1070–1080.

BARKER, R.D. & WORTHINGTON, P.F. 1973. Some hydrogeophysical properties of the Bunter Sandstone of northwest England. *Geoexploration*, **11**, 151–170.

BRUGGEMAN, D.A. 1935. Berechnung verschiedener physikalischer konstanten von heterogenen Substantzen. *Annuls Physik*, **24**, 636–664.

BUSS, S.R. 2000. *Attenuation of strong acids in the Birmingham Sherwood Sandstone aquifer*. PhD thesis, University of Birmingham.

BUSSIAN, A.E. 1983. Electrical conductance in a porous medium. *Geophysics*, **48**, 1258–1268.

CLAVIER, C., COATES, G. & DUMANOIR, J. 1977. The theoretical and experimental bases for the "Dual Water" model for the interpretation of shaly sands. *In: 52nd Annual Fall Technical Conference and Exhibition of the Society of Petroleum Engineers of AIME*, Denver, USA.

CAROTHERS, J.E. 1968. A statistical study of the formation factor relation to porosity. *Log Analyst*, **9**, 13–20.

DE LIMA, O.A.L. 1995. Water saturation and permeability from resistivity, dielectric, and porosity logs. *Geophysics*, **60**, 1756–1764.

DE WITTE, A.J. 1957. Saturation and porosity from electric logs in shaly sands. *Oil and Gas Journal*, **55**, 89–93.

DULLIEN, F.A.L. 1992. *Porous Media: Fluid Transport and Pore Structure*, 2nd edn. Academic Press, San Diego, CA.

ENDRES, A.L. & KNIGHT, R.J. 1991. The effects of pore scale fluid distribution on the physical properties of tight sandstones. *Journal of Applied Physics*, **69**, 1091–1098.

HANAI, T. 1960. Theory of the dielectric dispersion due to the interfacial polarisation and its application to emulsions. *Kolloid-Zeitschrift*, **171**, 23–31.

KELLER, G.V. 1953. Effect of wettability on the electrical resistivity of sand. *Oil and Gas Journal*, **51**, 62–65.

KELLER, G.V. 1966. Electrical properties of rocks and minerals. *In*: CLARK, S.P. (ed.) *Handbook of Physical Constants*. Geological Society of America Memoir, **97**, 553–578.

KNIGHT, R.J. 1991. Hysteresis in the electrical resistivity of partially saturated sandstones. *Geophysics*, **56**, 2139–2147.

KNIGHT, R.J. & DVORKIN, J. 1992. Seismic and electrical properties of sandstone at low saturations. *Journal of Geophysical Research*, **97**, 17 425–17 432.

KNIGHT, R.J. & ENDRES, A. 1990. A new concept in modelling the dielectric response of sandstones: Defining a wetted rock and bulk water system. *Geophysics*, **55**, 586–594.

KNIGHT, R.J. & NUR, A. 1987. The dielectric constant of sandstones, 60 kHz to 4 MHz. *Geophysics*, **52**, 644–654.

LANGMUIR, D. 1997. *Aqueous Environmental Geochemistry*. New Jersey: Prentice Hall, Englewood Cliffs, NJ.

OLD, R.A., HAMBLIN, R.J.O., AMBROSE, K. & WARRINGTON, G. 1991. *Geology of the Country Around Redditch*. Memoir of the British Geological Survey, Sheet 183 (England and Wales). HMSO, London.

MENDELSON, K.S. & COHEN, M.H. 1982. The effect of grain anisotropy on the electrical-properties of sedimentary-rocks. *Geophysics*, **47**, 257–263.

MITCHENER, R. 2002. *Hydraulic and chemical property correlations in the Triassic sandstone of Birmingham, UK*. Unpublished PhD thesis, University of Birmingham.

PATNODE, H.W. & WYLLIE, M.R.J. 1950. The presence of conductive solids in reservoir rock as a factor in electric log interpretation. *Transactions of the American Institute of Mining Engineers*, **189**, 47–52.

REVIL, A. 1999. Ionic diffusion, electrical conductivity, membrane and thermoelectric potentials in colloids and granular porous media: A unified model. *Journal of Colloid and Interface Science*, **212**, 503–522.

ROBERTS, J.J. & LIN, W. 1997. Electrical properties of partially saturated Topopah Spring tuff: Water distribution as a function of saturation. *Water Resources Research*, **33**, 557–587.

TAYLOR, S.B. 2000. *Electrical leak location and sandstone resistivity monitoring using a geophysical system permanently installed below a lined landfill site in the UK*. PhD thesis, University of Birmingham.

TAYLOR, S.B. & BARKER, R.D. 2002. Resistivity of partially saturated Triassic Sandstone. *Geophysical Prospecting*, **50**, 603–613.

TAYLOR, S.B. & BARKER, R.D. 2006. Modelling the DC electrical response of fully and partially saturated Permo-Triassic sandstone. *Geophysical Prospecting* (in press).

VINEGAR, H.J. & WAXMAN, M.H. 1984. Induced polarisation of shaly sands. *Geophysics*, **49**, 1267–1287.

WAXMAN, M.H. & SMITS, L.J.M. 1968. Electrical conductivities in oil-bearing shaly-sand. *Journal of the Society of Petroleum Engineering*, **8**, 107–122.

WAXMAN, M.H. & THOMAS, E.C. 1974. Electrical conductivities in shaly sands: I. Relation between hydrocarbon saturation and resistivity index. II. The temperature coefficient of electrical conductivity. *Transactions of the American Institute of Mining, Metallurgy and Petroleum Engineers*, **257**, 213–225.

WHITE, C.C. & BARKER, R.D. 1997. Electrical leak detection system for landfill liners: A case history. *Ground Water Monitoring and Remediation*, **27**(3), 153–159.

WILLS, L.J. 1976. *The Trias of Worcestershire and Warwickshire*. Report of the Institute of Geological Sciences, **76/2**. HMSO, London.

WINSAUER, W.O., SHEARIN, H.M., MASSON, P.H. & WILLIAMS, M. 1952. Resistivity of brine-saturated sands in relation to pore geometry. *AAPG Bulletin*, **36**, 253–277.

WORTHINGTON, P.F. 1973. Estimation of the permeability of a Bunter Sandstone aquifer from laboratory investigations and borehole resistivity measurements. *Water and Water Engineering*, **77**, 251–257.

WORTHINGTON, P.F. 1977. Influence of matrix conduction upon hydrogeophysical relationships in arenaceous aquifers. *Water Resources Research*, **13**, 87–92.

WORTHINGTON, P.F. 1982. The influence of shale effects upon the electrical resistivity of reservoir rocks. *Geophysical Prospecting*, **30**, 673–687.

WORTHINGTON, P.F. 1985. The evolution of the shaly-sand concepts in reservoir evaluation. *The Log Analyst*, **26**, 23–40.

WORTHINGTON, P.F. & BARKER, R.D. 1972. Methods for the calculation of true formation factors in the bunter sandstone of Northwest England. *Engineering Geology*, **6**, 213–228.

WYLLIE, M.R.J. & GREGORY, G.H.F. 1953, Formation factors of unconsolidated porous media: influence of particle shape and effect of cementation. *Transactions of the American Institute of Mining, Metallurgical and Petrological Engineers*, **198**, 103–110.

Flow and transport in the unsaturated Sherwood Sandstone: characterization using cross-borehole geophysical methods

PETER WINSHIP, ANDREW BINLEY & DIEGO GOMEZ

Department of Environmental Science, Lancaster University, Lancaster LA1 4YQ, UK
(e-mail: A.Binley@lancaster.ac.uk)

Abstract: Cross-borehole radar and resistivity measurements have been used to characterize changes in moisture content and solute concentration due to controlled injection of 1200 l of a saline tracer in the unsaturated zone of the Sherwood Sandstone at a field site in Yorkshire, UK. Borehole radar transmission profiles show the vertical migration of the wetting front during the tracer test. Three-dimensional cross-borehole electrical resistivity tomography was deployed to monitor changes over time in resistivity, caused by the increase in moisture content and pore-water salinity due to the tracer. The results show clearly the development of the tracer plume as it migrates towards the water table at a depth of 10 m. The tomographic results reveal the impact of a hydraulically impeding layer between a depth of 8 and 9 m. Geophysical and geological logs acquired at the site support this conceptualization. By combining the resistivity tomograms with cross-borehole radar tomograms, changes in pore-water concentration over time have been estimated. Changes in moisture content inferred from the geophysical results were compared with those produced by a three-dimensional unsaturated flow model. Using a sandstone effective hydraulic conductivity of 0.4 m day^{-1} in the model produced moisture profiles over time that were comparable with those inferred from the geophysical data during the early stages of the tracer test. Differences between modelled and field results were attributed to the impact of hydraulically impeding layers of finer sediments within the profile.

The ability to predict reliably the travel time of diffuse and point-source contaminants through the unsaturated zone of the Sherwood Sandstone is essential for the management of this nationally important water resource. Field characterization of flow and transport in the unsaturated zone is necessary in order to understand fully the natural processes that affect the fate of contaminants before they reach the water table. Traditionally, borehole-based sampling methods have been used to monitor transport processes in the subsurface. These methods are limited in that the measurement support volume is typically constrained to tens of cubic centimetres. In the unsaturated zone, the application of such methods is complicated by the need to extract pore-water samples at appropriate negative pressures (suctions) in order to obtain samples that truly represent the entire pore-size distribution.

Geophysical techniques have been widely used in hydrogeological studies for decades. For example, Rubin & Hubbard (2005) present the theoretical links between hydrological properties and geophysical parameters, and, through a wide range of case studies, highlight the potential hydrological value gained from geophysical surveys. Several methods, in particular resistivity and radar, allow high-resolution spatial and temporal sampling of the subsurface environment. The characterization of the shallow subsurface has been demonstrated by numerous applications of these methods; however, only recently have attempts been made to quantify directly hydrogeological properties using these techniques (Rubin & Hubbard 2005).

In 1998 a joint project between the universities of Lancaster and Leeds, funded by the UK Natural Environment Research Council and the UK Environment Agency, was initiated to examine, using geophysical methods, unsaturated flow and transport processes at two purposely developed field sites in the UK Sherwood Sandstone. This work, so far, has demonstrated: how cross-borehole (borehole to borehole) radar tomography can be used to monitor changes in moisture content in the unsaturated zone due to natural and forced (tracer) inputs (Binley *et al.* 2001); the evaluation of seasonal variation of moisture content profiles using high-resolution borehole resistivity and radar profiling (Binley *et al.* 2002*b*); initial attempts to utilise the geophysical data to develop numerical predictive models of unsaturated flow (Binley *et al.* 2002*a*; Binley & Beven 2003; Binley *et al.* 2004). In addition,

From: BARKER, R. D. & TELLAM, J. H. (eds) 2006. *Fluid Flow and Solute Movement in Sandstones: The Onshore UK Permo-Triassic Red Bed Sequence*. Geological Society, London, Special Publications, **263**, 219–231.
0305-8719/06/$15 © The Geological Society of London 2006.

petrophysical models relating geophysical data to hydrological properties have been developed (West *et al.* 2003). These articles have concentrated on monitoring and modelling moisture content variation. We report here on a recent joint hydrological–geophysical study of flow and transport in the Sherwood Sandstone at one of the field sites.

The two techniques used here are three-dimensional time-lapse electrical resistivity tomography (ERT) and time-lapse cross-borehole radar tomography and profiling. They provide geophysical measurements that can be related to the moisture content of the subsurface, and subsequently to the conductivity of that moisture content. They also yield data on a scale that is appropriate for numerical simulations of water movement in the subsurface. The two methods have been applied at a site (Lings Farm, Hatfield, near Doncaster, UK) on the outcrop of the Sherwood Sandstone (Fig. 1).

Cross-borehole radar and resistivity: basic concepts

In cross-borehole ERT, four-electrode resistance measurements are made using electrodes in two or more boreholes. Often surface electrodes are used to supplement the electrode array. Inversion of the resistance data is necessary in order to determine an image of resistivity between the boreholes. By discretizing the domain of interest into parameter cells, the objective of the inversion procedure is to compute the 'best' set of resistivity values that satisfies both the measured data set and any *a priori* constraints.

The inversion approach normally adopted uses regularization to stabilize the inversion and constrain the final image (e.g. LaBrecque *et al.* 1996). Cross-borehole ERT has been demonstrated in a wide range of environments. One of the earliest examples of hydrological applications of ERT is Daily *et al.* (1992) in a study of vadose zone moisture migration due to application of a tracer. Other examples of unsaturated zone studies using ERT include Slater *et al.* (1997), Ramirez & Daily (2001) and French *et al.* (2002). At the Hatfield site, Binley *et al.* (2002*a*) demonstrated how three- and two-dimensional ERT can be used successfully to monitor changes in moisture content in the unsaturated sandstone.

Borehole-to-borehole radar surveys may be conducted in two transmission modes in order to determine dielectric properties at the field scale. In both cases a radar signal is transmitted from one antenna placed in the first borehole and received by a second antenna in the other borehole. Measurement of the received electromagnetic wave permits determination of the first arrival and hence velocity of the wave (v). In one mode, using a multiple offset gather (MOG), the receiver is moved to different locations in one borehole whilst the transmitter remains fixed (Peterson 2001). The transmitter is then moved and the process repeated. Following collection of all data in this mode and determination of the travel time for each wave path-line it is possible to derive a tomogram of velocity within the plane of the borehole pair. In contrast, a zero offset profile (ZOP) may be determined by keeping both the transmitter and receiver at equal depth. By systematically lowering or raising the pair of antennae in the two boreholes it is possible to build a one-dimensional profile of average inter-borehole travel time over the entire borehole length.

Examination of the wave-form of the received signal allows the travel time, and hence the velocity of a radar wave, through the material between the boreholes to be determined. In low loss materials and at high frequency, the real part of the bulk dielectric constant (κ) is derived from:

$$\sqrt{\kappa} = \frac{c}{v} \qquad (1)$$

where c is the radar wave velocity in air (≈ 0.3 m ns^{-1}).

The increasing availability of commercial borehole radar systems and growing acceptance of radar in the hydrological community has led to a number of recent hydrogeological applications of the technique in unsaturated systems (e.g. Hubbard *et al.* 1997; Alumbaugh *et al.* 2002; Galagedera *et al.* 2003).

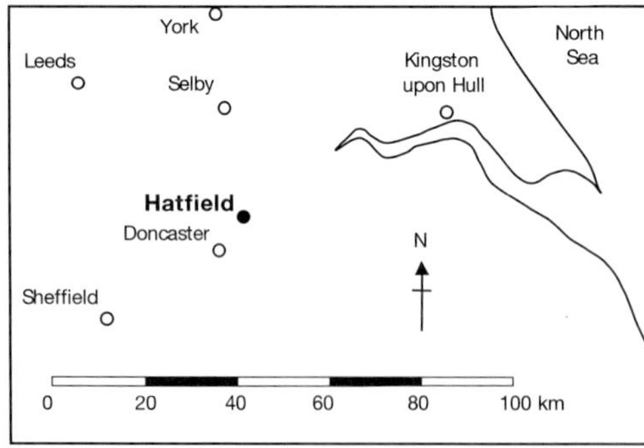

Fig. 1. Hatfield site location.

Site description

At the Hatfield site six boreholes were drilled in 1998 in order to monitor tracers injected into the sandstone (Fig. 2). Four of these boreholes (H-E1, H-E2, H-E3 and H-E4) were designed for resistivity measurements. These ERT boreholes contain 16 stainless steel mesh electrodes equally spaced at depths of between 2 and 13 m. Two boreholes (H-R1 and H-R2) were installed for radar measurements. These boreholes were drilled to a depth of 12 m and completed with 75 mm PVC casing. Both the ERT and radar boreholes have a weak sand–cement grout backfilling the gap between the host formation and installation. A tracer injection borehole (H-I2) was also installed within the centre of the borehole array (Fig. 2). The injection borehole is 3.5 m deep, with a 100 mm-diameter slotted section, and gravel pack between 3 and 3.5 m in depth.

Two cored boreholes were drilled at the site (Fig. 2) and logged by Leeds University (Pokar *et al.* 2001) (Fig. 3). The main lithology present in the core is medium-grained sandstone, interspersed with interlaminated fine- and medium-grained sandstones, particularly in the zone around 6 m depth, and between 8 and 9 m. Drift at the top of the section at the site is typically 2–3 m thick, and consists mainly of fluvio-glacial sands, derived from the underlying sandstones, with frequent large pebbles/cobbles.

In order to minimize disturbance, particularly from ingress of drilling fluids, cores were not extracted from the tracer array area. However, geophysical logs were obtained for all drilled boreholes using electromagnetic induction and natural gamma logging tools. Figure 4 shows example natural gamma logs for the boreholes H-E2, H-R2, H-R2 and H-E1. These logs reveal

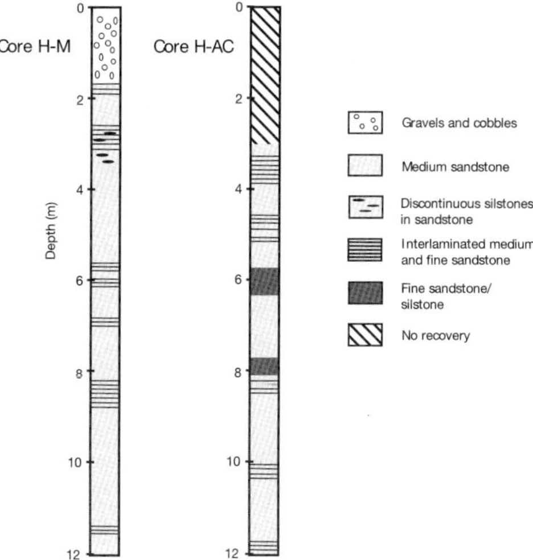

Fig. 3. Core logs (supplied by J. West, Leeds University).

subtle contrasts but support the conceptualization of repeated fine–medium sandstone layering.

Experimental procedures

During March 2003 a saline tracer was injected into the sandstone using borehole H-I2. Changes in bulk resistivity and dielectric constant of the sandstone were then monitored using radar and resistivity using the procedures described below.

Cross-borehole ERT

In order to compute a high-resolution image of the subsurface using cross-borehole ERT it is necessary to acquire a large number of four electrode measurements. During tracer tests the data capture time is critical, as each image should reflect a 'snapshot' of the subsurface. For this experiment, a six-channel Geoserve Resecs instrument was used, allowing the collection of 6372 measurements in about 2.5 h. The current and potential electrode pairs were chosen so that the dipoles they defined were horizontal, with one of the electrodes in each dipole being in one borehole and the other in any of the remaining three boreholes. The current and potential dipoles were restricted to being within 4.4 m of each other vertically, so that measured voltages were not too low. For all ERT surveys, reciprocal data (i.e. with current and potential electrodes swapped) were collected to assess

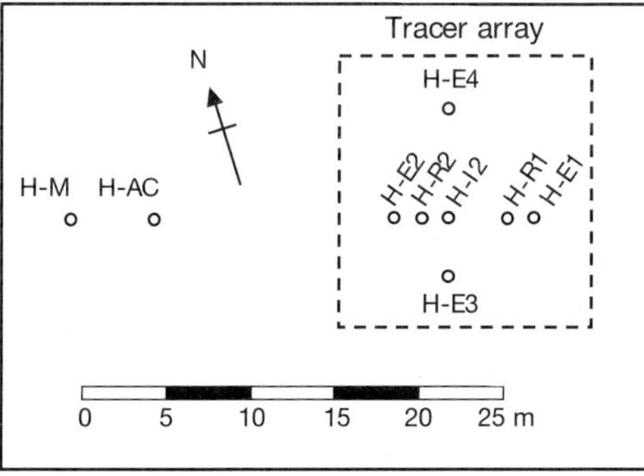

Fig. 2. Field site layout showing boreholes.

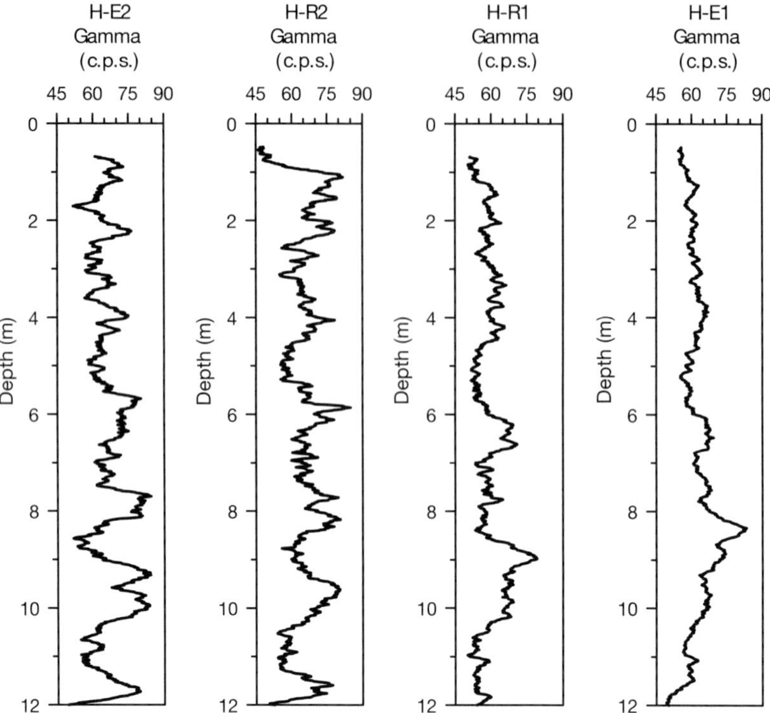

Fig. 4. Natural gamma logs in boreholes H-E2, H-R2, H-R1 and H-E1.

error levels (repeatability checks are often inadequate for this purpose: Daily *et al.* 2004). Thus, a maximum of 3186 measurements were used for data inversion. Inversion of the data in order to produce a three-dimensional resistivity tomogram was based on a regularized least-squares algorithm similar to that outlined in Morelli & LaBrecque (1996).

Resistivity values so obtained are assumed here to be related to hydrogeological parameters by Archie's Law (Archie 1942):

$$\rho = a\phi^{-m}\left(\frac{\theta}{\phi}\right)^{-n}\rho_w \qquad (2)$$

where ρ is the resistivity of the bulk material, ϕ is the porosity, ρ_w is the resistivity of the pore fluid, θ the volumetric moisture content, and a, m and n are formation constants. For resistivity measurements repeated at different times, then:

$$\frac{\rho_t}{\rho_0} = \frac{\theta_t^{-n}}{\theta_0^{-n}}\frac{\rho_{w,t}}{\rho_{w,0}} \qquad (3)$$

where the subscripts t and 0 refer to measurements at time t and time 0, respectively.

We recognize the limitation of using Archie's law in shaley sandstone (see, for example, Worthington 1977), but in the absence of appropriate petrophysical relationships the Archie model is adopted as a suitable first approximation.

Cross-borehole radar

Using boreholes H-R1 and H-R2, two radar data collection modes were adopted: zero offset profiling (ZOP) and multiple offset gathers (MOG). For both surveys a Sensors and Software Pulse EKKO PE100 system was used with 100 MHz antennae. For the ZOP surveys the antennae were lowered at 0.25 m increments. For the MOG surveys a 'complete' data set was not obtained due to time constraints imposed by the expected tracer movement. The MOG surveys carried out used transmitter locations at 1 m increments between depths of 1 and 10 m in H-R1, with receiver positions at 0.25 m increments between 1 and 10 m depth in H-R2 (also ensuring that the absolute vertical angle between transmitter and receiver did not exceed 45°). MOG data were inverted using the MIGRATOM code (Jackson & Tweeton 1994) to produce an image of radar velocity between

H-R1 and H-R2 and hence, using equation 1, an image of the bulk dielectric constant.

In order to describe the relationship between bulk dielectric constant and volumetric moisture content the complex refractive index method (CRIM) was used. The CRIM model can be stated as:

$$\sqrt{\kappa} = (1-\phi)\sqrt{\kappa_s} + \theta\sqrt{\kappa_w} + (\phi-\theta)\sqrt{\kappa_a} \quad (4)$$

where κ_s is the dielectric constant of the sediment grains, κ_w is the dielectric constant of water (assumed to be 81), κ_a is the dielectric constant of air (assumed to be 1) and ϕ is porosity. West $et\ al.$ (2003) carried out measurements of dielectric properties at different levels of water saturation in core samples extracted from the site. Based on these measurements we assume here that $\kappa_s = 5$ and $\phi = 0.32$. Note that the dielectric constant is independent of the electrical conductivity of the pore fluid.

Where measurements are taken at different times, equations 1 and 4 can be used to give the change in moisture content ($\Delta\theta$) as a function of the difference in observed radar wave velocity (Δv):

$$\Delta\theta = \Delta v\left(\frac{c}{\sqrt{\kappa_a} - \sqrt{\kappa_w}}\right). \quad (5)$$

The tracer experiment

The tracer consisted of 1200 l of water, dosed with NaCl to give an electrical conductivity of 2200 µS cm^{-1} (groundwater electrical conductivity at the site was measured as 650 µS cm^{-1}). The tracer was injected over a period of 3 days, from 14 to 17 March 2003, at a steady rate of approximately 17 l h^{-1}. A float valve in the injection borehole was used to control the head in the injection borehole, and hence the flow rate. Duplicate sets of background measurements of ERT were made on 6 and 13 March, and of radar measurements on 6 and 14 March. Tracer flow was monitored by means of a pressure transducer in a storage tank, which gave a way of calculating the cumulative injection volume over time. During the tracer test no rainfall was observed at the site. The water table was observed at approximately 10 m depth.

Results and analysis

Figure 5 shows the background (pre-tracer) ZOP results, converted to dielectric constant. Assuming that the dielectric constant is principally controlled by the moisture content, the

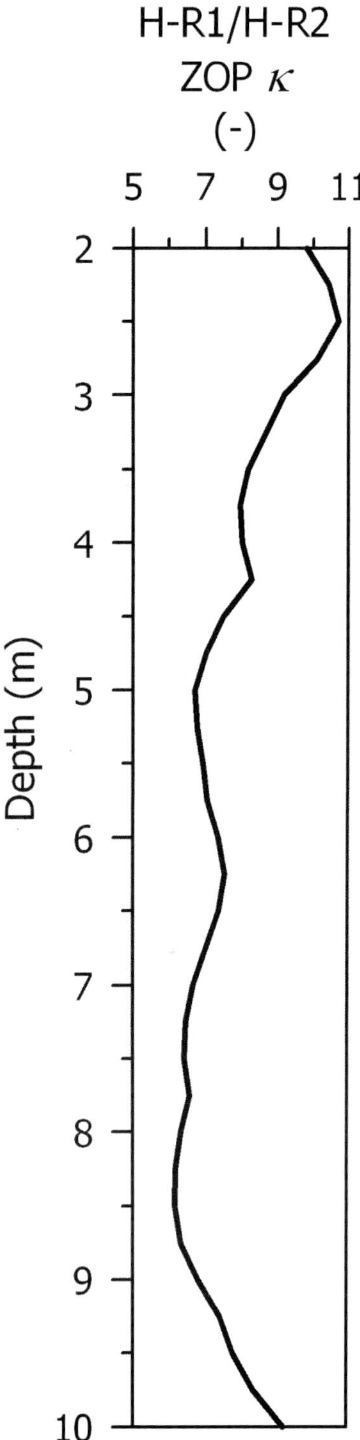

Fig. 5. Pre-tracer profile of dielectric constant determined from the average of ZOP surveys on 6 and 14 March 2003.

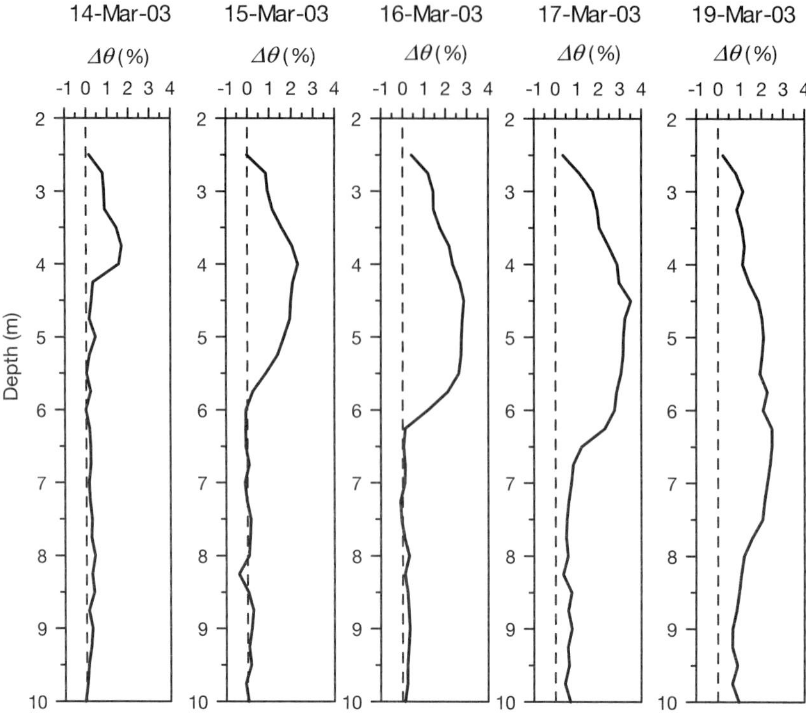
14-Mar-03
Δθ(%)
15-Mar-03
Δθ(%)
16-Mar-03
Δθ(%)
17-Mar-03
Δθ(%)
19-Mar-03
Δθ(%)
Depth (m)

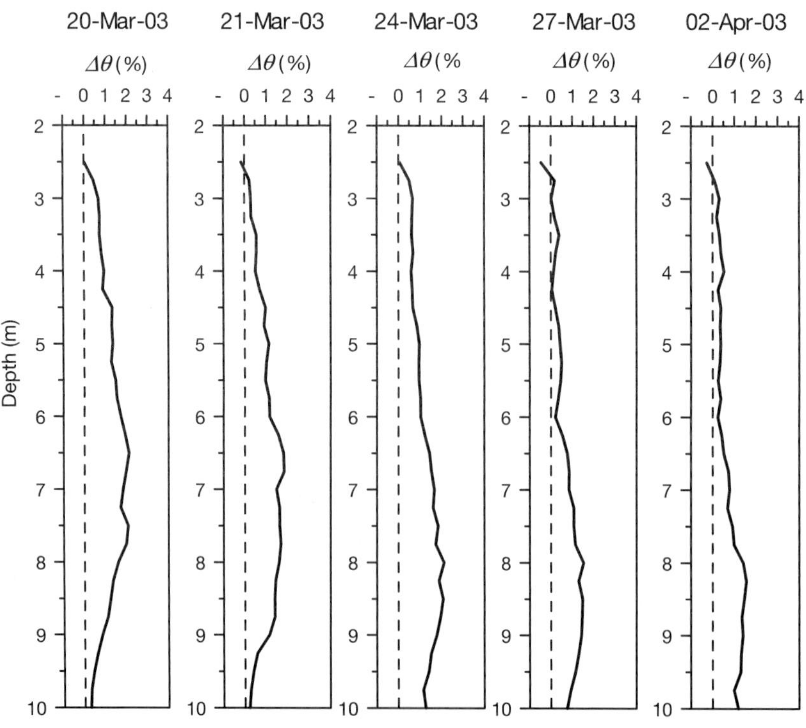
20-Mar-03
Δθ(%)
21-Mar-03
Δθ(%)
24-Mar-03
Δθ(%
27-Mar-03
Δθ(%)
02-Apr-03
Δθ(%)
Depth (m)

radar profile may be interpreted as follows. The high dielectric constant at 2.5 m depth is likely to be due to moisture retention at the base of the drift. At approximate depths of 4.2, 6.5 and 7.7 m increases in dielectric constant are seen, which are probably a result of further moisture retention by fine-grained units. These positions coincide with observed contrasts in the natural gamma logs shown in Figure 4 and the lithology of cores shown in Figure 3. Increases in the dielectric constant at depths greater than 9 m arc also likely to be the result of increased moisture retention but also will be due to the close proximity of the water table (10 m depth). Refraction of radar waves at the water table–capillary fringe can result in apparent high radar velocities as the first arrival may be a refracted wave, rather than the assumed direct wave.

Changes in moisture content inferred from radar measurements

As changes in radar velocity are not dependent on lithological parameters in the petrophysical model (see equation 5), changes in moisture content may be determined more reliably. Changes in moisture content from the pre-tracer conditions, inferred from the ZOP surveys, are shown in Figure 6. The development of the tracer plume during the injection (14–17 March) is clearly seen, as is the steady vertical migration of the wetting front. As this wetting front moves, the moisture 'bulb' grows and thus the volumetric change in water content observed by the radar decreases over time. The volume of the subsurface that is 'sensed' by the radar profile is described by the Fresnel zone for the particular radar wave frequency (Cervany & Soares 1992). The Fresnel zone is assumed to be an ellipse with a minor axis length of:

$$B = \left(\frac{\lambda^2}{4} + L \times \lambda \right)^{0.5} \qquad (6)$$

and a major axis length of

$$A = \left(\frac{\lambda}{4} + L \right) \qquad (7)$$

where L is the borehole separation (5 m) and λ is the wavelength (for a 100 MHz wave, with a velocity of about 0.1 m ns^{-1}, this is 1 m). For the

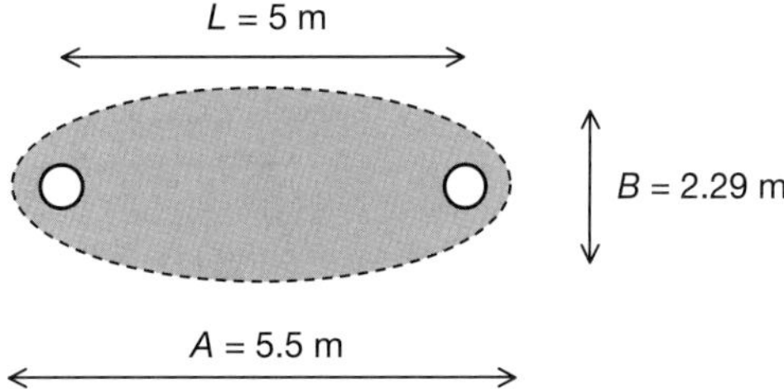

Fig. 7. Definition of the Fresnel zone for borehole radar measurements.

case reported here, $B = 2.29$ m and $A = 5.5$ m (Fig. 7).

The changes in moisture content, determined from the ZOP surveys, are shown in Figure 8 as hydrographs at particular depths. These time series reveal that approximately 230 h (about 9 days) after injection of the tracer was terminated (i.e. about 300 h after the start of tracer injection) moisture content at depths of 5 and 6 m return to near pre-tracer levels. At greater depths, however, the retention of moisture is observed for considerably longer. We infer this to be a result of fine-grained units between 8 and 9 m depth (Figs 3 and 4).

Changes in resistivity

The changes in moisture content determined from the radar profiles offer some insight into the mechanisms controlling unsaturated flow within the sandstone at the site. However, it is impossible to determine travel times of 'parcels' of water directly from these observations. Moisture already retained in the sandstone will be displaced by tracer water, but clearly 'new' and 'old' water cannot be differentiated. It is for this reason that electrical resistivity surveys were utilized. As already stated, changes in resistivity will be related to changes in moisture content and pore-water electrical conductivity (equation 3). With appropriate petrophysical relationships we may therefore use ERT and radar jointly to differentiate the 'new' tracer water from the existing formation water.

The changes in resistivity throughout the tracer test are shown in Figure 9. These are shown as isosurfaces of volumes with changes relative to the pre-tracer conditions above a certain threshold (in this case 7.5% for illustration of significant changes in moisture

---

Fig. 6. Changes in moisture content from pre-tracer conditions between boreholes H-R1 and H-R2 during tracer test, inferred from ZOP surveys.

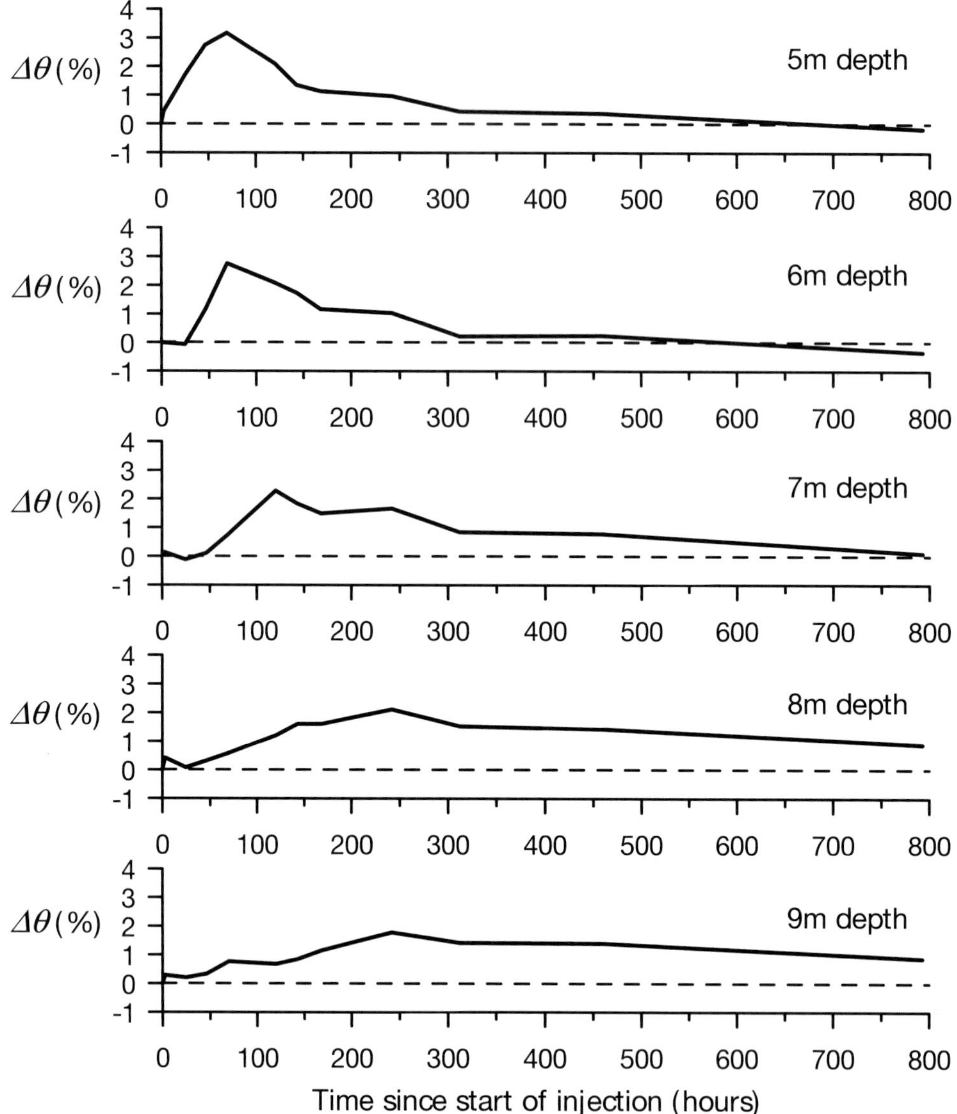

Fig. 8. Changes in moisture content during the tracer test at specific depths, inferred from ZOP surveys.

content). The images show clearly the development of the tracer 'bulb' during the injection and the subsequent vertical migration. Most striking is the obvious lateral spreading of the tracer between 8 and 9 m depth. These support the earlier hypothesis of a hydraulically retarding 'layer' at this depth. Note also, in Figure 9, that near the tracer-injection source (between depths of 3.5 and 6 m) the resistivity does not return to pre-tracer values even by 2 April. The volume apparently occupied by the tracer in this depth interval does shrink over the monitoring period but is still detectable 16 days after the tracer injection was stopped. To depths of 6 m, the moisture content has returned to pre-tracer levels by 27 March (Fig. 6); the change in resistivity is thus an indication that some fraction of the pore space has been replaced by the more electrically conductive tracer fluid.

Changes in pore-water solute concentration

If we assume that the solute concentration of pore water is linearly related to the fluid

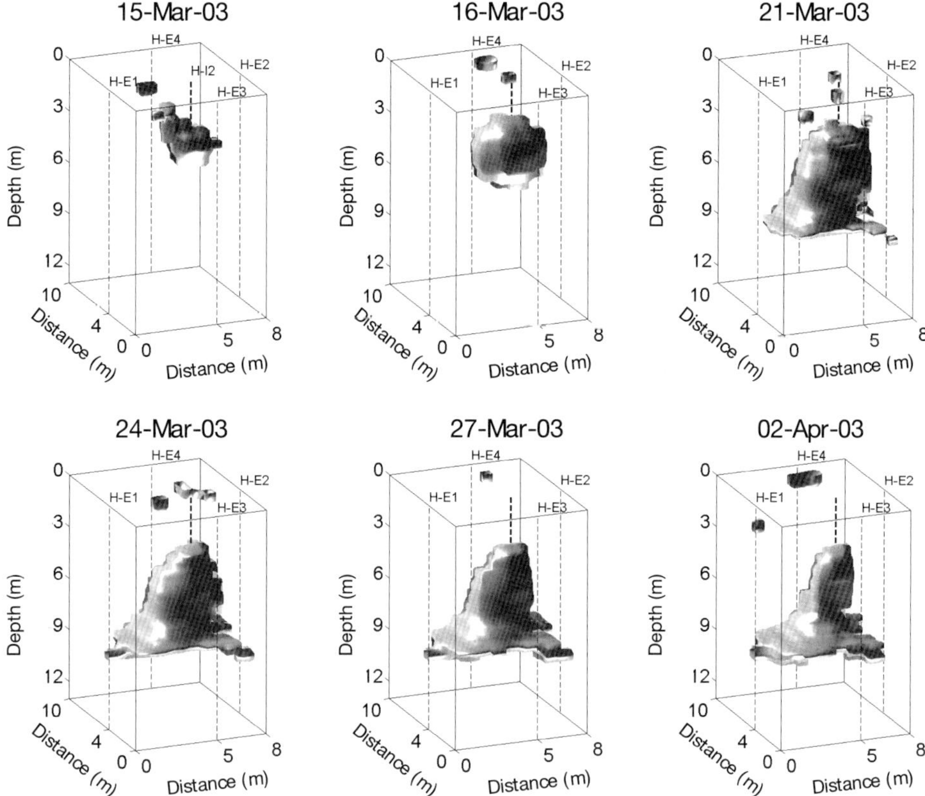

Fig. 9. Changes in resistivity during tracer test shown as isosurfaces of 7.5% reduction in resistivity relative to pre-tracer conditions. Shading is used to illustrate the shape of the moisture bulb that develops during the test.

electrical conductivity, i.e. inversely related to the fluid resistivity ρ_w, and given that κ_w will not change over time, then equations 3 and 4 can be combined to give an expression for the solute concentration relative to the background (pre-tracer) levels:

$$\frac{c_t}{c_0} = \frac{\rho_0}{\rho_t} \left(\frac{\sqrt{\kappa_0} + \phi(\sqrt{\kappa_s} - \sqrt{\kappa_a}) - \sqrt{\kappa_s}}{\sqrt{\kappa_t} + \phi(\sqrt{\kappa_s} - \sqrt{\kappa_a}) - \sqrt{\kappa_s}} \right)^n \quad (8)$$

where C_0 and C_t represents the solute concentration at time 0 and time t, κ_0 and κ_t are the dielectric constant values at time 0 and time t.

Assuming a value of $n = 1.13$ (Binley *et al.* 2002*b*) and other values defined as before, the resistivity quotients (ρ_0/ρ_t) were interpolated from the ERT images onto the vertical plane between radar boreholes H-R1 and H-R2. Then, using changes in dielectric constant obtained from the MOG radar inversions, the ratios of the pore water solute concentrations were

computed. The result is shown in Figure 10, from which it is apparent that the solute migrates at a much slower rate than the moisture front (as expected). Early transport is rapid to a depth of 6 m, at which point vertical transport is retarded somewhat – again supporting the hypothesis that the observed fine-grained units act as hydraulically impeding layers. We recognize that the results produced from application of equation 8 are subject to errors; increases in concentration *above* the tracer-injection zone, for example, are apparent in Figure 10. We also recognize that tomographic images are subject to inherent non-uniqueness. Nevertheless, this analysis offers some insight into unsaturated zone solute-transport processes that could not have been achieved without joint application of radar and resistivity.

Hydrological simulations

Binley *et al.* (2002*a*) applied a numerical model of unsaturated flow to tracer test data at the

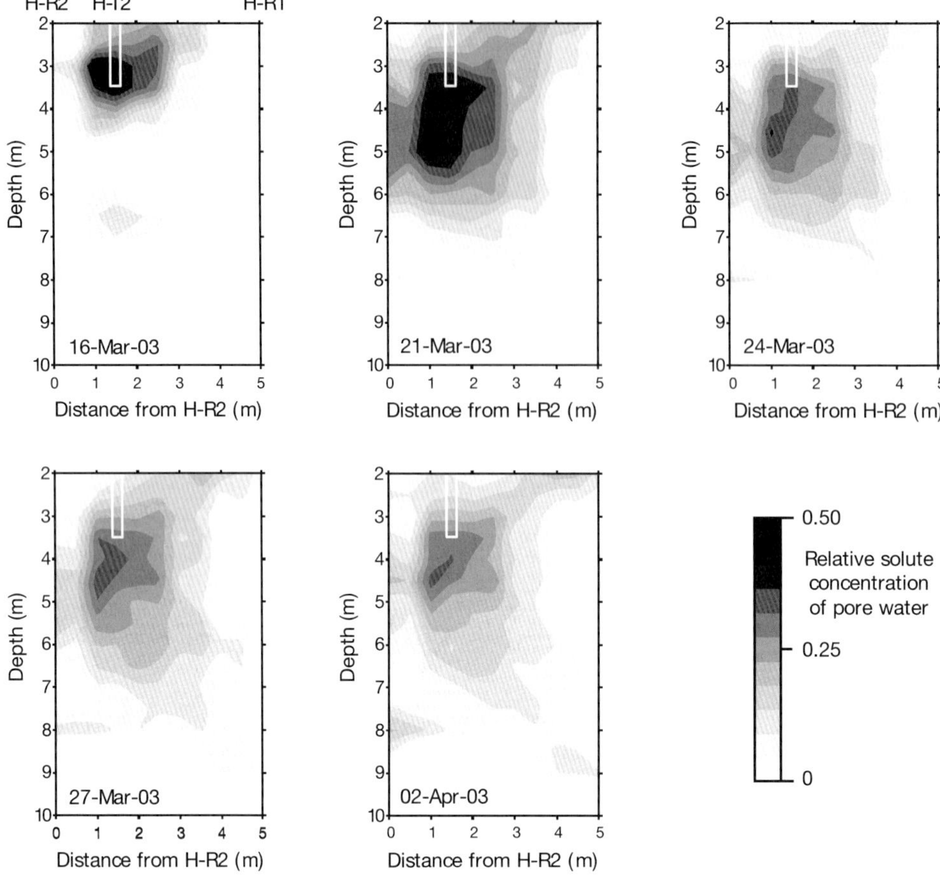

Fig. 10. Changes in pore-water solute concentration during the tracer test, inferred from radar and resistivity images.

Hatfield site. In their analysis the Richards equation was used, which can be written as:

$$\frac{\partial}{\partial x_i}\left(K(\psi)\frac{\partial h}{\partial x_i}\right) = \frac{\partial \theta(\psi)}{\partial t}, i = 1, 2, 3 \qquad (9)$$

where x_i are the co-ordinates (x_3 vertical co-ordinate), $K(\psi)$ is the hydraulic conductivity, ψ is the pressure head, h is the hydraulic head $= \psi + x_3$, $\theta(\psi)$ is volumetric moisture content and t is time.

In the analysis of Binley *et al.* (2002*a*) it was assumed that the unsaturated sandstone could be represented by a single effective hydrogeo-logical unit and attempts were made to deter-mine appropriate hydraulic parameters for the Hatfield site. The tracer test used was restricted in duration to approximately 200 h, i.e. 25% of that presented here. In an attempt to extend the

findings of this earlier study we apply here the same numerical parameterization as Binley *et al.* (2002*a*) and compare the simulated response with observations.

Modelling of the unsaturated zone was carried out using the three-dimensional (3D) finite-element model FEMWATER (Lin *et al.* 1997), which is based on a pressure head formulation. In FEMWATER, the widely used van Genuchten model (van Genuchten 1980) describing the unsaturated hydraulic relation-ships is adopted. With such an approach the unsaturated characteristics are described by:

$$\theta(\psi) = \theta_r + \frac{\theta_s - \theta_r}{[1+ \mid \alpha\psi \mid^\beta]^{\frac{\beta-1}{\beta}}} \qquad (10)$$

and

$$K(\psi) = K_s S_e^{0.5} \left[1 - (1 - S_e^{\beta/(\beta-1)})^{\frac{\beta-1}{\beta}} \right]^2 \quad (11)$$

where K_s is the saturated hydraulic conductivity, θ_r is the residual moisture content, θ_s is the saturated moisture content, S_e is effective saturation ($= (\theta - \theta_r)/\theta_s - \theta_r)$), and α and β are parameters.

As in Binley *et al.* (2002*a*), a model was set up to represent a parallelepiped of 11×11 m in plan (to allow specification of zero horizontal flow-boundary conditions) and 10 m in depth. The system was composed of three layers (Table 1): upper soil (Layer 1), sandy soil (drift) (Layer 2) and sandstone (Layer 3). The injection took place in Layer 3. The mesh model was composed of about 125 000 six-node prism elements and 65 000 nodes. The solution to the system of non-linear equations was achieved with a convergence threshold for hydraulic head equal to 0.001 m. The total simulation time was 500 h.

Figure 11 shows the observed and simulated change in volume of water in the system. In this figure the observed changes are inferred from the ZOP radar profiles. The total injected volume is 1.2 m³, but because radar transmission paths do not cover the entire volume invaded by the tracer a mass balance error results. This 'error' increases with time as the moisture bulb spreads laterally orthogonal to the radar transmission plane. The observed response shows a sharp increase in water volume (as the tracer is injected), followed by a slower decrease as the sampled volume drains. Note that some scatter is seen in the recession limb of the hydrograph, which is inevitable given that the observed changes in moisture content are very low (Fig. 6).

Equations 6 and 7 were used to define the appropriate sampling volume for the numerical simulations. From Figure 11 it is apparent that the model and observed responses match very well until approximately 200 h into the tracer test. After this point the model underpredicts the water volume within the Fresnel zone,

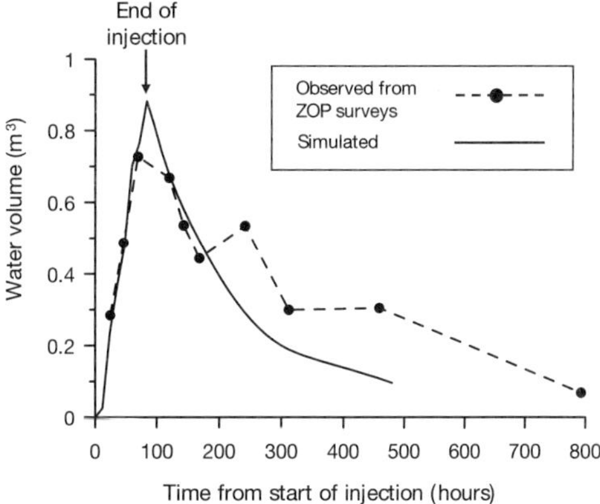

Fig. 11. A comparison of measured and modelled pore-water volumes between H-R1 and H-R2 during the tracer test.

implying that in the model drainage to the water table is too rapid. Interestingly, the optimization of effective hydraulic parameters by Binley *et al.* (2002*a*) was constrained to 250 h; it appears that, given extended data for the 2003 tracer test, parameterization of the deeper sandstone is inappropriate. This again supports the hypothesis that a low hydraulic conductivity unit exists deeper in the profile, for example at 8–9 m. It appears, therefore, that a single effective hydraulic conductivity value is not appropriate for the sandstone.

Conclusions

Cross-borehole radar and resistivity measurements have been used to characterize changes in moisture content and solute concentration due to controlled injection of a saline tracer in the unsaturated zone of the Sherwood Sandstone. Borehole radar transmission profiles show the vertical migration of the wetting front during the tracer test. Changes down to 1% volumetric moisture content appear detectable by the technique used, although we recognize

Table 1. *Hydraulic parameters for unsaturated flow modelling at the Hatfield site (after Binley* et al. *2002a)*

	θ_r	θ_s	β	α (m⁻¹)	K_s (m day⁻¹)
Layer 1 (0–0.5 m bgl)*	0.05	0.30	1.9	2	0.01
Layer 2 (0.5–3 m bgl)	0.04	0.32	2.2	2	0.048
Layer 3 (3–12 m bgl)	0.04	0.32	2.5	10	0.4[†]

* m bgl, metres below ground level.
[†] The K_s in this layer is based on the optimum value determined by Binley *et al.* (2002*a*).

that such signal sensitivity will not be achievable at all field sites. Three-dimensional cross-borehole electrical resistivity tomography was deployed to monitor changes in resistivity over time. The results show clearly the plume development and have revealed the impact of a hydraulically impeding layer above the water table. Geophysical and geological logs acquired at the site support this conceptualization. By combining the resistivity tomograms with cross-borehole radar tomograms we have estimated changes in pore-water concentration over time, albeit in a 2D vertical plane. Such information would not be obtainable without the joint application of radar and resistivity methods. By utilizing these in cross-borehole mode high-resolution imaging has been achievable.

In many previous hydrogeological studies geophysical techniques have been adopted in a purely qualitative manner. There is, however, hydraulic information that can be extracted from these techniques through appropriate integration with a hydrological modelling program. We have demonstrated how numerical models may be used jointly with geophysics and believe that further hydrogeophysical studies will show the immense value of geophysical data in constraining subsurface hydrological models. Our conceptual and numerical models of unsaturated flow and transport processes in the Sherwood Sandstone will continue to be refined and, we believe, ultimately help constrain predictive models used by water-resource managers and environment regulators.

We are grateful to the Environment Agency, UK, for continued support for our work. E. Mould and A. Walmsley (Environment Agency, UK) brought drilling expertise early on in the project. The work would not have been possible without agreement of site access by J. Cunliffe of Lings Farm, Hatfield. J. West supplied core logs and saturation–Dc resistivity data. This work was funded by the Natural Environment Research Council, UK, under NERC studentship grant NER/S/A/2001/06246.

References

ALUMBAUGH, D., CHANG, P.Y., PAPROCKI, L., BRAINARD, J.R., GLASS, R.J. & RAUTMAN, C.A. 2002. Estimating moisture contents in the vadose zone using cross-borehole ground penetrating radar: A study of accuracy and repeatability. *Water Resources Research*, **38**, 1309, doi:10.1029/2001WR000754.

ARCHIE, G.E. 1942. The electrical resistivity log as an aid to determining some reservoir characteristics. *Transactions of the American Institute of Mining Engineers*, **146**, 389–409.

BINLEY, A. & BEVEN, K. 2003. Vadose zone flow model uncertainty as conditioned on geophysical data. *Ground Water*, **41**, 119–127.

BINLEY, A., CASSIANI, G., MIDDLETON, R. & WINSHIP, P. 2002*a*. Vadose zone model parameterisation using cross-borehole radar and resistivity imaging. *Journal of Hydrology*, **267**, 147–159.

BINLEY, A., CASSIANI, G. & WINSHIP, P. 2004. Characterization of heterogeneity in unsaturated sandstone using borehole logs and cross-borehole tomography. *In*: BRIDGE, J.S. & HYNDMAN, D.W. (eds) *Aquifer Characterization by SEPM*. Society for Sedimentary Geology, Tulsa, OK, 129–138.

BINLEY, A., WINSHIP, P., MIDDLETON, R., POKAR, M. & WEST, J. 2001. High resolution characterization of vadose zone dynamics using cross-borehole radar. *Water Resources Research*, **37**, 2639–2652.

BINLEY, A., WINSHIP, P., WEST, L.J., POKAR, M. & MIDDLETON, R. 2002*b*. Seasonal variation of moisture content in unsaturated sandstone inferred from borehole radar and resistivity profiles. *Journal of Hydrology*, **267**, 160–172.

CERVANY, V. & SOARES, J.E.P. 1992. Fresnel volume ray tracing. *Geophysics*, **57**, 902–915.

DAILY, W., RAMIREZ, A., BINLEY, A. & LABRECQUE, D. 2004. Electrical resistance tomography. *The Leading Edge*, **23**, 438–442.

DAILY, W.D., RAMIREZ, A.L., LABRECQUE, D.J. & NITAO, J. 1992. Electrical resistivity tomography of vadose water movement. *Water Resources Research*, **28**, 1429–1442.

FRENCH, H.K., HARDBATTLE, C., BINLEY, A., WINSHIP, P. & JAKOBSEN, L. 2002. Monitoring snowmelt induced unsaturated flow and transport using electrical resistivity tomography. *Journal of Hydrology*, **267**, 273–284.

GALAGEDARA, L.W., PARKIN, G.W., REDMAN, J.D. & ENDRES, A.L. 2003. Assessment of soil moisture content measured by borehole GPR and TDR under transient irrigation and drainage. *Journal of Environmental Engineering and Geophysics*, **8**, 77–86.

HUBBARD, S.S., PETERSON, J.E., MAJER, E.L., ZAWISLANSKI, P.T., WILLIAMS, K.H., ROBERTS, J. & WOBBER, F. 1997. Estimation of permeable pathways and water content using tomographic radar data. *Leading Edge*, **16**, 1623–1628.

JACKSON, M.J. & TWEETON, D.R. 1994. *MIGRATOM – Geophysical Tomography Using Wavefront Migration and Fuzzy Constraints*. Bureau of Mines Report, **RI9497**.

LABRECQUE, D.J., MILLETO, M., DAILY, W., RAMIREZ, A. & OWEN, E. 1996. The effects of noise on Occam's inversion of resistivity tomography data. *Geophysics*, **61**, 538–548.

LIN, H.J., RICHARDS, D.R., TALBOT, C.A., YEH, G.T., CHENG, J. & CHENG, H. 1997. *FEMWATER: A Three-dimensional Finite Element Computer Model for Simulating Density-dependent Flow and Transport in Variably Saturated Media*. US Army Corps of Engineers and Pennsylvania State University Technical Report, **CHL-97-12**.

MORELLI, G. & LABRECQUE, D.J. 1996. Advances in ERT modelling. *European Journal of Environmental and Engineering Geophysics*, **1**, 171–186.

PETERSON, J. 2001 Pre-inversion corrections and analysis of radar tomographic data. *Journal of Environmental and Engineering Geophysics*, **6**, 1–18.

POKAR, M., WEST, L.J., WINSHIP, P. & BINLEY, A.M. 2001. *Proceedings of the Symposium on Applications of Geophysics to Engineering and Environmental Problems (SAGEEP2001)*. Environmental and Engineering Geophysical Society, Denver, CO.

RAMIREZ, A. & DAILY, W. 2001. Electrical imaging at the large block test – Yucca Mountain, Nevada. *Journal of Applied Geophysics*, **46**, 85–100.

RUBIN, Y. & HUBBARD, S.S. 2005. *Hydrogeophysics.* Springer, New York.

SLATER, L., ZAIDMAN, M.D., BINLEY, A.M. & WEST, L.J. 1997. Electrical imaging of saline tracer migration for the investigation of unsaturated zone transport mechanisms, *Hydrology and Earth System Science*, **1**, 291–302.

VAN GENUCHTEN, M.T. 1980. A closed-form equation for predicting the hydraulic conductivity of unsaturated soils. *Soil Science Society of America Journal*, **44**, 892–898.

WEST, L.J., HANDLEY, K., HUANG, Y. & POKAR, M. 2003. Radar frequency dielectric dispersion in sand and sandstone: Implications for determination of moisture content and clay content. *Water Resources Research*, **39**, 1026, doi:10.1029/2001WR000923.

WORTHINGTON, P.F. 1977. Influence of matrix conduction upon hydrogeophysical relationships in arenaceous aquifers. *Water Resources Research*, **13**, 87–92.

Non-reactive solute movement through saturated laboratory samples of undisturbed stratified sandstone

KHAIRUL BASHAR[1] & JOHN H. TELLAM

School of Geography, Earth and Environmental Sciences, University of Birmingham, Birmingham B15 2TT, UK

[1]*Present address: Department of Geology, Jahangirnagar University, Dhaka, Bangladesh (e-mail: J.H.Tellam@bham.ac.uk)*

Abstract: There has been much recent work on developing models of non-reactive solute migration in saturated stratified porous media. Almost all experimental results against which the models have been tested have been obtained using artificial media. The aim of the present study is to test the models against data from naturally stratified media. In this paper we report the results of the experiments carried out on samples of laminated, intact, saturated Triassic sandstone from the UK. Column experiments were performed at steady flow rates using samples with flow either parallel or perpendicular to the lamination. For flow parallel to the lamination, the breakthrough curves were asymmetrical. They were generally characterized by early breakthrough and tailing. Asymmetry and tailing increased with increasing flow rate. Column experiments in which flow was interrupted showed the presence of physical non-equilibrium. For flow perpendicular to the lamination the breakthrough curves were symmetrical. Simultaneous use of bromide and amino-G-acid, conservative tracers having very different diffusion coefficients, demonstrated the significance of diffusion particularly when the flow is parallel to the lamination. Thin-section analysis, dye staining and positron emission projection imaging (PEPI) techniques were used to study the spatial variations in hydraulic properties in the samples. Thin-section analysis indicated that the thickness of individual layers, each of different porosity and grain size, varies from less than 1 mm to several millimetres and occasionally exceeds 1 cm. The dye and PEPI experiments also identified stratification of flow when the flow is parallel to the lamination, but in the latter case the most obvious stratification was at a larger scale than for the former. No preferential flow was found for samples with flow perpendicular to the lamination. It is concluded that the dominant process in solute migration in the sandstone samples is stratification that is, at least, at two scales, a process which will result in a fractionation where two solutes of different diffusion coefficient are present.

Over the last two decades there has been much interest in non-reactive solute migration through saturated stratified porous media (e.g. Pickens & Grisak 1981; De Smedt & Wierenga 1984; Güven *et al.* 1984; Bhattacharya & Gupta 1986; Brusseau & Rao 1990; Li *et al.* 1994; Griffioen *et al.* 1998). The vast majority of this work has centred on experimental results using artificial media. Here we present the results of an investigation on the movement of solute through intact stratified sandstone samples. The paper deals exclusively with the qualitative results: a subsequent publication will analyse the data quantitatively. The samples used in the experiments are from the UK Triassic Sandstone, a red-bed aeolian–fluviatile sequence and the UK's second most-used aquifer.

Laboratory experiments

Approach

Several laboratory techniques were used for the study of mass transfer in the Triassic sandstone samples. Following batch experiments designed to test the suitability of various tracers, column experiments were performed in order to study the transport of dissolved solutes through the samples directly. Dye staining and positron emission projection imaging (PEPI) techniques were used for visualizing flow paths of dissolved materials within the sandstone. Thin-section analysis, dye staining and positron emission tomography techniques were used for the study of the heterogeneity in the hydraulic properties. Thin-section analysis and dye experiments are destructive methods; PEPI is a non-destructive method.

From: BARKER, R. D. & TELLAM, J. H. (eds) 2006. *Fluid Flow and Solute Movement in Sandstones: The Onshore UK Permo-Triassic Red Bed Sequence*. Geological Society, London, Special Publications, **263**, 233–251.

Batch experiments

Batch experiments were undertaken in order to test for any reaction between the tracers later used in the column experiments – amino-G-acid and bromide – and the rock: these tracers were initially selected because of their low reactivity (Levy & Chambers 1987; Trudgill 1987). Twenty grammes of disaggregated dry Triassic sandstone was placed in a conical flask with 200 ml of solution containing 5 mg l^{-1} amino-G-acid and 80 mg l^{-1} bromide. The flask was stirred continuously for 14 h and c. 5 ml samples were collected every hour. The concentrations of amino-G-acid and bromide were measured at a constant temperature using a fluorimeter (Perkin Elmer 204-A) and a temperature-compensated ion-specific probe with an Orion 290A meter. Batch experiments were also performed to test any reaction between Cu-EDTA and Triassic sandstones. Cu-EDTA was used for the PEPI experiments (Section 2.5). Thirty grammes of disaggregated Triassic sandstone was added to 250 ml of a solution containing 500 mg l^{-1} Cu-EDTA in a conical flask and the mixture was stirred continuously. Twelve millilitre samples were collected at 1 h intervals from both the flasks and the samples were filtered using a Millipore 0.45 μm filter. The concentration of Cu was measured using inductively coupled plasma-atomic emission spectrometry (ICP-AES).

Column experiments

Six cylindrical and one rectangular parallelepipedic, undisturbed, laminated samples of red Triassic Sandstone were used for miscible displacement experiments. Cylindrical samples were cut from borehole cores using a diamond core drill. The borehole cores were obtained when drilling at a site at Gatewarth [national grid reference SJ 5817 8722] in the Merseyside area in the NW of England. The rectangular parallelepipedic sample was collected from Sandy Lane Quarry, Bromsgrove [SO 954761], West Midlands, UK. Three cylindrical cores (columns 1–3) were cut in such away that the axis of the cylinder was parallel to the lamination. For the other three cylindrical cores (columns 11–13), the axis of the cylinder was perpendicular to the lamination. In the parallelepipedic sample (Slab 1), the long axis was aligned parallel to the lamination, and all sides were rectangular. After initial experimentation, Column 1 and Slab 1 were cut to smaller sizes (Column 1a and Slab 1a) in order to carry out experiments to investigate scale dependency. Table 1 shows the dimensions and the physical properties of the columns. The outer curved surface of the cylindrical samples was covered with silicon sealant, PTFE tape and heat-shrinking sleeve (polyolefin) to prevent any bypass of water through the side of the samples. The outer surfaces of Slab 1 were covered with Perspex sheets cemented in place (with Araldite). Manifolds were fitted on the shortest two of the four narrow sides in order to be able to perform column experiments (Fig. 1a). The samples were vacuum saturated with deaired water before starting the experiments. For the column experiments, the cylindrical samples were fixed in the sample holder with the help of rubber sleeves. The sample holder, which had two parts, was made of Perspex (Fig. 1b). The upper part of the sample holder held the sample tightly and allowed test solutions to flow

Table 1. *Dimensions and the physical properties of the columns**

Column	L (cm)	A (cm²)	V (cm³)	k (cm²)	Porosity (L^3/L^3)	Pore volume (ml)	Bulk density (g cm⁻³)
Column 1	6.9	10.75	74.19	1.55E-10	0.20	15.11	2.41
Column 1a	4.0	10.75	43.00	1.55E-10	0.20	8.70	2.41
Column 2	6.8	10.75	73.11	2.33E-10	0.22	14.94	2.48
Column 3	5.2	10.75	55.91	1.67E-10	0.22	10.95	2.164
Column 11	7.17	10.75	77.08	3.06E-12	0.19	14.96	2.44
Column 12	5.15	10.75	55.36	2.71E-12	0.18	10.13	2.41
Column 13	4.9	10.75	52.67	3.19E-12	0.20	10.61	2.38
Slab 1	30	31.5	945	1.08E-10	0.22	211	2.34
Slab 1a	15	31.5	473	1.08E-10	0.22	105	2.34
Trsand1	17.45	10.75	187.6	2.78E-9	0.39	72.97	1.59
Clysnd1	17.45	10.75	187.6	4.06E-9	0.41	77.67	1.53
sand2	17.45	10.75	187.6	4.87E-9	0.38	70.36	1.62

* L, length of the column; A, cross-sectional area of the column; V, column volume; and k, intrinsic permeability.

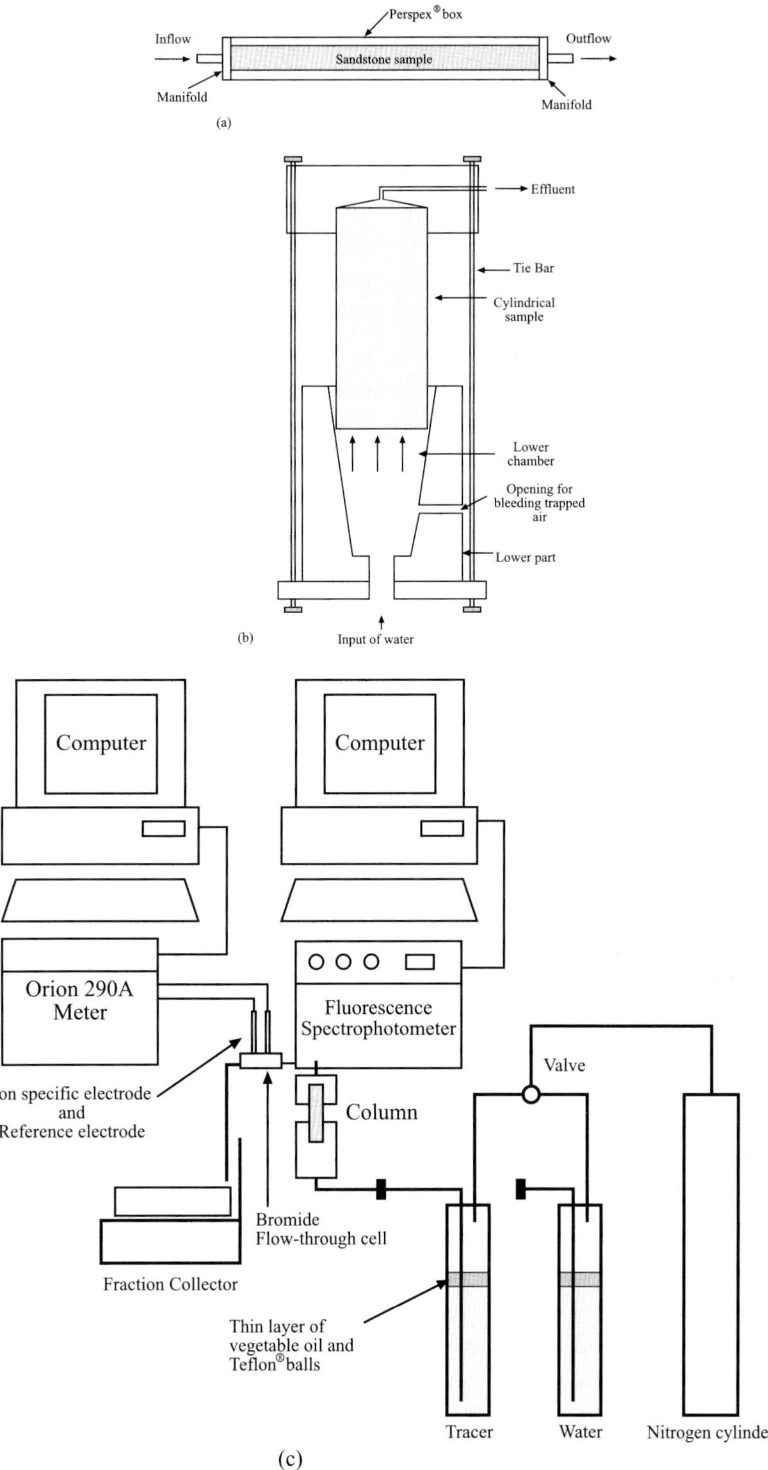

Fig. 1. (**a**) Cross-section of the parallelopepidic sample and holder. (**b**) Design of sample holder for column experiments. (**c**) Schematic of experimental set-up.

through thin tubing to the flow-through cell of a fluorescence spectrophotometer. Discharge water was collected with a fraction collector (Pharmacia Fraction Collector, Frac-100) at predetermined time intervals to measure the flow rate of water through the sample (Fig. 1c).

Column experiments were conducted using both amino-G-acid ($C_{10}H_9NO_6S_2$) and bromide ($CaBr_2$) as tracers. The test solutions were introduced at the base of the column. Pressurized nitrogen was used to force the test fluid through the samples. The liquids were degassed before being introduced into the reservoir by boiling them for few minutes. In order to prevent any N_2-rich test solutions from entering the columns, the reservoir always contained a buffer of unused fluid below the pressurized headspace. For the cylindrical samples, tracer solution was added to the column until full breakthrough occurred (i.e. it had a rectangular pulse input function). When a relative concentration of unity was achieved (i.e. $c/c_0 = 1$, where c is the effluent concentration, c_0 is the pulse input concentration) the injection solution was replaced by water without tracer to conduct a flush-out experiment. The solute pulse was displaced with water until c/c_0 was approximately 0. For slabs 1 and 1a, a short rectangular pulse of tracer was injected, and the relative concentration hence never reached 1. The concentration of amino-G-acid was measured by a fluorescence spectrophotometer (Perkin Elmer 204-A) fitted with a glass flow-through cell to determine the concentration of effluent continuously. The spectrophotometer was connected to a microcomputer to record the concentration of amino-G-acid in the flow line at a certain interval of time (typically twice per min). The response of the spectrophotometer to the influent concentration was measured at the beginning and at the end of the experiment. The overall machine drift (which was found to be linear by separate experiment) was taken into consideration in the calculation procedure. The concentration of bromide was measured using an Orion 290A digital concentration meter with double-junction and ion-specific probes. An ATC probe is also used to compensate the measured concentration change due to any change in the room temperature. The Orion 290A meter was connected to another microcomputer to record the effluent concentrations. The breakthrough curves produced by this experimental set-up comprises several hundred data points, the number being specified by the user: in diagrams in this paper only a small proportion of data

points are shown. As the water sample size is very small (<4% of the pore volume of samples), bias in the estimation of model parameters due to sample size is negligible (Schnabel & Richie 1987; Leij & Toride 1995). In some experiments both amino-G-acid and bromide ($CaBr_2$) were used simultaneously. In other experiments only amino-G-acid was used (Table 2).

Two reference columns were used, one containing loose fine to medium sand (column sand2) and the other containing fine sands and fired clay plates ('sand-with-clay-plates' column) (column Clysnd1). For the sand column, a cylindrical PVC pipe fitted with a fine mesh at each end was filled with loose fine sand. For the sand-with-clay-plates column, clay plates were prepared by firing dried, rectangular (1–2 cm), thin sheets (thickness 1 mm) of red clay in an oven until they became red hot. Then a cylindrical PVC column was packed with fine sands and the clay plates in such a way so that the clay plates remained parallel to the axis of the cylinder. Column experiments were also performed with disaggregated and repacked Triassic sandstone (column Trsand1). Disaggregation was achieved by using a rubber hammer to prevent any crushing of sand grains, the sandstone being weak. Table 2 gives the experimental conditions for all the breakthrough tests reported here.

Thin-section analysis

Thin sections were made in order to examine the heterogeneity of the sandstone (Murphy *et al.* 1977; Bullock & Thomasson 1979; Younger 1992). Before sectioning, the sandstone samples were impregnated with blue dyed Araldite resin to provide contrast between the void spaces and the framework components. The blue dye enabled the identification of different types of pores and also allowed visual estimation of porosity. The slices were cut in a direction perpendicular to the lamination. Thin sections were prepared for the sandstone samples used in the column, dye staining and PEPI experiments. Some of the dried slices obtained from dye experiments (see the next subsection on 'Dye staining') were cemented with 2-tonne clear epoxy cement in order to prevent the rhodamine WT (RWT) stains from dissolving during the thin-sectioning process. The sections were mounted on glass slides using the same cement. Thin sections were also prepared for the whole cross-section of sandstone block used in the positron emission tomography experiments to determine the distribution of zones of

Table 2. *Experimental system conditions for columns**

Column	Experiment	q (cm min^{-1})	$v = q/\theta$ (cm min^{-1})	T_{o}	Tracer[†]
1	1-1	0.003	0.013	3.723	AGA
1	1-2	0.014	0.067	3.891	AGA+Br–
1	1-3	0.026	0.131	5.351	AGA+Br–
1	1-4	0.085	0.418	4.717	AGA
1	1-5	0.118	0.585	7.064	AGA
1	1-6	0.142	0.702	6.592	AGA
1	1-7	0.199	0.984	9.642	AGA+Br–
1	1-8	0.225	1.120	13.683	AGA+Br–
1a	1a-1	0.070	0.347	4.458	AGA
1a	1a-2	0.155	0.772	5.551	AGA
2	2-1	0.012	0.055	3.671	AGA
2	2-2	0.045	0.206	4.909	AGA
2	2-3	0.062	0.285	5.415	AGA
2	2-4	0.165	0.756	5.683	AGA
3	3-1	0.014	0.064	3.360	AGA+Br–
3	3-2	0.046	0.208	4.104	AGA+Br–
3	3-3	0.075	0.341	4.691	AGA
3	3-4	0.176	0.799	4.750	AGA
3	3-5	0.313	1.422	4.911	AGA
11	11-1	0.007	0.037	1.790	AGA+Br–
12	12-1	0.006	0.034	2.757	AGA+Br–
13	13-1	0.008	0.039	2.595	AGA+Br–
Slab 1	Slab 1-1	0.005	0.023	1.549	AGA+Br–
Slab 1	Slab 1-2	0.043	0.191	1.943	AGA+Br–
Slab 1	Slab 1-3	0.177	0.789	3.637	AGA
Slab 1	Slab 1-4	0.269	1.200	4.018	AGA
Slab 1a	Slab 1a-1	0.063	0.284	1.699	AGA
Trsand1	Trsand1-4	0.123	0.316	2.103	AGA+Br–
sand2	Sand2-2	0.203	0.541	2.110	AGA+Br–
Clysnd1	Clysnd1-3	0.224	0.541	3.836	AGA+Br–

* q, Darcy velocity; θ, = porosity L^3/L^3; v, average linear velocity; and T_{o}, number of input pore volumes of tracer.
[†] AGA, amino-G-acid; Br–, bromide.

different permeability. Porosities of different layers were estimated using the point-counting technique.

Dye staining

Dye staining is a common technique for the study of heterogeneity in hydraulic conductivity. In this technique, dye is injected into a sample and then the sample is cut to determine the flow paths (Bouma & Dekker 1978; Omoti & Wild 1979; Seyfried & Rao 1987; Koch & Flühler 1994; Binley *et al.* 1996). It allows the relationship between water and solute flow paths and the structural features to be observed. The fundamental assumption underlying the interpretation of dye patterns is that the greater the amount of solution that passes a given point, the more darkly stained will be the rock at that point. The dye used in the experiments was rhodamine WT (RWT).

A total of 26 cylindrical Triassic sandstone samples were dye-tested, including some of those used for the column experiments. The samples were first vacuum-saturated with water. A solution of RWT with a concentration of 20 mg l^{-1} was then injected into the samples using the same method as for the column experiments. A steady flow rate of about 1 ml min^{-1} was used for samples where the flow direction was parallel to the lamination; because of the lower bulk permeability the lower flow rate of about 0.1 ml/min was used for samples where the flow direction was perpendicular to the lamination. After dye injection, slices were cut parallel to the flow direction (and perpendicular to the lamination) to observe the vertical profile of the dye front in the direction of flow.

The slices were dried in the oven and then photographed.

Positron emission projection imaging (PEPI)

Interpretation of physical property data sets using computerized tomography (CT) has provided a non-destructive and relatively non-intrusive technology for the measurement of the interior properties of porous media. The bulk density, solute transport, speed of flowing water and also the heterogeneities of water content in a porous medium can be estimated using CT technology. The usual application has been to resistivity data sets (i.e. resistivity tomography: Baker 1994; Binley *et al.* 1996) or to gamma or X-ray data sets (Petrovic *et al.* 1982; Hainsworth & Aylmore 1983; Crestana *et al.* 1985, 1986; Warner *et al.* 1989; Phogat *et al.* 1991).

In positron emission tomography (PET), the technique used in this study, a fluid labelled with a positron-emitting radionuclide is passed through a porous medium and images are taken using a positron camera, which consists of an extensive array of gamma detectors. This allows the quantitative measuring of the tracer as it migrates through a porous medium and hence enables depiction of the actual fluid flow paths (e.g. Hawkesworth & Parker 1991). In this study, a modified form of PET has been used – positron emission projection imaging (PEPI): this differs from PET only in that the image reconstruction assumes that tracer movement is in a two-dimensional (2D) plane (Loggia *et al.* 2004).

A positron is emitted as the tracer atom decays. This positron collides with and annihilates a nearby electron. This annihilation results in the production of two collinear 511 keV γ-ray photons that travel in diametrically opposed directions away from the annihilation site. The simultaneous detection of these particles by a pair of γ-ray detectors on either side of the decay site enables a line to be constructed that must pass through the point of origin of the γ-rays. By knowing the location of the sample, the origin of the annihilation can be calculated. The detection of a large number of events allows a mathematical reconstruction of the images of radionuclide-filled pore spaces. Each image represents the radioactivity recorded over a certain interval of time. As these γ-rays are very penetrating (50% will pass through 11 mm of steel) it is possible to image through a significant thickness of solid material.

The detected intensity of radioactivity at a point within the porous medium depends not only on the changing concentrations of tracer at that point but also on the changing quantities of the tracer (representing porosity) over a duration of time. Therefore, the contours of the transient images represent the effect of both the concentration gradient and changes in the storage of tracer throughout the system. The resolution of the system used was about 6 mm.

PEPI scanning was performed on Slab 1. Table 1 gives the physical properties and dimensions of Slab 1. Cu-EDTA solutions were prepared by using both radioactive copper (^{64}Cu) and non-radioactive copper (^{63}Cu). The concentration of Cu in both inactive Cu-EDTA and radioactive Cu-EDTA solutions was 500 mg l^{-1}. To remove the effect of density contrast between the displacing fluid and the displaced fluid the sample was flushed out with several pore volumes of inactive Cu-EDTA before the start of displacement experiments. Batch experiments showed that the test solutions behaved conservatively.

Each of the images obtained in this study represents the cumulative response of activity for 5 min intervals. Effluent solution was collected at the output using a fraction collector to obtain breakthrough curves. For both the image analysis and breakthrough curves compensation has been made for radioactive decay. The breakthrough experiments showed good repeatability.

Diffusion coefficient measurements

Diffusion coefficients for amino-G-acid and bromide were measured using the 'double reservoir' or 'through diffusion' method (Stenhouse *et al.* 1996) at $20 \pm 0.6\,°C$. Ten samples from the core of the Gatewarth borehole (see above) were cut into rectangular blocks of dimension $1 \times 5 \times 10$ cm: diffusion occurred across the 1 cm thickness of rock from the donor to the receiver cell. Five samples were cut so that diffusion could be measured in a direction parallel with lamination, and five so that it could be measured at 90° to lamination. Details are given by Bashar (1997).

Results and discussion

Reference column experiments: sand-only columns

For columns with uncemented fine- to medium-grained sands, the breakthrough curves are symmetrical in shape as expected for the case of a homogeneous porous medium (Fig. 2a). The breakthrough curves for amino-G-acid and

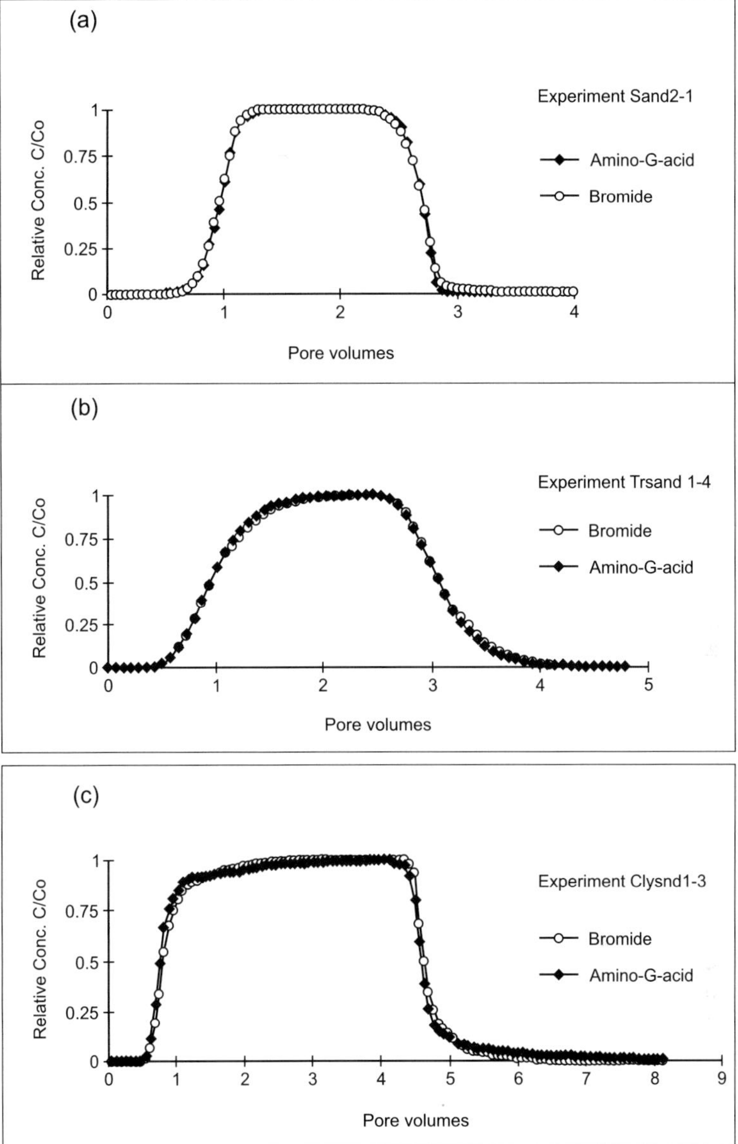

Fig. 2. Typical breakthrough curves for bromide and amino-G-acid in the synthetic media columns: (**a**) fine sand column; (**b**) disaggregated and repacked Triassic sandstone column; and (**c**) sand-with-clay-plates column.

bromide are identical. These experiments suggest that the apparatus does not affect the breakthrough curves in any significant way.

Batch experiments

No sorption on the sandstone of amino-G-acid, bromide or the Cu-EDTA could be detected.

It is concluded that amino-G-acid, bromide and Cu-EDTA may be treated as conservative tracers in the sandstone breakthrough experiments.

Disaggregated and repacked Triassic sandstone column breakthrough experiments

The breakthrough curves obtained for column experiments with disaggregated Triassic sandstone are symmetrical about $c/c_0 = 0.5$ (Fig. 2b), and the bromide and amino-G-acid break-through curves are very similar. This suggests that the sandstone grain geometries/pore geometries have little effect on the break-through curves.

Thin-section analysis

Thin sections were examined using a petrological microscope. It was found that in all cases the samples were laminated, with laminae of relatively large pore size, grain size and porosity alternating with laminae of relatively small pore size, grain size and porosity (Fig. 3). The thickness of the laminae generally varies from a fraction of a millimetre to several millimetres, but occasionally exceeds 1 cm.

Thin sections were also prepared for the whole cross-section of the slab sample used for positron emission tomography (see below). Examination of the thin sections showed that the whole cross-section exhibits a repetition of layers of: (i) fine- to medium-grained sand; and (ii) very-fine-grained sand–silt. Sixty-six layers of different grain size and pore size were recognized within the 15 cm-width of the slab. The thickness of the coarser grained layers varies between 3 and less than 1 mm, whereas the thickness of the finer grained layers are in the range of 2–<1 mm. The boundaries between layers are gradational. The layers with larger grain sizes have higher porosities. Coarse-grained layers dominate. The distribution of fine-grained and coarse-grained layers is not regular. In some parts of the section the concentration of fine-grained layers is much greater. Mica grains are oriented in the direction of stratification.

The degree of consolidation and the porosity of a sandstone are influenced by the shape of contact between grains and also the number of contacts per grain. It is observed that the contacts between grains range between tangential and sutured, presumably resulting from stress and the diagenesis of the sandstones related to the tectonic history of the rock. In general, the number of contacts between grains are much greater and more complex in low-porosity zones in comparison with the high-porosity zones.

Iron oxyhydroxides (probably mainly haematite), which impart the characteristic colour of these rocks, are present in a variety of forms including coating of detrital grains (often filling pits on the grain surfaces), as a pore lining on authigenic or detrital grain surfaces and as

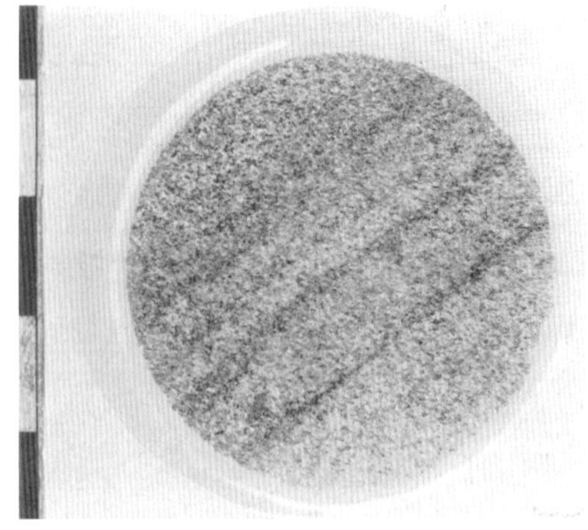

Fig. 3. Thin-section photomicrograph of Triassic sandstone sample showing low- and high-porosity laminae, in some cases occurring in groups of lower or higher values (scale bar marked in cm; plane polarized light; Column 10).

interstitial matrix where they are closely associated with clay minerals. The oxyhydroxides are more abundant in the lower porosity (and finer grained) zones than in higher porosity (and coarser grained) zones. Apart from iron oxide cements, carbonate cements are also present in some zones of low porosity. Quartz and feldspar overgrowths are also abundant in most of the samples. In some cases the contact between the overgrowths and the original grain surface is marked by iron oxide, indicating that the development of at least some of the over-growths is more recent than the development of at least some of the oxyhydroxides. All these cementing materials and overgrowths produce a complex pore architecture that presumably affects the permeability of the layers.

The thin-section examinations confirm the conclusion from examination of hand specimens that the rock is laminated. However, the thin-section study also demonstrates that the petrographic layering is associated with changes in pore size, pore volume and tortuosity. The laminae are therefore expected to have different permeabilities. The grouping of finer laminae and coarser laminae, particularly noticeable in the PEPI slab, suggest that a permeability layering may also occur on a larger scale. Orientation of platey minerals may impart an anisotropic character to the rock within individual layers. The complexity of the cements suggests the presence of dead-end zones into which diffusion may occur.

Sandstone column experiment results: flow parallel to the lamination

Figure 4 shows examples of breakthrough curves for flow parallel to the strata. The breakthrough curves for these and other experiments not illustrated here can be summarized as follows.

- The breakthrough curves are generally asymmetrical in shape (asymmetrical with respect to the $c/c_0 = 0.5$ point on the breakthrough curve).

- The relative concentration $c/c_0 = 0.5$ in the effluent is attained before 1 pore volume of tracer has passed through the sample. This phenomenon is known as early breakthrough in the soil science literature (Rao *et al.* 1980*a*, *b*).

- Marked tailing of the breakthrough curves is observed for all the experiments.

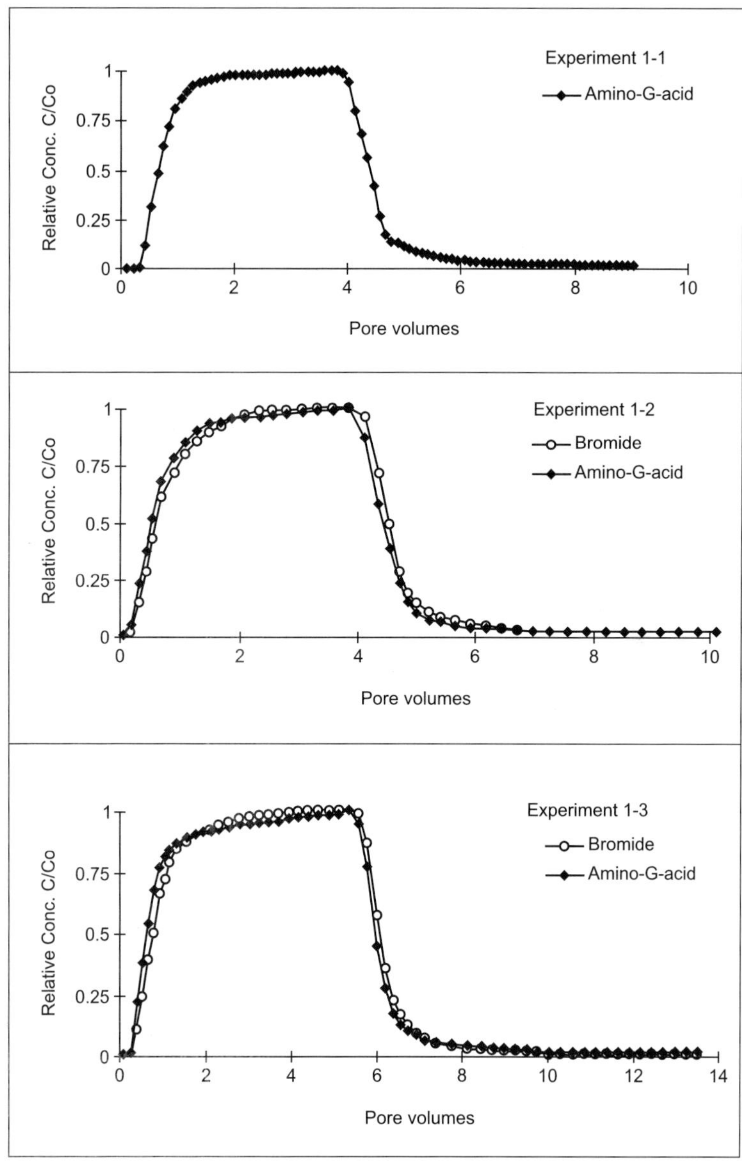

Fig. 4. Breakthrough curves for Column 1 (experiments 1-1, 1-2 and 1-3, flow parallel to laminations).

- Generally, asymmetry and tailing of the breakthrough curves increases with increasing flow velocity.
- The relative concentration of amino-G-acid reached $c/c_0 = 0.5$ much earlier than in the case of bromide in both transmission and flush-out phases of the experiments.
- The tailing of the breakthrough curves for amino-G-acid is much more pronounced than for bromide.
- The breakthrough curves for amino-G-acid are more asymmetrical than those of bromide.

The breakthrough curves for these experiments are very different from those obtained for the reference sand columns and the disaggregated sandstone columns, and presumably it is the layered structure seen in hand specimens and in thin section that causes the differences. The flow interruption technique (Brusseau *et al.* 1989; Koch & Flühler 1993) was also applied. Some of the column experiments were repeated and the flow was interrupted for a period Δt, after which the flow was continued. For transmission experiments flow was interrupted when the effluent concentration had reached at least 0.80 of the input concentration. Similarly, for flush-out experiments, flow was interrupted when the outflow concentration had fallen to at least 0.20 of the concentration of tracer. Breakthrough curves for the flow columns 1 and 2 (Fig. 5) showed significant drops in effluent concentrations during transmission experiments and corresponding rises in the flush-out experiments.

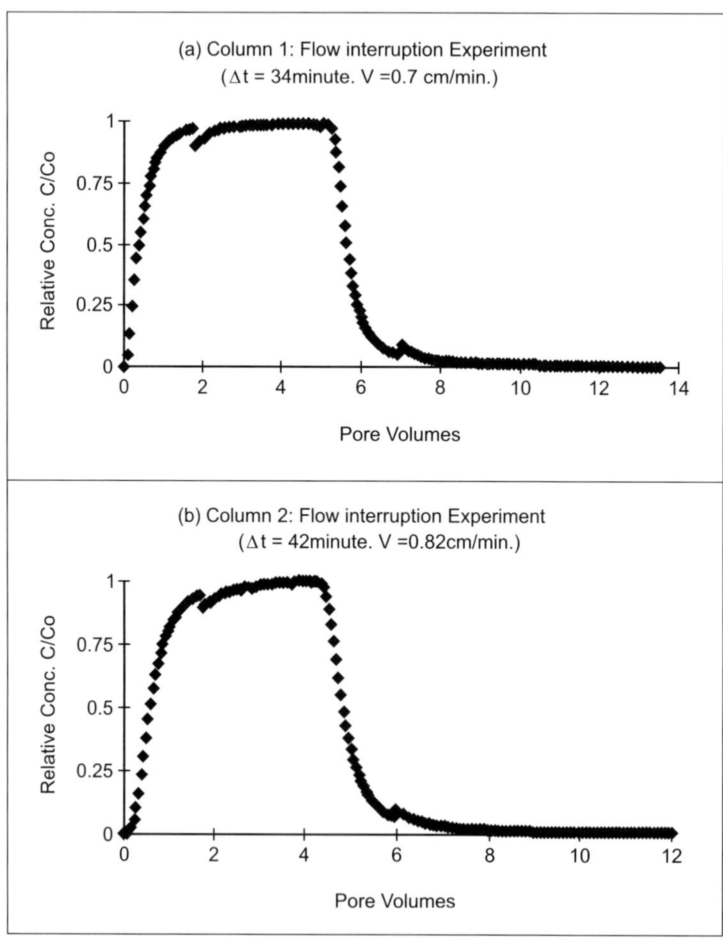

Fig. 5. Effect of flow interruption on the breakthrough curve: (**a**) Column 1; and (**b**) Column 2. Δt, = time interval flow stopped. Flow parallel to laminations.

During the period of no flow, the only concentration-changing process occurring will be diffusion. This implies that during the flushing of the rock there are variations in concentration within the pore water across the width of the columns. This, again, is consistent with layers of higher permeability and layers of lower permeability.

Sandstone column experiment results: flow perpendicular to the strata

Figure 6 shows examples of breakthrough curves for flow perpendicular to stratification. The breakthrough curves have the following characteristics.

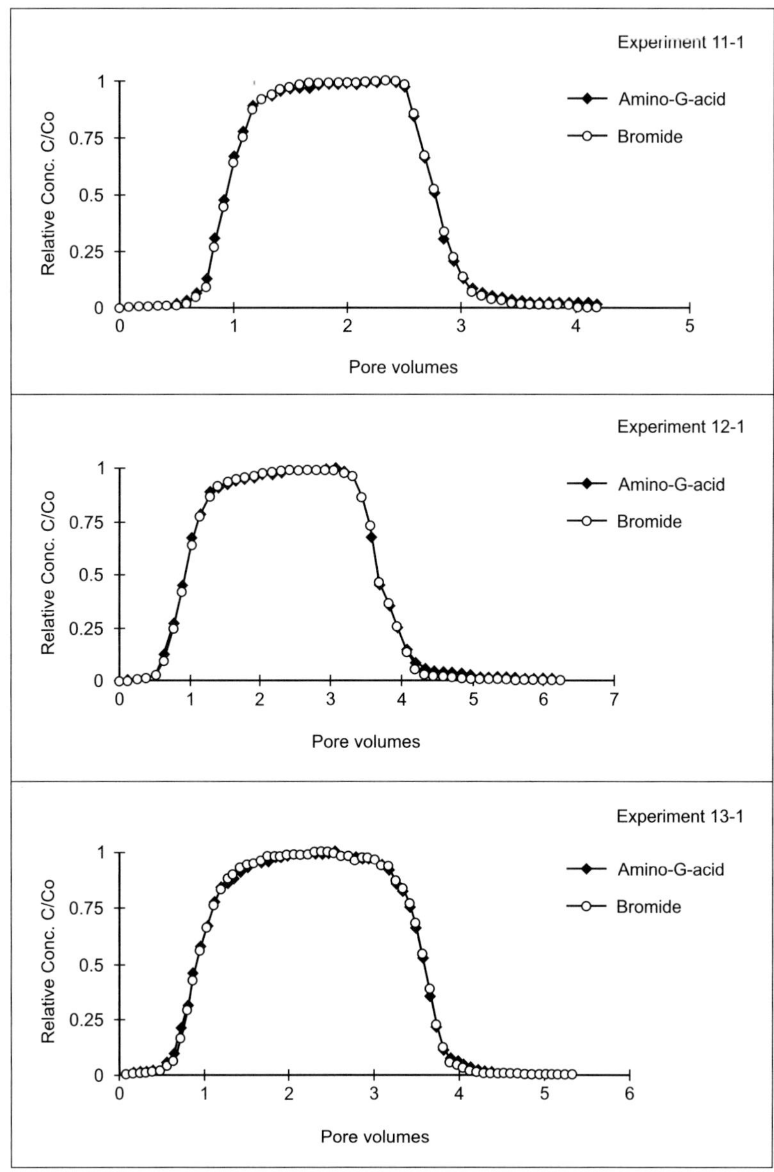

Fig. 6. Breakthrough curves for columns 11, 12 and 13 (experiments 11-1, 12-1 and 13-1; flow perpendicular to laminations).

- The breakthrough curves are slightly asymmetrical in shape (asymmetrical with respect to the $c/c_0 = 0.5$ point on the breakthrough curve).
- The relative concentration $c/c_0 = 0.5$ in the effluent is attained when about one pore volume of tracer has passed through the sample.
- No significant tailing of the breakthrough curves is observed in any of the experiments.
- Breakthrough curves for bromide and amino-G-acid are similar.

These results confirm that it is the layering, and not some intrinsic property of the pore architecture, that gives rise to the asymmetrical breakthrough curves.

The flow interruption technique was also applied. A change in concentration following the flow interruption was not observed.

It is concluded that the change in concentration seen in stopped flow experiments where flow is parallel with stratification is due to diffusion perpendicular to the strata.

Diffusion coefficients for amino-G-acid and bromide

The average intrinsic diffusion coefficients (D_i = –mass flux/(area × concentration gradient)) in the direction perpendicular to the strata are $1.3 \times 10^{-10} \text{ m}^2 \text{ s}^{-1}$ for bromide and $4.8 \times 10^{-11} \text{ m}^2 \text{ s}^{-1}$ for amino-G-acid. Similarly, the average diffusion coefficients in the direction parallel to the strata are 1.8×10^{-10} and $7.1 \times 10^{-11} \text{ m}^2 \text{ s}^{-1}$ for bromide and amino-G-acid, respectively.

The diffusion coefficient for bromide is at least twice that for amino-G-acid, and modelling has shown that this diffusion can explain the longer tailing of the breakthrough curves for amino-G-acid in the experiments where flow is parallel with stratification (Bashar 1997; cf. Hu & Brusseau 1995; Hölttä et al. 1996).

Reference column experiments: sand-with-clay-plates

Figure 2(c) shows the breakthrough curves for the sand-with-clay-plates column. The breakthrough curves are asymmetrical in shape. They exhibit early breakthrough and tailing. The relative concentration of amino-G-acid reached 0.5 much earlier than bromide in both the flush-in and flush-out phases of the experiments. Tailing of the breakthrough curves for amino-G-acid are more asymmetric than those of bromide. The style and breakthrough curves

obtained from the sand-with-clay plates experiments is very similar to that seen in the case of the sandstone column experiments where flow was parallel to the stratification. The layered nature of the sand-with-clay-plates columns appears to be a good analogue for the sandstone, despite the fact that the permeability difference between the sand and the fired clay plates is almost certainly much greater than the permeability difference between adjacent laminae in the sandstone columns. The sand-with-clay-plates experiments confirm that layering is likely to be the dominant mechanism, giving rise to the asymmetrical breakthrough curves in the sandstone experiments.

Dye-staining experiments

For Triassic sandstone samples with the flow direction parallel to the lamination, the dye experiments revealed that much more dye was present in some laminae relative to others (Fig. 7). Higher concentrations of dye were observed in the upflow parts of the samples and lower concentrations of dye were observed in the downflow parts. In some layers the solution moved much faster, producing finger-like protrusions of the stained area. At the contact of stained and unstained layers the concentration of dye gradually decreases in the direction transverse to the main flow. A few experiments were set up so that the breakthrough of the dye was observable as it emerged from the end of the sample. It was clear that breakthrough in any given layer does not occur simultaneously, but that the tracer appears in localized spots first, which then begin to coalesce. The breakthrough front location hence varies in all three dimensions.

In samples with the flow direction perpendicular to the lamination the dye front was nearly parallel to the lamination, indicating an almost uniform flow velocity of dye in the direction of flow.

Thin sections prepared from samples that were used for dye staining revealed that dye stains on the grains are present in layers with higher porosity and lower content of cementing materials (e.g. authigenic clays, carbonate cements and iron oxides). Both UV and plane-polarized light were used to exhibit the differences between these zones.

The dye-staining results confirm that some laminae/groups of laminae have significantly greater permeability than others. The reduction in colour longitudinally and transversely across laminae may indicate dispersion/diffusion. The thin sections indicate that this variation in permeability is influenced by the porosity, grain

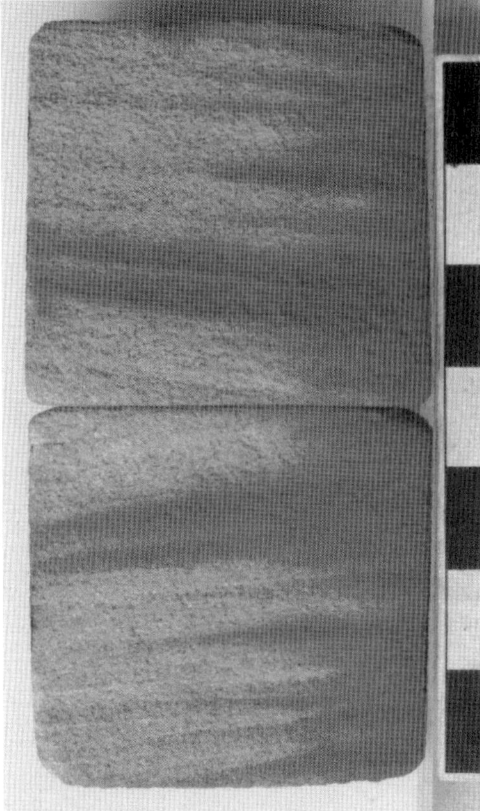

Fig. 7. The two halves of a column cut parallel to the axis after injection of dye. Dark coloured layers indicate the presence of the RWT; flow right to left, parallel with laminations.

size and the amount of cementing materials in the layers. It is observed that only a minor change in the porosity is associated with major change in permeability, the latter estimated from the length of the dye fingers. It would therefore appear that the hydraulic conductivities of the layers are influenced by pore interconnectivity and pore architectures, as well as by pore size. The lack of finger-like protrusions of dye in the sandstone samples where flow was perpendicular to the lamination is consistent with the uniform breakthrough of tracer indicated by the breakthrough curves.

The positron emission projection imaging results

Figure 8 shows some of the PEPI images of run 1. Each of the images shows contours of different activity indicating differences in mass of tracer (i.e. differences in concentration and/or total tracer volume). As the porosity does not vary a great deal over the whole cross-section and the thickness of the slab is also constant, the contours approximately represent the concentration gradient in the system.

The images shown in Figure 8 indicate the spread of the plume as it moves from the inlet manifold to the outlet. They show that the plume is extended in the direction of flow throughout the whole cross-section of the slab. The movement of tracer was faster in some layers in comparison with others, producing finger-like protrusions that extended to considerable distances. Development of at least three fingers or preferential flow paths can be observed from the beginning of the experiment.

The effluent was collected using a fraction collector and the concentration of radioactive Cu was measured from the specific activity in each fraction. The concentrations were later corrected for the radioactive decay. Figure 9 shows the breakthrough curve for radioactive Cu in Slab 1. The pore volume, as estimated from the PEPI images at maximum concentration, agreed to within 5% of the pore volume measured by saturation (211 cm^3). It is clear from Figure 9 that long tailing occurred during the breakthrough and flush-out phase of the experiment.

The growth of fingers in the direction of flow indicates non-uniform advection of tracer in the sample, which is presumably caused by variations in hydraulic conductivity of different layers. These fingers are associated with groups of laminae with higher porosities as revealed by thin-section examination.

The development of fingers is therefore another confirmation of the importance of layering, this time from a sample of sandstone of different shape (slab instead of cylinder), held in a different type of sample holder/manifold system and from a different formation 160 km distant from that where the sandstone columns were obtained. The scale of flow layering appears to be larger, but this is almost certainly because the resolution of the PEPI system is around 6 mm: the images are identifying groups of laminae rather than individual laminae, as was the case with the dye-staining experiments. As can be seen from Figure 8, the fingers not only grow in the longitudinal direction but also in the direction transverse to the main flow indicating some degree of transverse dispersion or diffusion. The contours in the direction of flow are generally more widely spaced than those in the direction perpendicular to flow. This feature is assumed to indicate greater dispersion, as expected, in the direction of flow.

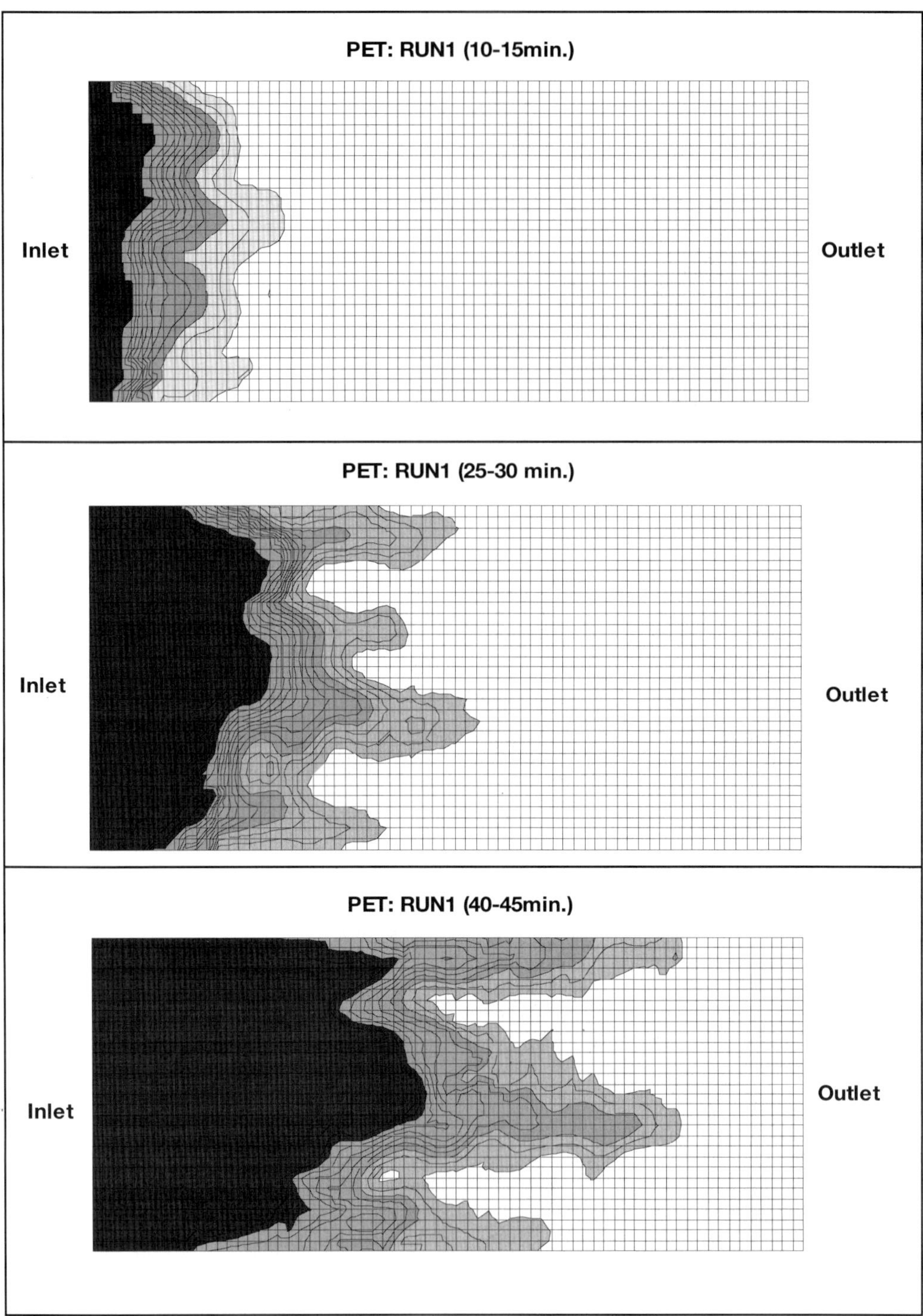

Fig. 8. (**a**) & (**b**) PEPI images for Slab 1 at the time intervals indicated.

PET: RUN1 (55-60min.)

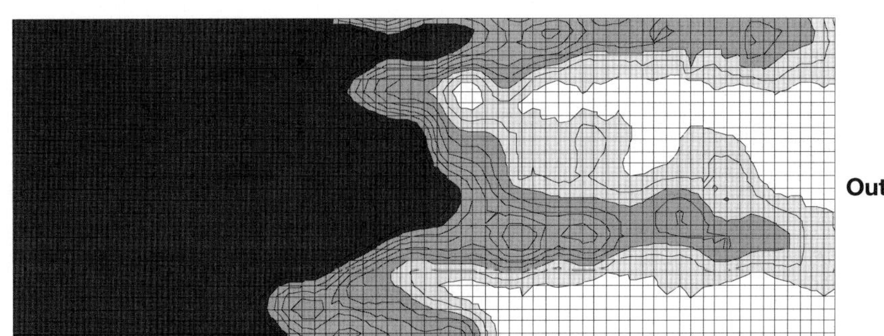

PET: RUN1 (85-90min.)

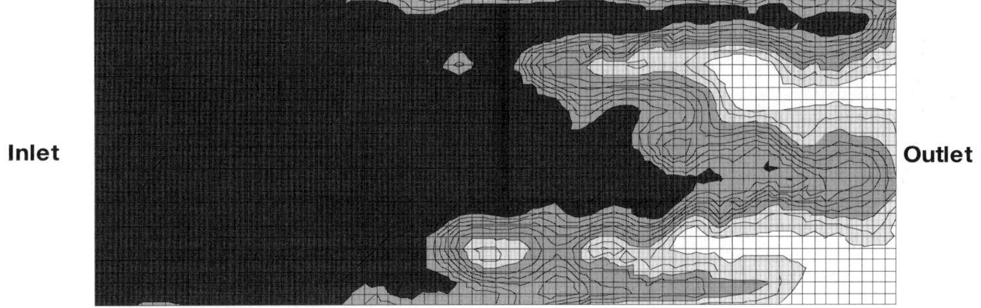

PET: RUN1 (115-120min.)

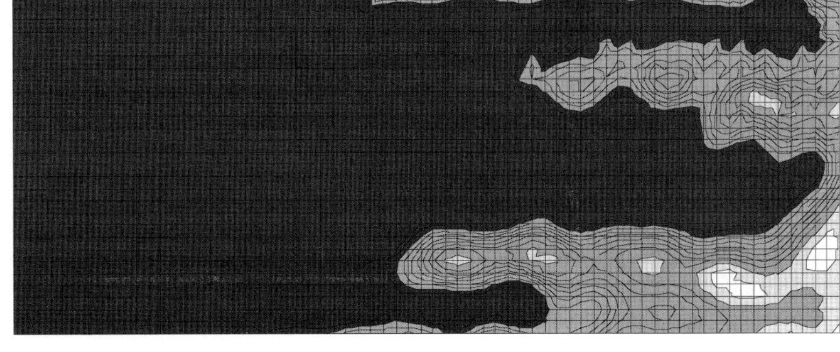

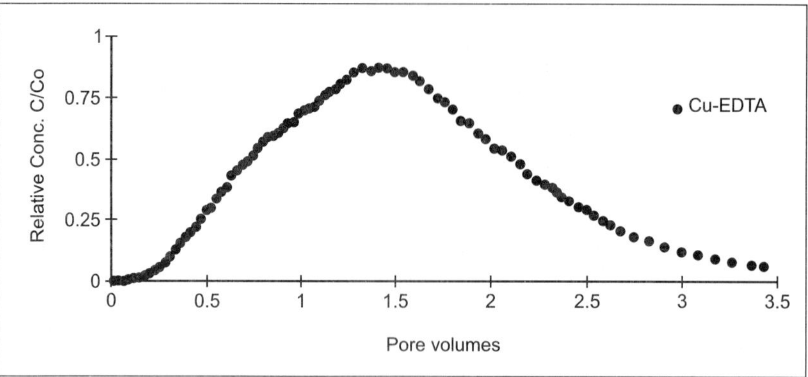

Fig. 9. Breakthrough curve for radioactive Cu^{2+} transport in Slab 1.

Conclusion

The aim of the work reported here was to establish that solute migration through the sandstone samples is dominated by stratification. To this end a wide range of techniques has been applied, including tracer breakthrough tests using amino-G-acid, bromide and Cu-EDTA, and tracer tests using staining dyes followed by sample sectioning, thin-section analysis and positron emission projection imaging (PEPI).

The sandstones appear laminated (i.e. stratified) in hand specimen and thin section. Comparison of breakthrough curves for disaggregated sandstone samples, intact samples with flow parallel with lamination and intact samples with flow at 90° to lamination demonstrate that the lamination affects solute migration, as significant departure from homogeneous system behaviour only occurs in the case of flow parallel to lamination. Dye-staining experiments and PEPI demonstrate that tracer moves much faster in some laminae/groups of laminae than in others. Stopped flow experiments and comparison of breakthrough curves using tracers with very different diffusion coefficient values indicate the presence of lateral concentration gradients in the sandstone columns when flow is parallel with lamination. All these features strongly suggest that the effect of the stratification is dominant for flow parallel with the lamination in these sandstones. However, limited evidence suggests that there might be some 'channelling' in the laminae, as breakthrough does not occur uniformly along each lamination. Figure 10 summarizes the conceptual model. For flow at 90° to stratification, the breakthrough curves are as expected from a homogeneous medium. The system is anisotropic in its dispersive properties (cf., for example, Dagan 1989). Bashar (1997) has shown that this conceptual model is quantitatively consistent with the breakthrough data: modelling results, including scale-dependency of model parameters, will be presented elsewhere.

The results presented here are of direct relevance to laboratory studies of reactive solute movement through sandstones similar to those used here: to avoid the problems of physical non-equilibrium, and the complexity that might arise if different laminae have different geochemical properties, it is suggested that experiments be run with flow perpendicular to the stratification. Although not the direct purpose of the study, the results do have implications for solute movement at the field scale. As the sandstones are bedded, with typical average bed size often in the region 0.5–1 m, similar asymmetric breakthrough may be expected at the field scale: however, the mass moved by diffusion will be much less in proportion to the mass within the pore volume of the bed, despite the much increased timescales.

The financial support for this research by the Commonwealth Scholarship Commission in the United Kingdom is gratefully acknowledged. The first author would like to express his gratitude to the authorities of the University of Jahangirnagar, Dhaka, Bangladesh for granting him study leave for this research. Thanks are due to Dr D. Parker and Dr P. McNeil of the School of Physics and Space Research of the University of Birmingham for running the PEPI scanning. R. Greswell is thanked for his help with many aspects of the laboratory work. Thanks are also due to Professor J. Lloyd for suggesting PEPI scanning.

(a)

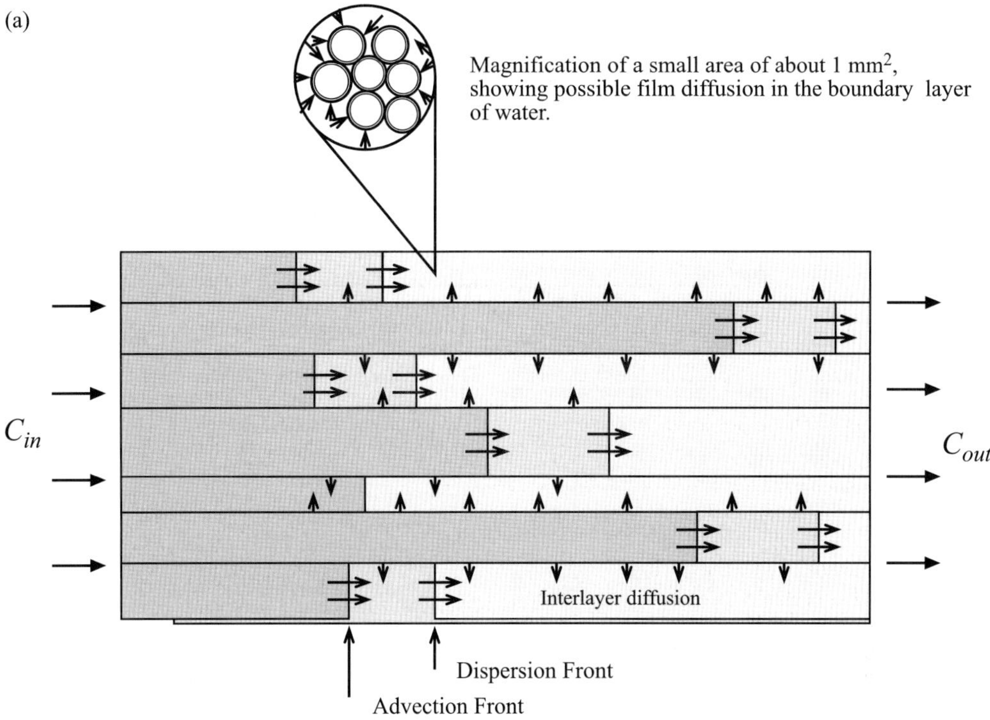

(b)

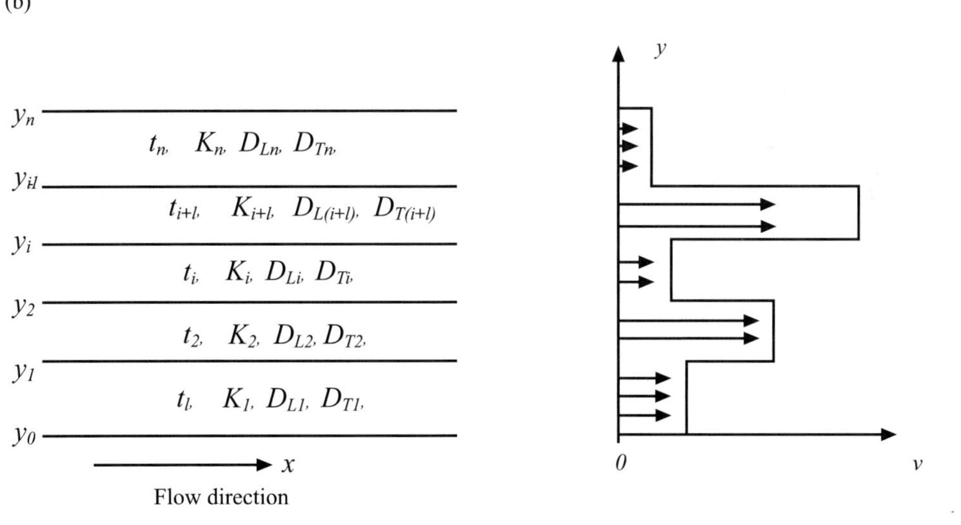

a. Layered structure

b. Stratified flow

Fig. 10. Diagrammatic representations of the mechanisms of mass transfer through sandstone samples with flow parallel to the lamination.

References

BAKER, A.C. 1994. *The measurement of hydraulic and solute transport parameters in clays.* PhD thesis, University of Birmingham.

BASHAR, K. 1997. *Developing a conceptual model of intergranular conservative solute transport processes for water flow through laboratory-scale samples of the U.K. Triassic sandstones.* PhD thesis, University of Birmingham.

BHATTACHARYA, R.N. & GUPTA, V.K. 1986. Solute dispersion in multidimensional periodic saturated porous media. *Water Resources Research,* **22**, 156–164.

BINLEY, A., HENRY-POULTER, S. & SHAW, B. 1996. Examination of solute transport in an undisturbed soil column using electrical resistance tomography. *Water Resources Research,* **32**, 763–769.

BOUMA, J. & DEKKER, L.W. 1978. A case study on infiltration into dry clay soil. I. Morphological observations. *Geoderma,* **20**, 27–40.

BRUSSEAU, M.L. & RAO, P.S.C. 1990. Modelling solute transport in structured soil. A review. *Geoderma,* **46**, 169–192.

BRUSSEAU, M.L., RAO, P.S.C., JESSUP, R.E. & DAVIDSON, J.M. 1989. Flow interruption: a method for investigating sorption nonequilibrium. *Journal of Contaminant Hydrology,* **4**, 223–240.

BULLOCK, P. & THOMASSON, A.J. 1979. Rothamsted studies of soil structure II. Measurement and characteristic of macroporosity by image analysis and comparison with data from water retention measurements. *Journal of Soil Science,* **30**, 391–413.

CRESTANA, S., CESAREO, R. & MASCARENHAS, S. 1986. Using a computed tomography miniscanner in soil science. *Soil Science,* **142**, 56–61.

CRESTANA, S., MASCARENHAS, S. & POZZI-MUCELLI, R.S. 1985. Static and dynamic three-dimensional studies of water in soil using computed tomographic scanning. *Soil Science,* **140**, 326–332.

DAGAN, G. 1989. *Flow and Transport in Porous Formations.* Springer, Berlin.

DE SMEDT, F. & WIERENGA, P.J. 1984. Solute transport through columns of glass beads. *Water Resources Research,* **20**, 225–232.

GRIFFIOEN, J.W., BARRY, D.A. & PARLANGE, J.-Y. 1998. Interpretation of two region model parameters. *Water Resources Research,* **34**, 373–384.

GÜVEN, O., MOLZ, F.J. & MELVILLE, J.G. 1984. An analysis of dispersion in a stratified aquifer. *Water Resources Research,* **20**, 1337–1354.

HAINSWORTH, J.M. & AYLMORE, L.A.G. 1983. The use of computer-assisted tomography to determine spatial distribution of soil water content. *Australian Journal of Soil Research,* **21**, 435–443.

HAWKESWORTH, M.R. & PARKER, D.J. 1991. Non-medical application of a positron camera. *Nuclear Instruments and Methods in Physics Research,* **A310**, 423–434.

HÖLTTÄ, P., HAKANEN, M., HAUTOJÄRVI, A., TIMONEN, J. & VÄÄTÄINEN, K. 1996. The effect of matrix diffusion in rock column experiments. *Journal of Contaminant Hydrology,* **21**, 165–173.

HU, Q. & BRUSSEAU, M.L. 1995. Effect of solute size on transport in structured porous media. *Water Resources Research,* **31**, 1637–1646.

KOCH, S. & FLÜHLER, H. 1993. Non-reactive solute transport with micropore diffusion in aggregated porous media determined by a flow-interruption method. *Journal of Contaminant Hydrology,* **14**, 39–54.

KOCH, S. & FLÜHLER, H. 1994. Lateral solute mixing in homogenous and layered sand column. *Geoderma,* **63**, 109–121.

LEIJ, F.J. & TORIDE, N. 1995. Discrete time- and length-averaged solutions of the advection dispersion equation. *Water Resources Research,* **31**, 1713–1724.

LEVY, B.S. & CHAMBERS, R.M. 1987. Bromide as a conservative tracer for soil–water studies. *Hydrological Processes,* **1**, 385–389.

LI, L., BARRY, D.A., CULLIGAN-HENSLEY, P.J. & BAJRACHARYA, K. 1994. Mass transfer in soils with local stratification of hydraulic conductivity. *Water Resources Research,* **30**, 2891–2900.

LOGGIA, D., GOUZE, P., GRESWELL, R. & PARKER, D.J. 2004. Investigation of the geometrical dispersion regime in a single fracture using positron emission projection imaging. *Transport in Porous Media,* **55**, 1–20.

MURPHY, C.P., BULLOCK, P. & TURNER, R.H. 1977. The measurement and characterisation of voids in soil thin sections by image analysis. Part I. Principle and techniques. *Journal of Soil Science,* **28**, 498–508.

OMOTI, U. & WILD, A. 1979. The use of fluorescent dyes to mark the pathways of solute movement through soils under leaching conditions: 2. Field experiments. *Soil Science,* **128**, 98–104.

PETROVIC, A.M., SIEBERT, J.E. & RIEKE, P.E. 1982. Soil bulk density analysis in three dimensions by computed tomographic scanning. *Journal of the Soil Science Society of America,* **46**, 445–450.

PHOGAT, V.K., AYLMORE, L.A.G. & SCHULLER, R.D. 1991. Simultaneous measurement of the spatial distribution of soil water content and bulk density. *Journal of the Soil Science Society of America,* **55**, 908–915.

PICKENS, J.F. & GRISAK, G.E. 1981. Scale-dependent dispersion in a stratified granular aquifer. *Water Resources Research,* **17**, 1191–1211.

RAO, P.S.C., JESSUP, R.E., ROLSTON, D.E., DAVIDSON, J.M. & KILCREASE, D.P. 1980a. Experimental and mathematical description of nonadsorbed solute transfer by diffusion in spherical aggregates. *Journal of the Soil Science Society of America,* **44**, 684–688.

RAO, P.S.C., ROLSTON, D.E., JESSUP, R.E. & DAVIDSON, J.M. 1980b. Solute transport in aggregated porous media: Theoretical and experimental evaluation. *Journal of the Soil Science Society of America,* **44**, 1139–1146.

SCHNABEL, R.R. & RICHIE, E.B. 1987. Elimination of time assignment bias in estimates of dispersion coefficient. *Journal of the Soil Science Society of America,* **51**, 302–304.

SEYFRIED, M.S. & RAO, P.S.C. 1987. Solute transport in undisturbed soil columns of an aggregated tropical soil: preferential flow effects. *Journal of the Soil Science Society of America*, **51**, 1434–1444.

STENHOUSE, M.J., MERCERON, T. & DE MARTINVILLE, E.S. 1996. Provision for diffusion coefficients for argillaceous media in support of preliminary safety assessment within the French HLW disposal programme. *Journal of Contaminant Hydrology*, **21**, 351–361.

TRUDGILL, S.T. 1987. Soil water dye tracing, with special reference to the use of rhodamine wt, lissamine FF and amino-G-acid. *Hydrological Processes*, **1**, 149–170.

WARNER, G.H., NIEBER, J.L., MOORE, D.D. & GEISE, R.A. 1989. Characterising macropores in soil by computed tomography. *Journal of the Soil Science Society of America*, **53**, 653–660.

YOUNGER, P.L. 1992. The hydrogeological use of thin sections: inexpensive estimates of groundwater flow and transport parameters. *Quarterly Journal of Engineering Geology*, **25**, 159–164.

Controls on dense non-aqueous-phase liquid transport in Permo-Triassic sandstones, UK

D. C. GOODDY & J. P. BLOOMFIELD

British Geological Survey, Maclean Building, Wallingford, Oxfordshire OX10 8BB, UK
(e-mail: dcg@bgs.ac.uk)

Abstract: The Permo-Triassic sandstones are geographically and lithologically diverse, and exhibit large variations in porosity (ϕ), pore-throat size, gas permeability (k) and mineral phases. These parameters influence considerably the flow and transport of contaminants. The aquifer is susceptible to contamination by a range of pollutants from both contemporary and historic sources as it outcrops in a number of large industrial towns. One such class of pollutants are chlorinated hydrocarbons, which have been used extensively for years in dry-cleaning and metal degreasing. The physical, chemical and biological properties of these dense non-aqueous-phase liquids (DNAPLs) make them a particular cause of groundwater contamination. Standard physical property measurements, along with a weak nitric acid extraction to determine dominant minerals, were carried out on consolidated sandstones. A laboratory method was developed to produce pressure–saturation curves for a DNAPL–water–sandstone system. Entry pressures were found to be lower than predicted from theoretical considerations, which could be explained in part by the relative amounts of calcite and dolomite present in the pore space. Calculations on relative permeability show that at roughly 60% DNAPL saturation no water flow occurs. Understanding the controls on DNAPL transport can greatly assist the need for appropriate remedial action.

Groundwater supplies face a threat from a wide range of synthetic organic chemicals (Lawrence & Foster 1991). The so-called chlorinated solvents, which are halogenated hydrocarbons, have been produced for more than a century, with the first production taking place in Germany in the late 19th century. It was not until World War II that the widespread use of chlorinated solvents in manufacturing industries began, and this increased markedly during the next three decades. Contamination of groundwater by these compounds went largely unnoticed until the late 1970s, as historically it was believed that chlorinated solvent released to the unsaturated zone of an aquifer would easily volatilize to the atmosphere. When disposed on dry ground, although a chlorinated solvent may appear to be lost entirely to the atmosphere, some will be transported into the subsurface by gaseous diffusion, by infiltration of contaminated water and, most significantly, as a moving non-aqueous phase. Once these contaminants reach the saturated zone, high volatility is of little assistance in removing the solvents.

The denser-than-water nature of liquid chlorinated solvents has led to their becoming known as 'dense non-aqueous-phase liquids' or DNAPLs. It is the physical, chemical and biological properties of these compounds that have made them a particular concern for groundwater contamination (Aurand *et al.* 1981;

Jackson & Dwarakanath 1999). Drinking water limits (or maximum concentration limits, MCLs) for the priority chlorinated solvents have been set at very low concentrations, for example under the revised EC Drinking Water Directive (Directive 98/83/EC) a maximum concentration of 10 µg l^{-1} is set for the sum of tetrachloroethene and trichloroethene. The justification for these standards has been that compounds like trichloroethene, trichloromethane (chloroform) and tetrachloromethane (carbon tetrachloride) are suspected carcinogens. There is also evidence that exposure to carbon tetrachloride and tetrachloroethene may result in kidney and liver damage. The solubilities of the chlorinated solvents are three–six orders of magnitude higher than their drinking water limits, and so the contamination resulting from a chlorinated solvent DNAPL will typically be greatly in excess of its MCL. In addition to the public-health issues related to chlorinated solvent contamination, the combination of high and increasing demand for water in the UK, and the increasing public awareness of water-quality-related issues, suggest that loss of groundwater resources as a result of pollution cannot be sustained. If neglected, the impact of DNAPL contamination on the costs of water supply may impose on UK industry significantly.

The Permo-Triassic sandstones are a typical consolidated sandstone aquifer, and in the UK

From: BARKER, R. D. & TELLAM, J. H. (eds) 2006. *Fluid Flow and Solute Movement in Sandstones: The Onshore UK Permo-Triassic Red Bed Sequence*. Geological Society, London, Special Publications, **263**, 253–264.
0305–8719/06/$15 © The Geological Society of London 2006.

they form the second most important aquifer supplying about a quarter of all licensed groundwater abstractions (Allen *et al.* 1997). They outcrop extensively in the Midlands and NE England, and a number of large industrial towns, including Manchester, Liverpool and Birmingham, obtain their water supplies at least in part from the aquifer. The aquifer is therefore susceptible to contamination by a range of pollutants from both contemporary and historic industrial sources. Many of the contamination problems occurring in Permo-Triassic sandstone aquifers within the UK result from the use of solvents several decades previously (Rivett *et al.* 1990; Burston *et al.* 1993). This delay, between spillage or release of solvents at the surface and their detection in a public-supply borehole, reflects both the likely lengthy travel times that can be anticipated in a very porous aquifer and the fact that routine groundwater monitoring for these compounds only commenced in the UK during the early–mid 1980s. Resistance of these compounds to chemical or biochemical breakdown in the subsurface environment (Aurand *et al.* 1981) has ensured that the chlorinated solvents persist in groundwater following their disposal and handling many years or even decades previously (Lawrence & Foster 1991; Lerner *et al.* 1993). A large body of work exists in the United States pertaining to the movement and fate of DNAPLs in aquifer systems (e.g. Pankow & Cherry 1996), although much of this work relates to unconsolidated rocks and so is not directly transferable.

The aim of this paper is to relate the behaviour that governs DNAPL entry and saturation of consolidated Permo-Triassic sandstones to the chemical and physical properties of the geological material. A series of standard chemical and physical measurements have been made on a range of Permo-Triassic sandstone samples, and a new laboratory method developed to investigate DNAPL entry pressures.

Transport and fate of DNAPLs in sandstones

DNAPL spillage and penetration

Solvent-transport mechanisms are complex in homogeneous media (Schwille 1988), but these complexities are compounded in heterogeneous media such as the Permo-Triassic sandstones (Eastwood *et al.* 1991). A spillage of a significant volume of DNAPL could result in the non-aqueous-phase solvent reaching the water table within weeks or months of the spillage. Conversely, slow release of an aqueous-phase solvent from contaminated soil as a result of leaching by precipitation could result in the aqueous-phase solvent not reaching the water table for many years.

In the unsaturated zone larger pores and fractures are generally drained and only the smaller pores retain water. DNAPL will enter the larger pores as it will tend to be repelled by water occupying smaller pores. The non-aqueous-phase solvent content of the unsaturated zone will accordingly be dependent on the antecedent moisture content of this zone. Once the main DNAPL body has passed through, some non-aqueous-phase solvent will remain, infilling some of the pore spaces. The fraction of the DNAPL that is retained is referred to as the residual saturation, which is the volume of NAPL trapped in the pores relative to the total pore volume.

Where a spillage is sufficiently large that the total volume of DNAPL spilled exceeds the saturation capacity of the unsaturated zone then the DNAPL will reach the water table and contaminate the aquifer directly. For a DNAPL to pass through the water table the pressure exerted by the DNAPL must exceed the capillary or excess pressure across the interface between itself and the water.

Wettability

For a system of two immiscible fluids (e.g. water and a DNAPL) in contact with a solid phase (e.g. a consolidated sandstone), both the cohesive forces within the fluids and the adhesive forces between the solid and each of the fluids are at work at the line of contact. If the adhesive forces between the solid and the water phase are greater than the cohesive forces inside the water itself, and greater than the forces of attraction between the organic phase and the solid, then the solid–water contact angle, α, will be acute and the water will preferentially wet the solid.

By convention, the wettability of a system is measured through the aqueous phase (Fig. 1). A system is considered to be water wet if the contact angle is less than 70°, and oil or solvent wetting if the contact angle is greater than 110°. If the contact angle is in the range 70°–110° then the system is described as neutral or intermediate wetting (Anderson 1987*a*). It has been observed in water–hydrocarbon systems that sandstones tend to be water or intermediate wetting, whereas carbonate rocks tend to be intermediate or oil wetting (Lake 1989; Taylor

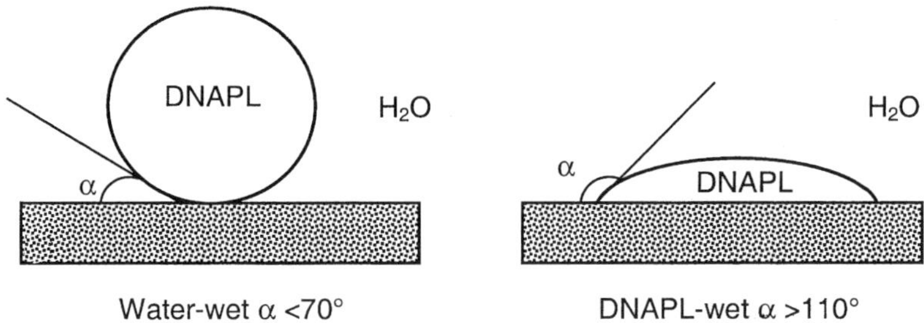

Fig. 1. Schematic diagram showing a DNAPL–water–solid system under water-wetting and DNAPL-wetting conditions.

et al. 2000). Most saturated media are preferentially wet by water. The contact angle provides the only direct measurement of wettability but is a difficult measurement to make under laboratory conditions without compromising the integrity of true environmental conditions. Recent work has shown the presence of subsurface chemical heterogeneities such as variations in aqueous-phase chemistry (Demond *et al.* 1994), mineralogy (Anderson 1987*a–c*; Bradford *et al.* 1998), organic matter distributions (Dekker & Ritsema 1994), and contaminant ageing (Harrold *et al.* 2001) can substantially alter the wettability, and both water– and organic–wet solid surfaces occur within the same porous medium.

Measuring capillary pressures

Capillary pressure causes porous media to draw in the wetting fluid and repel the non-wetting fluid because of the dominant adhesive force between the wetting fluid and the media solid surfaces. For a water–DNAPL system with water being the wetting phase, capillary pressure equals the DNAPL pressure minus the water pressure. The geometry of an interstitial pore space is obviously highly complex. Nevertheless, it is possible to conceive of a network of interstitial spaces connected by pore throats of a smaller characteristic dimension. The Laplace equation (equation 1) predicts the threshold value of the capillary pressure that must be exceeded for DNAPL to pass through a pore-throat of radius *r*:

$$\Delta P = \frac{2\sigma \cos \alpha}{r} \qquad (1)$$

where ΔP is the threshold pressure, σ is the interfacial tension and α is the contact angle. Thus, DNAPL will not enter an interstitial pore

until the capillary pressure exceeds the threshold value associated with the largest throat already in contact with the DNAPL. Once entry pressure has been achieved, the DNAPL moves into the pore. The water–DNAPL interfaces then position themselves across regions of the pore space that support radii of curvature consistent with the prevailing capillary pressure.

The pressure exerted by the DNAPL is:

$$H\Delta\rho g \qquad (2)$$

where H is the height of the DNAPL column, $\Delta\rho$ is the density difference between DNAPL and water, and g is acceleration due to gravity.

Equating these pressures for a circular capillary tube gives:

$$H = \frac{2\sigma \cos \alpha}{r\Delta\rho g} \qquad (3)$$

In the case of the solvent tetrachloroethene, the interfacial tension is reported by Dean (1979) as 0.044 N m⁻¹, although Harrold *et al.* (2003) measured a lower value of 0.038 N m⁻¹. As tetrachloroethene has a density of 1600 kg m⁻³, the density difference with water is 600 kg m⁻³. Mercer & Cohen (1990) also determined a contact angle of 33°–45° for a water–(fine–coarse) sand system. A contact angle of 40° and a density difference of 600 kg m⁻³ have been used in equation (3) to produce a plot showing the height of column of solvent required to enter various pore diameters for four different interfacial tensions (0.044, 0.038, 0.022 and 0.005 N m⁻¹), as shown in Figure 2.

Capillary pressure measurements are essential for determining the degree of penetration likely to occur in an aquifer system after the spillage of a DNAPL. A plot of capillary pressure against saturation for a rock core, called the capillary pressure curve, may be a

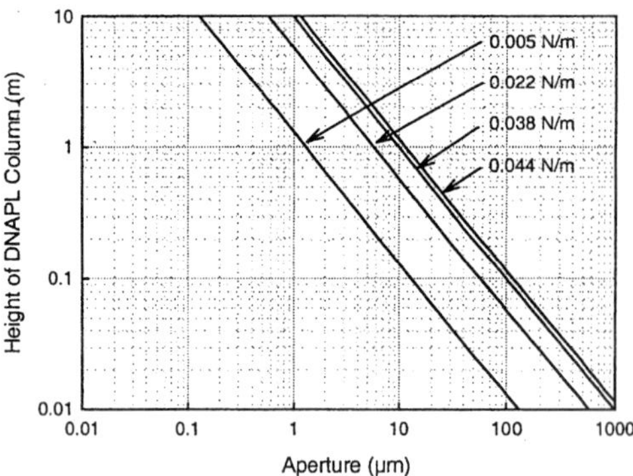

Fig. 2. Relationship between pore-throat radius and head of DNAPL required to invade a pore for a contact angle of 40° and $\Delta P = 600$ kg m^{-3} at different interfacial tensions.

useful way of estimating the potential risk to groundwater from any given DNAPL spill incident. These curves also provide data on the irreducible water saturation and the entry pressure of LNAPL or DNAPL into a water-saturated system.

Two main types of methods are available to determine capillary pressure–saturation, $P_c(S_w)$, relations in porous media: (i) displacement methods based on establishing successive states of hydraulic equilibrium; and (ii) dynamic methods based on establishing successive rates of steady flow of wetting and non-wetting fluids. Displacement methods are used more commonly than dynamic methods.

Three static (non-flow) methods of obtaining capillary pressure curves are the porous plate, mercury injection and centrifuge methods. The porous plate method may require weeks, whereas the centrifuge can take only days. Any fluid combination (gas–oil, gas–water, water–LNAPL, water–DNAPL, gas–water–LNAPL, gas–water–DNAPL) can be used in the centrifuge, it is non-destructive and the results are reproducible.

The seminal work on the centrifuge method was by Hassler & Brunner (1945). Slobod & Chambers (1951) demonstrated the simplicity, reproducibility and speed of the centrifuge method, and its good correlation with the porous plate method. More recently researchers have been focusing on using ultracentrifuges to obtain higher quality continuous data and to improve iterative methods for calculating capillary pressure curves (Ruth & Wong 1990; Kantzas *et al.* 1995; Nikakhtar *et al.* 1996). Employing a centrifuge method, Kinniburgh & Miles (1983) used an immiscible fluid to displace

water from consolidated and unconsolidated media, but were not interested in investigating pore entry pressure relationships. It is only very recently that the centrifuge has been used to measure pore entry pressure for solvents in consolidated media (Pantazidou *et al.* 2000; Harrold *et al.* 2001).

The simple model developed by Hassler & Brunner (1945) assumes that, at equilibrium, there is a network of DNAPL-filled pore throats that are connected to the surface layer of DNAPL. The aqueous phase is also assumed to form a similar nexus. The aqueous phase is only displaced from pore throats in which the driving pressure, resulting from the density difference between the two phases, exceeds the capillary pressure retaining the water in the pore throats.

The driving pressure at any point in the sample depends on the distance to the centre of rotation and varies continuously through the sample (Fig. 3). The average residual water saturation is a weighted average of the residual water saturation at each point in the sample.

At equilibrium, the driving pressure will be balanced by a capillary pressure, p_c. Consequently, the capillary pressure gradient at a point P distance r from the centre of rotation (Fig. 3) is related to the angular velocity, ω, and density difference between the fluid phases by:

$$\frac{dp_c}{dr} = \Delta\rho\omega^2 r \qquad (4)$$

where

$$\Delta\rho = \rho_{DNAPL} - \rho_{water} \qquad (5)$$

and

$$\omega = 2\pi v \qquad (6)$$

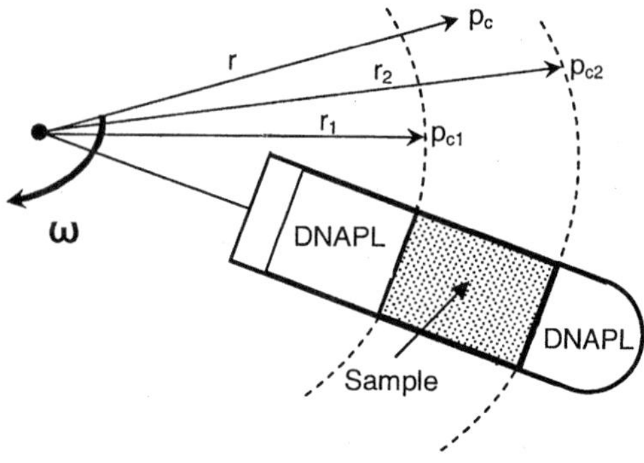

Fig. 3. DNAPL displacement of water using a fixed-angle rotor.

where ρ_{DNAPL} and ρ_{water} are the densities of the DNAPL and water, respectively, and v is the centrifuge speed in revolutions per s. Integrating equation 4 between the inner surface of the sample, distance r_1 from the centre of rotation, to point P and assuming that the capillary pressure at the inner surface, p_{c1}, is zero, gives the driving pressure at P:

$$p_c(r, \omega) = \int_{r_1}^{r} (\Delta\rho\omega^2 r)\mathrm{d}r = \frac{\Delta\rho\omega^2}{2}(r^2 - r_1^2). \quad (7)$$

In particular

$$p_{c2}(\omega) = \frac{\Delta\rho\omega^2}{2}(r_2^2 - r_1^2) \quad (8)$$

where p_{c2} is the capillary pressure at the base of the sample distance r_2 from the centre of rotation.

Empirical capillary pressure–saturation relationships

In materials where there is a range of pore sizes, the capillary behaviour of a material is a function of the material saturation and it can be described graphically using capillary pressure–saturation curves $[P_c(S_w)]$. A range of empirical functions has been proposed to describe the relationship between capillary pressure and saturation (Brooks & Corey 1966; van Genuchten 1980; Kool & Parker 1987; Lenhard & Oostrom 1998). Among the most commonly used is the van Genuchten (1980) equation. This was developed to describe capillary pressure–saturation phenomena in structured soils, but has been applied extensively to a range of materials:

$$P_c = P_0\left(S_e^{-\frac{1}{m}} - 1\right)^{1-m} \quad (9)$$

where P_0 is the characteristic entry pressure and m is a fitting parameter determined by the pore-size distribution. S_e is the normalized wetting fluid saturation defined as:

$$S_e = \frac{S_w - S_r}{S_m - S_r} \quad (10)$$

where S_r is the residual saturation (residual NAPL volume/total pore volume), S_m is maximum water saturation and S_w is the wetting phase saturation. The parameters m, P_0 and S_r in the van Genuchten equation are evaluated by fitting the equation to experimental data.

Relative permeability

Once a DNAPL has penetrated the water table, a two-phase system exists. The coexistence of another immiscible fluid in the pores reduces the area available for flow of either fluid and increases the tortuosity of the respective flow paths. The effective permeabilities are expressed as a product of the permeability and the relative permeability. The permeability, k, is considered to be a function only of the rock pore size, while the relative permeability, k_r, is a function of the fraction of fluid present in the rock pores and is a dimensionless property.

Permo-Triassic sandstones frequently exhibit cross-bedding and lamination that may result in the direction of maximum permeability being at a steep angle to the horizontal. This feature may enhance downward migration of DNAPLs in Permo-Triassic sandstones.

Determining relative permeability using the van Genuchten–Maulem equations

While it is possible to measure relative permeability functions in the laboratory, it is often found convenient to estimate them from $P_c(S_w)$ data. Burdine (1953) and Mualem (1976) developed models that relate relative permeability to the capillary pressure–saturation function. These models can be used to derive closed-form expressions for the relative permeability to the wetting phase saturation – $k_{rw}(S_w)$ – and the relative permeability to the non-wetting phase saturation – $k_{rnw}(S_w)$.

By combining the Mualem (1976) and van Genuchten models, a term for the relative of the wetting phase has been derived (van Genuchten 1980):

$$k_{rw} = S_e^{0.5}[1 - (1 - S_e^{1/m})^m]^2 \quad (11)$$

and Parker et al. (1987) used a similar procedure to derive a term for the relative permeability of the non-wetting phase:

$$k_{rnw} = (1 - S_e)^{0.5}(1 - S_e^{1/m})^{2m} \quad (12)$$

By definition, these functions range from zero to unity. The relative permeability to the wetting phase (water) is usually considered to be free of hysteresis. On the other hand, the saturations at which k_{rnw} is zero in a wetting or draining process are not usually the same because of entrapment of the non-wetting fluid (DNAPL) during the wetting process. It is usual to assume that k_{rnw} is greater than zero in a draining process for all $S_w < 1$. In reality there is a threshold saturation of non-wetting phase required to bring about an initial network of connected pore throats across the sample volume of interest.

Methods

Sample selection and preparation

The physical properties of the sandstones were characterized by measurements of the interconnected porosity and pore-size distributions, and a weak nitric acid extraction was used to characterize the surface chemistry. A centrifuge method was used to determine the capillary behaviour of the sandstone–DNAPL–water systems. The DNAPL used was tetra-chloroethene (PCE), a common dry-cleaning fluid.

On account of the diverse nature of the sandstones, it was not possible, within the scope of the present study, to sample systematically each of the main lithotypes from all the major sandstone formations. Instead, representative samples of consolidated sandstones were collected with a geographical spread and from a range of depths (poorly consolidated, friable sandstones were not investigated). A total of 110 samples were taken from 13 different localities (Fig. 4). All the samples were tested for porosity and extracted with weak acid. A smaller subset of 66 samples had previously been tested by mercury injection capillary pressure (MICP) to obtain pore-throat size

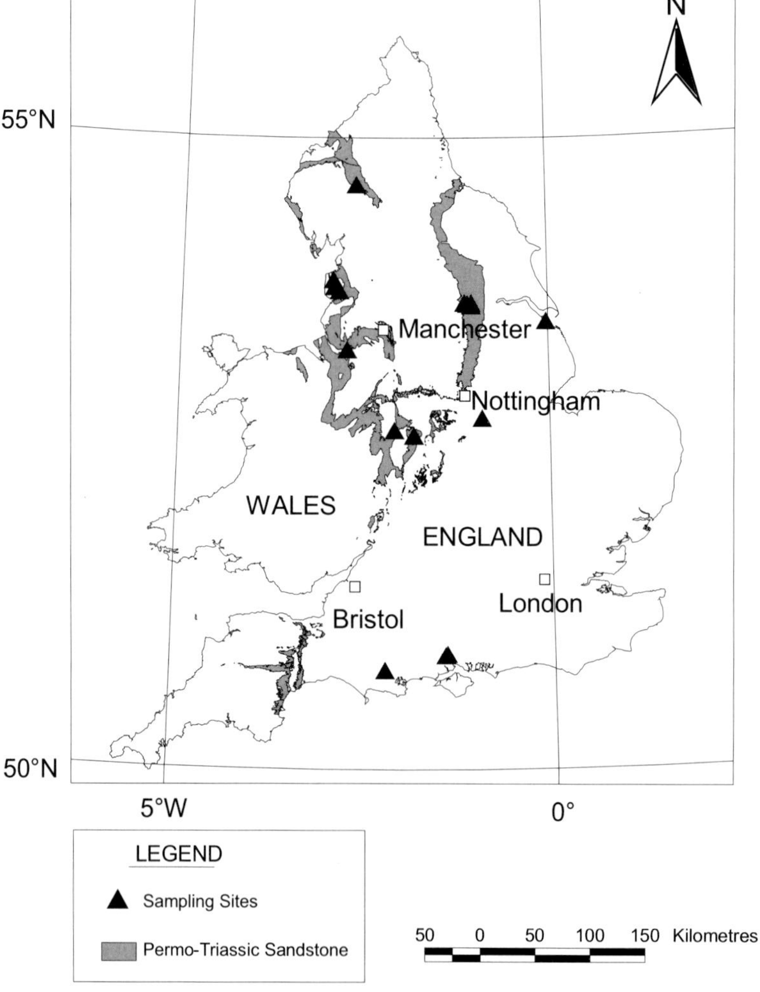

Fig. 4. Map of outcrops of the Permo-Triassic sandstones and location of sampling sites.

distributions (Bloomfield *et al.* 2001). A subset of a further 30 samples has been tested for this study by the centrifuge method to determine entry pressure. Samples for centrifuge studies were chosen on the basis of the porosity, chemical extraction and pore-size distribution data in conjunction with visual inspection of the core. Samples with exceptionally small pore sizes (<0.1 μm) were excluded from the study as the required entry pressures (approximately 120 m of PCE) would be considerably greater than experienced in any solvent spill scenario.

Measurement of solvent entry pressure in consolidated media

A centrifuge method was used to determine PCE entry pressures. A modified method for rapid determination of $P_c(S_w)$ curves in DNAPL–water systems, developed and described by Gooddy *et al.* (1999) and subsequently used by Harrold *et al.* (2001), has been used in this study. A fully water saturated sample is placed in a tube in a centrifuge rotor (Fig. 3). The tube is then filled with PCE in order to immerse completely the sample when the tube is in a horizontal plane. This arrangement permits the measurement of volumes of water expelled from the sample during centrifugation as the water floats to the top of the PCE in the centrifuge tube. Samples are centrifuged at increasing speeds and the volume of evolved water at each speed is measured. The applied macroscopic capillary pressure (equation 8) is calculated as a function of rotor speed and centrifuge and sample geometries (Kinniburgh & Miles 1983). The degree of sample saturation, S_e, and wetting phase saturation, S_w, are calculated on the basis of the evolved volume of water. The capillary pressure–saturation curves are fitted with the van Genuchten function to obtain a value for effective PCE entry pressure, P_0, the pore size distribution index, m, and the residual saturation, S_r.

Weak nitric acid extraction and chemical analysis

A weak acid extraction method has been used to characterize the surface chemistry of the sandstones. A 0.43 M HNO$_3$ solution is commonly used by soil scientists to extract weakly bound species (Boekhold *et al.* 1993; Gooddy *et al.* 1995) and as an alternative to neutral salt extractions such as calcium or barium chloride. The weak acid extraction method has the advantage over neutral salt

extractions in that no cation data are lost (for the conventional 0.01 M CaCl$_2$ extraction it is not possible to obtain any data on the bound calcium), although it may slightly overstate the weakly bound species by causing some mild dissolution of the matrix (Gooddy *et al.* 1995). Four grammes of ground, dry rock matrix was weighed into a preweighed 50 ml Oak Ridge polypropylene centrifuge tube to which was added 40 ml of 0.43 M HNO$_3$ (a final solid solution ratio of 1:10 g ml^{-1}). Duplicate sets of tubes were shaken on a box shaker for 2 h before centrifuging for 10 min at 17 000 rpm (a driving force of 40 000g) and the supernatant solution filtered through a 0.45 μm filter. Analysis of the filtered extracts was carried out using a Perkin-Elmer Optima 3300 DV ICP-AES (inductively coupled plasma-atomic emission spectroscopy) that was calibrated for simultaneous determination of 16 elements.

Results

DNAPL entry pressure

Figure 5 shows effective entry pressure, P_0, obtained by fitting the van Genuchten function (using a least-squares regression procedure) to the centrifuge pressure–saturation curves, plotted against median pore radius from the MICP tests (the figure also shows the equivalent height of PCE column on the right-hand y-axis). Superimposed on the same graph are a series of reference contours for IFTs (interfacial tension) calculated from equation 1 for a contact angle of 40°. This is based on Mercer & Cohen (1990) who cite contact angles of 33°–45° for

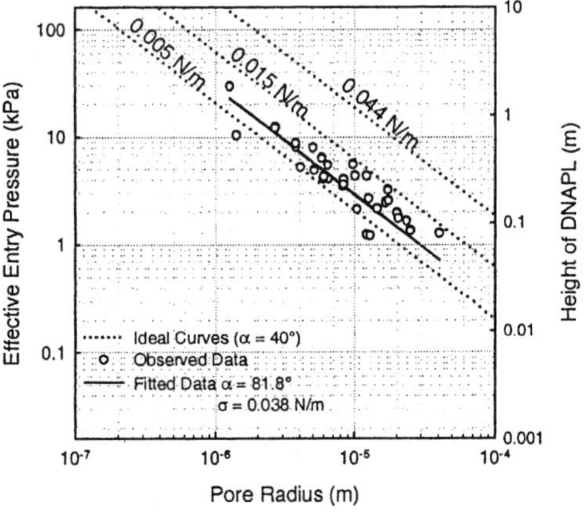

Fig. 5. The relationship between pore-throat radius and height of DNAPL required to overcome the entry pressure.

PCE–sand–water systems. It can be seen that most of the observations cluster between the 0.005 and 0.015 N m⁻¹ IFT contours. The interfacial tension, however, for PCE is 0.044 N m⁻¹ (Mercer & Cohen 1990, appendix B): consequently, the observed entry pressures are systematically and significantly lower (on average seven times lower) than those predicted by equation 1. This may be explained either by a lower than predicted value of IFT and/or a larger than predicted contact angle.

Given a contact angle of 40°, by fitting equation 1 to the observed data (optimizing on a least-squares basis) an apparent IFT of 0.007 N m⁻¹ is obtained ($R^2 = 0.79$). In a series of recent batch tests (Harrold *et al.* 2003) using PCE and dry Permo-Triassic sandstone, IFT values close to the 0.045 N m⁻¹ given by Mercer & Cohen (1990) were observed. The new study showed a small decrease in IFT from 0.044 to 0.040 N m⁻¹ over a period of 20 min and then a further decrease to 0.038 N m⁻¹ over the next 8 h. Given this observation it seems unlikely that an apparent IFT of 0.007 N m⁻¹ could account for the unexpectedly low entry pressures. Therefore, a larger than predicted contact angle is probably a more likely explanation for the low observed entry pressures. The best-fit line through the observed values ($R^2 = 0.79$) shown in Figure 3 based on an IFT of 0.038 N m⁻¹ requires an effective contact angle of 81.8°.

Figure 6 is a cross-plot of magnesium against calcium for the nitric acid extractions carried out on 110 Permo-Triassic sandstone samples from across the UK. The filled circles indicate samples used in this study and for which $P_c(S_w)$

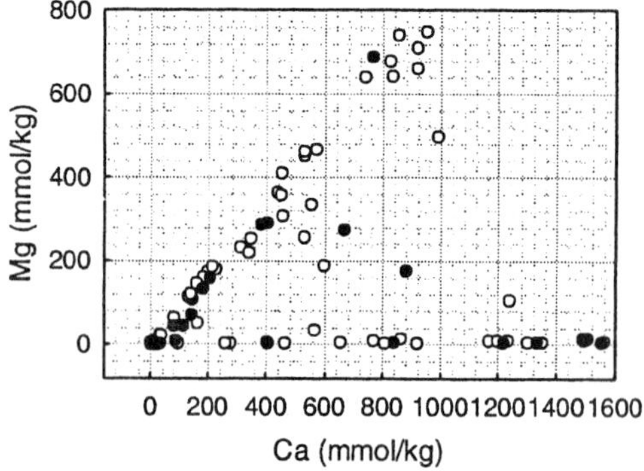

Fig. 6. Relationship between calcium and magnesium (assumed as proxies for calcite and dolomite) from 0.43 M HNO₃ extraction. Filled circles are samples for which DNAPL capillary pressure measurements have been made.

curves and MICP determinations have been carried out. It can be seen that the samples used in this study are representative of the larger data set. Figure 6 shows that the samples fall into one of two populations. Samples that are dominated by a calcite matrix [Ca(CO₃)] are depicted by the line running parallel to the *x*-axis, and those that are dominated by dolomite [CaMg(CO₃)₂] follow a 1:1 ratio. The samples rich in calcite account for about 40% of the population. Both calcite and dolomite are present as cements in the Permo-Triassic sandstones and are commonly found as euhedral pore-lining phases and as more diffuse cement phases, particularly in the pore throats of sandstones (Strong 1993).

It has been shown that the observed PCE entry pressures are significantly lower than would be expected on the basis of standard capillary theory, and that it is unlikely that this is because of changes in the IFT of the solvent. From this it has been inferred that the difference between observed and predicted entry pressures is due to unexpectedly large contact angles in the PCE–sandstone–water system. Barranco *et al.* (1997) found that contact angles increase near the pH of the zero surface charge of quartz (around pH 2); however, the system investigated in this study is pH neutral. Anderson (1987a) noted that surface-roughness effects will generally diminish the apparent contact angle, so even if surface-roughness effects are active in the samples that have been tested they will tend to reduce rather than increase the contact angle. Other workers have noticed significant changes in wettability and contact angles with time. Powers & Tamblin (1995) observed increases in contact angle for sands exposed to petroleum mixtures for between 14 and 60 days. Harrold *et al.* (2001) noted an increase in contact angle from 20° to 40° over 24 h for a trichloroethene–sand system. In the present study, however, the sandstones were only exposed to solvent for relatively short times, typically of the order of 2 h. On the basis of the observations of Harrold *et al.* (2001) and Powers & Tamblin (1995), it is unlikely that the length of exposure of the sandstones to PCE in the present study was sufficient to modify the contact angle significantly and hence to explain the observed entry pressures. Significant increases in wettability have also been observed with 'field', 'real-world' or 'dirty' solvents containing a cocktail of different solvents and impurities (Jackson & Dwarakanath 1999; Dwarakanath & Pope 2000; Harrold *et al.* 2001). The solvent used in the present system was an analytical reagent-grade PCE and so there are no significant quantities of impurities or surfactants in the PCE–sandstone–water system to affect the wettability.

In water–oil–rock systems it has been observed that sandstones tend to be water-intermediate wetting, whereas carbonate rocks tend to be intermediate or oil wetting (Lake 1989; Taylor *et al.* 2000). It is known that carbonate cements are present in the sandstones as shown in Figure 6, and this raises the possibility that the differences in observed and predicted entry pressures could be explained by changes in the surface mineral composition of the sandstones. A plot of calcium and magnesium per unit porosity (Fig. 7) has been constructed to test the observation that limestones (calcium and magnesium carbonates) tend to have more neutrally wetting properties than sandstone. Calcium and magnesium have been plotted against the percentage difference between observed and predicted entry pressures, as given by the following expression:

$$\text{Diff}_{P_\text{c}} = \frac{P_0 - P_\text{c}}{P_0} \times 100 \tag{13}$$

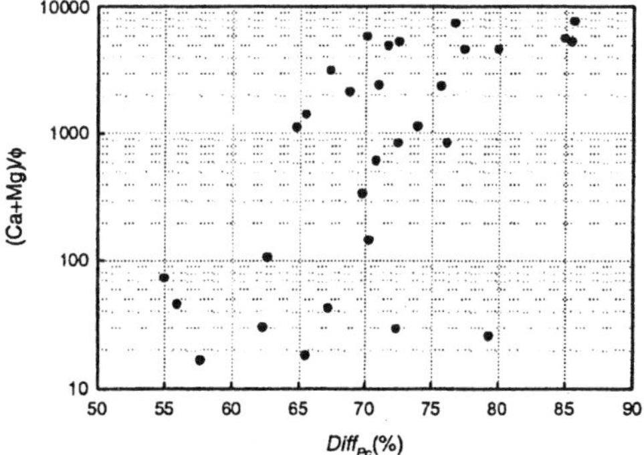

Fig. 7. Calcium and magnesium per unit porosity (mmol kg^{-1}) plotted against the percentage difference between calculated and observed entry pressures. See equation 13.

where P_0 is the effective entry pressure obtained by fitting the centrifuge data with the van Genuchten function (equation 9), and P_c is the predicted entry pressure based on ideal capillary behaviour, i.e. solution of equation 1 given a contact angle, α, of 40°, and an interfacial tension, σ, of 0.044 N m^{-1}. The plot shows a log-linear increase in the difference between the observed and predicted values of entry pressure with increasing calcium and magnesium per unit porosity. Higher calcium and magnesium contents are associated with lower than expected entry pressures. From this observation it is inferred that the surface chemistry of the sandstones, in particular the presence of calcite and dolomite cements, has a significant effect on the wetting properties of the rock, although were one to extrapolate to the origin in Figure 7 it clear that this cannot solely explain the inferred reduction in pore entry pressures.

This approach assumes a 'bundle of capillaries' model (Dullien 1992) of pore-size distribution. Indeed, the true pore-size distribution may be quite different from those inferred from standard analysis of MICP data. For example, a probability-based percolation model for interpreting MICP data (Gooddy 2003) suggests the presence of much larger pores. Therefore the observed entry pressures may either be interpreted in terms of an increase in contact angle (bundle of capillary tubes model) or the presence of more large and connected pores (percolation model) or indeed both. If there is a larger and more connected pore network than described by the bundle of capillary tubes model then the potential for pathogen and colloidal transport is considerably enhanced and the matrix of the Permo-Triassic sandstones is in fact more vulnerable than suggested by Figure 5.

A statistical summary of the key parameters measured during this study are shown in Table 1.

Table 1. *Statistical summary of key parameters measured*

	Porosity (%)	Gas permeability ($\times 10^{-15}$ m^2)	Median pore radius (μm)	Effective entry pressure, P_0 (kPa)	Ca (mmol kg^{-1})	Mg (mmol kg^{-1})
Number	30	30	30	30	108	108
Minimum	4.30	0.93	1.25	1.19	1.34	0.21
Maximum	32.90	7670	40.67	29.2	1560	749
Average	22.25	1400	10.43	5.22	423	84.2
Geometric mean	20.70	430	7.97	3.90	213	8.33
SD	6.42	2050	16.27	5.29	465	202

Table 2. *Statistical summary (based on 15 samples) of relative permeability to the wetting phase* (k_{rw}) *at four different non-wetting-phase saturations* (S_{nw})

S_{nw}	Mean	SD	Median	Maximum	Minimum
0.50	0.003	0.008	0.000	0.031	0.000
0.25	0.053	0.047	0.039	0.186	0.000
0.10	0.233	0.108	0.225	0.490	0.000
0.05	0.398	0.134	0.405	0.698	0.000

Relative permeability

Using the van Genuchten parameters of the fitted $P_c(S_w)$ curves, relative permeability to the wetting phase and non-wetting phase can be generated. Examples of such calculated curves are given in Figure 8. The example given is fairly typical in that $k_{rw} = 0$ (i.e. the point at which no flow of the wetting phase can occur) at a wetting phase saturation of roughly 0.36 (mean for all samples is 0.363 and the median is 0.368). Alternatively stated, when the core is 64% saturated with DNAPL, no water flow will occur.

Table 2 shows a statistical summary for the reduction in wetting-phase permeability for four different non-wetting-phase saturations, i.e. 50, 25, 10 and 5% DNAPL saturation. The table demonstrates that, although some flow of wetting phase is possible at 50% DNAPL saturation, this is on average less than 0.5% of the original. The table also shows how, at even 5% DNAPL saturation, the relative permeability is reduced, on average, to roughly 40% of its original value. In one case the relative permeability falls to zero at just 5% DNAPL saturation. Therefore, even a fairly small degree of DNAPL saturation can lead to significant reductions in groundwater flow.

Conclusions

DNAPL entry pressures have been found to be lower than predicted by theoretical considerations of capillary theory. It is suggested that changes in contact angle caused by the presence of more wetting carbonates in pore space is, in part, the reason for this.

On average the predicted relative permeability of the wetting phase falls to zero at a DNAPL content of roughly 60%, although a DNAPL content of just 5% can cause significant reductions in the permeability.

References

ANDERSON, W.G. 1987a. Wettability literature survey – Part 4: Effects of wettability on capillary pressure. *Journal of Petroleum Technology*, **39**, 1283–1300.

ANDERSON, W.G. 1987b. Wettability literature survey – Part 5: The effects of wettability on relative permeability. *Journal of Petroleum Technology*, **39**, 1453–1468.

ANDERSON, W.G. 1987c. Wettability literature survey – Part 6: The effects of wettability on waterflooding. *Journal of Petroleum Technology*, **39**, 1605–1622.

ALLEN, D.J., BREWERTON, L.J. ET AL. 1997. *The Physical Properties of Major Aquifers in England and Wales*. British Geological Survey Technical Report, **WD/97/34**.

AURAND, K., FRIESEL, P., MILDE, G. & NEUMAYR, V. 1981. Behaviour of organic solvents in the environment. *Studies in Environmental Science*, **12**, 481–487.

BARRANCO, F.T., DAWSON H.E., CHRISTENER, J.M. & HONEYMAN, B.D. 1997. Influence of aqueous pH and ionic strength on the wettability of quartz in the presence of dense non-aqueous phase liquids. *Environmental Science and Technology*, **31**, 676–681.

BLOOMFIELD, J.P., GOODDY, D.C., BRIGHT, M.I. & WILLIAMS, P.J. 2001. Pore-throat size distributions

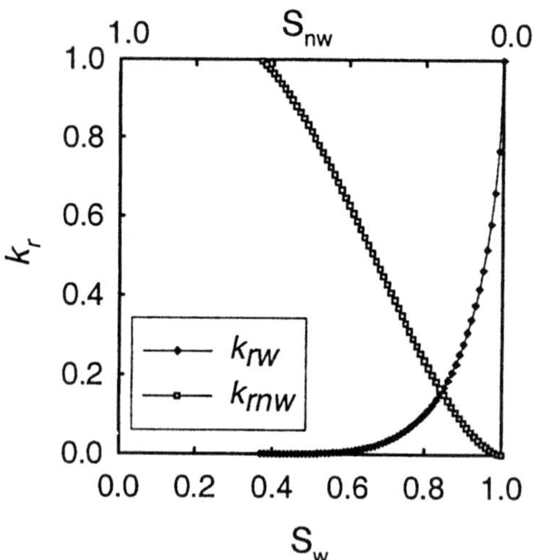

Fig. 8. Example of saturation–relative permeability relationship for a DNAPL–water–sandstone system (S_w is saturation of wetting phase; S_{nw} is saturation of non-wetting phase).

in sandstones and some implications for contaminant hydrogeology. *Hydrogeology Journal*, **9**, 219–230.

BOEKHOLD, A.E., TEMMINGHOFF, E.J.M. & VAN DER ZEE, S.E.A.T.M. 1993. Influence of electrolyte composition and soil pH on Cd sorption by an acid sandy soil. *Journal of Soil Science*, **44**, 85–96.

BRADFORD, S.A., ABRIOLA, L.M. & RATHFELDER, K.M. 1998. Flow and entrapment of dense nonaqueous phase liquids in physically and chemically heterogeneous aquifer formations. *Advances in Water Resources*, **22**, 117–132.

BROOKS, R.H. & COREY, A.T. 1966. Properties of porous media affecting fluid flow. *Journal of the Drainage and Irrigation Division, American Society of Engineering*, **92**, IR2, 61–88.

BURDINE, N.T. 1953. Relative permeability calculations from pore-size distribution data. *Petroleum Transactions of the American Institute of Mining and Metallurgy Engineers*, **198**, 71–77.

BURSTON, M.W., NAZARI, M.M., BISHOP, P.K. & LERNER, D.N. 1993. Pollution of groundwater in the Coventry region (UK) by chlorinated hydrocarbon solvents. *Journal of Hydrology*, **149**, 137–161.

DEAN, J.A. 1979. *Lange's Handbook of Chemistry*. McGraw Hill, New York.

DEKKER, L.W. & RITSEMA, C.J. 1994. How water moves in a water repellent sandy soil, 1, potential and actual water repellency. *Water Resources Research*, **30**, 2507–2519.

DEMOND, A.H., DESAI, F.N. & HAYES, K.F. 1994. Effect of cationic surfactants on organic liquid–water capillary pressure–saturation relationships. *Water Resources Research*, **30**, 333–342.

DULLIEN, F.A.L. 1992. *Porous Media. Fluid Transport and Pore Structure*. Academic, New York.

DWARAKANATH, V. & POPE, G.A. 2000. Surfactant phase behaviour with field degreasing solvent. *Environmental Science and Technology*, **34**, 4842–4848.

EASTWOOD, P.R., LERNER, D.N., BISHOP, P.K. & BURSTON, M.W. 1991. Identifying land contaminated by chlorinated hydrocarbon solvents. *Journal of the Institution of Water and Environmental Management*, **5(2)**, April, 163–171.

GOODDY, D.C. 2003. *Controls on the distribution of dense non-aqueous phase liquids in the matrix of Permo-Triassic sandstones*. PhD thesis, University of London.

GOODDY, D.C., BLOOMFIELD, J.P. & BRIGHT, M. 1999. *Development of a Centrifuge Method for the Rapid Determination of Solvent Pore-entry Pressures*. British Geological Survey Technical Report, **WD/99/6**.

GOODDY, D.C., SHAND, P., KINNIBURGH, D.G. & VAN RIEMSDIJK, W.H. 1995. Field-based partition coefficients for trace elements in soil solutions. *European Journal of Soil Science*, **46**, 265–285.

HARROLD, G., GOODDY, D.C., LERNER, D.N. & LEHARNE, S.A. 2001. Wettability changes in trichloroethylene contaminated sandstone. *Environmental Science and Technology*, **35**, 1504–1510.

HARROLD, G., GOODDY, D.C., REID, S., LERNER, D.N. & LEHARNE, S.A. 2003. Changes in the interfacial tension values for chlorinated solvents following their transport through various UK soil systems. *Environmental Science and Technology*, **37**, 1919–1925.

HASSLER, G.L. & BRUNNER, E. 1945. Measurement of capillary pressures in small core samples. *Petroleum Transactions of the American Institute of Mining and Metallurgy Engineers*, **160**, 114–123.

JACKSON, R.E. & DWARAKANATH, V. 1999. Chlorinated degreasing solvents: Physical–chemical properties affecting aquifer contamination and remediation. *Ground Water Monitoring and Remediation*, **19**, (4), 102–110.

KANTZAS, A., NIKAKHTAR, B., RUTH, D. & POW, M. 1995. Two phase relative permeabilites using the ultracentrifuge. *Journal of Canadian Petroleum Technology*, **34**, (7), 58–63.

KINNIBURGH, D.G. & MILES, D.L. 1983. Extraction and chemical analysis of interstitial water from soils and rocks. *Environmental Science and Technology*, **17**, 362–368.

KOOL, J.B. & PARKER, J.C. 1987. Development and evaluation of closed-form expressions for hysteretic soil hydraulic properties. *Water Resources Research*, **23**, 105–114.

LAKE, L.W. 1989. *Enhanced Oil Recovery*. Prentice Hall, Englewood Cliffs, NJ.

LAWRENCE, A.R. & FOSTER, S.S.D. 1991. The legacy of aquifer pollution by industrial chemicals: technical appraisal and policy implications. *Quarterly Journal of Engineering Geology*, **24**, 231–239.

LENHARD, R.J. & OOSTROM, M. 1998. A parametric model for predicting relative permeability–saturation–capillary pressure relationships of oil–water systems in porous media with mixed wettability. *Transport in Porous Media*, **31**, (1), 109–131.

LERNER, D.N., GOSK, E. ET AL. 1993. Postscript: summary of the Coventry groundwater investigation and implications for the future. *Journal of Hydrology*, **149**, 257–272.

MERCER, J.W. & COHEN, R.M. 1990. A review of immiscible fluids in the sub-surface: properties, models, characterization and remediation. *Journal of Contaminant Hydrology*, **6**, 107–163.

MUALEM, Y. 1976. A new model for predicting the hydraulic conductivity of unsaturated porous media. *Water Resources Research*, **12**, 513–522.

NIKAKHTAR, B., KANTZAS, A., DE WIT, P., POW, M. & GEORGE, A. 1996. On the characterising of rock/fluid and fluid/fluid intereactions in carbonate rocks using the ultracentrifuge. *Journal of Canadian Petroleum Technology*, **35**, (1), 47–56.

PANKOW, J.F. & CHERRY, J.A. 1996. *Dense Chlorinated Solvents and Other DNAPLs in Groundwater*. Waterloo Press, Waterloo.

PANTAZIDOU, M., ABU-HASSANEIN, Z.S. & RIEMER, M.F. 2000. Centrifuge study of DNAPL transport in granular media. *Journal of Geotechnical and Geoenvironmental Engineering*, **126**, 105–115.

PARKER, J.C., LENHARD, R.J. & KUPPANSAMY, T. 1987. A parametric model for constitutive properties governing multiphase flow in porous media. *Water Resources Research*, **23**, 618–624.

POWERS, S.E. & TAMBLIN, M.E. 1995. Wettability of porous media after exposure to synthetic gasolines. *Journal of Contaminant Hydrology*, **19**, 105–125.

RIVETT, M.O., LERNER, D.N. & LLOYD, J.W. 1990. Chlorinated solvents in UK Aquifers. *Journal of the Institution of Water and Environmental Management*, **4**, 242–250.

RUTH, D. & WONG, S. 1990. Centrifuge capillary pressure curves. *Journal of Canadian Petroleum Technology*, **29**, (3), 67–72.

SCHWILLE, F. 1988. *Dense Chlorinated Solvents in Porous and Fractured Media*. Lewis Publishers, Boca Raton, FL.

SLOBOD, R.L. & CHAMBERS, A. 1951. Use of centrifuge for determining connate water, residual water, and capillary pressure curves of small core samples. *Transactions of the American Institute of Mining Engineering*, **192**, 127–134.

STRONG, G.E. 1993. Diagenesis of the Triassic Sherwood Sandstone group of rocks, Preston, Lancashire, UK: a possible evaporitic cement precursor to secondary porosity? *In*: NORTH, C.P. & PROSSER, D.J. (eds) *Characterization of Fluvial and Aeolian Reservoirs*. Geological Society, London, Special Publications, **73**, 279–289.

TAYLOR, S.C., HALL, C., HOFF, W.D. & WILSON, M.A. 2000. Partial wetting in capillary liquid absorption by limestones. *Journal of Colloid and Interface Science*, **224**, 351–357.

VAN GENUCHTEN, M.TH. 1980. A closed form equation for predicting the hydraulic conductivity of unsaturated soils. *Soil Science Society of America Journal*, **44**, 892–898.

The arsenic concentration in groundwater from the Abbey Arms Wood observation borehole, Delamere, Cheshire, UK

DAVID G. KINNIBURGH[1], ANDREW J. NEWELL[1], JEFF DAVIES[1],
PAULINE L. SMEDLEY[1], ANTHONI E. MILODOWSKI[2],
JOHN A. INGRAM[3] & PHILIP D. MERRIN[4]

[1]*British Geological Survey, Wallingford, Oxfordshire OX10 8BB, UK*
(e-mail: dgk@bgs.ac.uk)

[2]*British Geological Survey, Keyworth, Nottingham NG12 5GG, UK*

[3]*Environment Agency, Knutsford Road, Warrington WA4 1HG, UK*

[4]*United Utilities, Lingley Mere Business Park, Great Sankey, Warrington WA5 3LP, UK*

Abstract: A 150 m observation borehole was drilled in Abbey Arms Wood, Delamere, Cheshire, UK in order to explore the local hydrogeological conditions and to understand better the source of the high concentrations of arsenic in some of the local groundwaters. The borehole was located on an outcrop of the Helsby Sandstone Formation (part of the Sherwood Sandstone Group) and was cored into the underlying Wilmslow Sandstone Formation. The aquifers in the area are unconfined and give rise to low-Fe groundwaters with As concentrations in the 10–50 µg l^{-1} range. The chemical composition of the sediments is quite uniform down to 150 m. The total arsenic content is in the range from 5 to 15 mg kg^{-1} and averaged 8 mg kg^{-1} ($n = 60$). There is no trend in sediment As concentration with depth, but pore water centrifuged from the core steadily increased in As concentration with depth. The As concentration ranges from 8 µg l^{-1} at 10 m (unsaturated zone) to 30 µg l^{-1} at 150 m. The source of the dissolved As remains unclear but the lack of evidence for discrete high-As minerals or zones of mineralization suggests that it is probably derived by desorption from rock-forming minerals in the sandstones, e.g. iron oxides. This may be in response to slightly higher pH (up to 8.0 at depth). If this trend applies throughout the area, restricting the screened interval for abstraction boreholes to the uppermost parts of the saturated zone may reduce As concentrations, but is likely to reduce yields and may also risk encountering groundwaters with high nitrate concentrations.

There has been an increased interest in the arsenic (As) concentration of groundwaters following the recent reduction of the European limit for As in drinking water from 50 to 10 µg l^{-1}. This has focused attention on certain groundwater sources that were previously, but are no longer, compliant. It is now known that naturally high-As groundwaters are found in many parts of the world (Smedley & Kinniburgh 2002) and the UK is no exception. It has been known for some time that groundwater sources from the Sherwood Sandstone aquifer in the UK, including some of those from the Delamere area, Cheshire, the study area of this paper, contain relatively high concentrations of As in the raw water (Edmunds *et al.* 1989; Edwards 2001). This has obliged the UK water industry to invest in As-removal plants in order to comply with the new drinking-water limit. A number of sources in the Delamere area now require treatment.

In March 2002, as part of a programme to enhance its water-level observation borehole network in the Delamere area, the Environment Agency (EA, North West Region), drilled a borehole in Abbey Arms Wood, near Delamere [SJ 56418 68133] (Fig. 1). This is within the high-As groundwater area and so provided an opportunity to investigate the nature and distribution of As in the sandstone and in the groundwater. In particular, it was not clear whether the As was derived from a few discrete horizons, perhaps related to the mineralization known to occur elsewhere in the Cheshire Basin, or was derived from a more diffuse and widespread source. Knowing the source and depth distribution of the arsenic would indicate whether it is possible to design abstraction boreholes that avoid the high-As source zones.

From: BARKER, R. D. & TELLAM, J. H. (eds) 2006. *Fluid Flow and Solute Movement in Sandstones: The Onshore UK Permo-Triassic Red Bed Sequence*. Geological Society, London, Special Publications, **263**, 265–284.
0305-8719/06/$15 © The Geological Society of London 2006.

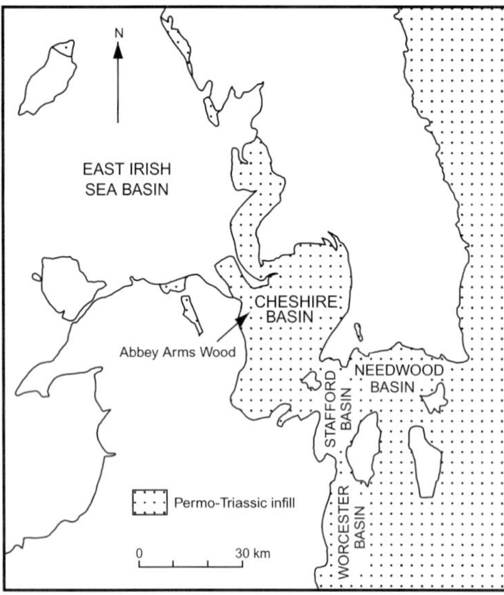

Fig. 1. Location of the Abbey Arms Wood borehole in relation to the location of the Cheshire basin and other nearby basins (after Plant *et al.* 1999).

Geological and hydrogeological background

Geology

The Abbey Arms Wood borehole is located in the northern part of the Cheshire Basin, which is one of the largest onshore basins in Britain (Fig. 1). The Cheshire Basin was formed in the late Permian (265–251 Ma) with the main phase of extension and subsidence occurring during the Triassic (251–205 Ma). Subsidence continued into the late Cretaceous, with the basin reaching its maximum depth of more than 6000 m in late Cretaceous or early Palaeocene times (60 Ma). Subsequent uplift and basin inversion led to the removal of more than 2000 m of Cretaceous and Jurassic strata in the depocentre, and the establishment of shallow, fresh groundwater systems in Triassic aquifers (Plant *et al.* 1999).

The site is situated within a small wood on the Helsby Sandstone Formation, part of a sequence of red-bed desert sandstones known as the Sherwood Sandstone Group. It is surrounded by rocks belonging to the younger Mercia Mudstone Group. The Cheshire Basin is infilled with a relatively thick Permo-Triassic

sequence that has a regional dip towards the ESE and thickens gradually toward the faulted eastern margin. The Sherwood Sandstone Group comprises a thick succession of Triassic sandstones, mudstones and conglomerates deposited in arid continental fluvial, aeolian and lacustrine environments (Bloomfield *et al.* 2005). The Abbey Arms Wood borehole penetrates both the Helsby Sandstone Formation and the Wilmslow Sandstone Formation. The Helsby Sandstone is a relatively silty, fine- to medium-grained flat-bedded sandstone, whereas the Wilmslow Sandstone consists of clean cross-bedded sandstones and mudstones. The boundary between these two formations occurs at a depth of 48.9 m and is well defined in the gamma-ray logs (Fig. 2). The Helsby Sandstone is pebbly in places. The original sediments were deposited between 245 and 235 Ma, after which the Cheshire Basin underwent a period of subsidence that continued until the beginning of the Tertiary period, some 65 Ma ago.

The borehole is 0.5 km west of the 'Overton–East Delamere Fault Zone', a major structural belt that defines the western margin of the Wem–Audlem Sub-basin. This fault zone is some 70 km in length and has an easterly downthrow that locally exceeds 1000 m at the top of the pre-Permian basement. The fault zone broadly forms the eastern margin of the 'Mid-Cheshire Ridge'. Sherwood Sandstone mainly outcrops to the west of the fault zone and Mercia Mudstone Group mainly to the east. An important feature of the faulting is its association with base-metal mineralization. At Bickerton, 15 km south of Delamere, there is a small sulphide mineral deposit hosted in Wilmslow Sandstone with Cu, Co, As, Ni, and Fe mineralization (Naylor *et al.* 1989). Mineralization within the sulphide ore deposits is believed by Plant *et al.* (1999) to have occurred during the late Triassic–early Jurassic when high-density metalliferous brines from the Mercia Mudstone Group flowed through the basin under the influence of gravity. These then encountered a small volume of reducing fluid in fault systems around the margins of the basin (Naylor *et al.* 1989; Tellam 1995).

Hydrogeology

A few metres of fluvio-glacial sands and gravels overlie the Sherwood Sandstone aquifer at the site of the borehole. These deposits can

---

Fig. 2. The gamma-ray log and lithostratigraphy of the Abbey Arms Wood borehole.

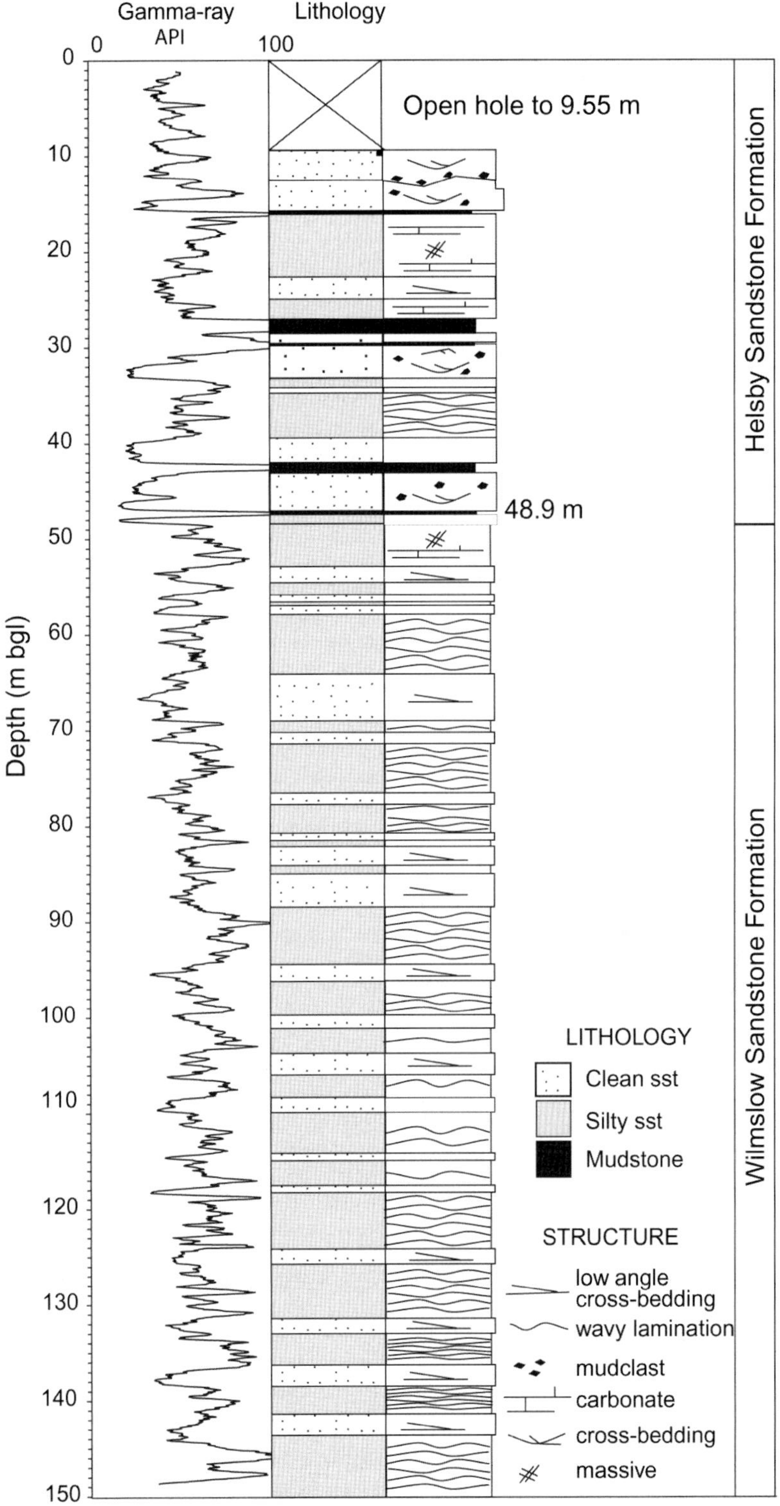
Gamma-ray
API
Lithology
0
100
0
Depth (m bgl)
Open hole to 9.55 m
10
20
30
40
48.9 m
50
60
70
80
90
100
110
120
130
140
150
Helsby Sandstone Formation
Wilmslow Sandstone Formation
LITHOLOGY
Clean sst
Silty sst
Mudstone
STRUCTURE
low angle
cross-bedding
wavy lamination
mudclast
carbonate
cross-bedding
massive

constitute a minor aquifer where they are thick enough and provide an important means of recharge to local wetlands, either directly or by providing continuity with surface waters. In the Cheshire region, the Helsby Sandstone Formation typically has a porosity of 20–33% and the Wilmslow Sandstone Formation has a porosity of 6–35% (mean 24%) (Allen *et al.* 1997). The high porosity and absence of sedimentary compaction are evidence that much of the primary cementation that existed initially has now been dissolved. Storage coefficients for the Permo-Triassic sandstones (undifferentiated) in the Cheshire region range from 1.2×10^{-5} (confined) to more than 0.09 (unconfined), with an approximately log-normal distribution having a geometric mean of 1.1×10^{-3} (Allen *et al.* 1997).

Transmissivities in the Sherwood Sandstone vary greatly and depend on the number, size and interconnectivity of fractures that are intersected. The recorded range of pumping-test determined transmissivities is from 0.9 to 4900 m^2 day^{-1} in the Sherwood Sandstone of the Cheshire region (Allen *et al.* 1997), with a geometric mean of 220 m^2 day^{-1} and a median of 250 m^2 day^{-1}. Pumping tests in the west Cheshire region, close to the Abbey Arms Wood borehole, indicate transmissivities ranging from 100 to 500 m^2 day^{-1}. Exceptionally high values of up to 12 000 m^2 day^{-1} have been recorded in the Delamere area with most of the transmissivity attributable to fractured zones. However, because the Wilmslow Sandstone is often poorly cemented and soft, the fractures do not always stay open and so do not necessarily contribute to the regional transmissivity.

The regional groundwater flow in the Sherwood Sandstone is towards the north and NW, although this is distorted locally by pumping from public-supply boreholes, in particular the United Utilities' Delamere Pumping Station which is about 0.4 km SW of the Abbey Arms Wood borehole. Groundwater residence times in the unconfined Sherwood Sandstone in other parts of the UK have been shown to be relatively short, as shown by modern ^{14}C dates and the presence of tritium and fertilizer nitrate (Bath *et al.* 1979; Foster *et al.* 1986). However, chemical evidence (Tellam 1994) and modelling (Furlong *et al.* 2000) from north Cheshire–south Lancashire and elsewhere shows that pre-industrial waters can exist at depth and in regions covered, even if not confined, by clay till.

Annual fluctuations in water levels are small, being often less than 4 m in the unconfined aquifer. In many parts of the Cheshire Basin, groundwater levels are rising due to reduced groundwater abstraction.

Core drilling and sampling

The borehole was constructed during February and March 2002 by rotary air-flush drilling. Steel casing (200 mm diameter) was installed and grouted to 9.55 m depth. The Helsby Sandstone was found at 6.35 m depth beneath Quaternary sand and gravel deposits.

Below 9.55 m, the sandstone was cored using a double-core barrel with rigid plastic core-lining tubes. This produced core of approximately 105 mm diameter. As each 3 m core run was completed, the core and liner were removed from the barrel and cut in half to give two 1.5 m sections. These were sealed at both ends with plastic caps and tape to reduce moisture loss. Cores were then boxed and transported to the BGS laboratory at Wallingford. Once in the laboratory, the plastic liner was opened with two longitudinal cuts and a lithological log of the core was prepared. Subsamples of core material about 10 cm long were then selected at approximately 2–3 m intervals. The outer 1 cm of material was discarded, the material sealed in a plastic bag and then mixed thoroughly.

Pore water was extracted from the core material by high-speed centrifugation using a Beckman J2–21C centrifuge. A set of six custom-made plastic liners with core inside was placed in a JA14 rotor and centrifuged at 14 000 rpm for 30 min to extract the water (Kinniburgh & Miles 1983). These liners were sealed at both ends with O-rings to minimize evaporation and degassing, although the liners were not designed to avoid access to the atmosphere completely and so some oxidation would have been possible during the preparation. The time delay between the coring and pore-water extraction was kept as short as possible, normally within 1 day and always within 1 week. The core was kept in an unheated store (5–10 °C) during this period.

The pore water was filtered through a 0.45 µm-membrane filter (Whatman) and split into subsamples for chemical analysis. The upper horizons of the unsaturated zone were quite dry and required repeated centrifugation to obtain sufficient pore water for analysis (15–20 ml). The gravimetric moisture content was determined by drying a representative subsample.

Following completion and flushing of the borehole, depth samples were collected from the unpumped water column at 65, 96, 110 and 140 m bgl (metres below ground level). The choice of depths is discussed below. Three aliquots were collected for each sample: one for cation analysis (acidified with concentrated, high-purity acid to 1% v/v HNO_3 final concentration), one for anions (no acidification) and a third for arsenic analysis (acidified with concentrated acid to 2% v/v HCl final concentration). Redox potential (E_H), pH, electrical conductivity and alkalinity were measured in the field using depth samples collected during logging. Major cations and sulphate were analysed by inductively coupled plasma-atomic emission spectroscopy (ICP-AES), anions by ion chromatography, and As by hydride generation-atomic fluorescence spectrometry (HG-AFS) with or without pre-reduction with KI to give As_T (total As) and As(III), respectively.

The borehole was logged using a variety of geophysical tools. A CCTV video of the borehole clearly showed the position of mudstone beds, bleached zones where iron oxide has been removed, and faults and fractures. A wet mudstone bed was observed in the unsaturated zone at 27 m depth. This bed was also evident from the core logging and gamma-ray log (Fig. 2) and appeared to be responsible for a perched water table. Various faults, fractures and cavities were observed that corresponded well with those recorded in the calliper log (Fig. 3). The greatest of these was a double cavity at 103–105 m bgl. The fluid conductivity and temperature logs in this depth range suggested that this was where a significant outflow of warm, low-conductivity water from deeper in the borehole occurred.

The core material was also logged lithologically and used for detailed physical (Bloomfield *et al.* 2006) and chemical characterization. Selected subsamples of the sediments were mixed thoroughly, air-dried, ground in a Tema mill and then subjected to a nitric–perchloric acid dissolution. The extracts were analysed by ICP-AES for major and minor elements including As. Arsenic was also analysed using HG-AFS using a pre-reduction with KI–ascorbic acid.

Water quality and aquifer characteristics

Regional scale: public and private supplies

The Sherwood Sandstone aquifer, which is the main aquifer in NW England, contains groundwaters that are typically of Ca–Mg–HCO_3 type with relatively high Mg/Ca ratios, indicative of a dolomitic source. Calcite, gypsum and anhydrite cements also contribute to the water quality. Chloride concentrations are typically less than 40 mg l^{-1} but not infrequently exceed those expected from atmospheric inputs, reflecting saline intrusion in coastal areas (in places induced by pumping) and mixing with deeper Na–Cl brines further inland (Tellam 1995). While low-nitrate sources (<2 mg l^{-1} NO_3–N) are common, nitrate concentrations exceeding 11.3 mg l^{-1} NO_3–N are found throughout the region.

Nearly 60% of the 2137 groundwater samples from 672 sources in the Environment Agency (EA) North-West region water-quality database fall below the reporting limit for As (normally 1 µg l^{-1} but sometimes 5 µg l^{-1}) (Fig. 4). These samples were derived from a variety of aquifer units, including the Sherwood Sandstone Formation, but not restricted to it. However, 11.5% of these samples exceeded 10 µg l^{-1}, the UK drinking water limit; 1.6% ($n = 11$) of samples exceeded 50 µg l^{-1} but these were mostly from farm and industrial sources. The three highest concentrations observed in public supply sources based on time-averaged raw water data were 81, 54 and 44 µg l^{-1}, all of these being from the Delamere area. The highest recorded As concentration was 261 µg l^{-1} from a market garden source near Irlam, 10 km west of Manchester. This borehole is 91 m deep and situated on the Wilmslow Sandstone Formation. Several other high-As (>30 µg l^{-1}) groundwater sources can be seen (Fig. 5), including a cluster near Ormskirk, west Lancashire. Most of the northern part of the area has low-As (<5 µg l^{-1}) groundwaters. The absence of groundwater sources to the east of Delamere reflects the lack of a suitable aquifer. A recent compilation of groundwater quality data in the Permo-Triassic sandstones of the Cheshire Basin found that As varied from less than 1 to 57 µg l^{-1}, with a median concentration of 6 µg l^{-1} ($n = 81$) (Griffiths *et al.* 2002).

Local scale: Delamere public supplies

There are nine United Utilities public-supply boreholes within 3 km of the Abbey Arms Wood borehole plus another, at Eaton, which is approximately 6 km to the south (Fig. 5). At this latter source, the aquifer is confined beneath Tarporley Siltstone Formation. Groundwater is abstracted from both the Helsby Sandstone and the underlying Wilmslow Sandstone which are in hydraulic continuity.

United Utilities public-supply well site at Delamere (Delamere PS) is about 0.4 km SW of the project borehole and Organsdale PS is about 1 km to the east. The N–S-trending East Delamere Fault lies about 0.5 km east of the

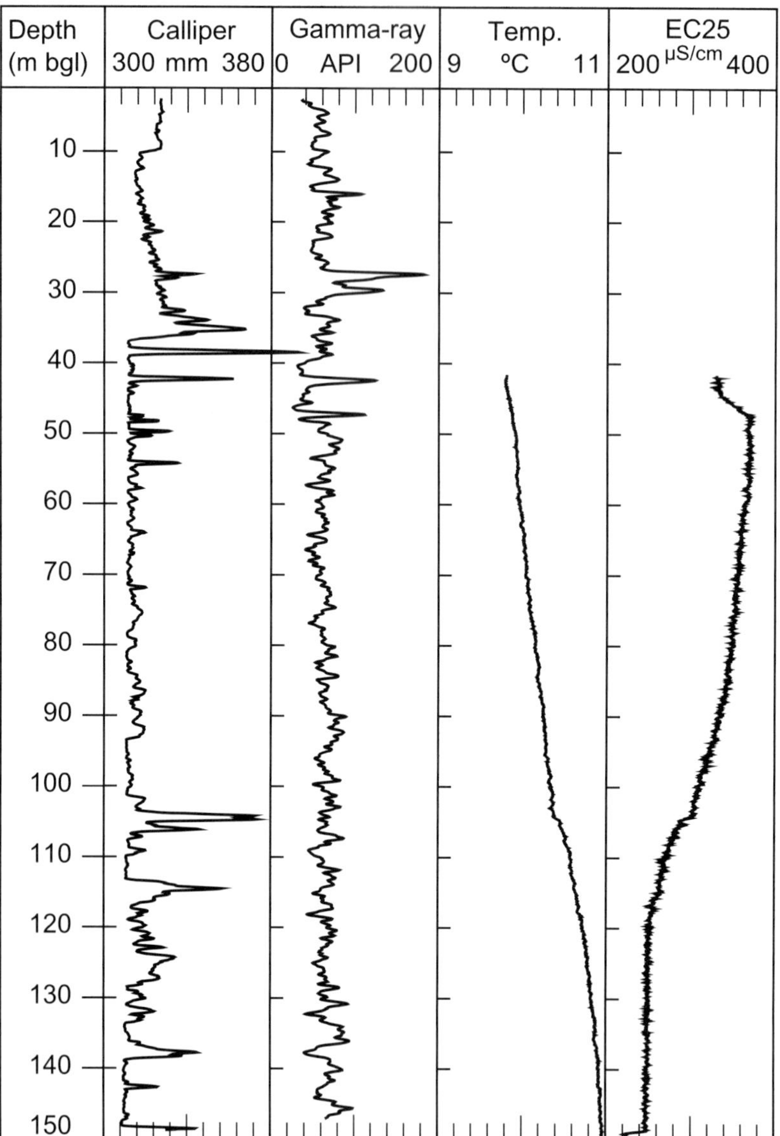

Fig. 3. Calliper, gamma-ray, temperature and specific electrical conductivity logs for the Abbey Arms Wood borehole.

Abbey Arms Wood borehole, and the Overton and Primrose Hill faults are within 2 km. The geological map shows that these boreholes are completed in the Helsby Sandstone Formation, although, as the drilling at Abbey Arms Wood showed, this can be overlain by unconsolidated sand and gravel.

In 1970 the average water level in the aquifer in this area was approximately 49 m AOD (metres above Ordnance Datum). During the late 1980s, average water levels were recorded at approximately 34 m AOD, a drop of 15 m, indicating that in this part of the aquifer system abstraction was exceeding recharge. Water levels in the early 1990s were at a minimum, but since 1994 have generally increased across the aquifer and are now similar to their 1970 levels. This regional rise in water level reflects the fact that recharge is now likely to be exceeding abstraction.

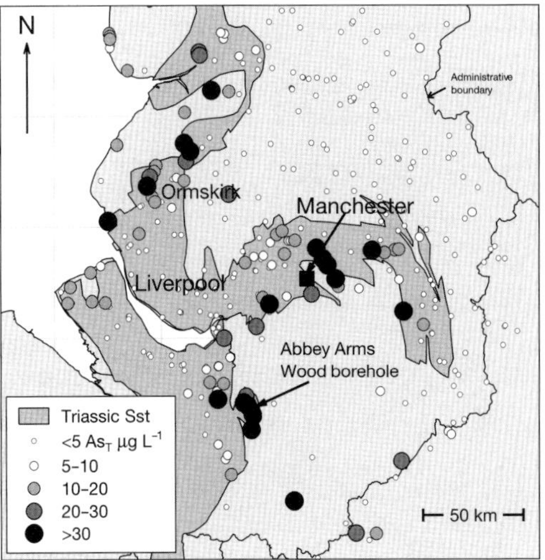

Fig. 4. Distribution of arsenic in groundwater from the EA North-West Region. The outcrop area of the Triassic Sandstone (ignoring Quaternary deposits) is shown (Environment Agency WIMS data).

A summary of the raw groundwater quality in the water-supply boreholes closest to Abbey Arms Wood is given in Table 1. The waters mostly have a near-neutral pH and are oxidizing, despite their considerable depth. This is indicated by their high E_H, high concentrations of DO (dissolved oxygen) and SO_4, and low concentrations of NH_4, Fe and Mn. The Delamere area is part of a Nitrate Vulnerable Zone (NVZ). Nitrate concentrations are mostly high (7–18 mg l⁻¹ NO_3–N) apart from at Eaton 2 (NO_3–N 0.47 mg l⁻¹).

Eaton 2 borehole has a number of distinctive features besides its low nitrate concentration, as it also has low concentrations of Cl (10.9 mg l⁻¹) and Br (40 µg l⁻¹), and is the source with the greatest As concentration (53 µg l⁻¹). It is the only one of the boreholes in which the Tarporley Siltstone Formation overlies the Helsby and Wilmslow formations. The Tarporley Siltstone Formation consists of marl with occasional sandstone bands and at Eaton 2 is 35.4 m deep. The marls may act as a partially confining layer and, judging from the low nitrate concentrations, appear to have protected the underlying groundwater from modern recharge. Eaton 2 also has a lower DO (more reducing) and slightly higher temperature than the other waters, which is consistent with a less active groundwater flow regime. While the chemistry is indicative of a relatively old groundwater, the $\delta^{18}O$ value (–7.7‰) (Table 1) is comparable to

that of modern recharge, precluding a major contribution from more isotopically depleted Pleistocene recharge (Smedley & Edmunds 2002). The isotopic data suggest that the groundwater from the borehole is of Holocene or younger age.

A second borehole at Eaton (Eaton 1, data not shown) sited within a few metres of Eaton 2 has a lower As concentration (29 µg l⁻¹) indicating some local variation. Eaton 1 also has a higher NO_3–N concentration (2.3 mg l⁻¹) and the concentration of most major elements is approximately 20% lower than Eaton 2. The two Eaton boreholes are overlain by a similar thickness of Tarporley Siltstone and are of a similar depth (243 m). It appears that Eaton 2 groundwater has been somewhat better protected from modern infiltration and recent groundwater flushing than Eaton 1. This may be one explanation for the difference in As concentrations.

Arsenic concentrations exceed 10 µg l⁻¹ in the raw water from all of the boreholes in the immediate Delamere area. There is no clear relationship between As concentration and borehole depth or borehole temperature (Table 2) for these boreholes. It appears that, if anything, there is a trend of increasing As concentration with borehole depth, although Eaton 2 has a large influence on this relationship and, as indicated above, this may reflect the influence of the overlying Tarporley Siltstone rather than simply reflecting a depth relationship. In order to comply with the new arsenic limit of 10 µg l⁻¹, United Utilities have invested significantly in water-treatment plants for As removal in the Delamere area.

When the As concentration is plotted against the pH for all of the sampled Sherwood Sandstone groundwaters in the west Cheshire–Wirral area (Fig. 6), there is a suggestion of increasing As concentration with increasing pH albeit with the highest pH only being about pH 8. This is significant in that As(V) desorption increases with increasing pH (Smedley & Kinniburgh 2002). However, Triassic Sandstone groundwaters in the pH range 7.5–8.0 are not unusual in themselves (Smedley & Edmunds 2002) and so other factors must also be involved.

Borehole scale: depth samples and pore water

The variation of gravimetric moisture content with depth in the Abbey Arms Wood borehole is shown in Figure 7. A perched water table is evident at about 27 m and is associated with a

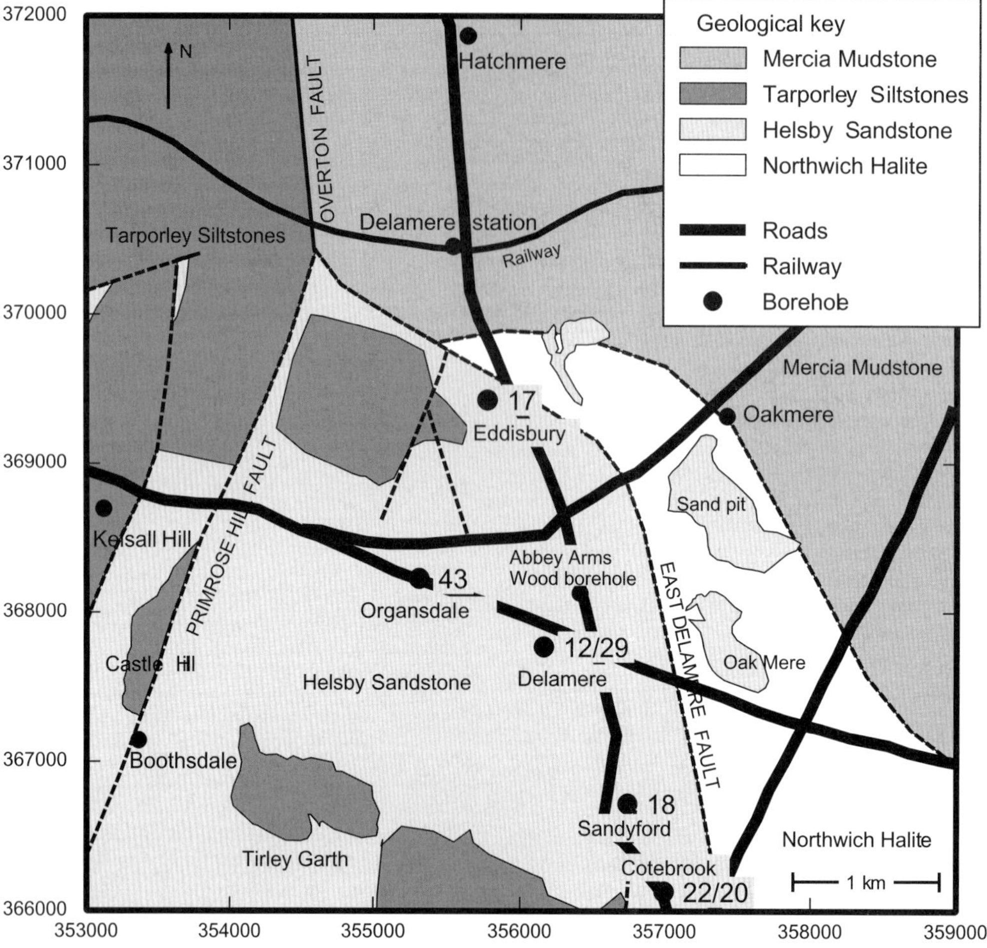

Fig. 5. Location of public-supply boreholes in the Delamere area, with the average arsenic concentrations shown (in $\mu g\ l^{-1}$) in the untreated groundwater (from EA data). Where two numbers are shown for a site, these are for separate boreholes. Eaton is about 6 km to the south of the Delamere pumping station.

fine-grained horizon. There is also a relatively high moisture content in the Wilmslow Sandstone compared with the more-cemented Helsby Sandstone. Most of the variation in the Wilmslow Sandstone reflects variations in the degree of cementation, with major changes particularly below 125 m. The moisture content increases from $20\ g\ kg^{-1}$ at 10 m to about $60\ g\ kg^{-1}$ at the water table, and has an average of about $80\ g\ kg^{-1}$ in the saturated zone. Given an average crystal density equivalent to that of quartz ($2650\ kg\ m^{-3}$), this corresponds to a moisture-filled porosity of 21%. This is in good agreement with independent laboratory measurements of porosity (Bloomfield *et al.* 2006). A steady increase in temperature with depth would normally be expected, and the

inflow and suspected downward movement of water to 105 m bgl has caused the temperature to remain relatively cool above the exit point. However, the temperature is not completely invariant with depth and there may be some outflow at other horizons or via the intergranular matrix. One possible outflow occurs at about 80 m bgl. There was also a difference in the fluid logs run on consecutive days (2 and 3 April 2002), which may reflect a change in pumping regime at the nearby Delamere pumping station borehole.

The locations of the depth samples at 65, 96, 110 and 140 m bgl were based on observed zones of inflow. The results of the field measurements and chemical analyses are given in Table 3 and show a decline in specific electrical

Table 1. *Water chemistry from public water-supply boreholes in the Delamere area prior to treatment*

Source and number	Cotebrook 40″	Cotebrook 15″	Delamere BH 3	Delamere BH 4	Eaton BH 2	Eddisbury SJ56/2	Organsdale SJ56/49	Sandyford
Laboratory number	S00-00972	S00-00971	S00-00969	S00-00968	S00-00973	88006805	88006811	S00-00970
Depth of borehole (m)	244	126	91	243	243	229	123	57
Easting	357150	357000	356150	356020	356690	355775	355310	356740
Northing	365760	366110	367780	367720	363410	369425	368230	366720
Date sampled	24-Oct-00	24-Oct-00	24-Oct-00	24-Oct-00	24-Oct-00	12-Jan-00	12-Jan-00	24-Oct-00
Temperature (°C)	10.9	10.6	10.1	10.1	12.1	9.4	10.1	10.8
pH	7.51	7.53	7.58	7.58	7.8	7.81	7.63	7.35
E_H (mV)	521	428	362	443	521			417
DO	6	8.6	10.9	7.7	1.7	8.2	8.8	7.9
SEC (μS cm^{-1})	582	546	554	484	331	515	584	736
Ca	75.7	74.3	82.5	72.4	37.2	54.5	89.2	94.3
Mg	18.9	19.8	6.16	5.43	16.6	23.6	8.47	14.7
Na	20.7	14.6	19.2	14.1	8.4	12.3	15.9	28.5
K	3.9	2.2	2	1.6	2.5	2.45	1.99	4.4
Cl	35.2	30	37.4	29.4	10.9	26.8	41.4	47.2
SO$_4$	41.6	34.9	30.8	21.7	24.7	30.3	25.2	37.2
Alk as HCO$_3$	201	227	157	158	153			218
NO$_3$ as N	8.36	7.13	14.5	10.2	0.47	15.1	17.5	17.7
Si	5.59	5.45	5	5.09	7.15	9.51	9.9	5.27
NO$_2$ as N	<0.001	<0.001	0.024	0.012	<0.001	<0.004	<0.004	0.008
NH$_4$ as N	0.005	0.008	<0.003	<0.004	0.004	<0.5	<0.5	<0.003
P	0.04	0.05	0.13	0.10	<0.02	<0.5	<0.5	0.10
TOC	1.4	2.3	3.5	1.7	1.1			2.9
DOC	2.9	4.3	1.9	1.7	2.5			4.9
F (μg l^{-1})	110	90	90	100	140	61	55	130
Br (μg l^{-1})	80	80	110	70	40	<50	83	100
I (μg l^{-1})	4	4	4	3	3			6
Al (μg l^{-1})	6	1	27	2	2		<10	4
As (μg l^{-1})	26	23	15	30	53	14	28	19
Ba (μg l^{-1})	122	135	499	624	125			299
Cu (μg l^{-1})	7.6	6.3	7.4	10.1	2.7	111	14.8	14.5
Fe (μg l^{-1})	<5	<5	7	<5	<5	<30	<30	<5
Li (μg l^{-1})	8	5	5	5	12			5
Mn (μg l^{-1})	<2	<2	<2	<2	<2	<10	<10	<2
Mo (μg l^{-1})	0.5	0.2	<.1	0.1	3			<0.1
Rb (μg l^{-1})	3.2	2.3	2.0	1.8	2.1			3.6
Sb (μg l^{-1})	0.2	0.15	0.13	0.17	0.1			0.1
Sc (μg l^{-1})	1.66	1.65	1.39	1.5	2.16			1.71
Se (μg l^{-1})	0.7	<.5	<0.5	<0.5	<0.5			<5
U (μg l^{-1})	2.5	1.5	0.2	0.9	7.5			0.5
V (μg l^{-1})	<1	<1	<1	<1	<1			<1
Zn (μg l^{-1})	11	15	11	8	8	150	12	10
δ^{13}C (‰)	−15.48	−15.82	−15.03	−15.52	−15.43			−15.93
δ^{18}O (‰)	−8.1	−7.4	−8.4	−8.6	−7.7			−7.6
δ^2H (‰)	−49	−53	−52	−55	−51			−53

* All data are in units of mg l^{-1} unless otherwise indicated. All data are for water prior to treatment. SEC, specific electrical conductivity (25 °C).

conductivity with depth, at least down to 110 m. This is similar to the trend shown by the electrical conductivity log (Fig. 3). E_H, Ca, Mg, Cl, NO$_3$–N, SO$_4$, alkalinity, As and Zn also decrease with depth. Only Mg increases with depth, possibly reflecting increasing dolomite dissolution. The 110 and 140 m samples are similar in terms of major-element chemistry and are consistent with a possible upwards movement of water to the large fracture at 105 m. The composition of the 96 m sample is substantially different and supports a discontinuity in flow between 96 and 110 m. The concentrations of As found in the depth samples are within the range found in public-supply boreholes in the area. About two-thirds of the arsenic was present as As(III) species (Table 3).

The measured pH values of the pore waters after centrifuging were in the range 6.92–8.25, but mostly in the range pH 7.8–8.2. The depth samples (pH 7.82–7.98) also lie within this narrower pH range (Table 3). Calcite saturation indices (not shown) for the pore waters increase gradually from –0.3 at 11 m depth to +0.2 at 50 m depth, and then remain approximately constant at this value to 150 m depth. This indicates possible dissolution of calcite in the Helsby Sandstone followed by a slight

Table 2. *Average arsenic concentrations in public-supply sources in the Delamere area and the depth of the boreholes*

Source	Depth (m)	Temperature (°C)	Average As (μg l^{-1})
Delamere 3	91	10.1	12
Delamere 4	243	10.1	29
Eaton 1	244	–	29
Eaton 2	243	12.1	51
Eddisbury	229	9.4	17
Cotebrook (15″)	126	10.6	22
Cotebrook (40″)	244	10.9	20
Organsdale	123	10.1	43
Sandyford	60	10.8	18

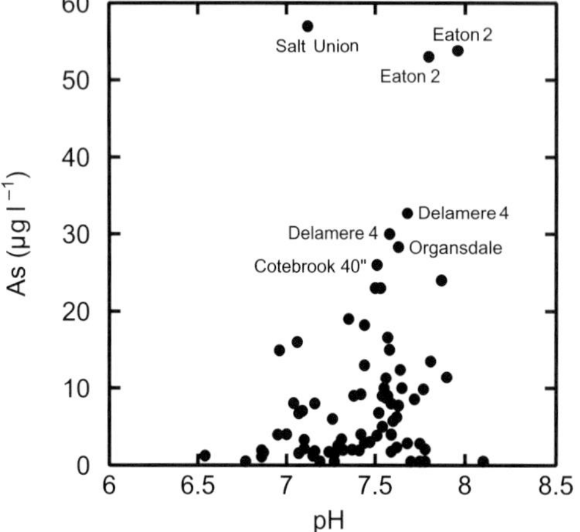

Fig. 6. Relationship between pH and total arsenic concentration in Sherwood Sandstone groundwaters from the west Cheshire and Wirral areas (from Griffiths *et al.* 2002).

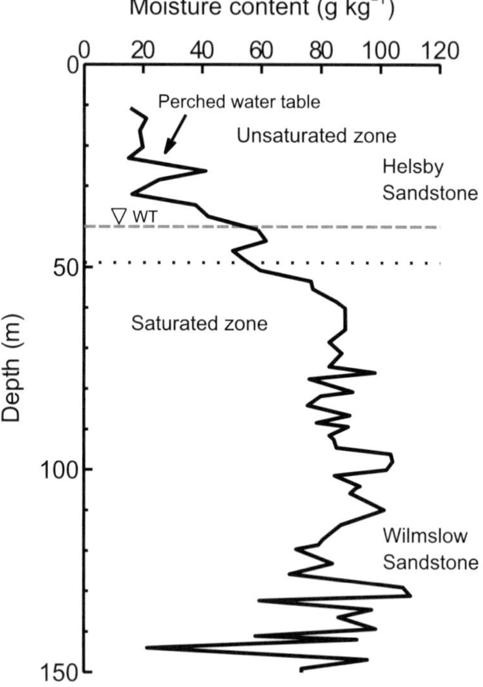

Fig. 7. Moisture content profile of the Abbey Arms Wood borehole.

supersaturation throughout the Wilmslow Sandstone. There is the possibility of some CO_2 degassing during pore-water extraction that would give artificially high pHs. However, all the indications are that the groundwater pH is unlikely to be much greater than pH 8.2 throughout the borehole depth. The depth variations of As, SO_4, Ca, Mg and NO_3–N are shown in Fig. 8.

The total As (As_T) pore-water profile shows an increasing trend in concentration with increasing depth, with three significant 'spikes', the greatest being 125 μg l^{-1} at 89.4 m. The other spikes were at 96.2 (89 μg l^{-1}) and 118.5 m (47 μg l^{-1}). These As spikes may be spurious because they were found in isolated samples and were unrelated to any visible change in sediment features, for example mineralization. Therefore, the original water samples were reanalysed. This showed that all three of the

total As concentrations had increased significantly with time (125–164, 89–100 and 47–107 μg l^{-1}, all over 203 days of storage). The most likely explanation is that a small amount of As(V)-rich colloid (<0.2 μm) was present in the sample and had slowly dissolved in the acidic solution in which the As samples were preserved. It is likely that the spikes are anomalous and that the background increase with depth observed represents the true As depth profile. This shows a steadily increasing

Table 3. *Chemical analysis of depth samples*

Parameter	Units	65 m	96 m	110 m	140 m
Sample number		S02-00777	S02-00778	S02-00779	S02-00780
pH		7.82	7.87	7.95	7.98
E_H	mV	388	403	174	159
SEC	μS cm^{-1}	425	357	270	280
Mg	mg l^{-1}	6.7	7.3	8.5	8.9
Ca	mg l^{-1}	63.6	52.3	34.5	34.0
Na	mg l^{-1}	14.7	11.9	9.0	9.0
K	mg l^{-1}	1.77	1.70	1.61	1.62
Cl	mg l^{-1}	38.3	29.1	16.5	16.1
NO_3–N	mg l^{-1}	7.0	5.8	3.8	3.7
SO_4	mg l^{-1}	15.7	11.8	7.8	7.7
Alkalinity as HCO_3	mg l^{-1}	141	135	119	111
Si	mg l^{-1}	5.3	5.3	5.4	5.4
As_T	μg l^{-1}	10.6	17.7	35.8	35.6
As(III)	μg l^{-1}	6.7	11.4	22.8	24.3
As(V)*	μg l^{-1}	3.9	6.3	13.0	11.2
Ba	μg l^{-1}	522	492	427	439
Cu	μg l^{-1}	1.2	1.8	0.4	1.3
Fe	μg l^{-1}	0.2	1.2	0.3	1.0
Li	μg l^{-1}	4.3	3.8	3.4	3.3
Mn	μg l^{-1}	1.4	0.4	<0.2	<0.2
Mo	μg l^{-1}	2.8	3.5	2.9	2.9
NO_2–N	μg l^{-1}	<3	<3	<3	<3
P	μg l^{-1}	82	78	66	59
Sr	μg l^{-1}	63	59	46	46
Zn	μg l^{-1}	67	135	27	14

* By difference: As_T – As(III).

concentration with depth from about 8 μg l^{-1} at the water table (40 m) to about 25–30 μg l^{-1} at 150 m. These results are broadly consistent with those from the depth samples, although the 110 m depth sample has a relatively high concentration (Fig. 8).

As(III) was not measured in all of the pore-water samples, but where measurements were made, this constituted the dominant species in those samples with background As_T concentrations. In contrast, As(V) (measured by difference) dominated the speciation in the two anomalous spikes. The presence of As(III) is unexpected given the oxidizing nature of the groundwaters and so needs further confirmation. The measured Fe concentration in pore-water samples from throughout the 150 m borehole was usually less than 5 μg l^{-1}.

Sulphate concentrations are low but variable in the unsaturated Helsby Sandstone pore waters. Some concentrations exceed 20 mg l^{-1}, although these do not coincide with the As spikes. The larger concentrations in the shallower parts of the profile could reflect a contribution from anthropogenic, fertilizer or drift (Quaternary deposit) sources. In the Wilmslow Sandstone there is a steady decline in sulphate concentration down to about 90 m and then a near-constant concentration of about 10 mg l^{-1}. There was a very strong correlation between SO_4 and Ba in the sediments (see below). There is no sign of the high concentrations of sulphate characteristic of extensive pyrite oxidation.

Calcium shows a steady decline in the pore waters from the Helsby Sandstone, a small increase between 50 and 70 m in the Wilmslow Sandstone, and then a steady decline between 70 and 100 m, after which it is constant at about 40 mg l^{-1}. Magnesium, on the other hand, has a low concentration in the Helsby Sandstone pore waters and steadily increases with depth in the Wilmslow Sandstone. There is a particularly sharp rise from 4 to 7 mg l^{-1} at 88.4 m, possibly related to the cavity observed at 87.15 m depth. There is also a jump in pore-water Mg concentration from 3 to 7 mg l^{-1} close to the water table. This goes back to 'background' a few metres deeper. The Mg concentrations in the centrifuged pore-water and depth samples are consistent (Fig. 8).

The top 30 m of the unsaturated zone shows high and erratic concentrations of NO_3 ranging from 1 to 17 mg l^{-1} NO_3–N. The upper part of the saturated zone contains approximately 7 mg l^{-1} NO_3–N but there is a tendency for concentrations to decrease with depth. These

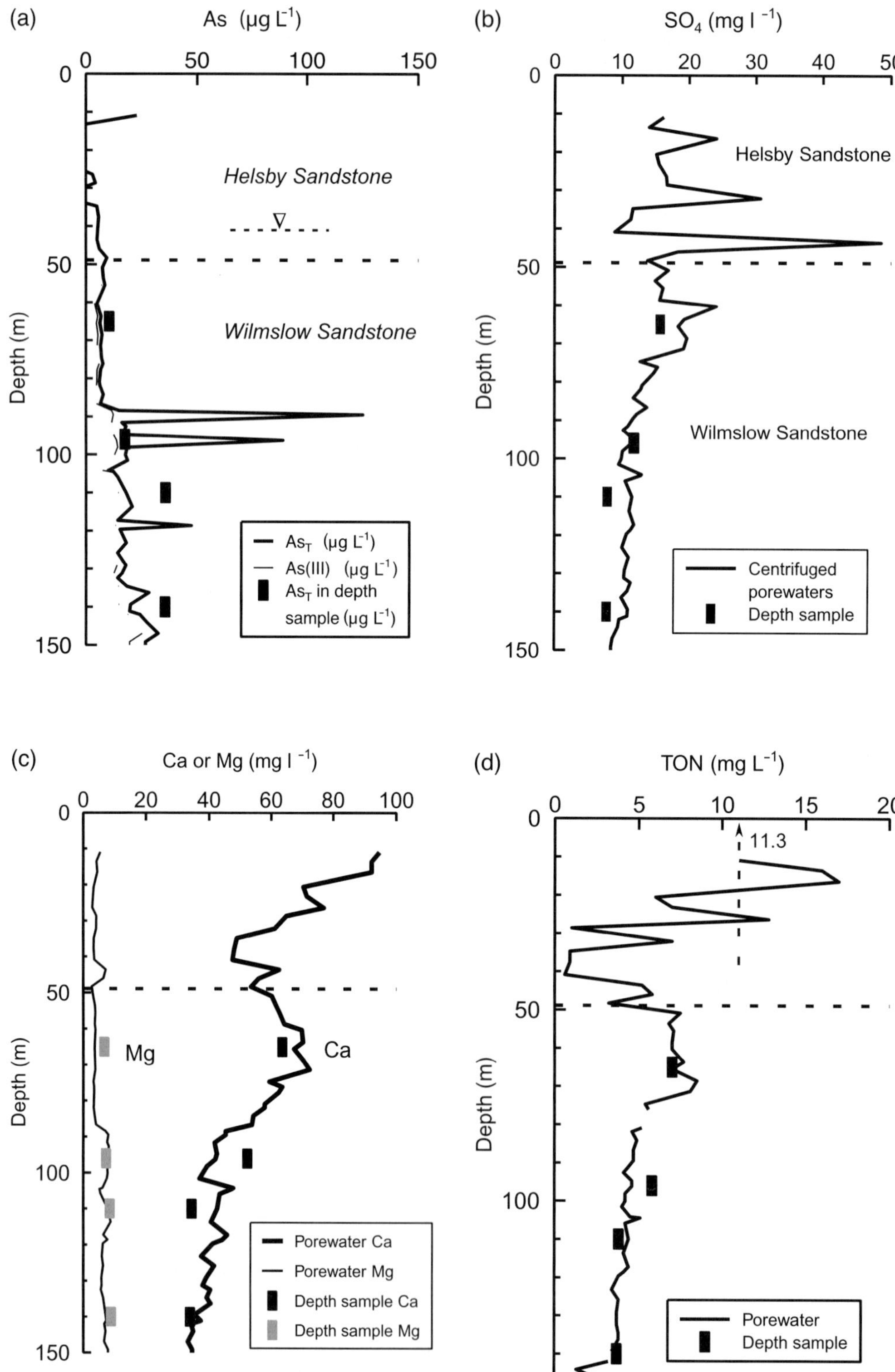
(a)
As (µg L⁻¹)
0
50
100
150
Depth (m)
0
50
100
150
Helsby Sandstone
Wilmslow Sandstone
As_T (µg L⁻¹)
As(III) (µg L⁻¹)
As_T in depth sample (µg L⁻¹)

(b)
SO₄ (mg l⁻¹)
0
10
20
30
40
50
Depth (m)
0
50
100
150
Helsby Sandstone
Wilmslow Sandstone
Centrifuged porewaters
Depth sample

(c)
Ca or Mg (mg l⁻¹)
0
20
40
60
80
100
Depth (m)
0
50
100
150
Mg
Ca
Porewater Ca
Porewater Mg
Depth sample Ca
Depth sample Mg

(d)
TON (mg L⁻¹)
0
5
10
15
20
Depth (m)
0
50
100
150
11.3
Porewater
Depth sample

data point to variable inputs of nitrogen. Given that the site is forested, the variation could be related to atmospheric variations in input (particularly the dry deposition of ammonium species), but it could also reflect temporal variations in the amount of soil N mineralized or soil N taken up by the trees and other vegetation. Forests are known to be more spatially variable than arable land in terms of nitrate leaching (Kinniburgh & Trafford 1996). There is no evidence for extensive denitrification in the profile. The environment is not sufficiently reducing. The trees could also scavenge sulphate aerosols and may explain the relatively high SO_4 concentrations seen in parts of the unsaturated zone.

The NO_3–N concentration at depth (greater than 100 m) is about 3–4 mg l^{-1}, which probably reflects inputs a century or more ago. These concentrations are considerably lower than currently pumped from nearby boreholes from an equivalent depth (e.g. Delamere 3 contains 14.5 mg l^{-1} NO_3–N), which points to the possibility of some drawdown of shallow, nitrate-rich groundwater at the pumping stations. Although woodland is often regarded as benign as far as nitrate leaching is concerned, the high concentrations of nitrate presently in the upper unsaturated zone, often exceeding the current drinking-water limit, indicate that 'nitrogen saturation' may currently exist within the forest. Again, the nitrate concentrations in the depth samples are consistent with those of the pore waters.

Petrography and mineralogy

Twelve samples of sandstone were analysed in detail. The samples represent material from the Helsby and Wilmslow sandstone formations and were selected to compare horizons in which high pore-water arsenic concentrations had been identified with horizons containing background pore-water As concentrations. Three of the sandstone samples, H675P1, H676P1 and H677P1, were from the horizons that produced pore-water As spikes. Samples H682P1, H683P1 and H684P1 were taken from horizons immediately adjacent to these pore-water As spikes. The remaining six samples were selected to represent horizons corresponding to 'background' pore-water As concentrations.

Hydrothermal and diagenetic As mineraliza-tion is known to occur in the Sherwood Sandstone Group and overlying Tarporley Siltstone Formation (Mercia Mudstone Group) in the Cheshire Basin (Carlon 1981; Plant *et al.* 1999). It is principally hosted in the upper parts of the Sherwood Sandstone Group, within the Helsby and Wilmslow Sandstone formations. The mineralization is closely associated with Cu, Fe, Pb, Zn, Co, Ni, Bi, Ag, Mo, Hg and Se sulphides, and barite. The mineralization is most extensively developed near Alderley Edge (Cheshire) and Clive (Shropshire) where it was formerly exploited economically.

Hydrothermal mineralization within the Sherwood Sandstone is also known from the nearby areas of Bickerton Hill (Carlon 1981) and Peckforton Hill in Cheshire. Most of the major primary sulphide mineralization in these areas has been altered to a complex secondary assemblage of Cu, Pb, Co and Zn carbonate, silicate, phosphate, oxide and oxyhydroxide phases associated with iron and manganese. Similar mineralization might be anticipated in the Abbey Arms Wood borehole. The objective of the petrographic analysis was therefore to examine the samples for any evidence of mineralization or for any other clues to the source(s) of the As in the groundwater.

The 12 sandstones were mineralogically very similar with respect to their detrital components. They are all subfeldspathic arenites–sublitharenites. Monocrystalline quartz is the major detrital component, with subordinate–minor amounts of polycrystalline quartz, lithic grains (which may include fine quartzite, schistose grains, fine siliceous volcanic grains, siltstone and fine sandstone), chert and K-feldspar (orthoclase and microcline). Sodic plagioclase is present in most samples as a very minor–trace component. Other accessory detrital minerals include very minor muscovite, ilmenite and magnetite, and traces of rutile, tourmaline, apatite, zircon and monazite.

All of the sandstones show the presence of thin coatings or pellicles of very fine illitic clay on the surfaces of detrital grains. These are impregnated with very-fine-grained disseminated hematite or other iron oxide. The iron oxide is responsible for the red colour of the rocks. The hematite is absent where reduction by later diagenetic mineralizing fluids has produced late calcite-cemented bleached bands (e.g. sample H680P1). Calcite is present as an

---

Fig. 8. Arsenic, sulphate, calcium, magnesium and total oxidized nitrogen (nitrate) concentrations versus depth in the porewater extracted from the project borehole and in four depth samples. The dashed line marks the boundary between the Helsby and underlying Wilmslow sandstone formations.

important cement in some of the sandstones but is absent from others (see more detailed sample descriptions given below). The calcite forms spherulitic micronodular concretions, which are often finely banded with thin layers of clay and hematite. These nodules may coalesce to form irregular cemented patches. The calcite cement preserves an 'expanded' or uncompacted grain fabric, and therefore predates sediment compaction. It is similar to early diagenetic nodular non-ferroan calcite and dolomite cements described elsewhere from the Sherwood Sandstone in the Cheshire Basin and other Permo-Triassic basins in the United Kingdom (see Plant *et al.* 1999 and references therein), and is interpreted as calcrete.

Minor weakly ferroan calcite cement also occurs as later diagenetic overgrowths seeded on the earlier calcite cement in some of the sandstones. The scanning electron microscopy (SEM) analysis of the heavily cemented 143.98–144.07 m sample showed that the calcite cement completely filled the pore spaces between the grains. In places, the calcite cement shows signs of corrosion and dissolution. Traces of authigenic quartz and K-feldspar are present locally as weak overgrowths on some detrital grains. Detrital iron and titanium oxides are sometimes partially altered to fine-grained authigenic titanium oxides. The Wilmslow Sandstone samples showed a greater amount of clay/oxide infilling material and a lower porosity than the Helsby Sandstone samples. The titanium oxide occasionally crystallized into coarser box-like crystals, which may be anatase or brookite.

In addition to secondary porosity formation caused by dissolution of calcite cement, there was also evidence for the dissolution of detrital plagioclase, K-feldspar and lithic grains (feldspathic fragments and chert grains). This secondary porosity contributed significantly to the overall porosity of some of the sandstones. No discrete arsenic minerals, sulphide minerals, or any other signs of hydrothermal or diagenetic mineralization were observed in any of the sandstones. In addition, arsenic could not be detected by SEM–EDX (energy dispersive X-ray) analysis in any of the minerals observed, either in randomly selected discrete point analysis of mineral grains, clay and iron oxide grain coatings, or cements, or by the more systematic searching employed using EDX analysis X-ray mapping techniques (Fig. 9). The EDXA detection limit for arsenic is of the order of 0.5 weight per cent As.

Calcite was detected in four of the 10 samples investigated. Three of these were from the Helsby Sandstone and contained about 5% by wt calcite by XRD analysis. Only the 143.98–144.07 m sample from the Wilmslow Sandstone contained detectable calcite and this sample was found to contain 26% calcite. All the samples contained small amounts of feldspar. XRD analysis of the clay fraction indicated that the clays were predominantly illite and expanding clays, possibly a randomly layered illite–smectite mixture (Plant *et al.* 1999), with a minor amount of chlorite. The samples only contained trace quantities (about 0.1%) of organic carbon.

The cation-exchange capacity (CEC) of the analysed samples was uniformly low from 2 to 13 meq kg^{-1} (Bellis 2002). The lowest CEC was in the heavily cemented 143.98–144.07 m sample. The variation in the CEC of the remaining samples was not obviously correlated with any other measured sample property.

Sediment chemistry and mineralogy

Measurement of As by direct aspiration using the multi-element ICP-AES method gave similar results to the more sensitive HG-AFS technique (Fig. 10). The results showed that the As content of the analysed sediments was 5–15 mg kg^{-1}. This is typical of sandstones (Smedley & Kinniburgh 2002) including UK red-bed sandstones (Haslam & Sandon 1991). There is no obvious trend of As concentration in the sediments with depth nor a clear difference in concentrations between the Helsby and Wilmslow sandstones (Fig. 10). The mean As content was 8 mg kg^{-1} overall, with a slightly lower average concentration in the Wilmslow Sandstone compared with the Helsby Sandstone (7.5 and 8.5 mg kg^{-1}, respectively). Arsenic does not correlate strongly with any of the other analysed elements.

Average concentrations for a number of chemical constituents in the Helsby and Wilmslow formations are given in Table 4, and a selection of sediment chemistry profiles is given in Figures 11 and 12. Some strong positive correlations ($r > 0.9$) exist between the element pairs: Al v. K, Al v. Mg, Al v. Na, Al v. P, K v. Na, K v. P, Mn v. Ca, Mg v. Fe and Ba v. SO$_4$. These are consistent with the overall mineralogy, especially the presence of feldspars and metal-rich coatings. However, the source of many of the minor elements (e.g. P) remains unclear. There was a weak negative correlation between Ca and Mg indicating that the presence of large amounts of dolomite is unlikely. It is also clear from the profiles for the major elements that calcite is probably rare or absent in the upper

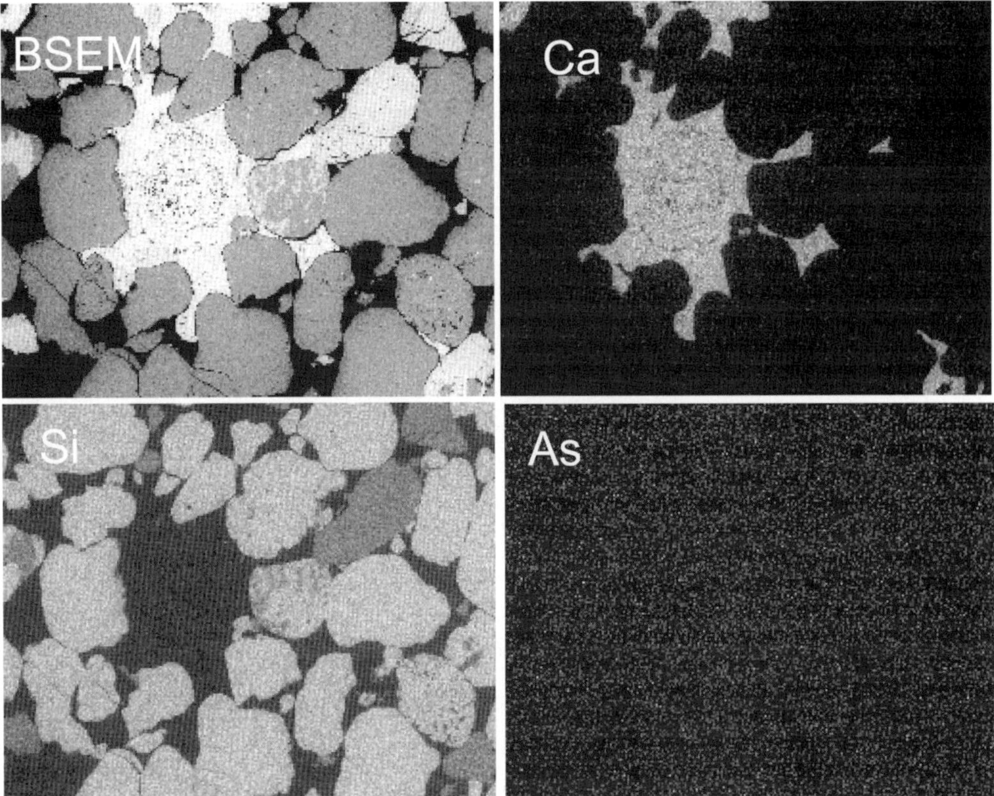

Fig. 9. Example of the EDX analysis X-ray elemental mapping. The figure shows (from top-left to bottom-right): back-scattered image showing bright nodular calcrete calcite cement (light areas), detrital quartz (dark grey) and K-feldspar (lighter grey) against a dark background. Light areas in individual element maps correspond to high concentrations. The Ca map shows calcite cement; the Si map shows high concentrations corresponding to quartz and lower concentrations corresponding to feldspars and clay coatings on grain surfaces; the As map shows only background noise. All images are for sample H684P1 (118.60–118.62 m). The images are each 2000 μm across.

part of the saturated Wilmslow Sandstone (50–100 m), but is present erratically above and below that. There is significantly more Na and K in the Wilmslow Sandstone than in the Helsby Sandstone, perhaps reflecting a higher concentration of feldspars or clays in the Wilmslow Sandstone. There is considerably more Al in the Wilmslow Sandstone, as well as more Fe and less Mn. The strongest correlation was Ba v. SO_4, which points to the occasional presence of barite in the Helsby Sandstone. The molar ratio in these high Ba–SO_4 horizons is close to unity and is consistent with this observation.

A stepwise multiple regression model was carried out on the sediment data (excluding those trace metals that were sometimes below the detection limit) to see which of the observed variables best predicted the As concentration of the sediment. Overall, the fit was highly significant in a statistical sense. With the seven variables, –depth, +Li, +Mg, –Mn, –Na, –SO_4, +Sr (the sign is the sign of the regression coefficient), it is possible to explain 58% of the variation in As concentrations. The three factors most indicative of a relatively large sediment As concentration were a large Mg concentration, a small Na concentration and a shallow depth. Iron was not selected as a significant factor, even though there was a weak positive correlation between Fe and As.

Note that despite the significant negative correlation with depth, there is not a clear decrease in As with depth. This is presumably due to compensating trends such as that of Mg or SO_4. The positive correlations with Li and Mg may point to a clay control on the sediment As

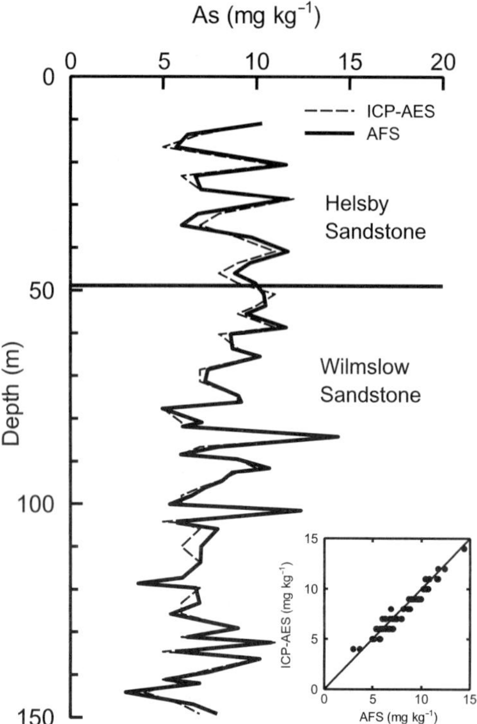

Fig. 10. Variation of arsenic concentration in the sediments with depth. A comparison of the results of the analysis of the sediment extracts measured by both HG-AFS and direct aspiration ICP-AES is also given.

concentration. The negative correlations are harder to explain and may reflect other unexplored factors.

Discussion and conclusions

Source and distribution of arsenic

The Delamere area is known for its occasional high-As groundwaters. Typically, these groundwaters have As concentrations in the range 10–50 $\mu g\ l^{-1}$, which is not high by international standards but relatively high for the UK. Both depth sampling and pore-water sampling in the Abbey Arms Wood borehole indicated that As concentrations increase steadily with depth from about 8 $\mu g\ l^{-1}$ at 10 m to about 30 $\mu g\ l^{-1}$ at 150 m. Arsenic speciation suggested that approximately two-thirds of the As was present as As(III). However, this is not consistent with the generally oxidizing nature of the groundwaters and so should be viewed with caution. Although higher As concentrations (up to

125 $\mu g\ l^{-1}$) were observed in three (out of 60) pore waters, these are believed to be an artefact, possibly due to the slow dissolution of an As(V)-rich colloid.

The trend of increasing groundwater As concentration with depth is the opposite to that observed for most of the major solutes (except Mg), which tend to decrease in concentration with depth. Considering that the location of the borehole was in a forested area, surprisingly high nitrate concentrations (up to 17 mg l^{-1} NO_3–N) were found in the unsaturated zone (top 40 m).

The sediment chemistry is fairly typical of that of red-bed sandstones (Haslam & Sandon 1991). Sediment As concentrations (5–12 mg kg^{-1}) are typical of sandstones elsewhere (Smedley & Kinniburgh 2002) and, if anything, showed a decrease with increasing depth. Erratic and high concentrations of pore-water sulphate in the unsaturated zone of the Helsby Sandstone suggested possible barite mineralization. No sulphide mineralization was observed petrographically in the core or from SEM–EDX analysis examination, nor were any specific As-bearing minerals observed. Possible source minerals are Ti–Fe oxides (magnetite, ilmenite, etc.). EDX analysis detected minor amounts of Cu and, possibly, Zn in some of these grains, and it might be possible that these minerals contained trace amounts of other heavy elements, including As. The early diagenetic fine-grained iron oxide (probably hematite) associated with the ferruginous clay coatings on detrital grains might also be another source of As. Iron oxides have a strong affinity for arsenic, and adsorbed concentrations can exceed 1000 mg kg^{-1} As (Smedley & Kinniburgh 2002). The early diagenetic hematite encountered in red-bed sandstones is considered to be a potential source of metals for red-bed type mineralization similar to that encountered elsewhere in the Cheshire Basin. However, a more detailed characterization of the surface chemistry of these minerals is needed to prove that the iron oxides are a potential source of As. Desorption of As from Fe oxides (hematite) was also suggested as the cause of relatively high As concentrations (up to 14 $\mu g\ l^{-1}$) in the East Midlands Triassic Sandstone aquifer, together with ageing and diagenesis (Smedley & Edmunds 2002).

The sediments involved are much older than those from the high-As areas found in most other parts of the world, which are typically Holocene–Pleistocene sediments with a relatively short history of freshwater flushing. Identifying the source of the arsenic in

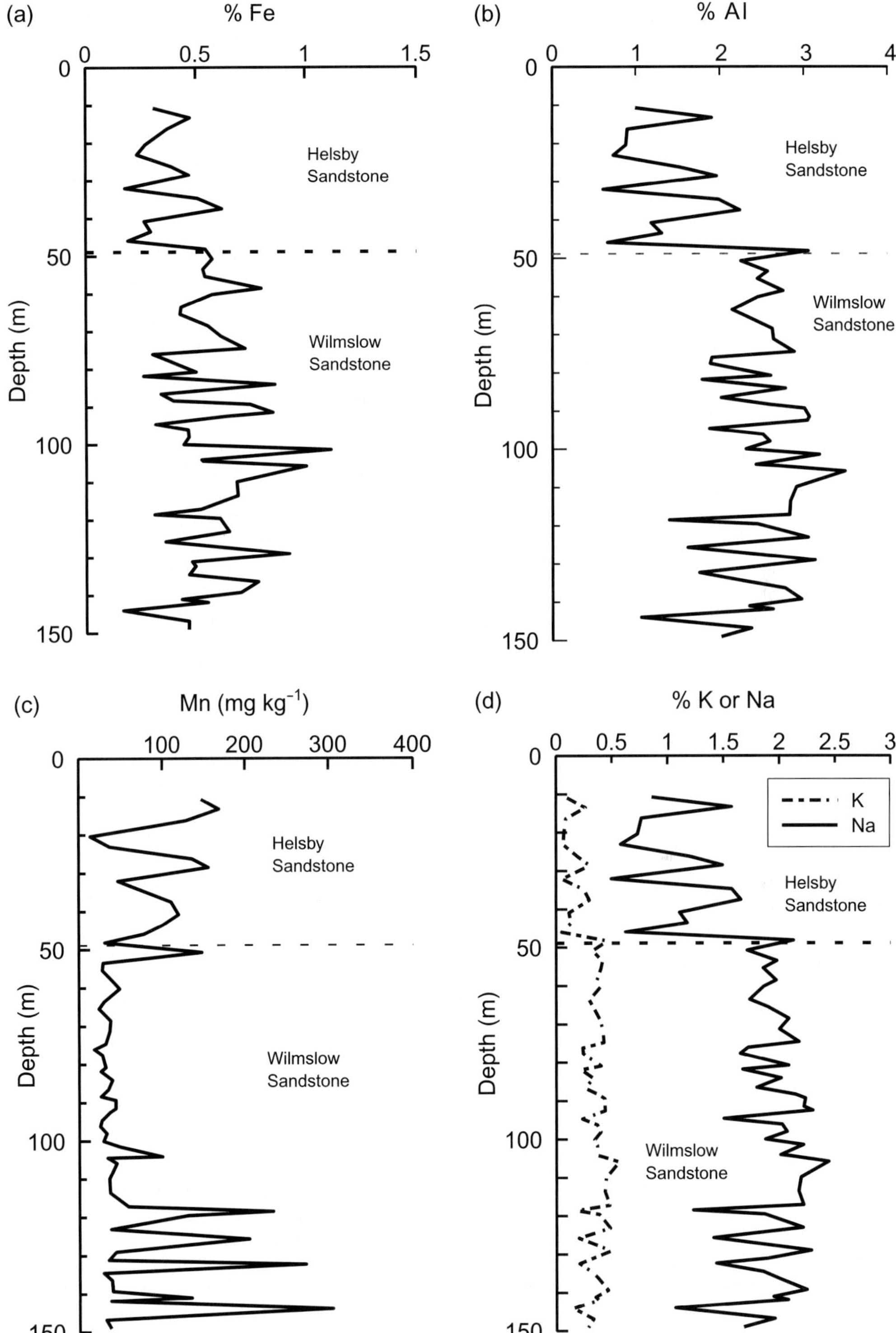

Fig. 11. Sediment chemistry profiles for iron, aluminium, manganese, sodium and potassium.

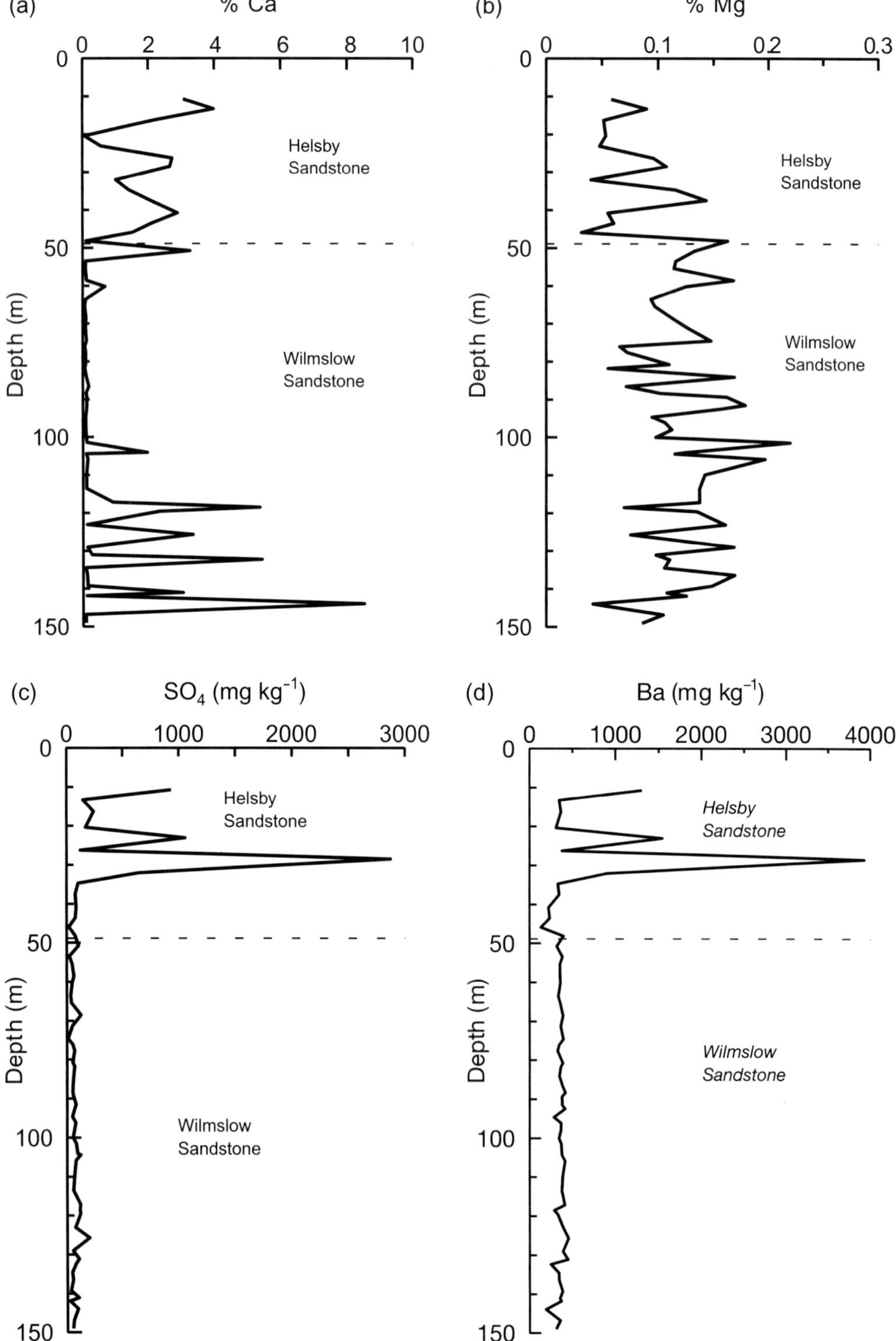

Fig. 12. Sediment chemistry profiles for calcium, magnesium, sulphur (as sulphate) and barium.

Table 4. *Average concentration of various major and minor elements in the sediments subdivided by formation*

Element*	Helsby Sandstone Formation	Wilmslow Sandstone Formation
n	14	46
%Al	1.43	2.47
%Fe	0.37	0.56
%Ca	1.89	0.85
%Mg	0.08	0.12
%K	1.14	1.94
%Na	0.17	0.36
As (by ICP-AES)	8.5	7.5
Ba	768	361
Cu	0.9	1.6
Li	11.4	12.3
Mn	96.7	62.5
P	171	293
Pb	<10	<10
SO$_4$	473	70
Sr	45	59
Zn	4.8	5.6

* Concentrations in mg kg^{-1} except where indicated as % (by weight).

groundwaters is always difficult as the quantities of As involved are so small. Given an average porosity of the sandstone (21%) and an average As content (8 mg kg^{-1}), it is only necessary to dissolve 0.04% of the As in the sandstone to give a groundwater As concentration of 30 μg l^{-1}, three times the existing EC drinking-water-quality limit for arsenic. This amount is extremely small and such dissolution clearly could not be detected by bulk chemical analysis. Indeed, given the sensitivity of groundwater to such small changes, it is perhaps surprising that high-As groundwaters are not more common.

The sediments were laid down around 245–235 Ma and have probably been exposed to freshwater flow for tens of millions of years or more, and so the major geochemical changes following deposition and uplift probably occurred long ago. Changes in the recent past must have been relatively small and over the last 100 000 years have probably been dominated by the ice ages and resulting sea-level changes. These would have affected groundwater flow patterns and so would have altered the composition of the groundwater and probably the As concentrations.

The extensive pumping of groundwater along with the possible introduction of pollutants, such as those derived from fertilizer leaching, are now likely to be the major drivers for water-quality change. Given that arsenic is currently

being removed from the aquifer in the abstracted groundwater, it is likely that it is still being desorbed from the aquifer, most probably from the iron oxides because these are known to bind arsenic very strongly. How long this might last is difficult to judge without more detailed information about the adsorption isotherms. There may also be very slow diagenetic changes taking place. Crystallization of the iron oxides may either reduce their surface area or change their structure, for example from hydrous ferric oxide to hematite, and this could be accompanied by the release of small amounts of arsenic. Other factors leading to increased arsenic concentrations could be an increase in the concentration of competing anions (like phosphate, silicate, bicarbonate or DOC) or a decrease in the concentration of co-ions aiding As adsorption, such as Ca^{2+}. Although no link with existing mineralization could be found in the affected sediments, the presence of known mineralization within a few tens of kilometres of the Delamere area could have provided a higher-than-usual source of As in the sediments and the secondary minerals (such as Fe oxides) that have evolved. This As could have been dispersed via the extensive fracture and faulting systems observed in the area. There are insufficient data on the As content of the Triassic sandstones in NW England to judge whether the concentrations observed in the affected sediments are different from those in Triassic sandstones elsewhere in the UK.

Edwards (2001) noted that the proximity to major faults was a possible indicator of high-As groundwaters in the study area, although the relationship proved not to be statistically significant. He also noted that there was a clear correlation between high-As groundwaters and the proximity to the mapped mudstone–sandstone boundary (the Tarporley Siltstone Formation–Helsby Sandstone Formation boundary). Clearly, both the source of the As and the groundwater flow are important factors and faulting can in principle affect both.

It should be stressed that the analyses of water samples referred to in this report are for raw groundwaters prior to any treatment taking place. The samples were not taken for the purpose of assessing water quality from each of the boreholes tested for drinking, cooking, or other domestic, agricultural or industrial purposes, and do not amount to certification of potability in respect of groundwater in the region. If such information is required, specific tests should be carried out for this purpose.

The main conclusion from the present study is that the As concentration in groundwater in the Delamere area appears to increase gradually with depth. This needs to be confirmed by detailed water-quality depth profiles in other boreholes from the area. If this trend is confirmed, it has obvious implications for water-resources management. In terms of As concentration, the shallower the borehole the better. This has to be set against the loss of yield that may occur with shallower boreholes and also possibly greater nitrate concentrations. For example, we observed a large, and potentially water-yielding, fracture at about 100 m. Therefore, completion just below this depth might provide an acceptable yield without excessive As. Nevertheless, the As concentration is likely to be still in the range 10–20 μg l^{-1} and so would still exceed the drinking-water limit of 10 μg l^{-1}. While the increase in As concentration with depth is somewhat erratic, it is sufficiently consistent to assume that the possibility of finding dramatically lower groundwater As concentrations by varying the depth of drilling appears unlikely.

We would like to thank P. Shand and K. Griffiths for an early sight of the 'baseline' data for the Delamere area. We also thank EA staff for their WIMS data, P. Williams for pore-water extraction, I. Neumann and J. Cobbing for the geophysical logging and depth sampling, and the BGS analytical staff for chemical analyses. We acknowledge G. Edwards and L. Bellis for providing a copy of their MSc report and thesis, respectively, and J. Tellam and an anonymous reviewer for their detailed and constructive comments. This paper is published with the permission of the Executive Director, British Geological Survey (NERC).

References

ALLEN, D.J., BREWERTON, L.J. ET AL. 1997. *The Physical Properties of Major Aquifers in England and Wales*. British Geological Survey Technical Report, **WD/97/34**.

BATH, A.H., EDMUNDS, W.M. & ANDREWS, J.N. 1979. Palaeoclimatic trends deduced from the hydrochemistry of a Triassic Sandstone aquifer, United Kingdom. *In: Isotope Hydrology 1978, Vienna*, International Atomic Energy Agency, Vienna, 545–568.

BELLIS, L. 2002. *A study of the variation in cation exchange capacity within the Triassic Sandstone of the Cheshire Basin, and its implications for risk assessment*. MSc thesis, University of Reading.

BLOOMFIELD, J.P., MOREAU, M.F. & NEWELL, A.J. 2006. Characterization of permeability distributions in six lithofacies from the Helsby and Wilmslow sandstone formations of the Cheshire Basin, UK. *In*: TELLAM, J.H. & BARKER, R.D. (eds) *Fluid Flow and Solute Movement in Sandstones: The Onshore UK Permo-Triassic Red Bed Sequence*, **263**, 83–101.

CARLON, C. 1981. The Gallantry Bank Copper Mine, Bickerton, Cheshire. *British Mining*, **16**, 1–50.

EDWARDS, G. 2001. *An investigation into the controls on arsenic distribution in groundwaters of north west England*. MSc Project Report, University of Birmingham.

EDMUNDS, W.M., COOK, J.M., KINNIBURGH, D.G., MILES, D.L. & TRAFFORD, J.M. 1989. *Trace-element Occurrence in British Groundwaters*. British Geological Survey Research Report, **SD/89/3**.

FOSTER, S.S.D., BRIDGE, L.R., GEAKE, A.K., LAWRENCE, A.R. & PARKER, J.M. 1986. *The Groundwater Nitrate Problem*. British Geological Survey Hydrogeological Report, **86/2**.

FURLONG, B.V., TELLAM, J.H., MACKAY, R., INGRAM, J.A. & HERBERT, A.W. 2000. Regional scale solute transport in a Permo-Triassic Sandstone aquifer. *In*: SILILO, O.T.N. ET AL. (eds) *Proceedings of the XXX IAH Congress on Groundwater: Past Achievements and Future Challenges*. Balkema, Rotterdam, 499–502.

GRIFFITHS, K.J., SHAND, P. & INGRAM, J. 2002. *Baseline Report Series: 2. The Permo-Triassic Sandstones of west Cheshire and the Wirral*. British Geological Survey Commissioned Report, **CR/02/109N**.

HASLAM. H. & SANDON, P.T.S. 1991. *The Geochemistry of Some Red-bed Formations in the United Kingdom*. British Geological Survey Technical Report, **WP/90/2**.

KINNIBURGH, D.G. & MILES, D.L. 1983. Extraction and chemical analysis of interstitial water from soils and rocks. *Environmental Science & Technology*, **17**, 362–368.

KINNIBURGH, D.G. & TRAFFORD, J.M. 1996. Unsaturated zone porewater chemistry and the edge effect in a beech forest in southern England. *Water, Air and Soil Pollution*, **92**, 421–450.

NAYLOR, H., TURNER, P., VAUGHAN, D.J., BOYCE, A.J. & FALLICK, A.E. 1989. Genetic studies of Red Bed mineralization in the Triassic of the Cheshire Basin, northwest England. *Journal of the Geological Society, London*, **146**, 685–699.

PLANT, J.A., JONES, D.G. & HASLAM, H.W. 1999. *The Cheshire Basin: Basin Evolution, Fluid Movement and Mineral Resources in a Permo-Triassic Rift Setting*. British Geological Survey, Keyworth.

SMEDLEY, P.L. & EDMUNDS, W.M. 2002. Redox patterns and trace-element behaviour in the East Midlands Triassic Sandstone aquifer, U.K. *Ground Water*, **40**, 44–58.

SMEDLEY, P.L. & KINNIBURGH, D.G. 2002. A review of the source, behaviour and distribution of arsenic in natural waters. *Applied Geochemistry*, **17**, 517–568.

TELLAM, J.H. 1994. The groundwater chemistry of the Lower Mersey Basin Permo-Triassic Sandstone aquifer system, UK: 1980 and pre-industrialisation-urbanisation. *Journal of Hydrology*, **161**, 287–325.

TELLAM, J.H. 1995. Hydrochemistry of the saline groundwates of the Lower Mersey Basin Permo-Triassic Sandstone aquifer, UK. *Journal of Hydrology*, **165**, 45–84.

Investigating rising nitrate concentrations in groundwater in the Permo-Triassic aquifer, Eden Valley, Cumbria, UK

ANDREW BUTCHER[1], ADRIAN LAWRENCE[1], CHRIS JACKSON[1], EMMA CULLIS[1], JENNIFER CUNNINGHAM[1], KAMRUL HASAN[2] & JOHN J. A. INGRAM[3]

[1]*British Geological Survey, Crowmarsh Gifford, Wallingford, Oxfordshire OX10 8BB, UK*
(e-mail: asb@bgs.ac.uk)

[2]*Environment Agency, National Groundwater and Contaminated Land Centre, Olton Court, 10 Warwick Road, Olton Solihull, West Midlands B92 7HX, UK*

[3]*Environment Agency, North West Region, PO Box 12, Richard Fairclough House, Knutsford Road, Warrington WA4 1HG, UK*

Abstract: Groundwater nitrate concentrations in the Permo-Triassic aquifer of the Eden Valley vary from less than 4 mg l^{-1} to in excess of 100 mg l^{-1} (as NO_3). A significant number of boreholes exhibit rising trends in nitrate concentration that either approach or exceed the CEC Directive 80/778 Maximum Admissible Concentration (MAC) of 50 mg l^{-1}. The main source of the nitrate is believed to be the nitrogen applied to grassland, both as slurry and as inorganic fertilizers.

The variability in groundwater nitrate concentrations is thought to be due in part to land use, particularly where low-yielding boreholes derive their water from a limited/localized area, and in part due to the variability in the travel times for water and solutes to migrate from the soil to the water table and then to the borehole. This variability in travel times is a function of surficial geology, depth to water table, depth of borehole and superficial deposit thickness, amongst other factors.

It is surprising, given the considerable storage within the saturated zone of the aquifer and the slow groundwater movement, that some relatively deep boreholes pump groundwater with nitrate concentrations in excess of 20 mgl^{-1}. Simple numerical modelling suggests that the fraction of modern water pumped is sensitive to the presence of fissures close to the abstraction boreholes and the location of the boreholes relative to superficial deposits. For some scenarios, using realistic superficial deposit geometries and aquifer hydraulic parameters, the proportion of modern water (water that is derived from infiltration that reached the water table since pumping started) could exceed 40% within 15 years of pumping.

The Eden Valley lies between two upland areas: the Pennines to the east and the Lake District to the west (Figs 1 & 2). It is aligned approximately NW–SE, and is some 56 km long and varies in width from 5 to 15 km. The valley floor is underlain by Permo-Triassic sandstones, which form the major aquifer in the region. The Permo-Triassic sandstones are overlain by superficial deposits ('drift') that can be up to 30 m thick, and include sands, gravels and clay-rich till. Groundwater in the sandstone aquifer is used by industry, for minor farm supplies and for public water supply. Groundwater resources are considerable and there is potential for further development of the aquifer.

Average annual rainfall is approximately 1000 mm year^{-1} in the Eden Valley and is in excess of 1500 mm year^{-1} on the surrounding higher ground. Recharge to the sandstone aquifer is estimated to average 150 mm year^{-1}. However, this recharge is distributed unevenly, with most occurring where superficial deposits are either absent or permeable. Runoff from the adjacent uplands drains to the River Eden, which flows northwards from Kirkby Stephen through Appleby and Penrith into the Carlisle Basin.

The Eden Valley is largely rural with a low population density of about 0.2 ha^{-1}. Agriculture, tourism and some industry are the major sources of income. Livestock rearing is the main agricultural activity; in recent years more intensive farming and higher stocking densities have resulted in greater applications of fertilizers to grassland and to fodder crops. The spreading of slurry wastes on grassland has increased and

From: BARKER, R. D. & TELLAM, J. H. (eds) 2006. *Fluid Flow and Solute Movement in Sandstones: The Onshore UK Permo-Triassic Red Bed Sequence*. Geological Society, London, Special Publications, **263**, 285–296.
0305–8719/06/$15 © The Geological Society of London 2006.

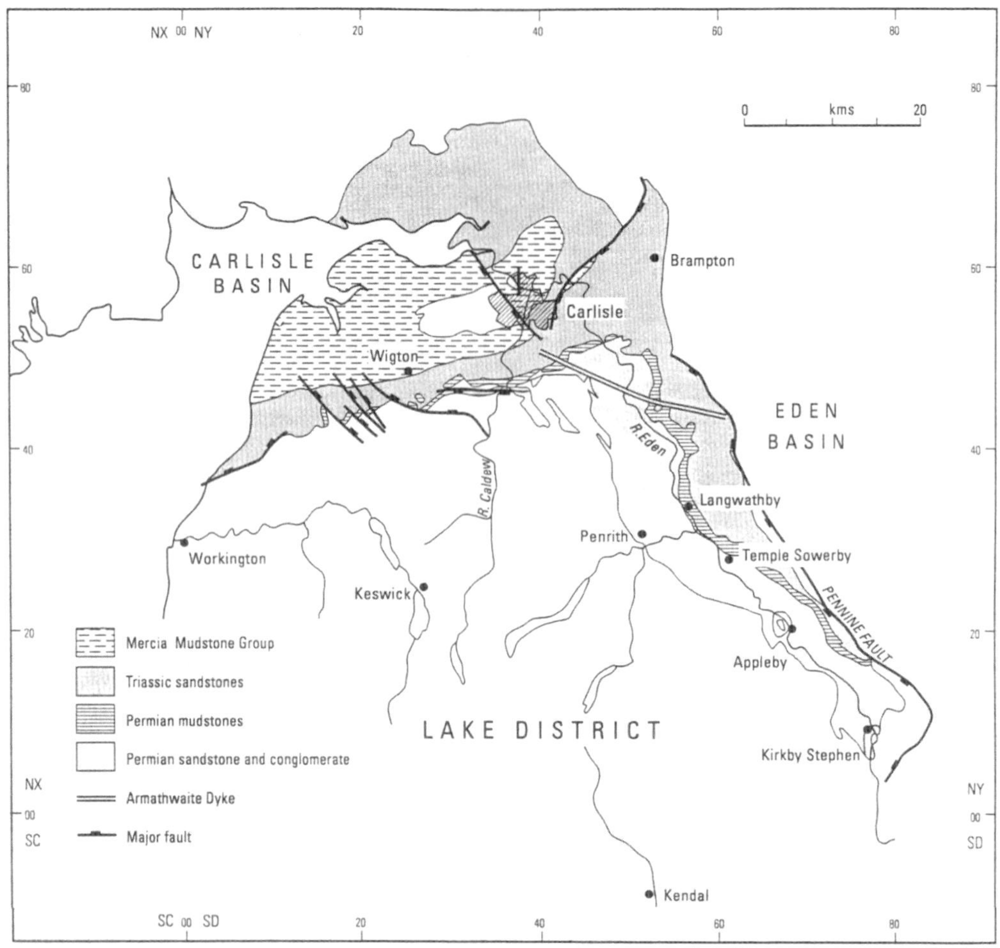

Fig. 1. Location map of the Eden Valley (modified from Allen *et al.* 1997). The valley is in the far NW corner of England.

both the timing and quantities applied are more dictated by the need to dispose of the slurry than to meet the crops' nutrient needs. However, within the Eden catchment there are also large areas of semi-natural habitat including unimproved grassland and woodland.

A significant number of boreholes that penetrate the Permo-Triassic aquifer have elevated nitrate concentrations (Fig. 3). These boreholes include both low-yielding farm supplies and major abstraction supplies.

The source of the nitrate is thought to be agriculture and a consequence of the intensification of farming and an increase in cattle stocking densities. A review of nitrate pollution and livestock farming (Hooda *et al.* 2000) suggested that the spreading of slurry on fields is probably the major source of nitrate in water. Because the spreading of slurry is usually considered a

solution to a waste disposal problem rather than a useful source of nutrients, rates of application frequently exceed crop requirements. Rates of nitrate leaching beneath intensively grazed grass can exceed those beneath intensive arable land (Parker *et al.* 1989; Chilton & Foster 1991). Intensification of agriculture has occurred during the past 20–30 years and so high nitrate groundwaters are likely to be associated with modern (post-1970s) recharge. Slurry pits are unlikely to be a major source of nitrate for large abstraction boreholes because of the considerable dilution within the borehole capture zone (Gooddy *et al.* 2001).

Hydrogeology

The Permo-Triassic sandstone aquifer comprises two formations, the Penrith Sandstone

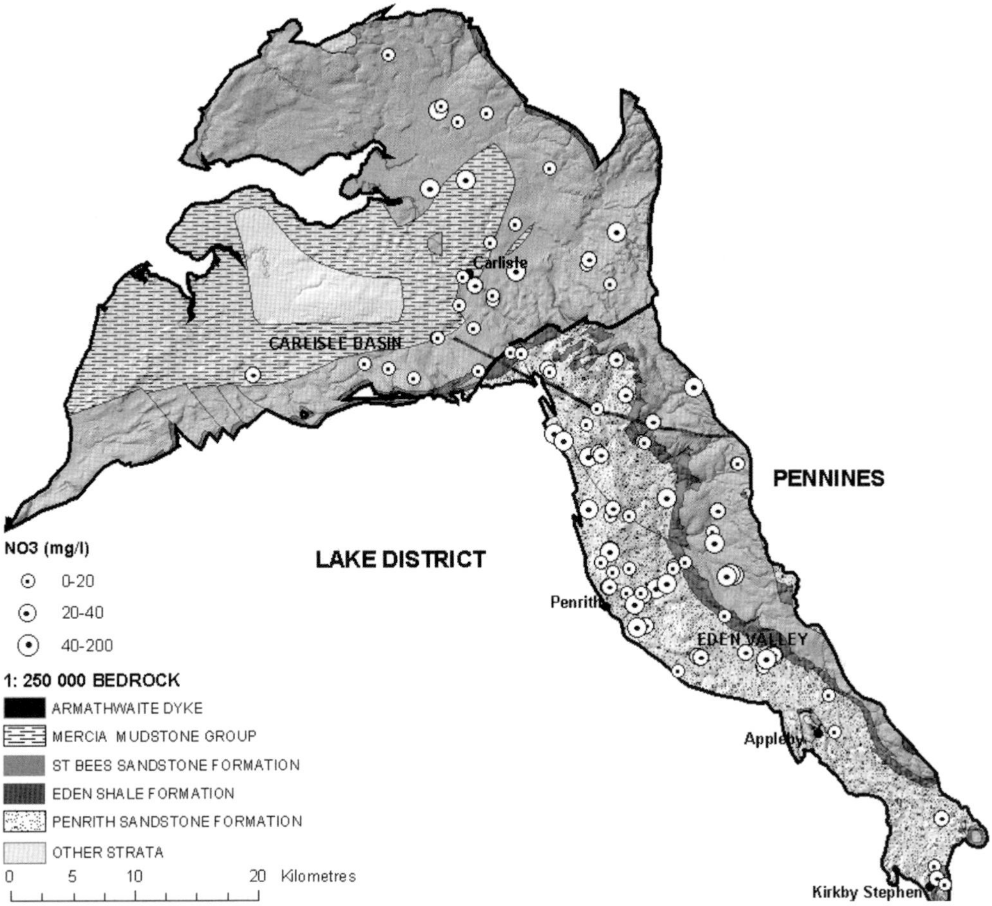

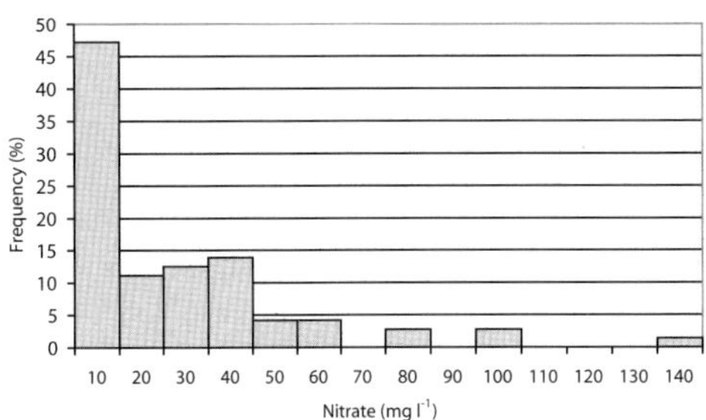

Fig. 2. Distribution of monitoring boreholes in the Permo-Triassic aquifers of the Eden Valley and Carlisle Basin indicating groundwater nitrate concentrations.

Fig. 3. Frequency distribution of groundwater nitrate concentration in monitoring boreholes in the year 2000.

(Permian) and the overlying St Bees Sandstone (mid-Triassic). These sandstones have moderate intergranular permeability (mean permeabilities are 0.8 and 0.24 m day^{-1} respectively: Allen *et al.*

1997), although pumping-test-derived hydraulic conductivities can be much higher because they include the fracture contribution to permeability (Lovelock 1972; Price *et al.* 1982).

Permeabilities used in regional models approach the intergranular value, which suggests that the influence of fracture permeability is probably more localized (Allen *et al.* 1997). As a consequence of this and the high porosity of the sandstone, rates of groundwater flow at a regional scale are low. Nevertheless, around abstraction boreholes groundwater velocities are locally higher because water movement is largely controlled by the fracture permeability (Worthington 1977).

Ingram (1978) estimated infiltration rates to the Penrith and St Bees sandstones, where these outcrop, as 315 and 350 mm year^{-1}, respectively. Infiltration rates to the sandstones are lower where they are overlain by superficial deposits, and were assumed by Ingram (1978) to be

90–100 mm year^{-1}. Using these infiltration rates, a good agreement was achieved between recharge and groundwater discharge to the River Eden, for part of the Eden catchment. However, the distribution of this recharge between areas where the sandstone is exposed and areas where it is covered by superficial deposits (approximately two-thirds of the catchment) is uncertain.

Vines (1984) estimated that the recharge rate through till to the Permo-Triassic sandstones was closer to 50 mm year^{-1}. His estimate was based on: (a) comparing water balance estimates for the Permo-Triassic sandstone aquifer in three adjacent catchments (in Lancashire) with different degrees of till cover; and (b) tritium profiles in the unsaturated zone of the sandstones in Cheshire.

If infiltration rates through less permeable superficial deposits are as low as 50 mm year^{-1} and, assuming the overall recharge rate to the sandstone aquifers of the Eden Valley remains the same, then clearly higher rates of infiltration are required in those areas of the Eden Valley where the superficial deposits are either permeable or absent. On this basis, rates of infiltration for the exposed sandstone could exceed 480 mm year^{-1}; the implications of this are discussed later.

The concept of higher recharge rates in areas where superficial deposits are either absent or permeable would appear reasonable, as runoff from adjacent areas where superficial deposits are thicker and less permeable is likely to contribute to, and significantly increase recharge in these former areas.

In the Eden Valley there is no obvious systematic distribution of the boreholes that pump higher nitrate groundwaters, the implication being that either the source of nitrate for these boreholes is localized (point source) or the travel times for water to move from the ground surface to the water supply boreholes are very variable. Long travel times may result in current pumped groundwaters originating as infiltration prior to the intensification of agriculture (which is the most likely source of nitrate) and thus be of low nitrate concentration.

Scope of the investigation

Initially, the water-quality data from the network of approximately 150 Environment Agency (EA) monitoring boreholes in the area around the Eden Valley were examined. Where boreholes had not been sampled and tested for groundwater nitrate concentration they were excluded. Borehole construction details held in the National Groundwater Archive were examined to determine whether these were comprehensive enough to use for further study. A subset of approximately 115 boreholes was finally reviewed in greater detail.

The groundwater nitrate concentration data available ranged from single measurements usually undertaken on completion of drilling, to datasets spanning longer periods. The data available from the EA cover the period from 1962 to 2002. However, the frequency of the groundwater nitrate data is generally very irregular. The data were plotted, and a brief assessment made of whether any reliable trends in the nitrate concentration could be recognized.

In the early stages of the study it was difficult to identify any pattern to the locations of the high groundwater nitrate boreholes (Fig. 2). However, after large-scale geology and land-use maps were produced the proximity of several of the boreholes, which had higher groundwater nitrate concentrations, to 'windows' in the superficial deposits became apparent.

An attempt was made to correlate groundwater nitrate concentrations with selected factors (Fig. 4). Although some correlations were observed (e.g. groundwater nitrate concentrations were generally higher and more variable where: (a) superficial cover was less than 10 m thick; and (b) boreholes were less than 100 m deep), no single controlling factor was apparent (Butcher *et al.* 2003).

Some deep boreholes (greater than 100 m depth) had relatively high nitrate concentrations. This is surprising as groundwater velocities within the Permo-Triassic sandstone aquifer at the regional scale are low, and so it might be anticipated that the residence time of water and solutes arriving at the borehole would be long and likely to originate as infiltration prior to the 1960s–1970s and before intensification of agriculture began.

The main purpose of this paper was to investigate how two factors, namely: (a) the development of a high-permeability fractured zone around the borehole; and (b) the influence of varying recharge rates associated with the presence of superficial cover, could influence the fraction of modern water pumped (and by implication the nitrate concentration). It is recognized that other factors, for example borehole and casing depth, unsaturated zone thickness and aquifer anisotropy, are likely to influence the fraction of modern water pumped, although these factors were not investigated at this stage.

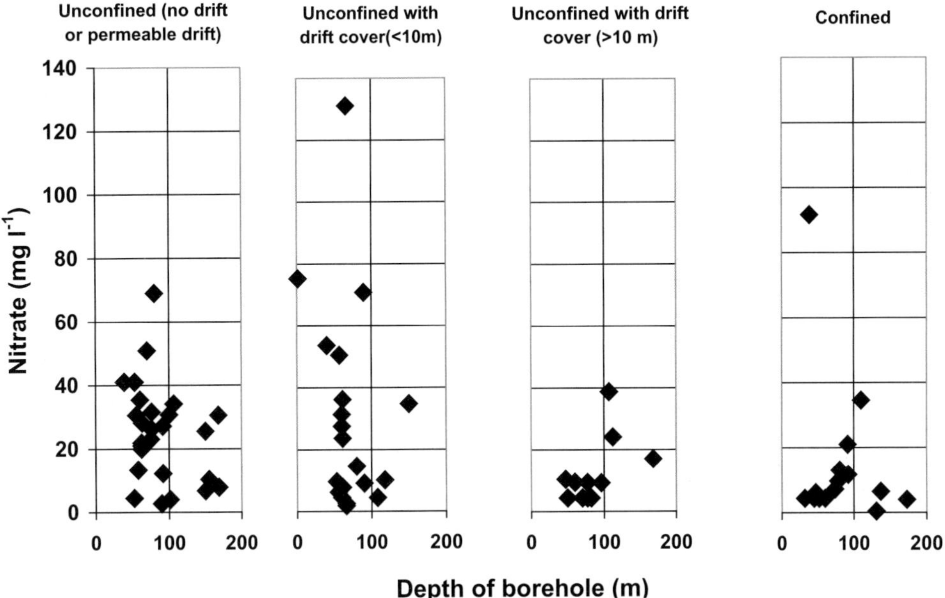

Fig. 4. Scatter plot of borehole depth and pumped nitrate concentration for the year 2000.

Numerical modelling

An attempt was made to assess how travel times for water to migrate from the water table to an abstraction borehole may vary. Accordingly, a relatively simple numerical model was developed to investigate the sensitivity of modelled travel times to two factors:

- the permeability of the sandstone and the development of local fissuring around abstraction boreholes;
- the distribution of superficial deposits relative to abstraction borehole.

The conceptual model that underpinned the numerical modelling was based on a review of previous studies and existing data (Monkhouse & Reeves 1977; Ingram 1978; Allen *et al.* 1997; ESI 1999).

The graphs presented below compare the percentage of modern water pumped against time since pumping started. Modern water is defined as recharge that reaches the water table after the abstraction well starts pumping. A high percentage of modern water in the borehole suggests that the nitrate concentration could also be high. Various scenarios were considered, which included a range of aquifer permeabilities and geometries of superficial deposits. The output from this modelling is used to show that

the percentage of modern water pumped is sensitive to these scenarios and provides a possible explanation as to how deep boreholes could pump water with a relatively high percentage of modern water.

Model structure

The model was constructed using the regional groundwater modelling code ZOOMQ3D (Jackson 2001), which incorporates unconventional local grid refinement. This has been used to simulate the abstraction borehole on a 50 m-mesh for improved accuracy. The model grid is shown in Figure 5. It is 8 km² and contains five layers. The horizontal hydraulic conductivity of the sandstone aquifer is set in the model as 1 m day⁻¹ and compares with a mean intergranular permeability of 0.8 m day⁻¹ (Allen *et al.* 1997). The vertical hydraulic conductivity is also defined in the model as 1 m day⁻¹, which is consistent with core permeability data (Allen *et al.* 1997). Introducing aquifer anisotropy into the model, by making the ratio of the horizontal to vertical hydraulic conductivies greater than unity (to simulate layering within the aquifer), would increase the fraction of modern water pumped by the borehole. This factor was not investigated by the modelling at this stage. The porosity value used in the model is 0.25 and

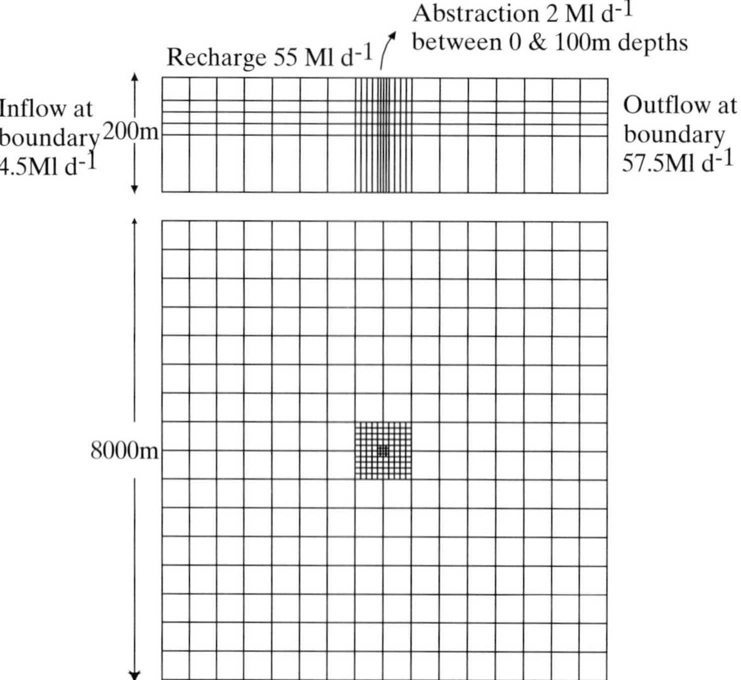

Fig. 5. Structure of the numerical model (cross-section over the plan view).

is based on core-porosity data (Allen *et al.* 1997). An abstraction well is located at the centre of the grid and pumps at a rate of 2000 m³ day⁻¹. The left-hand boundary inflow and right-hand bound outflow are specified to approximate a 1:50 regional hydraulic gradient. This gradient is consistent with the regional groundwater level contours observed in the Eden Valley by ESI (1999). For the model runs, the water pumped is derived predominantly from close to the borehole and therefore was insensitive to the distance to the boundary.

The model was used to simulate both an homogeneous intergranular aquifer and an aquifer that contains a horizontal fractured zone. The fractured zone is represented using a 5 m-thick layer, which has a hydraulic conductivity of 100 m day⁻¹. The modelled aquifer transmissivity varies from 200 (when no fracturing is present) to 700 m² day⁻¹ when a fracture zone is included. These values are consistent with field observations (Lovelock 1972; Lovelock *et al.* 1975; Allen *et al.* 1997). The fracture plane extends 500 m from the abstraction borehole in both horizontal Cartesian directions. The model layer thicknesses differ when simulating an intergranular or a fractured aquifer; however, the total thickness is always 200 m.

The total recharge to the aquifer is 55 Ml day⁻¹, its distribution varying depending on the extent of the superficial cover. Recharge rates through the sandstone and the superficial deposits were varied. Five recharge scenarios were considered; in all cases the total recharge to the model was set at 55 Ml day⁻¹. These were as follows:

- *Scenario 1*: there is no superficial cover and the recharge rate through the sandstone is 314 mm year⁻¹.
- *Scenario 2*: superficial deposits cover 35% of the modelled area and the recharge rate through these deposits is 119 mm year⁻¹. The recharge rate through the exposed sandstone is 416 mm year⁻¹.
- *Scenario 3*: superficial deposits cover 65% of the modelled area and the recharge rate through these deposits is 168 mm year⁻¹. The recharge rate through the exposed sandstone is 591 mm year⁻¹.
- *Scenario 4*: superficial deposits cover 35% of the modelled area and the recharge rate through these deposits is fixed at 50 mm year⁻¹ (corresponding to the recharge rate through till estimated by Vines 1984). The recharge rate through the exposed sandstone is 452 mm year⁻¹.

- *Scenario 5*: superficial deposits cover 65% of the modelled area and the recharge rate through these deposits is, again, fixed at 50 mm year^{-1}. The recharge rate through the exposed sandstone is 818 mm year^{-1}. This is significantly higher than can be expected realistically, except possibly in the case of focused runoff from an adjacent area covered by low-permeability superficial deposits: however, the model suggests that the travel time to the borehole was little affected by this recharge value as the borehole catchment (or capture zone) was restricted to the area overlain by superficial deposits.

These recharge scenarios are illustrated in Figure 6.

The simulations

In total 12 steady-state simulations were run, in which the different recharge scenarios were applied to the intergranular and fractured aquifer (Table 1). Particle tracking was then performed to determine the time of travel of water from the water table to the abstraction borehole. Particles were placed on the water table along a line through the centre of the model from the left to the right. A single line of particles was used, but, for more detail, particles could have been placed over the full areal extent of the borehole catchment (which varies between model runs). As will be shown, the use of one line of particles was sufficient to enable conclusions to be drawn regarding the influence of superficial cover and fracturing on the age of the abstracted water.

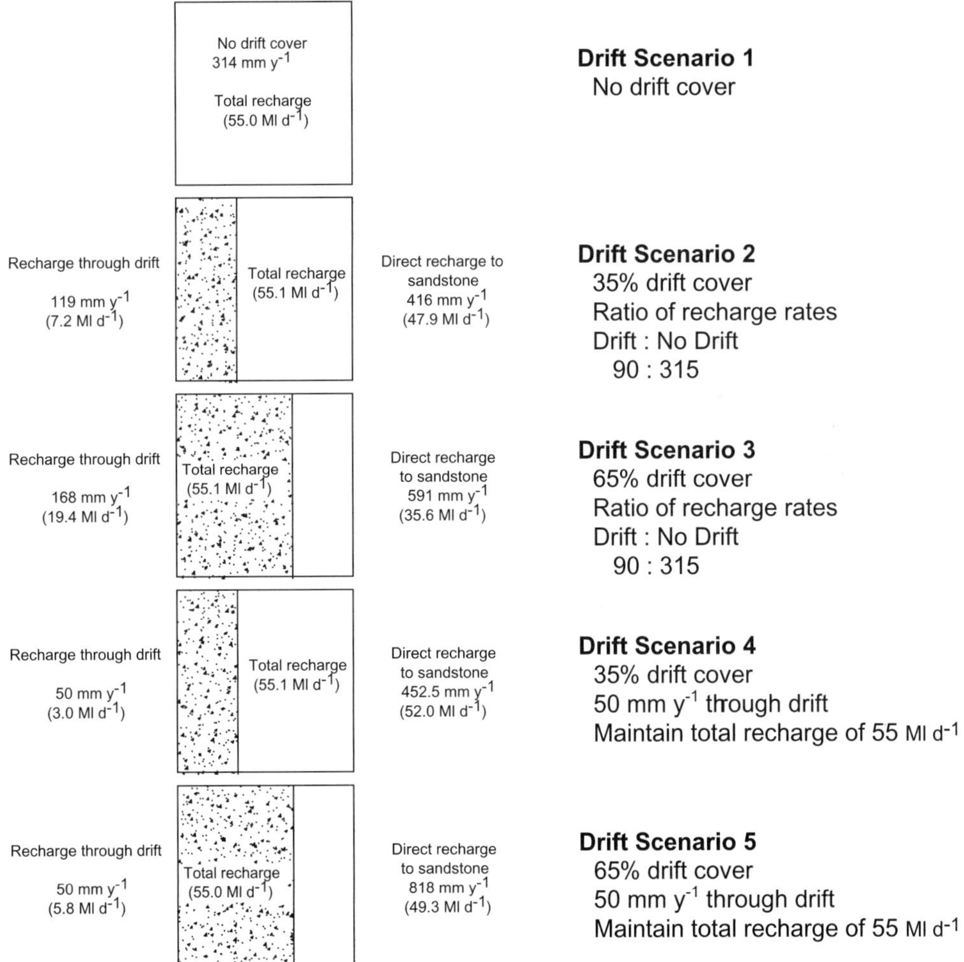

Fig. 6. Model recharge scenarios.

Table 1. *Details of model runs*

Model No.	Recharge scenario (% superficial deposit cover) [recharge through superficial deposits]	Intergranular or fractured aquifer	Hydraulic conductivity (1 m day⁻¹ unless given)	Porosity (25 % unless given)
1	1. (0)	Intergranular		
2	1. (0)	Intergranular	$K_x = 5$ m day⁻¹	
3	1. (0)	Fractured	$K_x = 100$ m day⁻¹ for fracture	
4	1. (0)	Fractured	$K_x = 100$ m day⁻¹ for fracture	5% for fracture
5	2. (35 %) [119 mm year⁻¹]	Intergranular		
6	3. (65 %) [168 mm year⁻¹]	Intergranular		
7	2. (35 %) [119 mm year⁻¹]	Fractured	$K_x = 100$ m day⁻¹ for fracture	
8	3. (65 %) [168 mm year⁻¹]	Fractured	$K_x = 100$ m day⁻¹ for fracture	
9	4. (35 %) [50 mm year⁻¹]	Intergranular		
10	5. (65 %) [50 mm year⁻¹]	Intergranular		
11	4. (35 %) [50 mm year⁻¹]	Fractured	$K_x = 100$ m day⁻¹ for fracture	
12	5. (65 %) [50 mm year⁻¹]	Fractured	$K_x = 100$ m day⁻¹ for fracture	

The spacing of the particles along the water table depends on the recharge rate. For comparisons to be made regarding the percentage of modern water arriving at the abstraction borehole over time between the different simulations each particle must represent the same *volumetric* recharge rate. In each model a particle is associated with 0.1075 m³ day⁻¹ of recharge per metre width of aquifer in the S–N direction. For example, if the recharge rate is 0.5 mm day⁻¹ the particle spacing will be 215 m.

Model output

Particle tracking is used to plot the pathline of each particle from the water table to the abstraction borehole. As an example, the particle paths for Model 1, an intergranular aquifer with no superficial deposit cover, are shown in Figure 7. The particle tracking model also calculates the travel time of individual particles. Particle travel times for Model 1 are plotted in Figure 8. Conclusions can be made regarding the effect of superficial deposit cover and aquifer fracturing by examining the particle travel times. For Model 1 (Figs 7 & 8), 33 particles arrive at the borehole from the water table. The first particle takes 0.81 years to travel from the water table to borehole. Consequently, it can be estimated that 0.81 years after the pump is switched on, approximately one thirty-third of the water abstracted is modern. The second particle arriving at the well takes 1.04 years to travel to the water table. Therefore, after 1.04 years we can assume that $2 \times 100/33\%$ of the abstracted water is modern. By applying this process, a graph of the percentage of modern water pumped against time can be drawn for each model simulation. For Model 1 this graph is shown in Figure 9.

Modelling results

The effect of fracturing on the percentage of modern water pumped over time. Comparison of models 1–4. Model 1 is taken as the base case

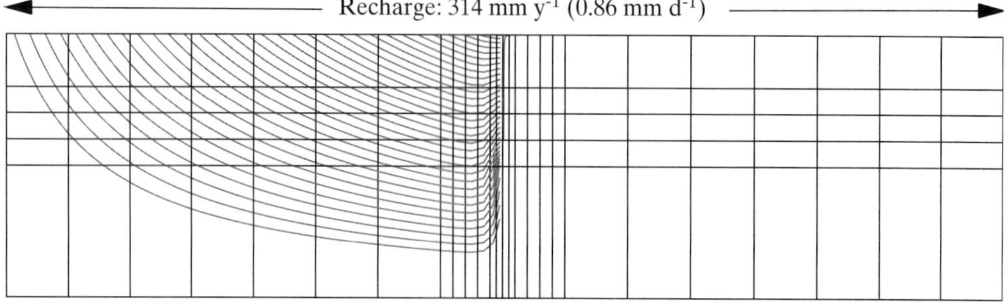

Fig. 7. Particle pathlines for Model 1 ($K_x = 1$ m day⁻¹, porosity = 0.25, no fracture, no superficial deposit cover). Particles down the hydrogeological gradient from the borehole are omitted.

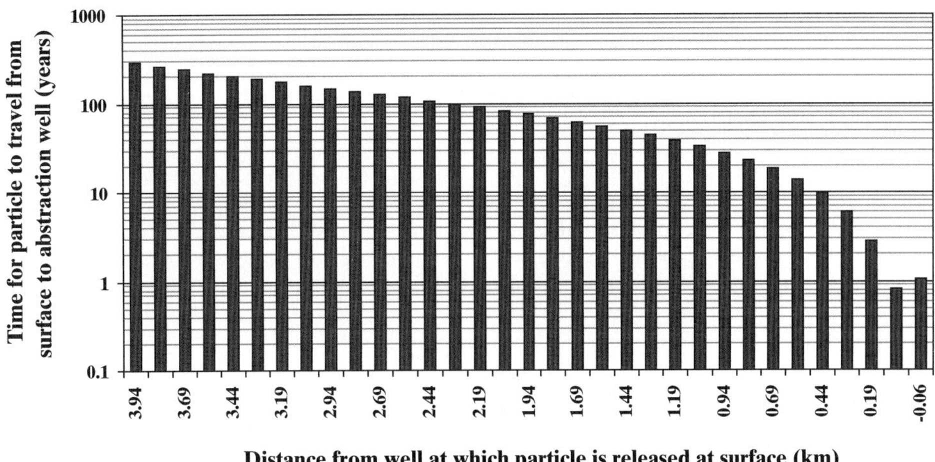

Fig. 8. Particle travel times for Model 1 ($K_x = 1$ m day^{-1}, porosity = 0.25, no fracture, no superficial deposit cover).

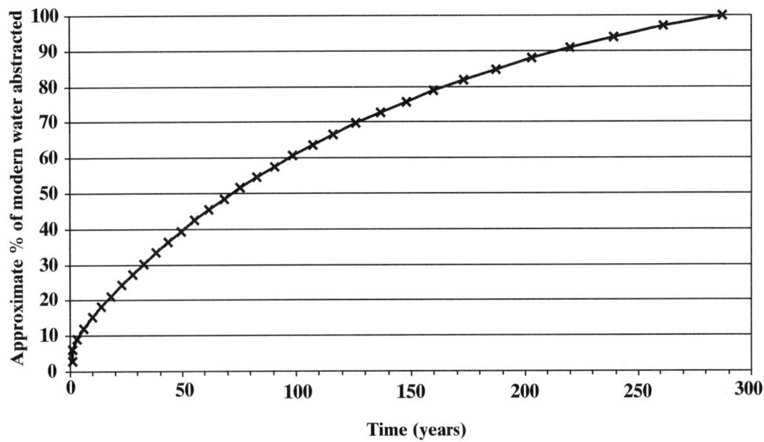

Fig. 9. Estimated percentage of modern water pumped since the start of abstraction, Model 1 ($K_x = 1$ m day^{-1}, porosity = 0.25, no fracture, no superficial deposit cover).

for comparison with the other simulations. Model 1 represents an intergranular aquifer with uniform recharge (no superficial deposit cover).

The influence of fracturing close to the abstraction borehole on the proportion of modern water pumped is indicated in Figure 10. This shows that when fracturing is present, the proportion of modern water pumped is normally higher. For example, after 15 years of pumping a borehole in the fractured aquifer pumps about 40% modern water compared with 20% for a borehole where no fracturing is developed.

The fractured zone transports water rapidly to the borehole. Particles starting approximately 500 m from the well (i.e. over an interval that is similar to the horizontal extent of the fracture zone) travel vertically down to the fracture zone and then rapidly to the well. Consequently, a greater percentage of modern water is pumped at earlier times in the fractured aquifer models (Model 3 and Model 4) than in the homogeneous aquifer model (Model 1).

The particles that arrive at the well after approximately 20 years travel for a significant time through the unfractured part of the aquifer (i.e. from greater than 500 m from the well). The introduction of the fracture has less of an impact on the travel times of these particles than those starting closer to the well.

The effect of varying cover of superficial deposits on the percentage of modern water pumped over time in an intergranular aquifer. Comparison of models 1, 5, 6, 9 and 10. The influence of superficial cover on the proportion of modern water pumped from a non-fractured intergranular aquifer is shown in Figure 11.

It is clear from Figure 11 that the higher recharge rate (through the exposed sandstones) close to the borehole (recharge scenario 2, model run 5) and recharge scenario 4 (model run 9) allowed a greater fraction of modern water to reach the borehole more rapidly. This is because the borehole preferentially sources its water from the exposed sandstone where recharge is higher. There is more vertical flow close to the borehole and consequently younger water arrives at the borehole more rapidly. The curves of the percentage of modern water pumped against time rise steeply compared with

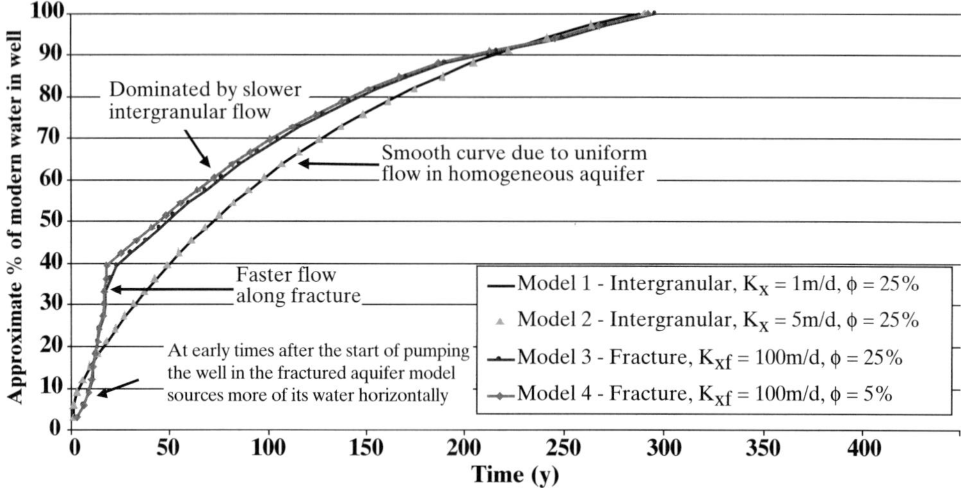

Fig. 10. The effect of the presence of a horizontal 500 × 500 m fracture on the breakthrough of modern water at the pumped well. No superficial deposit cover. K_{xf}, horizontal permeability of fracture; ϕ, porosity.

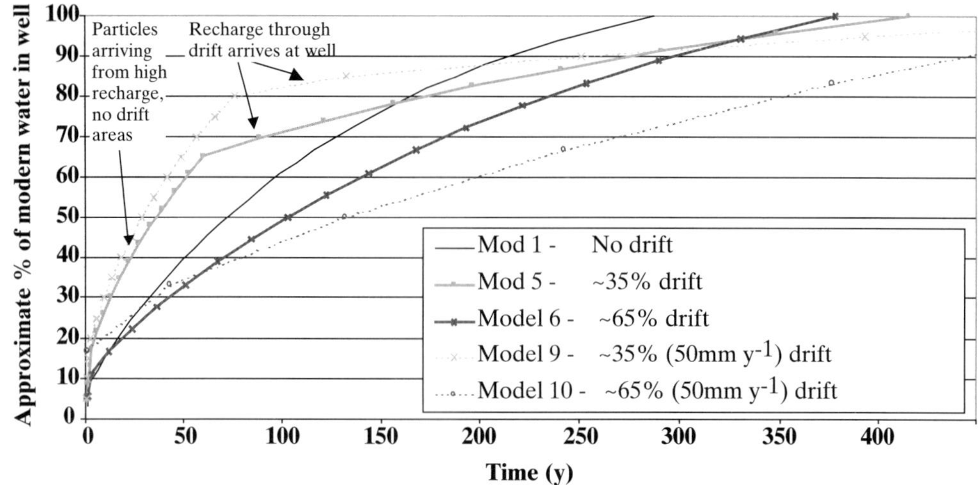

Fig. 11. The effect of superficial ('drift') deposits on the breakthrough of modern water at the pumped well. All runs are with an intergranular aquifer of K_x = 1 m day^{-1} and porosity 25%. In models 6 and 10, the borehole catchment is entirely beneath the low recharge drift covered area. The curve for Model 10 is unreliable for early times because of the low number of particles applied.

the base case (Model 1) for approximately 60–70 years. After this time water begins to arrive from the areas covered by superficial deposits and the slope of the curve reduces.

In the scenarios where superficial deposits cover 65% of the modelled area the abstraction borehole is located beneath these deposits. The larger borehole catchment, due to the reduced recharge rate, means that modern water takes significantly longer to arrive at the borehole.

Discussion

Modelling suggests that two factors:

- focused recharge through permeable windows in the superficial cover
- localized fissure flow to the abstraction borehole

could have a significant influence on the proportion of modern water pumped from a borehole. The scenarios modelled used realistic parameters that fitted current understanding of the groundwater flow system in the Permo-Triassic sandstone aquifers in the Eden Valley.

In general, the modelling suggests, not surprisingly given the high porosity of the aquifer, that it would take many decades (even centuries) for all the pore water within the borehole catchment to be flushed out by modern water.

The model predicted that groundwater pumped from a deep borehole beneath extensive superficial deposits would have only a small percentage of modern water (during the first 50 years of pumping) and is therefore likely to have low nitrate concentrations. Field data show that the major abstraction boreholes located beneath thick (more than 10 m) superficial deposits do, indeed, pump groundwater of low nitrate concentrations. The residence times of the groundwaters are yet to be determined.

The results of the modelling also suggest that, for boreholes located in a 'recharge window' or where fracturing is developed close to the borehole, the proportion of modern water pumped could increase rapidly with time initially but levels off at later times. An implication for pumped nitrate concentrations is that these may increase rapidly once pumping commences but the upwards trend may decline later. Thus, nitrate concentrations currently observed in the Eden Valley are strongly influenced by the pumping history, and maximum nitrate concentrations have probably not yet been reached. Tellam & Thomas (2002) reached a similar conclusion for the sandstone aquifer beneath Birmingham.

This investigation suggests that even when land-management practices are introduced to reduce nitrate leaching from soils, the timescales for reversing upwards trends in groundwater nitrate concentrations are likely to be many decades. This has clear implications for the implementation of the Water Framework Directive.

Modelling has provided a possible explanation both as to how some relatively deep boreholes can pump water containing a significant percentage of modern water and why there can be differences in nitrate concentration between boreholes in different hydrogeological environments that cannot be explained by differences in land-use alone.

Conclusion

Groundwaters in the Permo-Triassic sandstone aquifers in the Eden Valley show a considerable variation in nitrate concentration, from less than 4 to more than 100 mg l^{-1}. The principal source of the elevated groundwater nitrate concentrations is believed to be the spreading of animal slurry to grassland, which may be applied at rates in excess of the crop nutrient requirements.

A number of factors appear to influence groundwater nitrate concentrations observed in boreholes. These include, amongst others, surficial geology, the depth of the borehole, the thickness and nature of superficial deposits, and the development of horizontal fracturing around the borehole. However, the controls on groundwater nitrate concentration are complex and no single factor dominates.

One surprising observation is the relatively high nitrate concentrations measured in some deep boreholes. Simple numerical modelling suggests that the proportion of modern water pumped by a deep borehole is sensitive to: (a) the development of horizontal fracturing close to the borehole; and (b) the location of the borehole relative to windows in the low-permeability superficial deposits.

One implication from the modelling is that the rising trend in pumped nitrate concentration observed in some abstraction boreholes may level off with time provided that nitrogen loadings to the aquifer do not increase, but that these timescales may be very long indeed.

References

ALLEN, D.J., BREWERTON, L.J., ET AL. 1997. *The Physical Properties of Major Aquifers in England and Wales*. British Geological Survey Technical Report, **WD/97/34**. Environment Agency R&D Publication, **8**.

BUTCHER A.S., LAWRENCE, A.R., JACKSON, C.R., CUNNINGHAM, J., CULLIS, E., HASAN, K. & INGRAM, J. 2003. *Investigation of Rising Nitrate Concentrations in Groundwater in the Eden Valley, Cumbria: Project Scoping Study.* Joint British Geological Survey and Environment Agency R&D Publication, **NC/00/24/148**.

CHILTON, P.J & FOSTER, S.S.D. 1991. Control of ground-water nitrate pollution in Britain by land-use change. *In*: BOGARDI, I. & KUZELKA, R.D. (eds) *Nitrate Contamination.* Springer, Heidelberg, 333–347.

ESI (Environmental Simulations International) 1999. *Improvements to Groundwater Protection Zone Network: Gamblesby Model. Hydrogeological Setting, Conceptual Model and Model Idealisation.* Report to Environment Agency North West Region, report reference Environmental Simulations International report to Environment Agency, North West Region, 6061\GAM\R1D2.

GOODDY, D.C., HUGHES, A.G., WILLIAMS, A.T., ARMSTRONG, A.C., NICHOLSON, R.J. & WILLIAMS, J.R. 2001. Field and modelling studies to assess the risk to UK groundwater from earth based stores for livestock manure. *Soil Use Management*, **17**, 128–137.

HOODA, P.S., EDWARDS, A.C., ANDERSON, H.A. & MILLER, A. 2000. A review of water quality concerns in livestock farming areas. *Science of the Total Environment*, **250**, 143–167.

INGRAM, J.A. 1978. *The Permo-Triassic Sandstone Aquifers of North Cumbria.* Hydrogeological Report, Environment Agency, Warrington, UK.

JACKSON, C.R. 2001. *The Development and Validation of the Object-oriented Quasi Three-dimensional Regional Groundwater Model ZOOMQ3D.* British Geological Survey Internal Report, **IR/01/144**.

LOVELOCK, P.E.R. 1972. *Aquifer properties of the Permo-Triassic sandstone aquifers of the United Kingdom.* PhD thesis, University of London.

LOVELOCK, P.E.R., PRICE, M. & TATE T.K. 1975. Groundwater conditions in the Penrith Sandstone at Cliburn, Westmorland. *Journal of the Institute of Water and Engineering Science*, **29**(4), 157–174.

MONKHOUSE, R.A. & REEVES, M.J. 1977. *A Preliminary Appraisal of the Groundwater Resources of the Vale of Eden, Cumbria.* Central Water Planning Unit, Reading, Technical Note, **11**.

PARKER, J.M., CHILTON, P.J. & MCKITTRICK, R. 1989. *Nitrate Leaching to Groundwater from Grassland on Permeable Soils.* British Geological Survey Report, **WD/89/40c**.

PRICE, M., MORRIS, B.L. & ROBERTSON, A.S. 1982. A study of intergranular and fissure permeability in Chalk and Permian aquifers, using double-packer injection testing. *Journal of Hydrology*, **54**, 401–423.

TELLAM, J.H. & THOMAS, A. 2002. Well water quality and pollutant source distributions in an urban aquifer. *In*: HOWARD, K.W.F. & ISRAFILOV, R.G. (eds) *Current Problems of Hydrogeology in Urban Areas*, **139–158**.

VINES, K.J. 1984. *Drift Recharge.* North West Water Hydrogeological Report, **145**.

WORTHINGTON, P.F. 1977. Permeation properties of the Bunter Sandstone of northwest Lancaster, England. *Journal of Hydrology*, **32**, 259–303.

The capillary characteristic model of petroleum hydrocarbon saturation in the Permo-Triassic sandstone and its implications for remediation

K. D. PRIVETT

SRK Consulting, Windsor Court, 1–3 Windsor Place, Cardiff CF10 3BX, UK

Present address: Hydrock Consultants, Over Court Barns, Over Lane, Almondsbury, Bristol BS32 4DF, UK (e-mail: kevinprivett@hydrock.com)

Abstract: Up to 9×10^6 l of light non-aqueous-phase liquid (LNAPL) is present in a 400 m-wide zone in the Permo-Triassic sandstone 30 m below a working industrial site. Remediation by skimmer wells failed to meet the expectations of the regulatory authorities. A detailed study has concluded that this form of remediation is not possible in this formation.

Initial estimates of the volume of LNAPL in the sandstone had been made by applying the concept of correcting for an 'exaggerated thickness' in the wells and multiplying by the porosity. Regulatory requirements were to remove most, preferably all, of the LNAPL. Site observations and measurements from high-quality core samples showed that the fine pores of the rock obey the capillary characteristic model. This was derived some 50 years ago in the oil production industry but has, in the opinion of the author, not been fully appreciated by the contaminated groundwater remediation sector in the UK. The model has been used to make a more reliable estimate of LNAPL volume and to demonstrate to the regulatory authorities the processes involved and how these control the ability to clean the site.

The concept of 'apparent volume' is introduced as an overall measure of apparent thickness in the site-wide monitoring borehole network. Apparent volume is shown to be broadly negatively correlated with groundwater-level fluctuations and so the measure of changes in apparent thickness is not a sound basis for regulatory compliance. It was found that LNAPL thickness in wells is less than the thickness of the contaminated zone in the sandstone and, importantly, the degree of saturation is no more than 30%. This leads to a reduction in effective permeability of two orders of magnitude (from the fully saturated condition), which has serious implications for remediation. Most of the LNAPL is immobile and so pumping is ineffective; it does not reduce the volume of aquifer impacted nor does it have any significant effect on reducing the source zone for dissolved-phase generation, the size of any dissolved phase plume or the longevity of such a plume.

The Permo-Triassic sandstone is a major UK aquifer and contamination by petroleum hydrocarbons (an EC List I substance) presents a significant challenge to the ability of that aquifer to supply clean water. The capillary characteristic model describes the behaviour of light non-aqueous-phase liquids (LNAPL), such as petroleum hydrocarbons, within the pores of a granular medium. It can be used to make predictions about the volume of LNAPL present, its migration through the pores and its ability to be removed by any remediation programme involving pumping.

The model has been understood in the field of petroleum production for some 50 years, but has been slow to be appreciated by practitioners in the field of contaminated land and groundwater. Indeed, there is only passing reference to it in the Environment Agency technical guidance on the subject of LNAPL monitoring (Erskine *et al.* 1998).

This paper highlights experience gained from an LNAPL contamination remediation project where failure to recognize the significance of the capillary properties of the Permo-Triassic sandstone meant that initial remedial objectives were founded on incorrect assumptions.

Case study background

The remediation project that forms the case study concerns an industrial site with a working factory covering some 17 ha. At the time of writing, the details of the site remain confidential but the main features are as follows. Bedrock lies below 5 m of terrace gravel deposits and is stratigraphically the lowest part of the Permo-Triassic sandstone, a dune-bedded sequence of cemented, jointed, fine- to medium-grained sandstone with dual permeability, dipping at 5°–7°. Aquifer properties were measured by the consultants responsible for the

From: BARKER, R. D. & TELLAM, J. H. (eds) 2006. *Fluid Flow and Solute Movement in Sandstones: The Onshore UK Permo-Triassic Red Bed Sequence.* Geological Society, London, Special Publications, **263**, 297–309.

initial work. A range of hydraulic conductivity from 3.9 to 50 m day^{-1} (mean 20.5 m day^{-1}) was calculated from eight falling head tests, and the secondary hydraulic conductivity was calculated to be between 178 and 289 m day^{-1} from a single 24-h constant-rate pumping test. There is no information available as to the variation across the site, or with depth. The same consultants recorded porosity in the range 7–17%, although the method they used is not known. Porosity measurements carried out for the present study utilized a helium expansion porosimeter that yielded values in the range 22.3–33.3%. The groundwater is approximately 30 m below ground level, fluctuating by approximately 1 m year^{-1}, and with a very low hydraulic gradient (estimated to be approximately 0.001). The regional flow direction is from the NE, but there is a radial component because the site lies close to a meander loop of a large river.

The factory was constructed on a greenfield site in 1958 and extended up until 1970. The industrial process uses kerosene (and other similar LNAPLs) that appear to have been lost into the ground since the process started. However, this leakage was only discovered in the early 1990s after an oil-budget audit was performed. Consultants were employed, who drilled some 50 boreholes across the site and identified a pool of LNAPL approximately 400 m across. Two main leakage points in the plant were identified, which corresponded to peaks in the measured thickness of LNAPL floating on the groundwater in the monitoring boreholes, the maximum thickness being approximately 3.5 m (points X and B in Fig. 1). The edge of this LNAPL accumulation lies 600 m from the river, thought to be in hydraulic connection with the groundwater, but there are no nearby groundwater abstraction points.

Textbook conceptual models of LNAPL transport (see for example Freeze & Cherry 1979) show downward flow through the unsaturated zone leaving a residual, sorbed phase, with the mobile, free phase, spreading laterally on the capillary fringe just above the phreatic surface. The weight of the LNAPL may depress the capillary fringe and phreatic surface slightly, producing what is often referred to as a 'lens' of LNAPL contamination. Within the saturated zone, a dissolved phase can form as groundwater flows across the base of the LNAPL lens, and this takes some of the more soluble fractions into solution. A dissolved-phase plume will develop down-gradient from the LNAPL lens in response to groundwater flow and other transport mechanisms. Because LNAPL is only sparingly soluble, the free phase and sorbed

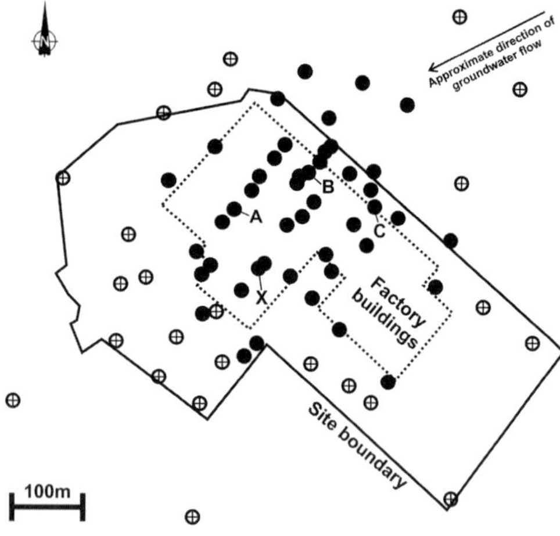

Fig. 1. Monitoring borehole layout. The solid symbols denote free LNAPL present. The points corresponding to the higher and lower peaks in Figure 3 are boreholes X and B, respectively. Cored boreholes used in this study are A, B and C.

phase will act as a source zone for the dissolved phase for a substantial period.

It is normally assumed that the thickness of LNAPL in a borehole is greater than the thickness of LNAPL-contaminated ground through which the borehole passes (see, for example, CONCAWE 1981). This is because the borehole destroys the capillary fringe locally and LNAPL can sit directly on the water at the piezometric level. The weight of LNAPL depresses the LNAPL–water interface, further increasing the thickness of LNAPL in the borehole. The measured thickness in the borehole, known as the 'apparent thickness', can be used to calculate the 'true thickness' in the ground, usually by application of an 'exaggeration factor' from a variety of published literature. The volume of LNAPL in the ground is then calculated by multiplying by the porosity. This has become known as the 'sharp interface model' because it assumes pores within the LNAPL-contaminated zone are fully saturated with LNAPL, and that there are sharp interfaces between the air and LNAPL and between the LNAPL and water *in the ground* (Fig. 2). The pores in the unsaturated zone above the LNAPL pool (and its capillary fringe) are mostly air-filled and those below (in the saturated zone) are mostly water-filled, although there may be some small residual air content. This is a simplification that does not take into account LNAPL in the unsaturated zone.

Estimates of the likely LNAPL volume beneath the case study site, based on the sharp

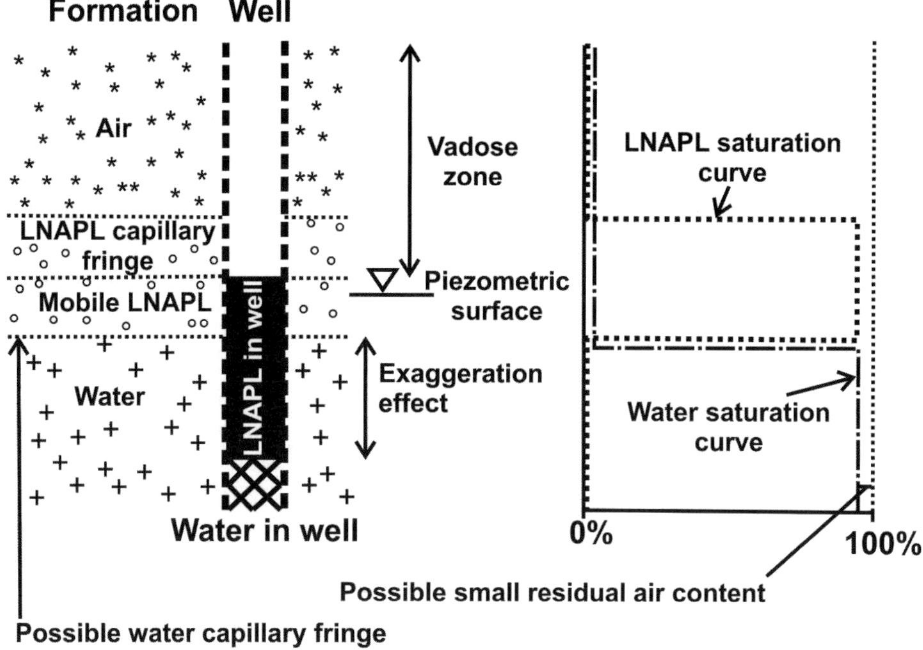

Fig. 2. The sharp interface model of LNAPL contamination. Cross-section through a monitoring well plus LNAPL and water depth–saturation curves. (After Erskine *et al.* 1998.)

interface model, varied wildly depending on the method and assumptions made. An uncorrected apparent thickness, and porosity of 10–30%, produced estimates ranging between 11.6×10^6 and 34.7×10^6 l. The methods of de Pastrovich *et al.* (1979) and Kemblowski & Chiang (1990) gave between 0.26×10^6 and 331×10^6 l depending on a range of porosity, density and capillary thickness. Four separate consultants who considered the problem based on their experience of exaggeration factor estimated between 1.8×10^6 and 10.4×10^6 l. From an assessment of the results, it was concluded that the most likely range was between 3×10^6 and 7×10^6 l in the pore space (with a further 0.5×10^6 l in the fractures), but this is not a very sound conclusion based on the wide range of estimates mentioned above.

A remedial scheme was designed and commissioned in 1998, in consultation with the regulatory authorities, which involved removing the LNAPL source using skimmer pumps in deep boreholes across the site. The stated remedial objectives were: first, to stop on-going LNAPL leakage from the factory plant by the end of 2000 (phased refurbishment being allowed to maintain production from the factory); secondly, to achieve a maximum 1 m thickness of LNAPL in the monitoring boreholes by the end of 2001; and, thirdly, to reduce this to 0.2 m by the end of 2003. The ultimate desire was to remove all LNAPL from the aquifer.

This paper is concerned primarily with the LNAPL contained in the primary (intergranular) porosity. The vast majority of the LNAPL resides in the intergranular pores and this is far less mobile than the LNAPL in the discontinuities. A new remedial strategy was developed for the site, key issues of which were the assessment of the practicalities of removing the LNAPL and of the risks associated with any residual LNAPL (that could not be removed).

Reassessment of the problem

By the end of 1999 approximately 150 000 litres had been recovered by the pumping system (from 12 boreholes). There was little, if any, change in the apparent thickness measurements and the Environment Agency was concerned about the lack of progress towards the stated remedial objectives. SRK Consulting was employed to reassess the problem, putting it for the first time into a risk-based framework using the source–pathway–receptor concept of pollution linkages.

Part of this work involved a risk assessment of the free-phase LNAPL during which the borehole records of LNAPL thickness and LNAPL–water interface level were examined in detail. In order to use the data from all the wells, a series of calculations have been made of what is called 'apparent volume'. The apparent thickness of the oil in all the wells has been contoured for each of these time snapshots and the volume under the contours calculated (using the 'Surfer' program). This is the 'apparent volume', which is an arbitrary concept, and does not represent an estimate of the volume of oil present. The scale on the apparent volume is given so the reader can see the relative changes over time. It is a device for converting a series of one-dimensional thicknesses (in the wells) into a single three-dimensional volume by integrating over the area of the oil accumulation.

It should be noted that the same pattern is produced if the apparent thickness variation in each borehole is plotted, although this is a very time-consuming exercise. Typical annual variation in apparent thickness is between 0.2 and 0.8 m for boreholes with an average apparent thickness above 0.5 m. The annual variation for those with less than 0.5 m is much smaller (less than 0.1 m). The advantage of the apparent volume plot is that a single graph can be used to represent the whole site and can be used to see seasonal variation in the whole data set. The apparent volume is not used to calculate the actual volume of LNAPL in the ground.

Figure 3 shows the apparent volume surface for three dates, the peaks corresponding to the main leakage zones in the factory above. Visual comparison of the surfaces indicates that the volume has increased between July 1995 and October 1998. This initially seems reasonable as leakage was still occurring and pumping had not yet started. By January 2000 the volume appears to have reduced slightly and this could be interpreted as a response to pumping having started in 1998.

It had been remarked upon by the previous consultants prior to 1999 that LNAPL apparent thickness was affected by seasonal groundwater fluctuations but the full implications of this were not investigated. Part of the reassessment work focused closely on the historic data in order to understand why the remediation scheme was not meeting expectations. Figure 4 is a plot of groundwater level and apparent volume on a quarterly basis since records began. Monthly data are available but quarterly data are presented for clarity.

The borehole hydrograph shows the annual fluctuation of approximately 1 m, superimposed

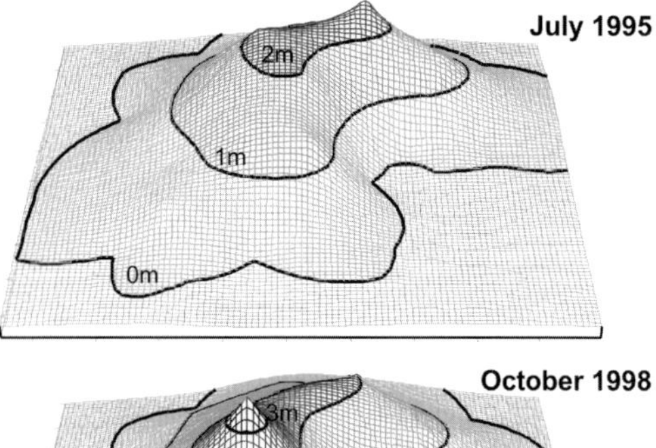

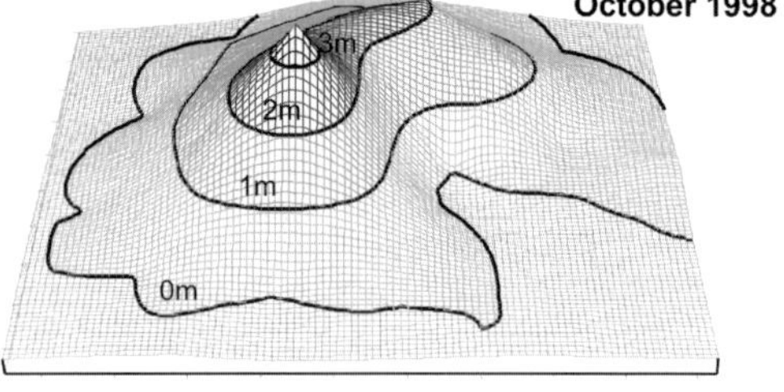

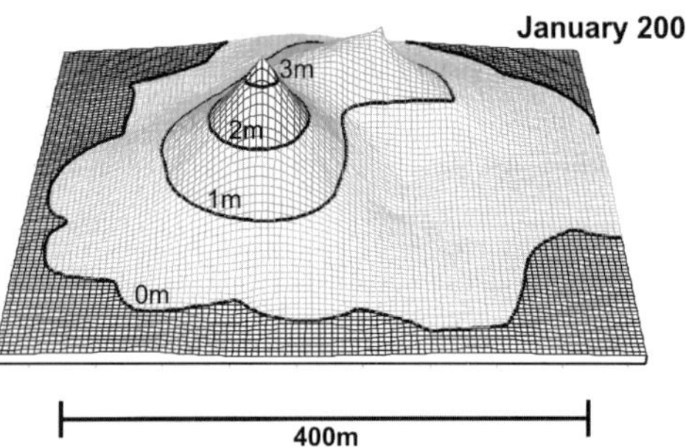

Fig. 3. The apparent volume of LNAPL on three dates. Surfaces produced by contouring apparent thickness in monitoring wells.

on a longer wavelength trend, representing variation over something like a 6-year cycle. The hydrograph is from one of the monitoring boreholes that does not contain free LNAPL. This hydrograph was compared to a number of others across the site, including some containing LNAPL, whose piezometric surface was calculated by taking into account the depression in the LNAPL–water interface. All the data gave the same picture and it is concluded that the hydrograph in Figure 4 is a true representation of the groundwater regime across the site. A visual assessment of Figure 4 shows the tendency for peaks in the apparent volume curve to coincide with troughs in the groundwater level curve, both in the annual and long-term cycles. Figure 5 is a cross-plot of groundwater elevation v. apparent volume, which shows that the correlation coefficient for the whole data set is only 0.25. The coefficients

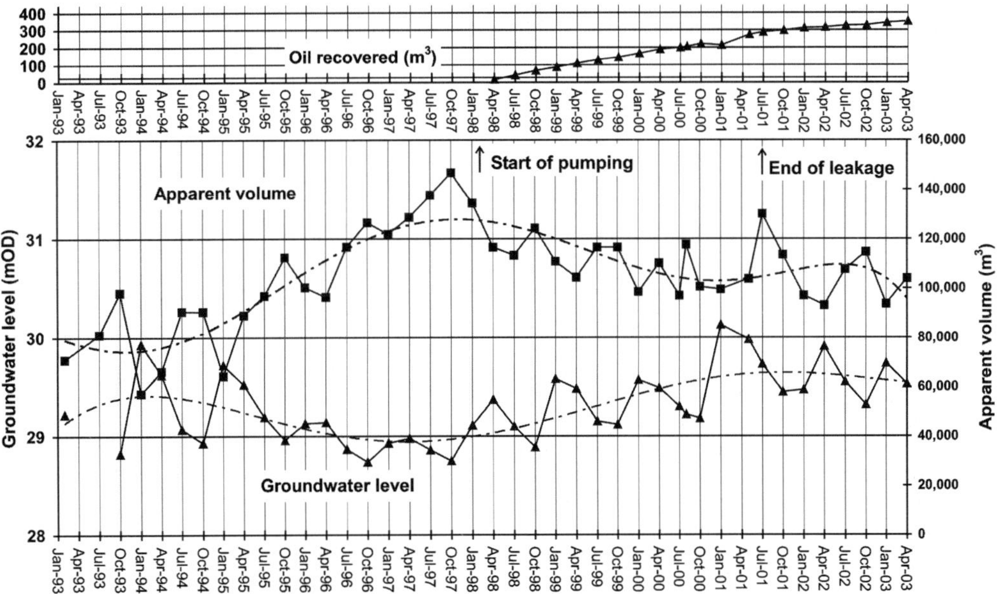

Fig. 4. The variation of apparent volume over time in relation to groundwater-level fluctuations. Polynomial trend lines show long-term trends.

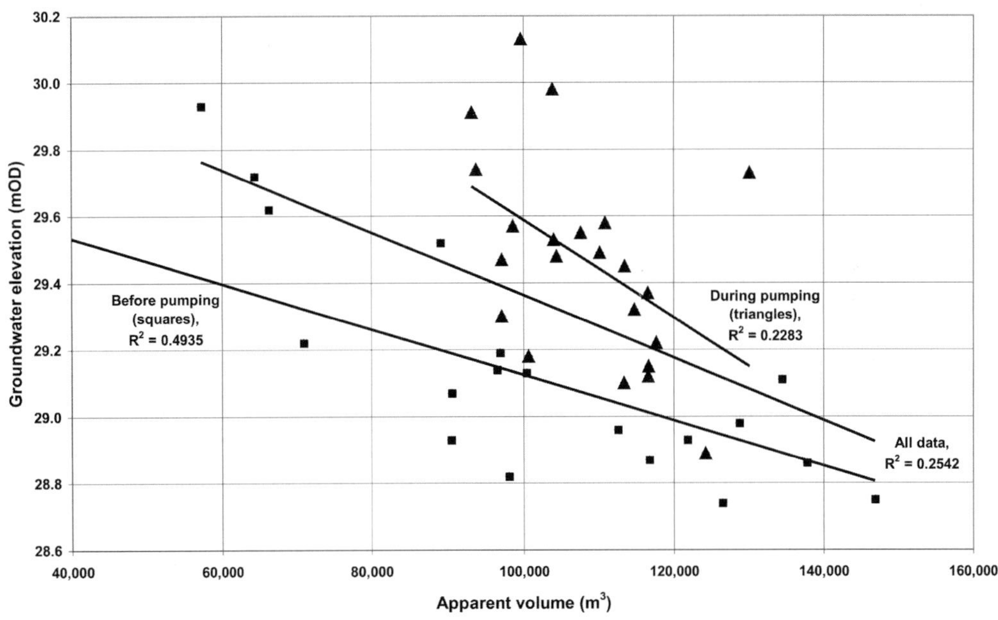

Fig. 5. Correlation between groundwater elevation (metres above Ordnance Datum) and apparent volume.

for the data before pumping started and during pumping are 0.49 and 0.23, respectively. This demonstrates two things: first, there is significant scatter in the data despite the visual recognition of a pattern, and this is interpreted as a result of time lags between groundwater level changes and the piston effect driving apparent thickness changes; and, secondly, there appears to be an effect caused by the pumping that is overprinting the broader picture.

The volumes of LNAPL being removed by pumping are small in relation to the total volume (Fig. 4, upper curve). Despite continued pumping, the apparent volume begins to increase at the beginning of 2001, and continues to increase after the final leaks were halted in July 2001, whilst at the same time the groundwater level falls. The conclusion reached is that the change in the apparent volume is *not* caused primarily by pumping but is an artefact of longer term groundwater-level fluctuation, which coincidentally started to rise at the end of 1997 just before pumping commenced. The relationship between groundwater level and apparent volume was evident in the data collected before pumping started and continues throughout the data set. It is clear that measurement of apparent LNAPL thickness in boreholes can be misleading and is, therefore, not considered a sound basis for regulatory compliance.

The capillary characteristic model

These observations cannot be explained satisfactorily using the sharp interface model of LNAPL saturation, but can be explained by the capillary characteristic model. It is not the intention here to fully explain all the concepts. The reader is directed to a number of publications where the model is used to calculate oil volumes and estimate oil mobility in porous media, which have a bearing on the findings of the current investigation (see Farr *et al.* 1990; Lenhard & Parker 1990; Huntley *et al.* 1994*a, b*; Beckett & Lundegard 1997; Lundegard & Mudford 1998; Huntley 2000; Lundegard & Beckett 2000; Huntley & Beckett 2002).

The key feature of the model (Fig. 6) is the recognition that the pores are not fully saturated with LNAPL and there is no clearly separate, floating pool or lens of free product. Instead, there is a two-phase (water–LNAPL) zone broadly co-incident with the floating LNAPL layer that appears in boreholes. Above this is a three-phase (water–LNAPL–air) zone. The degree of LNAPL saturation in the two-phase zone is dependent on the capillary characteristics of the pore spaces, in particular the pore throats, and on the fluid properties of the non-wetting (LNAPL) and wetting (water) fluids.

It is easier for LNAPL to replace water than it is for water to replace LNAPL when the fluids are moving through the intergranular porosity. This is controlled by the surface tension and the pore entry pressure. The LNAPL saturation profile, as a proportion of the total pore volume, of the aquifer is not constant (compare Fig. 2). Under conditions of vertical equilibrium (i.e. no groundwater fluctuation), the profile starts

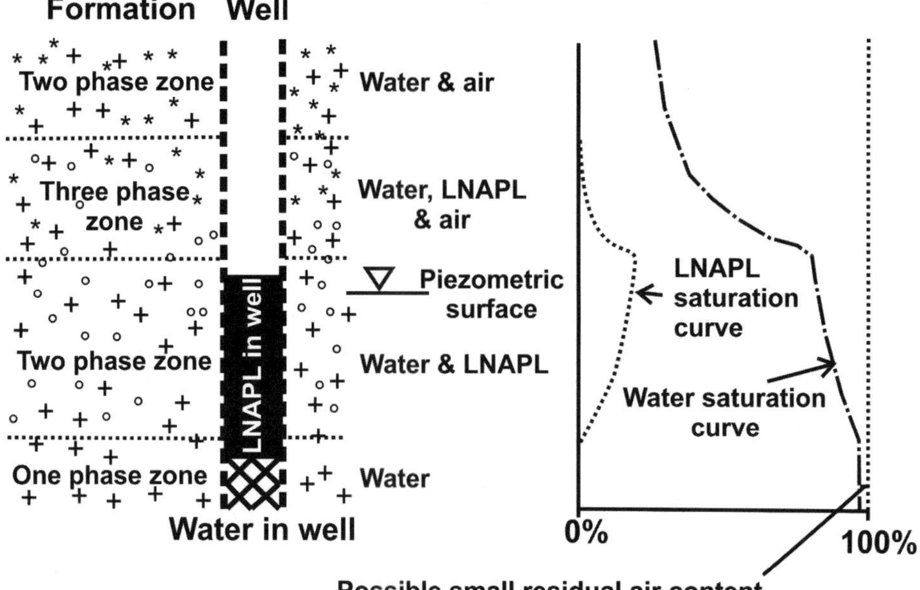

Fig. 6. The capillary characteristic model of LNAPL contamination. Cross-section through a monitoring well plus LNAPL and water depth–saturation curves. Compare with Fig. 2. (After Farr *et al.* 1990.)

above the level of the top of the LNAPL in the borehole and rapidly increases to a maximum value at the top of the LNAPL column. The saturation profile declines to zero at the level of the base of the LNAPL column in the borehole (Fig. 6).

The maximum LNAPL saturation varies with particle size. Figure 7 is an example from Huntley & Beckett (2002) for a 2 m apparent thickness of petrol in wells in different soil types. The maximum saturation for a silty clay might be less than 20% of the pore volume, rising to 90% for a medium sand. Clearly, only in sediments coarser than medium sand does the LNAPL saturation approach the classic sharp interface model of Figure 2.

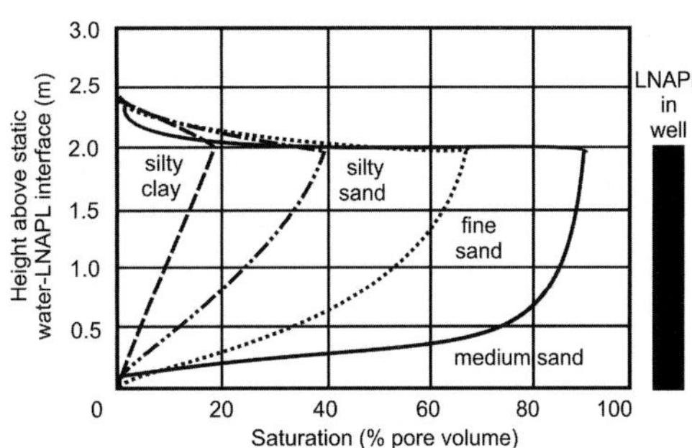

Fig. 7. LNAPL saturation curves for a range of formation grain sizes for 2 m apparent thickness of petrol in a monitoring well. (After Huntley & Beckett 2002.)

Cored boreholes

As part of the reassessment of the case study site, three new boreholes were drilled to obtain high-quality cores through the LNAPL zone (labelled A, B and C in Fig. 1). The boreholes were completed as wells for future LNAPL apparent thickness measurement. Six metres of 100 mm-diameter core were obtained from each hole using a polymer flush and plastic core liner. This was a technique originally developed by the contractor for obtaining pore-fluid samples for nitrate in groundwater studies. The cores were immediately subsampled, the outer drilling-induced smear zone removed, and the subsamples securely wrapped in cling film and aluminium foil to avoid volatilization, and then shipped to the laboratory. Determinations of LNAPL saturation, water saturation, porosity, dry density and grain specific gravity were made (at vertical separations typically between 0.17 and 0.58 m). The core samples were unwrapped in the laboratory and a section broken out from the centre of each core, placed in an extraction thimble and weighed. The pore fluids were extracted with dewatered xylene for 24 h in a Dean Stark apparatus.

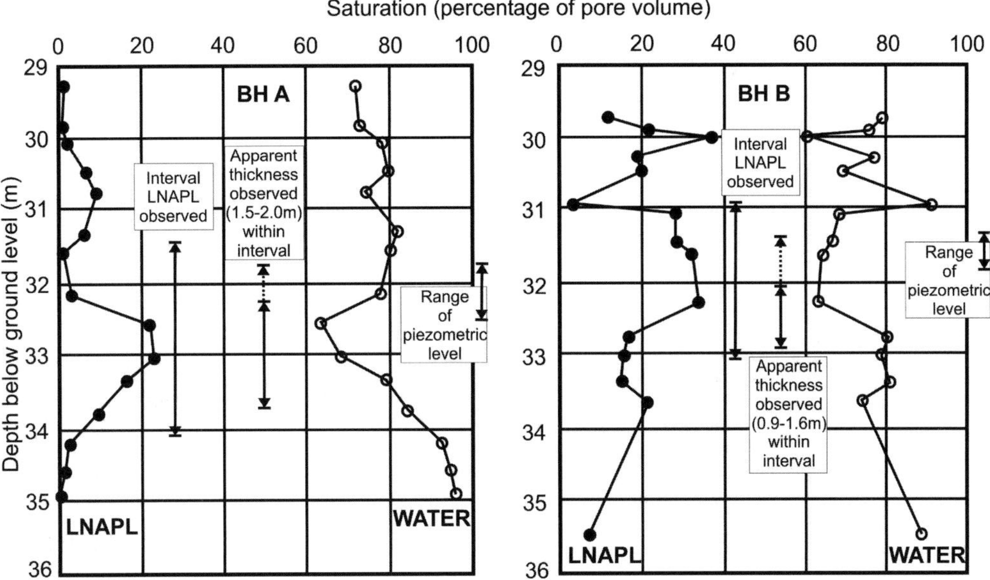

Fig. 8. Observed water and LNAPL saturation curves for two boreholes in the Permo-Triassic aquifer, as a percentage of pore volume. The bars represent subsequent measurements of the range of apparent thickness, the maximum vertical interval over which LNAPL was observed and the piezometric level.

The measured oil and water saturation profiles for two of these borehole cores are shown in Figure 8. Also indicated are the total vertical interval over which free LNAPL has been observed subsequently in each well, the maximum and minimum apparent thickness (in that interval), and the calculated range of piezometric levels (based on apparent thickness, LNAPL–water interface level and measured LNAPL specific gravity of 0.804). These observations are taken from the monthly recordings once equilibrium had been reached, defined here as the first true peak in apparent thickness, being 11 and 22 months after drilling of boreholes A and B, respectively.

Comparison with Figure 7 shows that the LNAPL saturation curve conforms to the classic shape of the capillary characteristic model and that the maximum saturation in Borehole A is 20%. This is very similar to a very-fine-grained soil (the silty clay example in Fig. 7), and is attributed to the combined effect of a fine-grained matrix and cement in the Permo-Triassic sandstone. The smaller peak in the LNAPL saturation curve above the LNAPL column in the borehole is the result of smear caused by groundwater-level fluctuations and shows that the true state of the aquifer is more complex than that described by the equilibrium model. Borehole B (Fig. 8) is broadly similar but displays a maximum LNAPL saturation of about 30% and a more pronounced smear effect above. In this case the thickness of the subsequent LNAPL column in the well is significantly less than that of the LNAPL-contaminated zone.

Revised LNAPL volume calculation

Far from there being an exaggerated thickness of LNAPL in the monitoring boreholes, as in the classic theory (Fig. 2), the above investigation shows that it is *thinner* than the contaminated zone. Importantly, LNAPL is present in the aquifer at a much lower degree of saturation than previously expected.

The shape of the saturation profile is similar to that predicted by the van Genuchten equations (van Genuchten 1980 as modified by Lundegard & Mudford 1998) and it is, therefore, assumed that these equations can be of use in estimating the true volume of LNAPL in the sandstone. One of the variables in these 'closed-form' equations is the height above the LNAPL–water interface. It is, therefore, possible to calculate saturation profiles for boreholes where only LNAPL apparent thickness is known, if the other variables in the equation can be determined from laboratory tests or site observations.

The equations are based on the height of the LNAPL in the well above the LNAPL–water interface, and so this point with zero LNAPL saturation is the base of the calculated profile.

The shape of the profile is generated by one set of equations over the full height of the oil in the well (between the elevations of the LNAPL–water interface and the LNAPL–air interface). The maximum oil saturation is generally at the top of the oil layer and the equations produce this inflection point at the top of the LNAPL. Water saturation (S_w) is calculated by:

$$S_\mathrm{w} = (1 - S_\mathrm{r})\left[\frac{1}{1+(\alpha_\mathrm{ow}P_\mathrm{cow})^n}\right]^{(1-1/n)} + S_\mathrm{r} \quad (1)$$

where α_ow and n are the van Genuchten capillary parameters for the LNAPL–water system, S_r is the residual water saturation and P_cow is the LNAPL–water capillary pressure head. $P_\mathrm{cow} = (1 - \rho_\mathrm{r})h_\mathrm{ow}$, where ρ_r is the LNAPL relative density (the ratio of the LNAPL and water densities ($\rho_\mathrm{o}/\rho_\mathrm{w}$)) and h_ow is the height above the LNAPL–water interface. LNAPL saturation (S_o) is calculated using:

$$S_\mathrm{o} = 1 - S_\mathrm{w}. \quad (2)$$

A second set of equations complete the profile in the zone above the LNAPL in the well, closing at zero saturation at the top of the profile. The total liquid saturation ($S_\mathrm{o} + S_\mathrm{w}$) is calculated by using:

$$S_\mathrm{o} + S_\mathrm{w} = (1 - S_\mathrm{r})\left[\frac{1}{1+(\alpha_\mathrm{ao}P_\mathrm{cao})^n}\right]^{(1-1/n)} + S_\mathrm{r} \quad (3)$$

where α_ao is the van Genuchten parameter for the air–LNAPL system and P_cao is the air–LNAPL capillary pressure head. $P_\mathrm{cao} = \rho_\mathrm{r}h_\mathrm{oa}$, where h_oa is the height above the air–LNAPL interface. LNAPL saturation above the air–LNAPL interface in the well is then calculated from equation 3 minus equation 1.

The area under the saturation profile curve is known as the 'specific volume' and has units of 'm^3 of LNAPL per m^2 of aquifer' (the area of aquifer being in plan view). It is possible to integrate the specific volume data for all the wells in the area of LNAPL contamination by contouring the values and calculating the volume under the contour surface, in much the same way as the apparent volume was calculated earlier. The results of this volume calculation, undertaken here using Surfer, represents an estimate of the total LNAPL volume in the pore space. A correction can then be applied to take into account the additional oil in the sandstone, in the smear zone caused by non-equilibrium effects of fluctuating groundwater level (seen in Fig. 8).

The three field profiles from the borehole cores have be used to back-calculate the other unknown parameters necessary to solve the van Genuchten equations. The method of Lundegard & Mudford (1998) was used to define the LNAPL saturation profiles. The closed-form equations were set up in a spreadsheet and a graphical output prepared that plots observed data and calculated data on the same axes. The equations were verified by using the data published by these authors. The three van Genuchten parameters plus the residual water saturation, S_r, were varied to obtain a best fit between the observed and the calculated profiles. The optimized fitting was undertaken using the method of least squares (using the 'Solver' facility in Microsoft Excel).

Figure 9 shows the observed and calculated profiles for Borehole B (the x-axis scale has been expanded compared to Fig. 8). The upper graph shows a very good fit between the calculated profile and the core data. Good fits were also obtained for the other two boreholes, but the four variables derived from the three boreholes covered a range of values. In order to define a set of 'best-estimate' parameters, which could be taken as representative of the whole site area, it was necessary to average the values. This was achieved through a sensitivity analysis, using some judgement, rather than a simple mean. Boreholes A and B gave similar results, but Borehole C was on the edge of the LNAPL accumulation zone and possessed a poorly developed profile compared with the other two. For that reason, it was omitted and the 'best-fit' parameters were derived from boreholes A and B. The arithmetic means of the parameters from the two boreholes were used as the starting point for another Solver optimization, and the final results were taken as being appropriate for all the boreholes across the site. This is undoubtedly a sweeping generalization, but it is considered a far better approximation than any of the originally tried methods based on the sharp interface model.

Figure 9b shows the comparison between the observed profile for Borehole B and the recalculated profile based on the best-estimate parameters. Although not as good a match as

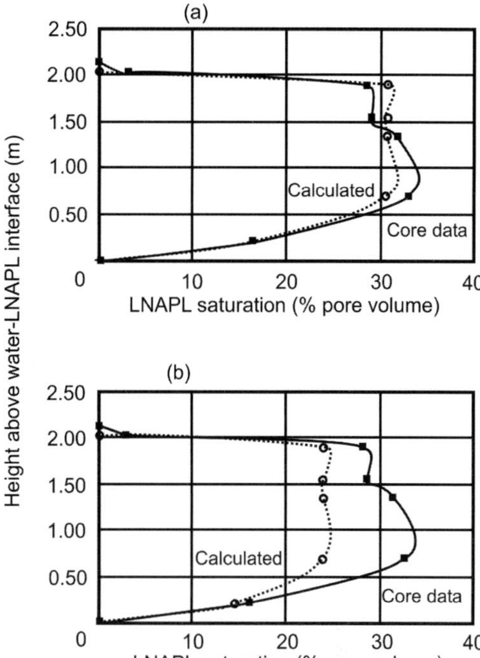

Fig. 9. Comparison of measured (solid line) and calculated (dotted line) LNAPL saturation curves for Borehole B. (**a**) The calculated curve is based on back-calculated parameters for the actual borehole and and demonstrates the goodness of fit between observed and calculated profiles. (**b**) The calculated curve is based on the 'best-fit' parameters derived from two boreholes and is judged appropriate for the whole site.

the borehole-specific calculated profile (Fig. 9a), it is considered a reasonable fit and, therefore, a reasonable basis for a volume calculation. It is accepted that the result will not be precise, but it is an estimate based on site-specific observations that can be used by the site owner and the regulatory authorities is assessing the magnitude of the contamination problem.

The total volume of LNAPL was calculated, by integrating the estimated saturation profiles derived for all the boreholes across the site, to be 5.9×10^6 l. This is an equilibrium condition as it includes only the portion of the saturation curve in Figure 9 and not the smear effect shown in Figure 8. Examination of the three observed saturation profiles led to the conclusion that there could be as much as 50% more LNAPL in the smear zone, which extended both above and below the column of LNAPL in the boreholes. If this was typical of the site as a whole, the maximum LNAPL volume would be nearer to 9×10^6 l.

There are clearly a number of assumptions behind this calculation but it is considered a reasonable estimate in the circumstances, not least because it is based on an understanding of the mechanisms operating at the site.

Effects of the capillary characteristic model

When LNAPL saturation is less than 100%, the permeability with respect to LNAPL is reduced because LNAPL has to flow around water-filled pores where it cannot displace the water. This 'relative permeability' depends on grain size (hence pore size) and apparent thickness. As we have seen, the silty clay example of Huntley & Beckett (2002) in Figure 7 bears a close resemblance to the observed behaviour of the Permo-Triassic sandstone. Reference to Figure 10 shows that if there is 3 m of LNAPL in a borehole in a silty clay, the permeability of the LNAPL in the ground may be only 1% of the value it would have been were the ground 100% saturated with LNAPL. This starts to explain why the originally designed pumping scheme did not meet its expectations. The wells are rapidly emptied and take a long time to refill. Most of the oil recovered probably originates from the fracture system. Furthermore, the permeability falls dramatically as the apparent thickness is reduced. Figure 11 is a further example from Huntley & Beckett (2002), this graph being in absolute units, and shows a fall of two orders of magnitude in permeability for the silty clay example when the apparent thickness falls by one order of magnitude. If the head in a well is permanently reduced by pumping, the effective permeability (at equilibrium) in the zone of influence falls dramatically.

A number of effects have been noted by various authors (see for, example, Farr et al. 1990; Lenhard & Parker 1990; Huntley et al. 1994a, b; Beckett & Lundegard 1997; Lundegard & Mudford 1998; Huntley 2000; Lundegard & Beckett 2000; Huntley & Beckett 2002) and are summarized in Table 1. These have implications for the management of LNAPL-contaminated aquifers with fine pore size, such as the Permo-Triassic sandstone.

The change in apparent thickness is seen as a key indicator that the capillary characteristic model applies to a given site. The explanation (Huntley et al. 1994a) is that when the groundwater-level falls in the summer, the LNAPL

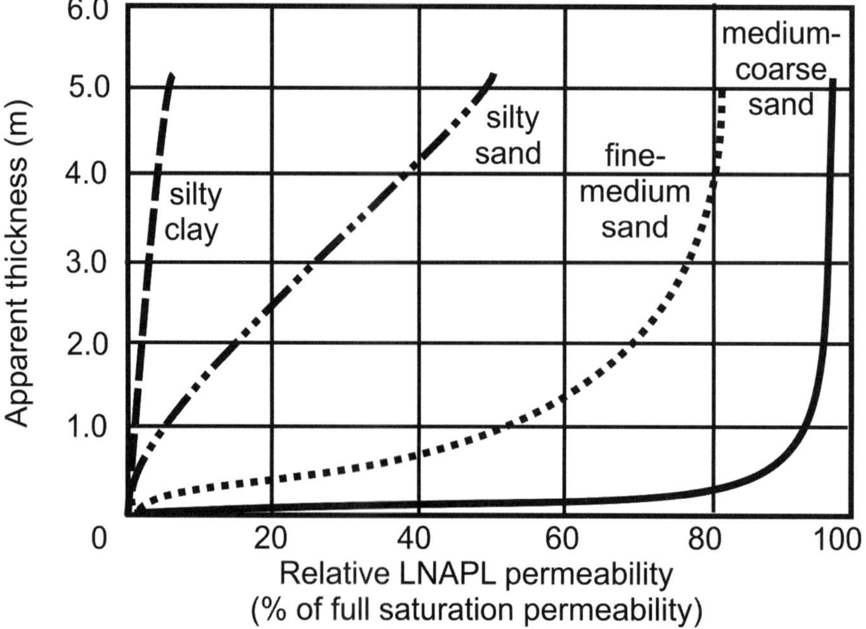

Fig. 10. The effect of formation grain size on LNAPL relative permeability, for a range of apparent thickness of LNAPL in a monitoring well. The permeability with respect to LNAPL for partially saturated materials is generally much lower than it would be if the material were to be fully saturated with LNAPL. (After Huntley & Beckett 2002.)

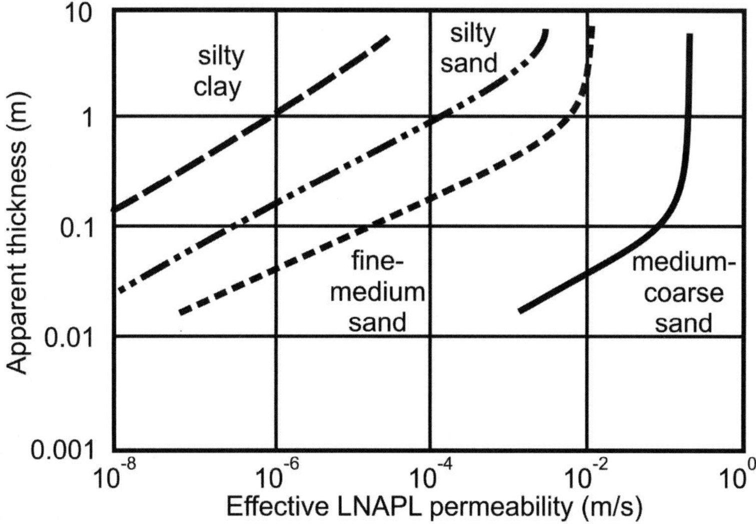

Fig. 11. The effect of formation grain size on LNAPL effective permeability, for a range of apparent thickness of LNAPL in a monitoring well. (After Huntley & Beckett 2002.)

Table 1. *Summary of the effects of the capillary characteristic model*

Apparent LNAPL thickness depends on the capillary properties of the aquifer:
* higher LNAPL saturation leads to greater apparent thickness, all else being equal;
* finer grain size (i.e. pore size) leads to greater apparent thickness, at the same degree of saturation;
* LNAPL volume may be predicted from the van Genuchten, or similar, equations but assumes vertical equilibrium (i.e. no smear).

Permeability decreases exponentially with reduced LNAPL saturation:
* at 10% saturation, permeability is two orders of magnitude less than it would have been at 100% saturation at the case study site;
* LNAPL generally occurs at low saturation except in very coarse sediments;
* even at 20 or 30% saturation, LNAPL in the sediments may be effectively immobile as far as remediation by pumping is concerned;
* there is a 'residual saturation' of 20 or 30% (may be as high as 60% in fine soils and rocks) that is impossible to remove by pumping;
* lateral LNAPL spread at the margins of the accumulation zone tends to be self-limiting as low saturation at the margin and low apparent thickness leads to low permeability.

The sharp interface model is only applicable to very coarse soils and rocks:
* in the Permo-Triassic sandstone there is no 'exaggeration effect', and the LNAPL extends above and below the column of LNAPL in the boreholes – it is stranded in the pores as groundwater-level fluctuates and cannot be displaced by water;
* the LNAPL in the aquifer above and below the column in the boreholes is immobile (but will still contribute to the generation of a dissolved-phase plume);
* of the LNAPL in the aquifer adjacent to the column in the boreholes, only a small proportion is easily mobilized, as demonstrated by the poor performance of pumping and the change in apparent thickness in response to groundwater fluctuation.

cannot keep pace with it. This causes the LNAPL to drain locally into the well and the apparent thickness increases. When the groundwater-level rises again in the winter, some of the LNAPL has, by then, moved down and it gets trapped in the pores by the rising groundwater. LNAPL from the well is pushed above the LNAPL level in the aquifer and LNAPL flows back into the aquifer, thereby reducing the apparent thickness. This is a hysteresis loop and leads to smear and permanently trapped LNAPL in the long term.

Implications for remediation

The above-mentioned effects have serious implications in any attempt to remediate the Permo-Triassic aquifer using skimmer pumps.

- The apparent thickness of oil measured in wells cannot be used as a basis for regulatory compliance because of the large variations that can be induced by groundwater-level fluctuations.
- Oil recovery reduces exponentially as saturation falls, with distance from the well, and as time proceeds.
- Oil recovery does not reduce the volume of aquifer impacted because oil remains locked in its original position.
- Oil recovery has no effect on reducing the size of a dissolved-phase plume and only a very limited effect on longevity, because only a small proportion can be removed.

Conclusions for the case study site

The regulatory authorities originally demanded complete removal of LNAPL to return the aquifer to pristine condition, as a strategic groundwater resource. This process was commenced by skimmer pumping, and apparent thickness was to be a measure of success. Since pumping is a long-term remedial option, it is inevitable that trends in apparent thickness were used to gauge progress. It has been shown that the capillary pores are important as they hold most of the LNAPL spill. A study of the mechanisms involved has shown that most of the LNAPL cannot be removed and because groundwater-level changes are the major driver for apparent thickness in the wells, this is not a satisfactory indicator for regulatory compliance.

Regulatory compliance at the case study site is now based on risk assessment. The LNAPL is relatively static and immobile to pumping, and migration is not perceived as a serious risk. The LNAPL pumping exercise continues as it removes some of the free product from the discontinuities. However, it is primarily viewed as a monitoring exercise and not as a clean-up technique. Long-term observations of LNAPL recovery rates will help to assess the mobility of the LNAPL. Yield has fallen in the 4 years since the operation started in 1998, from an average of approximately 250 to 150 l day^{-1}, with a total of only 350 000 litres recovered.

The dissolved-phase plume aspect of the risk management is being addressed by monitored natural attenuation and is the topic of a separate paper (Rees 2006).

The author expresses his thanks to P. D. Lundegard of Unocal Corporation for his valuable assistance, in particular with advice on solving of the van Genuchten equations.

References

BECKETT, G.D. & LUNDEGARD, P.D. 1997. Practically impractical – the limits of LNAPL recovery and relationship to risk. *Proceeding of the NGWA/API Conference on Petroleum Hydrocarbons and Organic Chemicals in Ground Water*, National Ground Water Association, Westerville, Ohio, 442–445K.

CONCAWE. 1981. *Revised Inland Oil Spill Clean-up Manual*. Report, **7/81**. CONCAWE, The Hague.

DE PASTROVICH, T.L., BARADAT, Y., BARTHEL, R., CHIARELLI, A. & FUSSELL, D.R. 1979. *Protection of Groundwater from Oil Pollution*. CONCAWE, The Hague.

ERSKINE, A.D., GREEN, H.R. & HEATHCOTE, J.A. 1998. *Review of LNAPL Monitoring Techniques in Groundwater*. Environment Agency Technical Report, **P148**.

FARR, A.M., HOUGHTALEN, R.J. & MCWHORTER, D.B. 1990. Volume estimation of light nonaqueous phase liquids in porous media. *Ground Water*, **28**, 48–56.

FREEZE, R.A. & CHERRY, J.A. 1979. *Groundwater*. Prentice-Hall, Englewood Cliffs, NJ.

HUNTLEY, D. 2000. Analytical determination of hydrocarbon transmissivity from baildown tests. *Ground Water*, **38**, 46–52.

HUNTLEY, D. & BECKETT, G.D. 2002. *Evaluating Hydrocarbon Removal from Source Zones and its Effect on Dissolved Plume Longevity and Magnitude*. American Petroleum Institute, Washington, DC, Publication, **4715**.

HUNTLEY, D., HAWK, R.N. & CORLEY, H.P. 1994*a*. Nonaqueous phase hydrocarbon in a fine-grained sandstone: 1. comparison between measured and predicted saturations and mobility. *Ground Water*, **32**, 626–634.

HUNTLEY, D., WALLACE, J.W. & HAWK, R.N. 1994*b*. Nonaqueous phase hydrocarbon in a fine-grained sandstone: 2. effect of local sediment variability on the estimation of hydrocarbon volumes. *Ground Water*, **32**, 778–783.

KEMBLOWSKI, M.W. & CHIANG, C.Y. 1990. Hydrocarbon thickness fluctuations in monitoring wells. *Ground Water*, **28**, 244–252.

LENHARD, R.J. & PARKER, J.C. 1990. Estimation of free hydrocarbon volume from fluid levels in monitoring wells. *Ground Water*, **28**, 57–67.

LUNDEGARD, P.D. & BECKETT, G.D. 2000. Practicability of LNAPL recovery – implications for site management. *In: Proceedings of the 2nd International Conference on Remediation of Chlorinated and Recalcitrant Compounds, 2000*, Battelle Press, Columbus, Ohio.

LUNDEGARD, P.D. & MUDFORD, B.S. 1998. LNAPL volume calculation: parameter estimation by nonlinear regression of saturated profiles. *Ground Water Monitoring and Remediation*, **18**, 88–93.

REES, S.B. 2006. Investigaton and management of a kerosene leakage into a Permo-Triassic sandstone aquifer in the UK. *In*: TELLAM, J.H. & BARKER, R.D. (eds) *Fluid Flow and Solute Onshore UK Permo-Triassic Red Bed Sequence*. Geological Society, London, Special Publications, **263**, 311–324.

VAN GENUCHTEN, M.T. 1980. A closed form equation for predicting the hydraulic conductivity of unsaturated soils. *Soil Science Society of America Journal*, **44**, 892–899.

Investigation and management of a kerosene leakage into a Permo-Triassic sandstone aquifer in the UK

S. B. REES

Geotechnology, Ty Coed, 5 Cefn-yr-Allt, Aberdulais, Neath, Swansea SA10 8HE, UK
(email: benrees@geotechnology.net)

Abstract: Leakage over a 35-year period, starting in the 1960s, of approximately 9×10^6 l of kerosene from an industrial complex located above fractured Permo-Triassic sandstone aquifer resulted in the potential for a significant dissolved phase. However, a phased, risk-based approach identified no significant risks in terms of dissolved-phase petroleum hydrocarbons, BTEX, or volatile or semi-volatile organic compounds. This is considered to be due to biodegradation of potential contaminants via metabolic pathways using electron acceptors primarily provided by the groundwater recharge (oxygen, nitrate and sulphate). Contaminant concentrations have remained below regulatory limits for protection of the receptor since monitoring was initiated in 2000. This case study demonstrates the potential for biodegradation within the Permo-Triassic sandstone aquifer of the UK: the risk-based approach described and the management plan developed could be applicable at other sites.

Between the mid 1960s and 2002, an estimated 9×10^6 l of predominantly kerosene oil leaked from a series of point sources at an operational industrial complex located in Shropshire, UK, on gravels overlying Permo-Triassic sandstone. Specific site details are not given for confidentiality reasons, but further background information is given in Privett (2006).

Investigations by previous consultants, comprising installation of groundwater monitoring boreholes, showed that free-phase oil had migrated vertically through the sand and gravel and was in contact with groundwater approximately 30 m below the site. Monthly monitoring of the oil and groundwater levels has indicated that over 2 m apparent depth of free-phase oil occurs in some areas (see Privett 2006). Figure 1 illustrates the approximate extent of aquifer impacted by the free-phase oil during the period 1993–2003.

Between December 1999 and November 2000, a detailed review of all historic groundwater-quality monitoring information was undertaken. The aim of the review was to assess the risk posed by the potential development of a dissolved-phase plume. This case study describes the approach adopted at the time, the risk-based management plan developed in response to the findings and the monitoring results for the subsequent period.

Site hydrogeology

The site is on level high ground within a meander loop of the nearby river, and is immediately underlain by a sequence of alluvial terrace sands and gravels approximately 5 m thick. Fractured Permo-Triassic sandstone underlies these terrace deposits.

The water table is approximately 30 m below ground level and coincident with the water level in the river (some 400 m to the south of the site boundary). Seasonal fluctuations in groundwater level are typically less than 1 m with little impact on the very shallow (0.001), approximately N–S-directed, hydraulic gradient estimated for the site. Privett (2006) describes the influence of the oil on the local groundwater regime. The site also has a licensed soakaway that causes localized groundwater mounding in the SE corner. Contours on the surface of the groundwater are given in Figure 2a and b. The effect of the kerosene NAPL mound has been to depress the water surface across much of the site, except in the south where the water table mound below the soakaway can be seen.

During the installation of the existing groundwater monitoring network by previous consultants, several tests were undertaken to assess aquifer properties. These tests were only made at a few locations and little information exists about lateral or vertical variation within the aquifer. Eight falling head tests indicated that hydraulic conductivity varies from 3.9 to 50 m day^{-1}, with a mean of 20.5 m day^{-1}. A single 24-h constant-rate pumping test showed hydraulic conductivity to be between 178 and 289 m day^{-1}. The same consultants recorded porosity in the range 7–17%, although the

From: BARKER, R. D. & TELLAM, J. H. (eds) 2006. *Fluid Flow and Solute Movement in Sandstones: The Onshore UK Permo-Triassic Red Bed Sequence*. Geological Society, London, Special Publications, **263**, 311–324.
0305–8719/06/$15 © The Geological Society of London 2006.

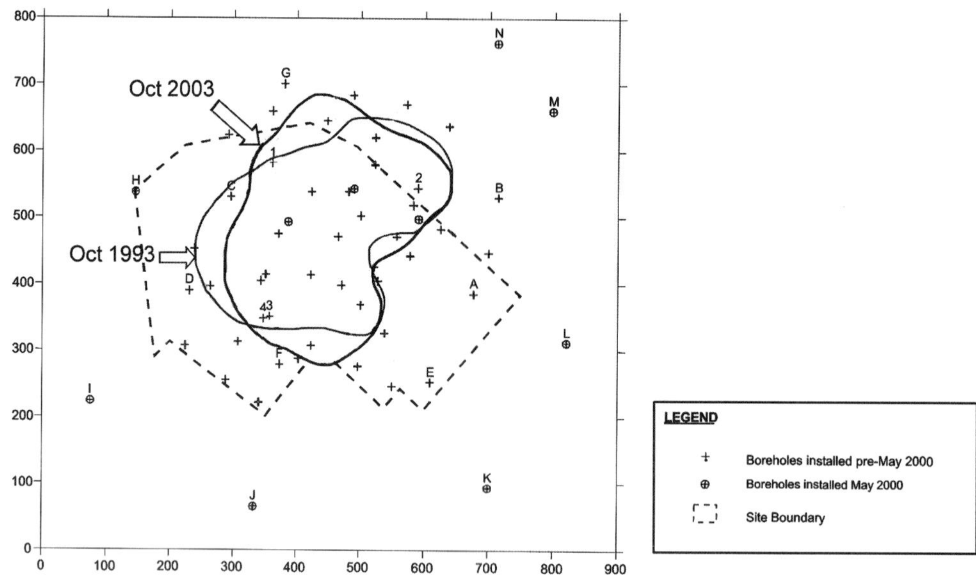

Fig. 1. Location of 0.5 m apparent oil-thickness contour for the month of October for 1993 and 2003.

method used is unknown. Porosity measurements carried out on aquifer material during the present study utilizing a helium-expansion porosimeter yielded values in the range 22.3–33.3% (Privett 2006).

Study approach

Risk assessment methodology

The approach adopted for the risk assessment followed the UK Environment Agency methodology as described by Marsland & Carey (1999). This enabled a phased approach, ensuring the efficient allocation of resources through a structured decision-making framework incorporating cost–benefit consideration and progressive data collection. Essentially, the methodology advocates four successive tiers, each one involving increasing degrees of data collection and complexity.

- *Tier 1* considers whether the source-zone pore-water contaminant concentrations are sufficient to impact on the receptor – dilution, dispersion and attenuation are ignored.
- *Tier 2* involves the comparison of the observed contaminant concentration in groundwater below the site and the target concentration for the receptor.

- *Tiers 3 and 4* consider whether natural attenuation (including dispersion, retardation and degradation) of the contaminant as it moves to the receptor is sufficient to reduce the concentration to an acceptable level.

Tiers 3 and 4 are distinguished by the sophistication of the modelling and prediction processes.

Initially, the receptor was considered to be the groundwater beneath the site. However, groundwater would also act as a pathway for contaminant movement towards the river, which is used for public water supply several kilometres downstream. Ultimately, therefore, it was agreed with the Environment Agency that the river was the principal receptor, although further contamination of the groundwater down-gradient of the site would not be allowed. The assessment of risk posed by the free-phase oil is described by Privett (2006).

Site investigation

In May 2000, seven 50 mm boreholes, identified as H–N in Figure 2b, were installed. These boreholes were considered necessary as the existing groundwater monitoring network did not include up-gradient (background) or down-gradient (off-site) 'sentinel' monitoring locations. These factors limited assessment of

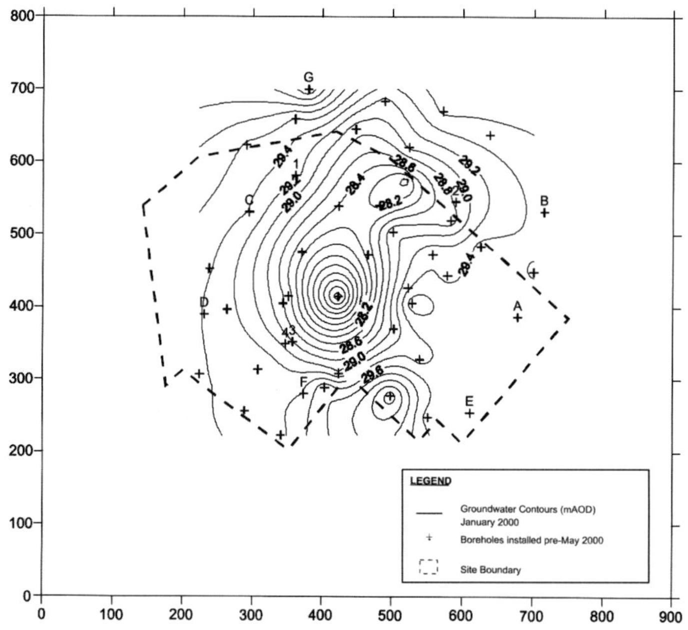

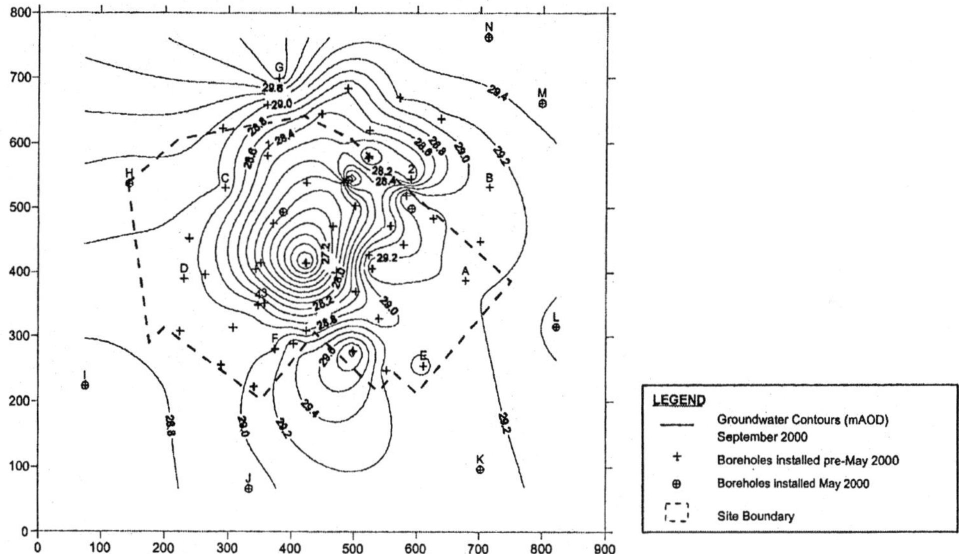

Fig. 2. Contours on the groundwater surface (heads where there is no oil present – see Fig. 1 – and the oil–water interface where it is present). (**a**) January and (**b**) September 2000. Additional monitoring boreholes are included for September 2000 (see text for details of site investigation in May 2000).

the dissolved phase and the development of a conceptual groundwater model.

Boreholes were drilled in locations with available access for the drilling rig and long-term access for monitoring. All new monitoring boreholes were completed with UPVC well pipe with slotted casing in the bottom 6 m, within which the seasonal water table fluctuates.

Similar borehole construction was used during the installation of the earlier monitoring network, although precise records detailing the length of the screened intervals are incomplete.

Chemical analysis and testing

The assessment of the dissolved phase was hindered by the fact that the historical data set was based on several different, not easily comparable, analytical techniques for determination of dissolved hydrocarbon in groundwater. As a result, a standardized approach was developed that involved the analysis of: (a) speciated total petroleum hydrocarbons (TPH) as described by Vorhees *et al.* (1999); and (b) volatile organic compounds. The analytical programme is summarized in Table 1.

One of the concerns during the positioning of the new boreholes in May 2000 was the ability to be able to assess groundwater-contaminant concentrations along profiles from the area closest to the oil accumulation to the new sentinel boreholes. Two distinct techniques were used to assess the maximum likely concentration of dissolved hydrocarbons beneath the oil, as this would represent the contaminant source concentration:

- direct sampling of groundwater beneath the oil layer;
- mixing uncontaminated groundwater from upgradient (borehole M) with oil obtained from wells 1, 2 and 4 and analysing the subsequent aqueous phase in a series of elution tests.

Groundwater samples from below the oil were collected using a disposable bailer following purging of the oil and groundwater with a Grundfos MP-1 pump. During sampling, the oil from the surrounding aquifer formation flowed back into the borehole. Consequently, the integrity of the groundwater samples was impossible to maintain, although no oil emulsion was visible in the samples collected.

For the purposes of the elution test, a 1:5 (oil: water) dilution was selected following experimenting with various mixtures between 1:100 and 1:10. The aim was to identify a ratio that balanced minimizing the formation of an emulsion with a volume of water that would allow for the analytical detection of soluble oil components at low levels. The oil–water mixture was shaken for 5 min in stoppered separation funnels and allowed to stand until separated. Following separation, subsamples of water were drawn off and analysed in accordance with the analytical programme (Table 1). The oil was also analysed for the same parameters.

Monitoring

High-quality sampling and analysis of groundwater using consistent sample collection and analytical techniques was fundamental to the risk-assessment process. Quarterly sampling of boreholes A–N was initiated in May 2000. Hydrocarbons are susceptible to microbially mediated biodegradation that may result in changes to the concentration and distribution of inorganic groundwater parameters (Fetzner 1998; Fang *et al.* 2000). Therefore, dissolved

Table 1. *Analytical programme*

Analysis	Includes	Method[†]	Detection limit
Speciated total petroleum hydrocarbons (TPH)	Carbon groups as follows: Aliphatics: C5–C6, >C6–C8, >C8–C10, >C10–C12, >C12–C16, >C16–C35* Aromatics: C6–C7, >C7–C8, >C8–C10, >C10–C12, >C12–C16, >C16–C21, >C21–C35	GC-FID	0.01 mg l^{-1}
Semi-volatile organic compounds (SVOCs)	Semi-VOCs from general scan from library match	GC-MS	0.01 mg l^{-1}
Volatile organic compounds (VOCs)	Pollutants listed under USEPA Methods 624/8260 plus tentatively identified compounds from laboratory library database	GC-MS	0.001 mg l^{-1}
BTEX	Benzene, toluene, ethyl benzene, xylene are included in speciated TPH analysis and in VOC scan	GC-FID as part of speciated TPH. GC-MS as part of VOC scan	0.01 mg l^{-1} as part of speciated TPH. 0.001 mg l^{-1} as part of VOC scan

* Since July 2001 these are reported as a two separate ranges: C16–C21 and C21–C35.
[†] GC-FID, gas chromatography–flame ionization detection; GC-MS, gas chromatography–mass spectroscopy.

oxygen (DO), temperature, redox potential (E_H), alkalinity, electrical conductivity, pH, NO_3, Fe^{2+}, Mn, SO_4 and CH_4 are also monitored, as recommended by Carey *et al.* (2000).

Boreholes are first purged of approximately 3 well volumes using a Grundfos MP-1 pump, and then sampled with a disposable bailer once DO, E_H and pH are stabilized to within 10% of the previous reading in the purged water. Agitation of the water and the potential loss of volatiles during sample collection is minimized through the use of a low-flow valve attached to the base of the bailer. Field measurements are made using calibrated Hanna portable field equipment and, since early 2001, a flow-through cell that prevents contact of the water with the atmosphere. Alkalinity is measured on filtered samples by colorimetric titration using standardized 1.6 N H_2SO_4 to a pH 4.5 end point. All samples analysed for dissolved hydrocarbons are collected in opaque glass jars and vials ensuring there is no free head space. Blind trip blanks containing ultra pure water and blind field blanks containing distilled water are also routinely submitted for analysis of dissolved hydrocarbons. All samples are stored in cooler boxes and analysed within 48 h of collection at a United Kingdom Accreditation Service accredited laboratory.

Dissolved phase assessment

Chemical data

Full chemical results from the groundwater monitoring analysis, oil analysis and the elution tests for May 2000 are given in Tables 2 and 3. Volatile organic compounds (VOCs) and semi-volatile organic compounds (SVOCs) were found to be below the analytical detection limits (0.001 mg l^{-1}) within the groundwater.

Source concentration

The approximate oil phase analyses given in Table 3 indicate that aliphatic fractions (specifically C10–C16) dominate the TPH. Other VOC and SVOC components were also detected as indicated.

Higher dissolved hydrocarbon concentrations were found in the groundwater samples from below the oil than from the elution test samples (Table 2). The recorded concentrations of C8–C35 aliphatics in the groundwater were several orders of magnitude greater than their maximum solubility (Vorhees *et al.* 1999). Coupled to a similarity in speciated TPH distribution with the free-phase oil, this suggests that these samples contained entrained free product from sample collection as suspected. Alternatively, the results may reflect increased solubility of these fractions within the oil mixture relative to those quoted by Vorhees *et al.* (1999).

Beyond the oil accumulation

Taking the up-gradient groundwater to represent background water quality, it appears that background TPH accounted for between 0.03 and 0.13 mg l^{-1} (Table 2). Groundwater down-gradient of the 0.5 m free-phase oil apparent thickness contour only contained aliphatic fractions C5–C6, >C6–C8 and >C8–C10, and aromatic fractions >C8–C10 at maximum concentrations of 0.03, 0.09, 0.12 and 0.18 mg l^{-1}, respectively, in May 2000. The TPH concentration beyond the oil was between 0.02 and 0.34 mg l^{-1}, with the highest value detected in borehole B, very close to the edge of the oil accumulation and considered slightly up-gradient, as suggested in Figure 2. TPH concentrations are plotted as a function of distance from the 0.5 m free-phase apparent thickness contour in Figure 3.

The concentrations of dissolved hydrocarbons detected beyond the oil accumulation in May 2000 were therefore largely within the range found in the background water quality. Continued monitoring of the same monitoring network on a quarterly basis following the same sampling and analytical techniques has substantiated this observation. There are no discernible trends in TPH concentration or fractionation since May 2000 (Table 4). This suggests that there is no detectable dissolved phase in terms of TPH and that conditions have not changed significantly during the 3 years of monitoring.

In terms of BTEX, VOC and SVOCs in May 2000, toluene was detected in borehole E at 0.002 mg l^{-1}, xylene in boreholes E and G at 0.001 and 0.002 mg l^{-1}, and naphthalene in borehole G at 0.002 mg l^{-1}. Since May 2000 naphthalene has been detected in two boreholes at 0.002 mg l^{-1} in June 2002 and benzene at 0.002 mg l^{-1} in one borehole in July 2001. Other VOCs have not been detected.

Biodegradation indicators

Hydrocarbon biodegradation here refers to biologically mediated oxidation reactions involving the transfer of electrons from an organic contaminant to an electron acceptor (Stumm & Morgan 1981). Micro-organisms preferentially mediate reactions yielding greatest energy, resulting in an order of electron

Table 2. *Groundwater chemistry May 2000*

Borehole	M	N	A	B	C	D	E	F	G	H	I	J	K	L	1	2	3	4	1	2
Location	Up-gradient	Up-gradient	Close to oil pool	Close to oil pool	Close to oil pool	Close to oil pool	Close to oil pool	Close to oil pool	Close to oil pool	Down-gradient	Down-gradient	Down-gradient	Down-gradient	Down-gradient	Below oil	Below oil	Below oil	Elution test	Elution test	Elution test
Speciated TPH (mg l⁻¹)																				
Aliphatics																				
C5–C6	0.01	–	–	0.02	0.03	0.03	0.03	–	0.02	0.01	–	–	–	–	0.02	0.02	0.04	–	–	–
>C6–C8	0.07	0.03	0.02	0.02	0.09	0.02	0.04	0.02	0.01	0.04	–	0.02	0.02	0.01	0.17	0.92	1.06	0.06	0.04	0.05
>C8–C10	0.02	–	–	0.12	–	–	0.01	–	0.02	0.01	–	0.02	–	0.02	12.99	32.60	19.26	0.12	0.22	0.34
>C10–C12	–	–	–	–	–	–	–	–	–	–	–	–	–	–	1.03	2.47	2.91	–	0.14	0.05
>C12–C16	–	–	–	–	–	–	–	–	–	–	–	–	–	–	3.79	6.43	10.48	–	0.09	0.08
>C16–C35	–	–	–	–	–	–	–	–	–	–	–	–	–	–	1.03	1.28	5.88	–	0.13	0.02
Total aliphatics	0.10	0.03	0.02	0.16	0.12	0.05	0.08	0.02	0.06	0.06	–	0.04	0.02	0.03	19.04	43.73	39.63	0.18	0.62	0.54
Aromatics																				
C6–C7	–	–	–	–	–	–	–	–	–	–	–	–	–	–	–	–	–	–	–	–
>C7–C8	–	–	–	–	–	–	–	–	–	–	–	–	–	–	–	–	–	–	–	–
>C8–C10	0.03	–	–	0.18	–	0.01	0.02	–	0.03	0.02	–	0.02	–	0.03	0.46	1.63	1.21	0.23	0.38	0.61
>C10–C12	–	–	–	–	–	–	–	–	–	–	–	–	–	–	–	–	–	–	0.03	–
>C12–C16	–	–	–	–	–	–	–	–	–	–	–	–	–	–	–	–	–	–	0.02	–
>C16–C21	–	–	–	–	–	–	–	–	–	–	–	–	–	–	–	–	–	–	–	–
>C21–C35	–	–	–	–	–	–	–	–	–	–	–	–	–	–	–	–	–	–	–	–
Total aromatics	0.03	–	–	0.18	–	0.01	0.02	–	0.03	0.02	–	0.02	–	0.03	0.46	1.63	1.21	0.23	0.43	0.61
TPH (= sum ali + aro C5–C35)	0.13	0.03	0.02	0.34	0.12	0.06	0.10	0.02	0.09	0.08	–	0.06	0.02	0.06	19.50	45.36	40.83	0.41	1.06	1.15
BTEX and naphthalene (mg l⁻¹)																				
Benzene	–	–	–	–	–	–	–	–	–	–	–	–	–	–	–	–	0.006	–	0.005	–
Toluene	–	–	–	–	–	0.002	–	–	–	–	–	–	–	–	–	–	–	–	0.022	–
Ethylbenzene	–	–	–	–	–	–	–	–	–	–	–	–	–	–	0.003	0.001	–	–	0.008	–
p/m-Xylene	–	–	–	–	–	–	0.001	–	–	–	–	–	–	–	0.006	0.010	0.028	–	0.013	–
o-Xylene	–	–	–	–	–	–	–	–	0.002	–	–	–	–	–	0.010	0.033	0.050	0.029	0.065	0.021
Naphthalene	–	–	–	–	–	–	–	–	0.002	–	–	–	–	–	–	–	–	–	–	–
Biodegradation indicators																				
Temp. (°C)	12.3	12.5	13.9	12.4	12.5	13.5	13.7	13.9	12.9	11.6	14.1	12.5	11.7	12.8	12.9	13.4	15.1	–	–	–
pH	6.61	6.91	6.22	6.6	6.65	6.61	7.2	6.55	7.01	7.14	7.07	6.81	7.02	6.67	7.18	6.68	6.79	–	–	–
E_H (mV)	311	355.8	220	303	326	215	305	247	308	283	242	287	330.8	318	70	26	60	–	–	–
Electrical cond. (µS cm⁻¹)	337	425	351	138	174	754	668	742	698	486	1144	599	311	639	799	1407	720	–	–	–
Alkalinity (mg l⁻¹ as CaCO₃)	134	121	123	155	83	371	212	235	76	182	192	135	166	125	300	405	425	–	–	–
DO (mg l⁻¹)	10.8	16.3	3.16	3.2	5.16	4.5	6.4	2.34	4	10.46	9.37	10.29	10	7.02	1.99	2.9	1.7	–	–	–
Nitrate (mg l⁻¹)	24.1	30	12.8	25.8	22	71.8	39.9	19.2	44.4	13.7	87.5	207.2	32	26.3	12.1	11.1	6.5	–	–	–
Sulphate (mg l⁻¹)	48	52	20	51	13	100	80	62	88	46	118	51	62	57	43	11	41	–	–	–
Methane (mg l⁻¹)	–	–	–	–	–	–	–	–	–	0.002	–	–	–	–	–	0.01	–	–	–	–
Fe(II) (mg l⁻¹)	–	–	0.24	–	–	–	–	–	–	–	–	–	–	–	0.32	13.82	0.32	–	–	–
Mn (mg l⁻¹)	–	–	–	–	–	–	–	–	–	–	–	–	–	–	9.48	0.58	–	–	–	–

Detection limits: Fe^{2+} and Mn detection limit is 0.05 mg l⁻¹; BTEX and naphthalene 0.001 mg l⁻¹; speciated TPH 0.01 mg l⁻¹.

Table 3. *Approximate oil phase analyses (May 2000)*

	Borehole	
Determinand	1	2
Speciated TPH		
Aliphatics		
C5–C6	–	–
>C6–C8	10	7
>C8–C10	38	34
>C10–C12	457 661	427 445
>C12–C16	481 703	472 842
>C16–C35	253 434	115 779
Total aliphatics	1 192 847	1 016 106
Aromatics		
C6–C7	0.2	0.1
>C7–C8	1	1
>C8–C10	88	76
>C10–C12	8244	13 112
>C12–C16	17329	21 116
>C16–C21	2234	1628
>C21–C35	1327	9789
Total aromatics	29 224	45 721
TPH	1 222 070	1 061 828
VOCs		
Benzene	1	–
Toluene	9	6
Chlorobenzene	36	–
Ethylbenzene	26	21
p/m-Xylene	111	97
o-Xylene	81	70
Isopropylbenzene	19	17
Propylbenzene	38	32
1,2,4-Trimethylbenzene	110	114
1,3,5-Trimethylbenzene	378	403
sec-Butylbenzene	52	57
tert-Butylbenzene	57	75
n-Butylbenzene	381	540
Naphthalene	447	515

All concentrations in mg l^{-1}.

acceptor preference, typically $O_2 > NO_3 >$ Mn(IV) > Fe(III) > $SO_4 > CH_4$. During this process, the groundwater becomes increasingly reducing and alkalinity increases due to oxidation of the hydrocarbon carbon content.

Figure 4 illustrates the variation of selected biodegradation indicators reported in Table 2, including groundwater samples taken from below the oil. Relative to up-gradient and down-gradient groundwater, DO, NO_3 and SO_4 are at a lower concentration below the oil and immediately down-gradient, while alkalinity (and Fe^{2+} and Mn) are elevated below the oil. Such variations suggest that biodegradation is occurring via several metabolic pathways, as noted at other sites where hydrocarbon

biodegradation has been confirmed (Fetzner 1998; Thornton *et al.* 2001*b*).

Monitoring of boreholes A–N between May 2000 and 2003 indicates that the concentration levels of the biodegradation indicators up-gradient and down-gradient of the oil accumulation has not significantly changed (Table 5). This suggests the processes affecting their distribution have not significantly changed.

Electron acceptors typically occupy a discrete area of the aquifer in relation to the contaminant plume and in accordance with their relative metabolic energies (Wiedemeier *et al.* 1995; Fang *et al.* 2000). The detection of both electron acceptors and metabolic by-products in groundwater from below the oil collected from boreholes with screened intervals of up to 6 m either suggests that there was analytical error or very pronounced aquifer heterogeneity and limited vertical dispersivity. The opportunity for analytical error is considered low as all field instruments are calibrated routinely and the same laboratory analytical techniques have always been used. There is, however, the possibility that during the measurement of DO in May 2000 the water came into contact with the atmosphere during measurement and oxygen transfer occurred. This situation was resolved in early 2001 when a flow through cell was used for field measurements.

Tier 2 risk assessment

Figure 3 suggests that no significant dissolved TPH was present above background levels in May 2000. As discussed above, continued monitoring has confirmed this situation, suggesting that there is no detectable TPH dissolved phase associated with the oil leakage. These high background levels may be the result of indigenous substances in the groundwater, such as humic and fulvic acids or analytical errors. Consequently, following discussion with the Environment Agency, it was agreed that there was no requirement to move to a Tier 3 risk assessment and that no active remedial action was required with respect to TPH fractions. The situation was the same with respect to VOCs and SVOCs, as no contaminants were found above any applicable regulatory values in the groundwater in May 2000, as summarized in Table 6.

Biodegradation capacity

The total biodegradation capacity (TBC)

The lack of a detectable dissolved phase and the distribution of biodegradation indicators

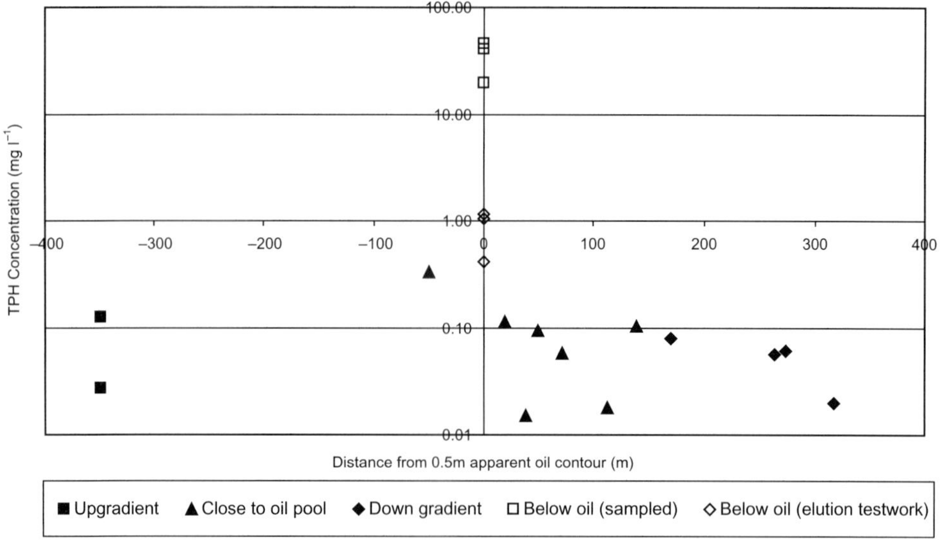

Fig. 3. TPH concentrations in May 2000 as a function of distance from the 0.5 m apparent oil-thickness contour. Concentrations below the detection limit of 0.01 mg l⁻¹ have been ignored.

suggests that contaminants have been biodegraded to levels below analytical detection. The total biodegradation capacity (TBC) is considered to comprise two components: groundwater-sourced and aquifer-sourced.

Groundwater biodegradation capacity (GBC)

The expanded monitoring network established in May 2000, and the subsequent monitoring has improved understanding of groundwater flow and its influence on biodegradation. Regional groundwater recharge occurs to the NE of the site, with groundwater generally flowing to the south. As illustrated in Figure 2, the free-phase oil suppresses the local groundwater beneath the site such that flow directions are locally complicated. There appears to be an element of radial groundwater flow from the southern boundary of the oil accumulation towards the sentinel wells I–K.

As described in detail by Privett (2006), the pores of the aquifer beneath the site have been shown to be partially saturated with free-phase oil at the oil–water interface. At the interface, there is a two-phase (water–oil) zone that is broadly co-incident with the oil that is monitored in the boreholes. Above this is a three-phase (water–oil–air) zone. As a consequence, seasonal variations in water level will serve to supply groundwater-sourced electron acceptors

(primarily O_2, NO_3 and SO_4) to the areas where oil dissolution would be envisaged to be occurring, particularly in the lower portion of the two-phase zone.

The importance of contaminant biodegradation utilizing groundwater-supplied electron acceptors can be evaluated by estimating the potential GBC, outlined in principle by Carey *et al.* (2000). This is determined by dividing the background groundwater electron acceptor concentrations by a contaminant-specific utilization factor (UF) (Wiedemeier & Rifai 1999). The UF refers to the stoichiometric mass of electron acceptor consumed during degradation of a given mass of a contaminant along a specific biodegradation pathway. The UF can be estimated from the stoichiometry of the relevant balanced biodegradation reactions, as illustrated in Table 7. The GBC (for the surrogate compound: see Table 7) can subsequently be calculated. This has been carried out for the present study, and the results for C_5H_{12} are presented in Table 8. This calculation can be repeated for each TPH fraction. This has not been presented, but the calculation clearly demonstrates the excess GBC available within the aquifer.

Sulphide produced during sulphate reduction may be re-oxidized back to sulphate. There is also evidence that CH_4 may be oxidized by sulphate (INAP 2003). Both processes may therefore result in the underestimation of the contribution of GBC to TBC from sulphate

Table 4. *Hydrocarbon monitoring data for the period 2000–2003*

Borehole	Date	Ali C5–C6	Ali C6–C8	Ali C8–C10	Ali C12–C16	Ali C16–C21	Ali C21–C35	Total aliphatics	Aro C8–C10	Aro C10–C12	Aro C12–C16	Aro C16–C21	Aro C21–C35	Total Aromatics	TPH	Naphthalene	Benzene	Xylene
H	06/06/2000	0.01	0.035	0.014				0.059	0.021					0.021	0.08			
	26/03/2002											0.016		0.016	0.016			
I	17/07/2001											0.016	0.039	0.055	0.055			
J	06/06/2000		0.02	0.015				0.035	0.022					0.022	0.057			
	17/07/2001											0.014	0.024	0.038	0.038			
K	16/06/2000		0.019					0.019						0.019				
	17/07/2001											0.019	0.03	0.049	0.049			
	27/03/2002										0.361	0.108		0.469	0.469			
	03/09/2002	0.017	0.048					0.083	0.014	0.012				0.026	0.109			
L	16/06/2000		0.014	0.02				0.034	0.03					0.03	0.064			
	18/07/2001											0.021	0.034	0.055	0.055		0.002	
	16/06/2003					0.037	0.071	0.108							0.108			
M	22/05/2000	0.007	0.069	0.02				0.096	0.029					0.029	0.125			
	17/07/2001											0.039	0.037	0.076	0.076			
	26/03/2002											0.018		0.018	0.018			
	24/06/2002					0.018		0.018						0.018		0.002		
	16/06/2003					0.139	0.375	0.514				0.034		0.034	0.548			
N	23/05/2000		0.027					0.027						0.027				
	17/07/2001											0.151	0.235	0.386	0.386			
	26/03/2002											0.036		0.036	0.036			
	24/06/2002															0.002		
	16/06/2003					0.103	0.232	0.335							0.335			
D	03/05/2000	0.029	0.018					0.047	0.011					0.011	0.058			
	17/06/2003					0.101	0.226	0.327							0.327			
G	03/05/2000	0.024	0.014	0.022				0.06	0.034					0.034	0.094	0.002		0.002
	18/07/2001				0.109	0.292	0.752	1.153			0.297	0.658	0.305	1.26	2.413			
F	08/06/2000		0.018					0.018							0.018			
	27/03/2002					0.14		0.14							0.14			
	11/12/2002					0.045		0.045							0.045			
	17/06/2003					0.015	0.048	0.063							0.063			
B	03/05/2000	0.022	0.021	0.117				0.16	0.175					0.175	0.335			
	18/07/2001											0.035	0.039	0.074	0.074			

All values in mg l^{-1}. Blanks indicate hydrocarbon below analytical detection limit (0.01 mg l^{-1} for TPH and 0.001 mg/l for VOCs).

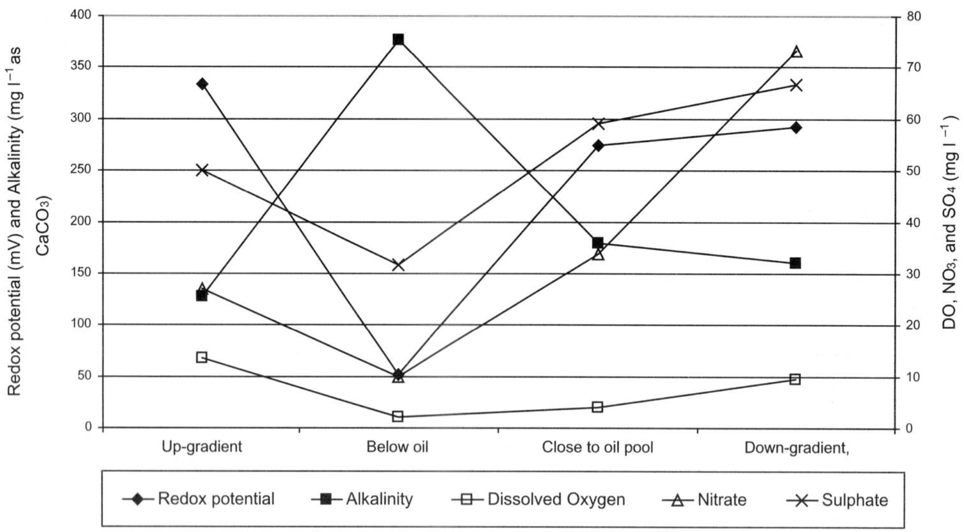

Fig. 4. Variation of biodegradation indicators (May 2000) across the site.

Table 5. *Statistical summary of biodegradation indicators for the period 2000–2004*

	DO (mg l⁻¹)	Temp. (°C)	E_H (mV)	EC (µS cm⁻¹)	pH	Field alkalinity (mg l⁻¹ as CaCO₃)	Nitrate (mg l⁻¹)	Sulphate (mg l⁻¹)
Upgradient boreholes (M and N)								
Minimum	7.4	10.6	162	245	6.23	54	21	39
Maximum	16.3	15.5	356	425	7.82	140	38	52
Count	16	17	15	17	17	14	17	17
Average	9.3	13.3	244	301	6.75	91	29	44
Standard Deviation	2.2	1.18	61	51	0.44	30	5.4	3.8
Boreholes close to oil pool (A–G)								
Minimum	0.2	8	68	92	5.15	76	11	13
Maximum	7.7	15.6	560	909	8.7	373	97	100
Count	43	42	39	43	43	34	39	39
Average	2.4	12.7	217	585	6.63	217	36	61
Standard Deviation	1.8	1.36	91	174	0.71	83	25	21
Down-gradient boreholes (H–L)								
Minimum	4.4	10.3	32	119	5.92	42	4.2	22
Maximum	13	14.9	371	1230	8.11	350	207	118
Count	50	50	45	50	50	40	45	45
Average	8.4	12.6	223	551	6.84	149	55	53
Standard Deviation	2.0	1.2	75	296	0.50	73	42	20

reduction on the basis of absolute concentrations alone. Natural organic material present in the aquifer would also be susceptible to oxidation, although only 0.06 wt% was detected in three samples (Privett 2006).

Aquifer biodegradation capacity (ABC)

The detection of increased concentrations of the metabolic by-products Fe^{2+} and Mn at 14 and 9 mg l⁻¹, respectively, below the oil (Table 2)

Table 6. *Tier 2 assessment for BTEX and naphthalene (concentrations in mg l^{-1})*

Chemical[1]	EQS Freshwater[2]	WHO Health[3]	EU Drinking Water[4]	UK Drinking Water 1989[5]	UK Drinking Water 2000[6]	Observed concentration beyond oil accumulation (May 2000)	Observed concentration beyond oil accumulation (June 2000–March 2003)
Benzene	0.05	0.1	0.001	–	0.001	<0.001	0.002
Toluene	0.05	0.7	–	–	–	0.002	
Ethyl benzene	–	0.3	–	–	–	<0.001	
Xylene	0.03	0.5	–	–	–	0.001–0.002	0.002 xylene
Naphthalene	0.01	–	–	0.012 (0.005 on odour)[7]		0.002	0.002

[1] BTEX guidelines from Environment Agency Internal Guidance on Regulation 15, Appendix C (Environment Agency, December 1999), citing [3–5] below.
[2] Environmental Quality Standards are annual average concentrations.
[3] World Health Organization (WHO) Guidelines for Drinking Water Quality, 1984: 'Health' value represents the concentration that does not result in any significant risk to the consumer over a lifetime of exposure.
[4] EU Drinking Water Standards, 1998 (98/83/EC).
[5] UK Water Supply (Water Quality) Regulations 1989.
[6] UK Water Supply (Water Quality) Regulations 2000, which took effect from 1 January 2004.
[7] Naphthalene guidelines from Bates *et al.* (1997). Proposed reference level for potable water is 0.012 mg l^{-1} on health-based criteria, 0.005 mg l^{-1} on odour criteria.

Table 7. *Biodegradation of pentane (surrogate for the aliphatic C5–C6 TPH fraction)*

Balanced biodegradation reaction	UF* (mg mg^{-1})	Biodegradation pathway
$\underline{C_5H_{12}} + \underline{8O_2} \rightarrow 5CO_2 + 6H_2O$	3.6	Aerobic oxidation
$\underline{C_5H_{12}} + 6.4\underline{NO_3^-} + 12.8H^+ \rightarrow 3.2N_2 + 5CO_2 + 9.2H_2O$	5.5	Nitrate reduction
$\underline{C_5H_{12}} + \underline{4SO_4^{2-}} + 8H^+ \rightarrow 4H_2S + 5CO_2 + 6H_2O$	5.3	Sulphate reduction
$\underline{C_5H_{12}} + 16MnO_2 + 32H^+ \rightarrow \underline{16Mn^{2+}} + 5CO_2 + 22H_2O$	12.4	Manganese reduction
$\underline{C_5H_{12}} + 32Fe(OH)_3 + 64H^+ \rightarrow \underline{32Fe^{2+}} + 5CO_2 + 86H_2O$	25	Ferric iron reduction
$\underline{C_5H_{12}} + \underline{4SO_4^{2-}} + 8H^+ \rightarrow 4H_2S + 5CO_2 + 6H_2O$	5.3	Sulphate reduction

* Stoichiometric mass ratio of underlined electron acceptor or metabolic by-product to underlined organic compound.

Table 8. *Worked example of biodegradation capacity calculation for C_5H_{12} (May 2000)*

Electron acceptor or by-product	Utilization factor (mg mg^{-1})	Background concentration (mg l^{-1})	Biodegradation capacity (mg l^{-1})
Oxygen	3.6	13.27	3.7
Nitrate	5.5	26.89	4.9
Sulphate	5.3	49.96	9.5
Total biodegradation capacity mg l^{-1}			18.1

indicates that biodegradation is not only utilizing groundwater-sourced electron acceptors. Bacterially mediated reduction of ferric iron and manganese oxide coatings and minerals within the aquifer sediment is also considered to be occurring. These processes have been demonstrated to be important contributions to the TBC elsewhere (Heron & Christensen 1995).

Oxidation of the aliphatic C5–C6 TPH

fraction via ferric iron and manganese reduction is shown in Table 7. Calculation of the potential available ABC would require knowledge of the total Fe and Mn contents of the aquifer and the potential bioavailability of each fraction. As this information is not known, the only method for assessing the relative importance of the ABC is to calculate the biodegradation that has occurred. This is possible as both processes generate metabolic by-products. Therefore, deducting the up-gradient (background) concentration of each by-product from the concentration detected below the oil and dividing the difference by the contaminant-specific UF, the amount of biodegradation that has occurred may be evaluated.

UF values for C5–C6 TPH aliphatics for ferric iron and manganese reduction are 25 and 12.4 mg mg^{-1}, respectively. Therefore, as less than 14 mg l^{-1} of each associated metabolic by-product was detected below the oil above background concentrations, the contribution of these reduction processes to TBC appears small. Similar conclusions were reached by Thornton *et al.* (2001*a*) for a similar UK Permo-Triassic sandstone aquifer contaminated with over 4000 mg l^{-1} of phenol.

The proportion of Fe reduction, and to a lesser extent Mn reduction, is likely to have been underestimated. Dissolved Fe may rapidly react with any dissolved sulphide produced during sulphate reduction to form insoluble sulphide. Iron removal may also occur via auto-catalytic oxidation, which has a short timescale at near-neutral pH (Wehrli 1990). Mineralogical controls on the bioavailability of ferric iron and manganese are also considered to affect the contribution of ABC to TBC, as discussed by Thornton *et al.* (2001*a*).

The assessment of the TBC only provides a semi-quantitative estimate of each contributory component. The calculations do, however, indicate significant excess TBC. Coupled with the low oil saturation described by Privett (2006), these two factors are considered to provide an explanation for the lack of an extensive detectable dissolved phase. The situation has not changed since May 2000, as indicated by the stable contaminant concentrations and biodegradation indicator concentrations.

Further possible evidence for the occurrence of biodegradation at the site came during the coring of several holes within the centre of the oil accumulation as part of an assessment of the free phase (Privett 2006). In several discrete sections of core lengths, a black sooty substance was observed coating sand grains and fractures. This was considered the residue of bacterially mediated contaminant degradation, although additional investigation has not been undertaken to confirm this.

Conclusions

The assessment of the groundwater chemistry in May 2000 as summarized above indicated that no remedial action was required within a risk-based framework. However, as part of the overall risk-management strategy for the site, a long-term programme of monitoring and assessment to confirm compliance was agreed with the Environment Agency. The programme is outlined in Figure 5 and involves:

- maintenance of the groundwater monitoring network;
- quarterly monitoring, assessment and regulatory reporting;
- annual reviews and meetings;
- commitment of sufficient managerial and financial resources.

At all stages, the monitoring results are interpreted within an iterative risk-based management framework (e.g. points 2 and 4 in Fig. 5) that includes contingency elements (e.g. points 5b, 6 and 8 in Fig. 5). This approach provides continued protection of the groundwater down-gradient of the site and, ultimately, of the river.

The hydrogeological properties of the Permo-Triassic sandstone have proved suitable for the degradation of a kerosene leakage in this locality, such that no significant dissolved phase has been detected. This is considered to be due to the age of the oil leakage, acclimatization of the aquifer microbial communities to the contaminant concentrations, and the continuous supply of electron acceptors, particularly O_2, NO_3 and SO_4. There are no current reasons to suggest that these conditions will change in the near future and it is concluded that biodegradation of the dissolved phase will continue.

Many thanks are due to K. D. Privett for improvements to the manuscript and for discussing ideas, and to G. Thomas for assistance with the figure preparation. The views expressed are those of the author at the time of submission.

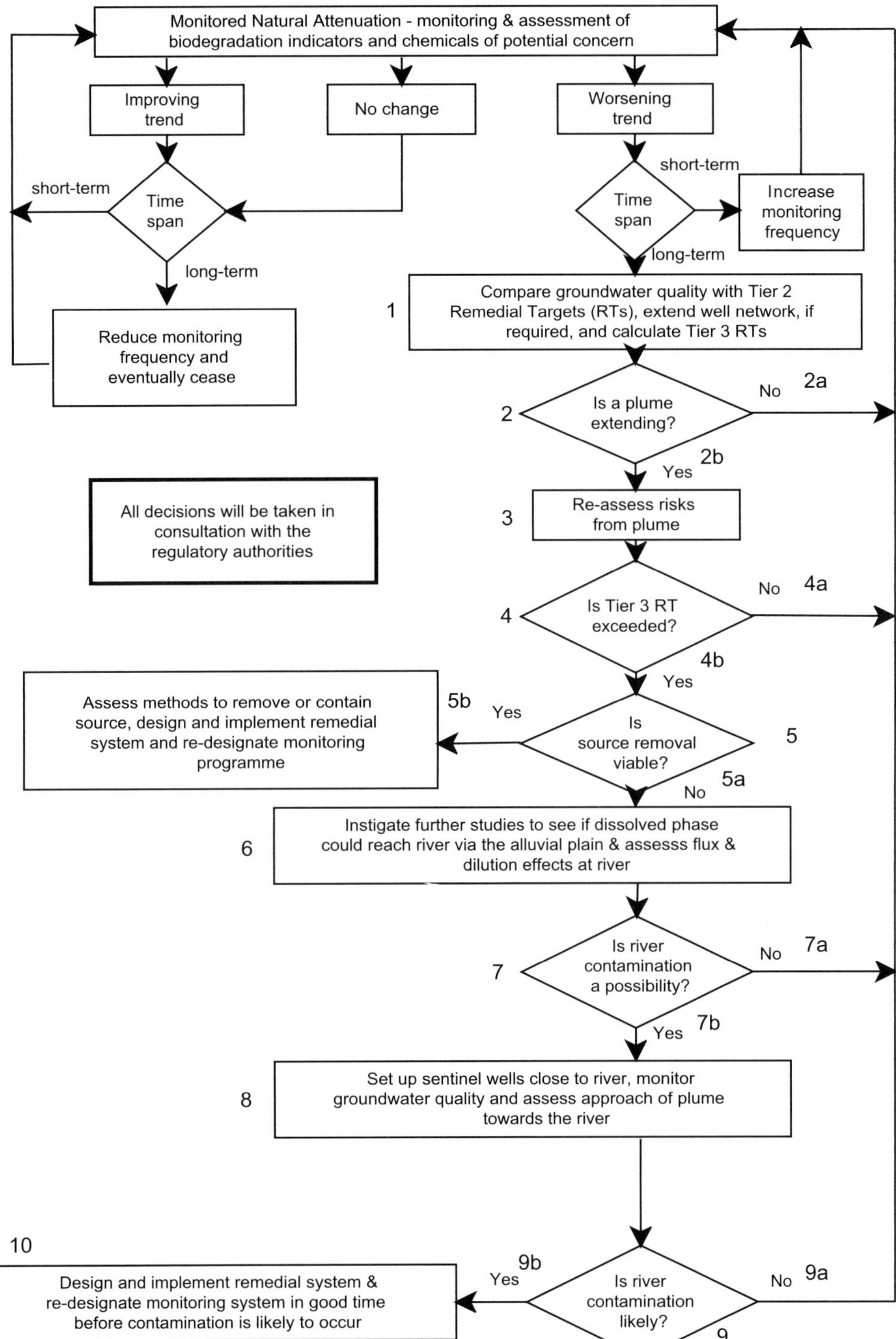

Fig. 5. Risk-based management and assessment plan.

References

BATES, K., YOUNG, W. & SUTTON, A. 1997. *Proposed Environmental Quality Standards for Naphthalene in Water*. Environment Agency R&D Technical Report, **P54**.

CAREY, M.A., FINNAMORE, J.R., MORREY, M.J. & MARSLAND, P.A. 2000. *Guidance on the Assessment and Monitoring of Natural Attenuation of Contaminants in Groundwater*. Environment Agency R&D Publication, **95**.

FANG, J., BARCELONA, M.J., KRISHNAMURTHY, R.V. & ATEKWANA, E.A. 2000. Stable carbon isotope biogeochemistry of a shallow sand aquifer contaminated with fuel hydrocarbons. *Applied Geochemistry*, **15**, 157–169.

FETZNER, S. 1998. Bacterial degradation of pyridine, indole, quinoline, and their derivatives under different redox conditions. *Applied Microbiology and Biotechnology*, **49**, 237–250.

HERON, G. & CHRISTENSEN, T.H. 1995. Impact of sediment-bound iron on redox buffering in a landfill leachate polluted aquifer (Vejen, Denmark). *Environmental Science and Technology*, **29**, 187–192.

INAP. 2003. *Treatment of Sulphate in Mine Effluents*. International Network for Acid Prevention. www.inap.com.au

MARSLAND, P.A. & CAREY, M.A. 1999. *Methodology for the Derivation of Remedial Targets for Soil and Groundwater to Protect Water Resources*. Environment Agency R&D Publication, **20**.

PRIVETT, K.D. 2006. The capillary characteristic model of petroleum hydrocarbon saturation in the Permo-Triassic sandstone and its implications for remediation. This Volume.

STUMM, W. & MORGAN, J.J. 1981. *Aquatic Chemistry*, 2nd edn. Wiley, New York.

THORNTON, S.F., LERNER, D.N. & BANWART, S.A. 2001*a*. Assessing the natural attenuation of organic contaminants in aquifers using plume-scale electron and carbon balances: model development with analysis of uncertainty and parameter sensitivity. *Journal of Contaminant Hydrology*, **53**, 199–232.

THORNTON, S.F., QUIGLEY, S., SPENCE, M.J., BANWART, S.A., BOTTRELL, S. & LERNER, D.N. 2001*b*. Process controlling the distribution and natural attenuation of dissolved phenolic compounds in a deep sandstone aquifer. *Journal of Contaminant Hydrology*, **53**, 233–267.

VORHEES, D.J., WEISMAN, W.H. & GUSTAFSON, J.B. 1999. *Human Health Risk-based Evaluation of Petroleum Release Sites: Implementing the Working Group Approach*. Total Petroleum Hydrocarbon Criteria Working Group Series, **5**. Amherst Scientific Publishers, Amherst, MA.

WERHLI, B. 1990. *Redox Reactions of Metal Ions at Metal Surfaces. In:* STUMM, W. (ed.) *Aquatic Chemical Kinetics*. Wiley, Chichester, 311–336.

WIEDEMEIER, T. & RIFAI, H. 1999. *Natural Attenuation of Fuels and Chlorinated Solvents in the Subsurface*. Wiley, Chichester.

WIEDEMEIER, T.H., SWANSON, M.A., WILSON, J.T., KAMPBELL, D.H., MILLER, R.N. & HANSEN, J.E. 1995. Patterns of intrinsic bioremediation at two U.S. Air Force bases. *In:* HINCHEE, R.E., WILSON, J.T. & DOWENEY, D.C. (eds) *Intrinsic Bioremediation*. Battelle Press, Columbus, OH, 31–51.

Combined isotopic and modelling approach to determine the source of saline groundwaters in the Selby Triassic sandstone aquifer, UK

S. H. BOTTRELL, L. J. WEST & K. YOSHIDA

School of Earth Sciences, University of Leeds, Leeds LS2 9JT, UK
(e-mail: S.Bottrell@earth.leeds.ac.uk; j.west@earth.leeds.ac.uk)

Abstract: Groundwater abstraction from the Triassic Sherwood Sandstone aquifer in the Selby area, Yorkshire, UK, has caused decline of the water level in the aquifer and groundwater quality problems such as high chloride and sulphate concentrations. Geochemical and isotopic analysis of groundwaters and groundwater modelling of an approximately 25 × 30 km area around Selby was carried out to understand the groundwater flow conditions and identify the source of saline water. Isotopic compositions (δ^{34}S and δ^{18}O) of seawater sulphate, Coal Measures brine and Permian evaporite sources do not match that associated with salinity in the Selby wellfield. Rather, the source of saline groundwater in the Selby wellfield matches the isotopic composition of Triassic evaporites in the overlying Mercia Mudstones.

Steady-state groundwater flow modelling demonstrates that that majority of water abstracted from the Selby wellfield is balanced by recharge from the west of Selby at Brayton Barff, and from the Escrick and York moraines to the north. A small proportion of the abstracted water originates from leakage from the River Ouse through the confining layer as a result of drawdown in the Selby area. The development of a cone of drawdown centred on Selby has created a new E–W hydraulic gradient to the east of Selby, allowing water ingress from the east. Capture-zone analysis indicates that the four abstraction boreholes contaminated by saline water collect groundwater from the north to NE (i.e. from the direction of the boundary between the Sherwood Sandstone and the Mercia Mudstone), confirming that the source of salinity is likely to be the Mercia Mudstone evaporites.

Groundwater is an important water resource. In the UK, 80% of the total groundwater abstraction is from only two aquifers, the Cretaceous chalk (54%) and the Triassic sandstone (26%) (Owen *et al.* 1991). Incursion of saline waters into aquifers is a very common cause of lowered groundwater quality and, ultimately, loss of groundwater resources (Todd 1980). Marine incursion is the usual cause of increased groundwater salinity in coastal aquifers in hydraulic continuity with the sea, but in other situations a variety of sources of salinity can be involved. Brines from different sources often have similar chemistry (Tellam & Lloyd 1986; Tellam 1995), making distinction between different brine sources difficult from chemical analyses alone.

The town of Selby in East Yorkshire has a number of flour and feed mills and other industrial sites that have private abstraction wells in the Sherwood Sandstone aquifer. Abstraction at boreholes in the Selby wellfield wells has taken place for over 100 years at increasing rates, and in the 1970s some wells began to be affected by increases in salinity. This increase was severe enough in some cases to mean that wells were abandoned. In order to constrain the probable causes of increased salinity of these waters, a dual approach has been adopted. Isotopic compositions of sulphate from wells affected by salinity increases have been compared with those of sulphate associated with different potential sources of salinity, as a possible 'fingerprint' of the source involved; and to test whether, and under what conditions, the sources identified might realistically contribute salinity at the Selby wellfield we have also modelled groundwater flows within the aquifer.

Geology and hydrogeology of the Selby area

The geology of the study area is shown in Figure 1. Quaternary deposits ('drift') cover the Triassic Sherwood Sandstone, except for a small sandstone outcrop area at Brayton Barff. The area where drift is absent is relatively high ground. The ground elevation of the study area rises from 10 m AOD (metres above Ordnance Datum) in the SE to over 30 m AOD in the NW and at Brayton Barff. To the east of study area, the Sherwood Sandstone Group is overlain by the Triassic Mercia Mudstone Group (approximately 14 km from Selby town). The Sherwood

From: BARKER, R. D. & TELLAM, J. H. (eds) 2006. *Fluid Flow and Solute Movement in Sandstones: The Onshore UK Permo-Triassic Red Bed Sequence.* Geological Society, London, Special Publications, **263**, 325–338.
0305–8719/06/$15 © The Geological Society of London 2006.

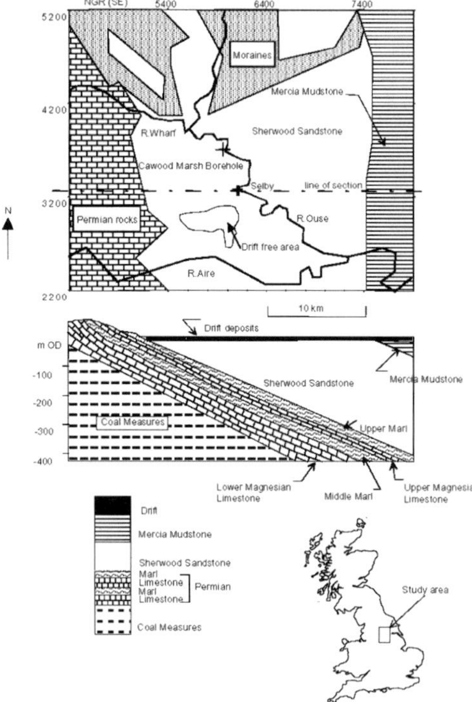

Fig. 1. Geological map and section for the Selby area, UK. NGR, national grid reference (data supplied by Environment Agency).

Sandstone is underlain by the Permian Marls and the Permian Magnesian Limestones, which crop out to the west of the study area. The Permian rocks themselves unconformably overlie Carboniferous Coal Measures strata and, until 2004, coal was deep-mined from these strata in the Selby coalfield.

The sandstone dips towards the east at a regional angle of about 2°–4° and increases in thickness from its feather-edge in the west to 400 m in the east. The thickness at Selby is around 180 m. The sandstone aquifer is confined below by the Upper Permian Marl. Pumping test transmissivity values for the sandstone aquifer in the York, Selby and Goole areas are 50–400 m² day⁻¹; analysis of the trend of transmissivity data with borehole depth suggests hydraulic conductivity values in the range of 1.8–4 m day⁻¹ (Allen *et al.* 1997). Hydraulic conductivity values estimated from core samples of the sandstone of the Yorkshire area range from 0.18 to 2.2 m day⁻¹ (Reeves *et al.* 1975; Lovelock 1977), which suggests that perhaps half of the flow into some of the more transmissive boreholes is via fractures. A significant proportion of this flow may occur via E–W

faults that are known to be present in the area (L. Brown pers. comm. 2003).

Recharge to the aquifer can take place as infiltration of precipitation through the drift (or directly into the sandstone at Brayton Barff) or by leakage from the rivers Aire and Ouse that cross the area. The majority of Quaternary drift covering the area is comprised of lacustrine clays and silts with low hydraulic conductivity, formed in pro-glacial and moraine and sea ice-dammed lakes during the Devensian glacial period (Cooper & Gibson 2003). However, to the north and west of the area (see Fig. 1) a complex pattern of sand and gravel deposits associated with the York (Askham) and Escrick moraines and intervening kames provides a far more permeable recharge pathway to the aquifer.

Owing to heavy abstraction in the Selby wellfield and in the Brayton Barff recharge zone, water levels have fallen significantly in recent decades (Fig. 2). The groundwater head dropped more than 8.0 m between 1984 and 1992. As a result, a huge cone of depression has been created by the abstraction at Selby, which appears to influence the groundwater flow direction within the whole study area. This hydrographic change has been accompanied by a dramatic increase in salinity (chloride and sulphate concentrations) in some of the wells within the Selby wellfield area.

Sources of salinity

Possible sources of salinity

The hydrogeology of the aquifer means that there are several possible sources for the salinity affecting the Selby wellfield: the brines and evaporites in the Permo-Carboniferous rocks below the Sherwood Sandstone Group; the

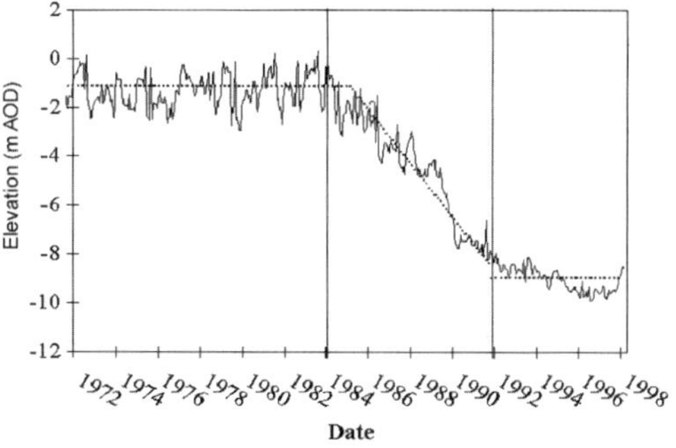

Fig. 2. Hydrograph of an observation borehole in Selby. m AOD, metres above Ordnance Datum.

brines and the evaporites in the Triassic sequence; and Quaternary or Recent marine incursion.

Sources in Permo-Carboniferous rocks below the Sherwood Sandstone

Upwards migration of brines from Permian evaporites or Coal Measures strata that underlie the Sherwood Sandstone (Fig. 1) have been invoked as possible sources of the saline waters (Aldrick 1976; Stagg 1995), possibly related to deep coal-mining activity and subsidence-related faulting. This migration is a possibility if the aquifer were in hydraulic continuity with groundwaters in the underlying strata, as depression of the groundwater head in the vicinity of the Selby wellfield could induce upconing of underlying denser brines.

Deep brines derived by dissolution of underlying Permian evaporites do occur in the Sherwood Sandstone further north in Cleveland, where hydraulic continuity is provided by fracturing of the intervening marl units (e.g. Bottrell *et al.* 1996). The Triassic and older sediments of the Selby area are affected by a series of ENE–WSW-trending faults that could possibly provide hydraulic continuity if permeable; none of these faults have sufficient throw to juxtapose Sherwood Sandstone directly against Permian evaporites. Strongly saline groundwater 'formation water' is also found in sandstones within the Carboniferous Coal Measures (e.g. Edmunds 1975; Sheppard & Langley 1984) and is a possible source of salinity in the Triassic aquifer. However, mine dewatering has reduced groundwater head in the underlying Coal Measures strata, which could act against any tendency for upward migration of brines.

Sources in Triassic rocks

Continental evaporite deposits are known from the Triassic Sherwood Sandstone; for example, elevated sulphate concentrations in the confined parts of the Birmingham aquifer are attributed to dissolution of gypsum and/or celestobarite that formed as authigenic phases under evaporative conditions (Jackson & Lloyd 1983; Hughes *et al.* 1999; cf. Sullivan & Koppi 1995). Much more significant deposits of evaporites occur within the overlying Mercia Mudstone, from which gypsum is worked commercially (Taylor 1983; Harvey & Stewart 1998). Saline groundwater could enter the Sherwood Sandstone from the Mercia Mudstone in the eastern, confined part of the aquifer.

Quaternary or Recent marine incursion

The river Ouse is tidal within part of the area under study, but is not saline or brackish, so saline incursion into the aquifer by leakage of estuarine water is unlikely at the present day. However, infiltration of seawater into the eastern part of the aquifer could have taken place under higher sea-level conditions than at present during previous Quaternary interglacials.

Isotope composition of sulphate as a tracer of source of salinity

Sulphate molecules contain oxygen and sulphur, both of which exhibit measurable natural variation in their stable isotopic composition and can be used as a 'fingerprint' of different sulphate sources (e.g. Robinson & Bottrell 1997; Hughes *et al.* 1999). At environmental temperatures and near-neutral pH of most groundwaters, sulphate is resistant to isotopic exchange on timescales of several 100 000 years (Lloyd 1968). Only bacterial sulphate reduction can modify sulphate isotopic compositions markedly, so, provided groundwater redox potentials have remained sufficiently high to prevent this from occurring, sulphate isotopic compositions can act as a powerful tracer for different sulphate, and thus brine, sources. In the Selby aquifer redox potential (E_H) measurements are not made as part of routine monitoring and were not made specifically for this study. However, a compilation of existing data for the Selby area groundwaters by Stagg (1995) shows that groundwater E_H has small positive values (+2 to more than +100 mV). Sulphate reduction typically occurs at E_H around –400 mV and thus sulphate in the Sherwood Sandstone groundwaters is unlikely to have been affected by sulphate reduction.

Methods

Samples collected and analyses performed

Permian marl evaporites were collected as chipping samples from an RJB Mining proving borehole at Cawood Marsh, North Yorkshire ([NGR SE5880 3718], see Fig. 1). Coal Measures water samples were collected at water ingress points into mines in the Selby coalfield. Groundwater samples were also collected from boreholes into the Magnesian Limestone used for water supply at some mines, as well as observation and producing wells in the Sherwood Sandstone aquifer and the Selby wellfield, an observation well in the Mercia Mudstone and shallow monitoring wells in Quaternary drift deposits. Samples for $\delta^{13}C$–DIC (dissolved organic carbon) analysis were collected according to Bishop (1990) and DIC recovered as $SrCO_3$ for analysis by the method of McCrea (1950). Evaporite-bearing marl and shale samples were leached with distilled water for 7 days to extract soluble ions. Anion concentrations of leachate and groundwater samples were measured using a Dionex DX-100 ion chromatograph. Cation compositions of groundwater samples were measured by ICP-OES (inductively coupled plasma-optical emission spectroscopy). Sulphate was recovered from leachates and groundwaters by precipitation as $BaSO_4$ at pH 3 and approximately 70 °C and analysed for $\delta^{18}O$ (McCarthy *et al.* 1998) and $\delta^{34}S$ (Halas *et al.* 1982) using a VG SIRA 10 isotope ratio mass spectrometer. Isotopic data are reported in standard delta (δ) notation as per mil (‰) deviations from the Cañon Diablo Troilite (CDT) standard for sulphur, Standard Mean Ocean Water (SMOW) standard for oxygen and Pee Dee Belemnite (PDB) for carbon.

Groundwater modelling

Model design. In order to understand the present groundwater flow conditions and identify a saline water source, groundwater modelling was carried out using the MODFLOW code (McDonald & Harbaugh 1988), followed by particle-tracking analyses using the MODPATH code (Pollock 1989). The groundwater level of the sandstone aquifer in the Selby cone of drawdown dropped more than 8 m between 1984 and 1992 after which it stabilized (Fig. 2). There were not enough data available to construct a model to simulate the pre-1984 water-level conditions, so a steady-state model was developed to simulate the post-1992 conditions, using average water levels measured by the UK Environment Agency in 14 observation wells between 1992 and 1996. Preliminary modelling used two layers to represent the Quaternary drift and the Sherwood Sandstone Group, respectively. This modelling showed that, under steady-state conditions, the drift layer is not permeable enough to permit significant lateral flow and simply transmits recharge to the underlying sandstone. Hence, the further modelling work reported here did

not explicitly model flow in the drift layer, although Quaternary deposit characteristics were used to infer recharge rates to the sandstone.

An approximately 30 km E–W by 25 km N–S area of the aquifer was modelled; Figure 3 shows the model boundaries. The northern model boundary is where the highest groundwater levels are observed and is thought to be a groundwater divide, and hence is defined as a no-flow boundary. The eastern boundary represents the edge of the overlying Mercia Mudstone. In steady-state flow conditions, groundwater flow across this boundary will be very small, because of the deep confined nature of the aquifer to the east. Hence, it was assumed to be a no-flow boundary. In fact, a small amount of water is likely to have been drawn across this boundary from the confined zone during the period of transient drawdown prior to 1992. The western model boundary, and the lower boundary of the model, represent the contact between the Sherwood Sandstone and the underlying Upper Permian Marl. As the Upper Permian Marl has a relatively low permeability, and faults passing through it have insufficient throw to juxtapose the sandstone against the Upper Magnesian Limestone below, this boundary was assumed to be a no-flow boundary. The southern model boundary is defined by the position of the River Aire, and a constant head boundary, based on piezometric levels measured in Environment Agency observation wells, was assumed.

The topology of the top of the Sherwood Sandstone was obtained by kridging data from 70 boreholes penetrating the superficial deposits; the topology of its base was inferred from 35 boreholes penetrating to the base of the group, and from the position of its feather edge.

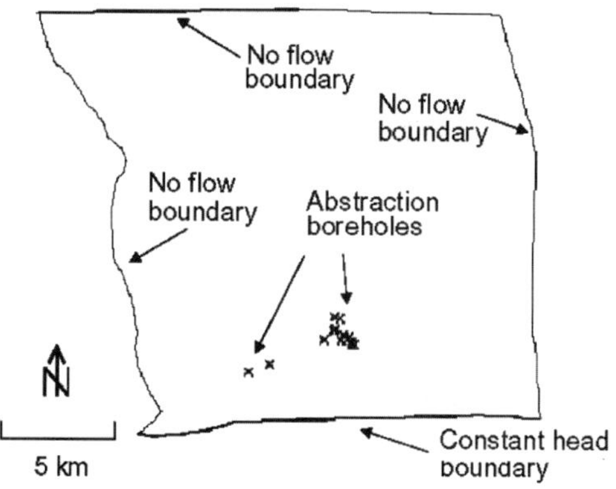

Fig. 3. Model boundary conditions.

These data indicate that the Sherwood Sand-stone Group was up to 400 m thick in the eastern part of the modelled area. However, a plot of transmissivity v. effective aquifer thickness (i.e. borehole depth) shows that transmissivity tends to increase with thickness up to 100 m, but then remain approximately constant. Furthermore, a few very high transmissivities (>600 m^2 day^{-1}), thought to be associated with fracturing, are seen in boreholes less than 100 m deep. These data suggest that most flow occurs in the upper 100 m of the aquifer (Allen *et al.* 1997), so the model layer thickness was restricted to a maximum of 100 m.

The characteristics of the Quaternary deposits overlying the sandstone were used to define three recharge zones, labelled A, B and C in Figure 4. Zone A is the area around Brayton Barff where the sandstone is not covered by drift. Here, it was assumed that all available water flows into the sandstone. A recharge rate of 250 mm year^{-1}, was calculated from rainfall minus actual evaporation data supplied by the UK Meteorological Office. In drift-covered areas recharge rates will depend on the permeability of the drift sequence, which controls the proportion of recharge lost to field drainage systems. Two zones were defined according to the nature of the superficial deposits, as described on geological maps and sheet memoirs (Edwards *et al.* 1950; Institute of Geological Sciences 1973) and seen in observation well logs supplied by the Environment Agency. In Zone C, in the northern sector of the modelled area, the superficial deposits consist of glacial sands and gravels associated with the

Escrick and York moraines and intervening kame deposits. In Zone B, which covers the majority of the study area, the superficial deposits comprise mainly glacio-lacustrine silts and clays with interbedded sands. River alluvium overlies these glaciolacustrine deposits in the river valleys. It was assumed that direct recharge was zero in Zone B, because of: (i) the low-permeability nature of the silt and clay deposits; and (ii) the extensive nature of the agricultural deep drainage, which has lowered the hydraulic head in much of the drift in the area to below that in the aquifer during the summer months. The recharge rate in Zone C (sands and gravels of the York and Escrick moraines) was found from model calibration to be 16 mm year^{-1}. Although this recharge rate seems low compared to that in the drift-free area at Brayton Barff (250 mm year^{-1}), it provides a significant proportion (33%) of the recharge because of the relatively large area of Zone C.

The rivers Wharf and Ouse cross the model area as shown in Figure 4. Over most of the modelled area, these rivers are floored by fine-grained alluvium, so their hydraulic connection with the aquifer is likely to be weak. However, in Selby the River Ouse has been dredged to allow navigation, and piezometric levels are well below river levels due to the cone of drawdown. Furthermore, chemical data discussed in the results section below (see Table 3 later) suggest that recharge from the overlying Quaternary drift deposits makes up, on average, 6% of abstracted water in the Selby wellfield. To incorporate this recharge into the model, river leakage via the drift was specified such that Ouse river water comprised 6% of the abstracted well water. The river conductance required was consistent with a hydraulic conductivity for the intervening drift layer of 0.01 m day^{-1}, assuming 10 m drift thickness and a 15 m-wide river channel. However, the sensitivity of the flow model and particle-tracking analyses to the magnitude of river leakage was low.

The hydraulic conductivity of the sandstone aquifer was assumed to be uniform and was used as a calibration parameter; a value of 3 m day^{-1} provided the best match when aquifer flowing thickness was restricted to 100 m. This value is consistent with aquifer transmissivity values measured in the area but higher than most core measurements, which suggests that fracture flow may be significant. Abstraction data for the Selby wellfield and the wells in the Brayton Barff area were supplied by the Environment Agency; these data show that

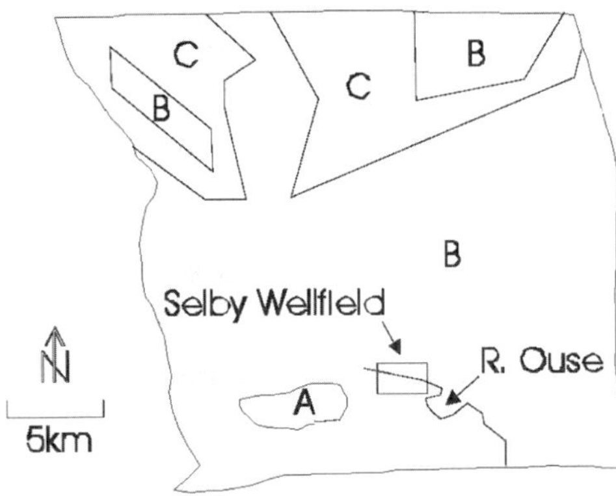

Fig. 4. Model recharge zones. Zone A, Brayton Barff recharge window (unconfined sandstone); Zone B, Vale of York glaciolacustrine deposits and till (mainly silts and clays); and Zone C, Escrick and York moraines and associated kame deposits (sands and gravels).

abstraction was approximately constant over the modelled period. Changes in hydraulic head at the northern and eastern boundaries over the modelled period were less than the seasonal variation (*c.* 0.5 m).

MODFLOW's performance is primarily dependent on the application of a sufficiently fine grid (Haitjema *et al.* 2001). In this model a finer grid spacing was used around wells than in other areas. The grid spacing ranged from 50 m close to wells to 500 m at the edge of the model. The total number of cells was 79 279 and the total model area was approximately 574 000 000 m². Mesh-refinement exercises showed that this discretization was sufficiently fine.

Model calibration

Calibration was performed using measured water levels in 14 observation boreholes penetrating the sandstone. In this study, sensitivity analyses to Zone C recharge and hydraulic conductivity were carried out to match the average water level values for the 1992–1996 period. Modelled v. target head values are shown in Figure 5, and the groundwater head contours and residuals are shown in map form in Figure 6. In general, mismatches between the observed and target values are less than 2 m, which is similar to the range of fluctuation seen in many of the boreholes for data collected between 1992 and 1996. Thus, it is considered that the model adequately represents the aquifer flow.

Water budget. The water budget for the steady-state model is summarized in Table 1. The sources of most inflow are recharge from the Brayton Barff outcrop area (40%), recharge through the York and Escrick moraines (33%),

leakage from the river Ouse in the Selby area where the groundwater heads are depressed (6%) and inflow from the constant head boundary in the SE where the hydraulic gradient is northwards as a result of the Selby abstractions (21%). The main discharges are well abstraction (92% of total discharge); the remaining discharge occurred across the constant head boundary in the SW, directly south of the Brayton Barff recharge area.

Well capture-zone analysis. To identify the source zones of the abstraction boreholes, capture-zone analysis was carried out using the MODPATH particle tracking code (Pollock 1989). The MODPATH code uses the cell-by-cell flow terms that are created by the MODFLOW simulation to determine particle movement directions and rates. Particle travel time analysis also requires a flowing (kinematic) porosity and a travel time to be specified. The flowing porosity of the Sherwood Sandstone is much lower than the total porosity (Allen *et al.* 1997). This is because a significant proportion of flow may be concentrated in fractures or faults, or relatively coarse-grained sandstone units that make up a low volumetric proportion of the sequence (Truss 2004). Here, a value of 5% was assumed. The particle travel time was specified as 100 years, on the grounds that abstraction has been going on for at least this length of time in the Selby wellfield.

Results

Groundwater chemistry

Groundwater chemical data are summarized in Table 2. The compositions of pumped waters from the Selby wellfield (Prefix P in Table 2) have been modelled in terms of mixtures of

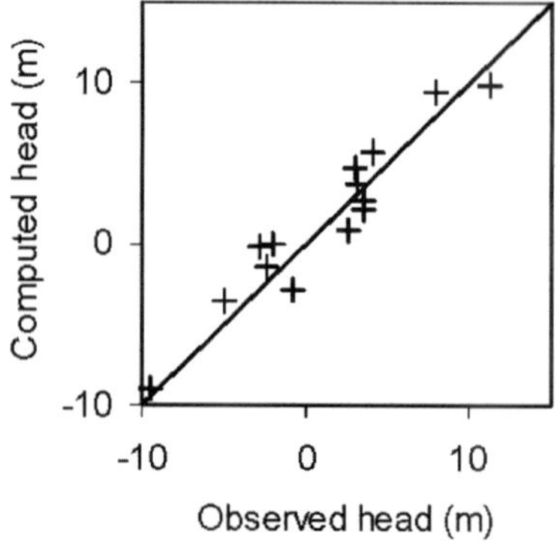

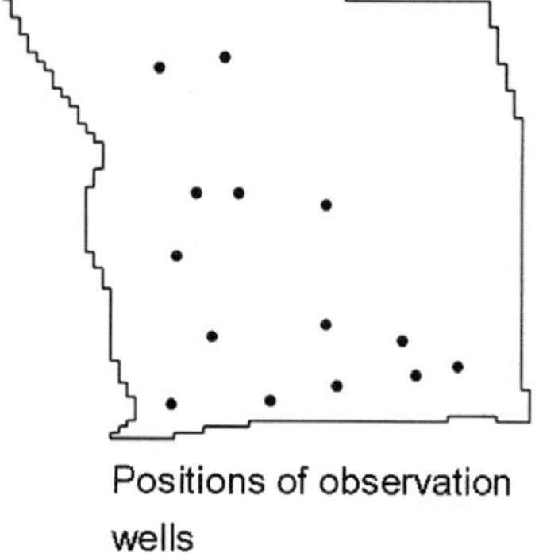

Fig. 5. Relationship between observed and computed hydraulic head, and positions of observation wells.

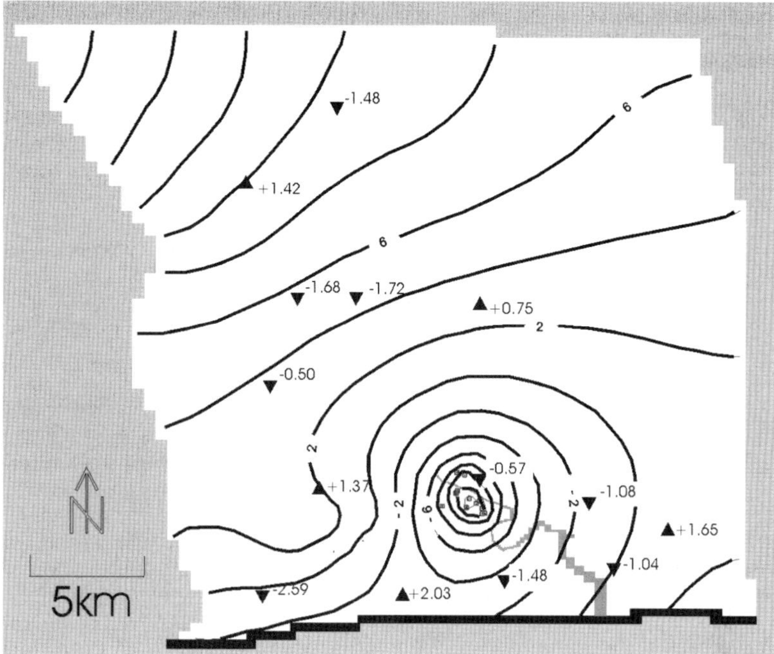

Fig. 6. Contour map of modelled hydraulic heads (m AOD) and differences from calibration targets (m, observed – computed).

Table 1. *Water balance for steady-state model*

Inflows	m³ day⁻¹	%	Outflows	m³ day⁻¹	%
Zone A (Brayton Barff)	5970	40	Abstraction	−13 590	92
Zone C (York and Escrick moraines)	4940	33	Southern boundary	−1220	8
Southern boundary	3090	21			
River Ouse	810	6			
Total	14 810	100		−14 810	100

three water types: aquifer water outside the cone of depression of the Selby wellfield; groundwater from wells in Quaternary drift; and Triassic evaporite brine. End-member compositions are based on representative analysed compositions of these water types (Table 2). Calculations used the NETPATH code (Plummer *et al.* 1991). The chemistry of many of the Selby wellfield waters can be explained as a simple mixture of aquifer water (sampled outside the wellfield area) and a small brine component supplying excess chloride and other components to affected wells (Table 3). However, drawdown resulting from heavy pumping abstractions in the Selby wellfield has markedly increased the head difference between drift water and Sherwood Sandstone aquifer in the wellfield area (Fig. 2). This has

induced water leakage from the overlying drift deposits into Sherwood Sandstone aquifer and some wells require a significant additional component of drift water to generate the observed compositions (Table 3). The model predicts the $\delta^{13}C$ of the DIC in the calculated mixture, and this is compared with the analysed $\delta^{13}C$ of the DIC in Table 3 as a test of model fit: the majority of predictions are in good to reasonable agreement with analysed compositions.

Sulphate stable isotopes

Sherwood Sandstone. Figure 7 shows sulphate isotopic compositions plotted against chloride concentration for abstraction wells in the Selby wellfield. Wells with low chloride exhibit a very

Table 2. *Chemical analyses of water types and pumped waters from the Selby wellfield. All units are mg l^{-1}, except pH*

Site	N*	Na	Ca	K	Sr	Cl	HCO$_3$	SO$_4$	NO$_3$	pH
Aquifer water	11	15	84	2	0.04	20	379	5	b.d.	7.3
Drift water	4	30	242	3	0.43	117	250	430	73	7.1
Brine	1†	70 000	1900	140	28	105 000	24	3600	0.43	7.93
P1	1	101	154	7	1.06	576	296	173	10	7.2
P2	1	23	146	4	0.52	37	490	297	8	7.7
P4	1	9	93	2	0.10	39	338	133	3	7.3
P6	1	58	268	5	0.05	675	248	128	11	7.3
P7	1	105	188	6	0.38	576	372	298	5	7.6
P8	1	93	224	5	0.29	891	310	205	10	7.2
P10	1	9	74	2	0.18	67	323	22	1	7.3
P11	1	10	71	2	0.13	40	343	43	1	7.7
P12	1	10	58	2	0.02	69	249	10	5	7.5
P13	1	18	132	3	0.13	65	378	324	6	7.1

* Number of sites from which representative analysis was taken.
† Brine from Brookhouse Farm borehole, Lower Mersey Basin (Tellam 1995). This brine is a high-concentration end-member of the saline waters in the Permo-Triassic sandstones of the Lower Mersey Basin aquifer: their source is the Mercia Mudstone Group evaporites.
b.d., below limit of detection.

Table 3. *Calculated water mixtures at Selby wells. Mixtures are based on representative analysed compositions of three water types: aquifer water outside the cone of depression of the Selby wellfield; groundwater from wells in Quaternary drift; and Triassic evaporite brine. Calculations used the NETPATH code (Plummer et al. 1991) and full details are given in Yoshida (2000). Note that the chemistry of sample P-8 is the only one requiring a large component of drift water, explaining the difference in its sulphate isotopic composition from the other high-chloride waters (P1, P6 and P7), which have zero or small drift-water component*

Site	Predicted fraction in groundwater			Computed δ^{13}C-DIC ‰ V-PDB	Measured δ^{13}C-DIC ‰ V-PDB
	Aquifer water	Drift water	Brine		
P1	0.97	0.00	0.03	−16.7	−15.1
P2	0.98	0.00	0.02	−13.1	−13.0
P4	0.89	0.11	0.00	−16.9	−17.3
P6	0.92	0.00	0.08	−16.6	−15.8
P7	0.90	0.06	0.04	−17.0	−16.5
P8	0.76	0.17	0.05	−16.7	−14.7
P10	0.90	0.10	0.00	−17.0	−17.3
P11	0.91	0.09	0.00	−17.0	−17.3
P12	0.76	0.24	0.00	−17.5	−15.9
P13	0.62	0.38	0.00	−17.6	−15.2

wide range of sulphate isotopic compositions, but those with high chloride are all distinctly enriched in ^{34}S and ^{18}O. The wide range of sulphate isotopic compositions in low-chloride samples reflects a variety of sulphate sources in the aquifer and drift groundwaters, but only two samples overlap the isotopic signature of the saline waters (P4 and P12 for δ^{34}S and P4 for δ^{18}O). Three of the saline water samples have δ^{34}S in the range +17.2–+17.5‰ and δ^{18}O in the range +14.6–+15.9‰; the fourth sample (P8) has δ^{34}S = +8.5‰ and δ^{18}O = +13.7‰. The saline water is thus associated with an isotopically heavy sulphate source, but one sample (P8 Fig. 7) has lighter compositions than the rest. The isotopic composition of this sample can be explained by mixing with a sulphate source with lighter δ^{34}S and δ^{18}O. The light sulphate source is likely to be the oxidation of pyrite as oxygenated water is drawn through the drift deposits, as this sample contains a significant drift-water component according to the mixing model calculations summarized above in Table 3. The isotopic composition of the group of

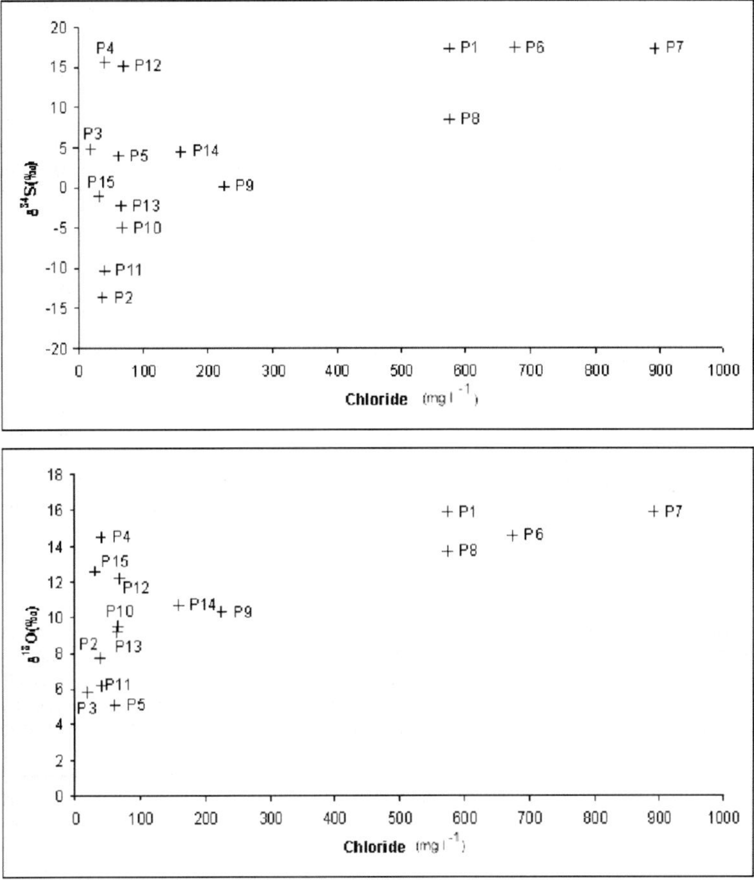

Fig. 7. Relationship between isotopic composition and chloride concentration for Sherwood Sandstone groundwater samples. **(a)** $\delta^{18}O$ of sulphate and Cl concentration; and **(b)** $\delta^{34}S$ of sulphate and chloride concentration.

three saline samples is thus taken to be indicative of the sulphate in the saline water source.

Mercia Mudstone. Table 4 presents the sulphate isotopic data from samples taken from the only well that intercepts Mercia Mudstone evaporite horizons in the study area. Table 4 also includes isotopic data on sulphate: in groundwater from an evaporite horizon in the Mercia Mudstone of the Cheshire Basin (Lymm Marina borehole) (Barker 1996); in groundwater from the sandstone sequence in the Mersey Basin (Gatewarth Observation borehole, Warrington, and data from Kimblin 1995); and solid samples from the Winsford Halite Mine and elsewhere in the Cheshire Basin (Taylor 1983; Hughes 1998).

Permian Marl. Figure 8 shows chloride and sulphate concentrations in Permian Marl drill-chipping leachate samples. Chloride concentrations are generally low, except around 210 m in the Middle Permian Marl. Downhole geophysical logging of the Cawood Marsh borehole identified halite around 200–205 m and the origin of the elevated chloride is likely to be the halite. Much chloride will, however, have been lost by dissolution of halite into the drilling fluid before chippings could be sampled. Sulphate is low (similar to the overlying Sherwood Sandstone) in the uppermost 10 m of the Upper Permian Marl but increases substantially below this, reaching values of 50 000–70 000 ppm in the rock chippings. These high concentrations are found throughout the Middle Permian Marl. Figure 8 also shows $\delta^{34}S$ and $\delta^{18}O$ of sulphate leached from the Permian marls. The upper, low-concentration samples show a wide variation in isotope composition from –4.3 to

Table 4. *Isotopic compositions of sulphate in Mercia Mudstone groundwaters and evaporite minerals*

Sulphate source	δ^{34}S‰ CDT	δ^{18}O‰ SMOW
Mercia Mudstone groundwater east of Selby (this study)	+17.3	+13.7
Sulphate associated with Traissic halite, Winsford Mine, Cheshire: Mean (SD) of six samples (Hughes 1998)	+17.9 (0.2)	+12.4 (0.2)
Gatewarth borehole 160 m (Mercia Mudstone evaporite-derived brine present in the sandstone, Merseyside) (Barker 1996)	+16.6	+12.5
Lymm Marina borehole 160 m (Mercia Mudstone evaporite brine, Merseyside) (Barker 1996)	+15.5	+12.1
Mercia mudstone evaporitic gypsum (Taylor 1983)	+17.6	+13.9
Groundwater influenced by Mercia Mudstone evaporites	+17.5	+13.7
Kimblin (1995)	+17.9	+14.2

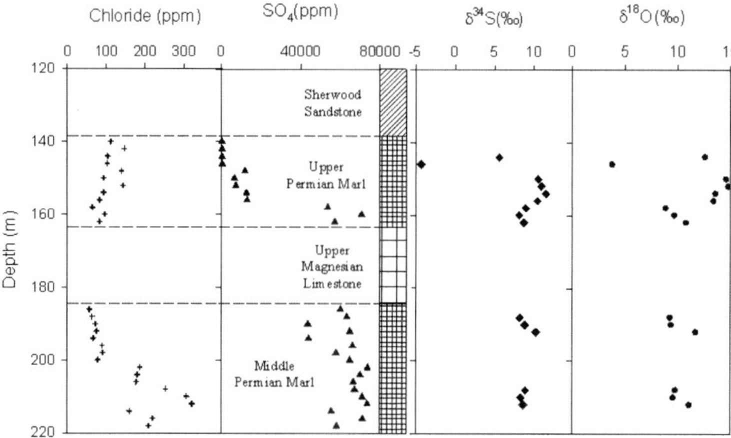

Fig. 8. Depth profiles of leachable chloride and sulphate from Permian marl units in the Cawood Marsh borehole and corresponding sulphate isotopic compositions. ppm, mg kg^{-1} in the solid phase.

+11.5‰ δ^{34}S and from +3.8 to +14.8‰ δ^{18}O. The deeper, higher concentration samples have a much more restricted range of δ^{34}S (+8.2–+10.2‰) and δ^{18}O (+9.3–+11.7‰). As these compositions are associated with elevated chloride they are used below to represent the sulphate isotopic 'fingerprint' of this source of salinity.

Coal Measures brines. Mine-water samples collected between 650 and 750 m bgl in the Selby area have a range of concentrations: chloride ranges from 3 to 10 g l^{-1} and sulphate from 54 to 57 g l^{-1}. Figure 9 shows the sulphate isotopic composition of these mine-water samples: δ^{34}S ranges from +3.9 to +11.7‰, while δ^{18}O ranges from +2.6 to +11.9‰.

Groundwater modelling

Flow conditions. Modelled head values (Fig. 6) suggest that the cone of drawdown beneath Selby has created an E–W hydraulic gradient to the east of Selby. This gradient is important from a contamination point of view, as it allows much older waters from the east of Selby to be drawn in, rather than simply flowing southwards to discharge into the River Aire, as would be the case under natural hydraulic conditions with no pumping. The modelled values also suggest the piezometric surface reached a minimum level of around –16 m AOD in the centre of the cone of drawdown. This is very close to the top of the sandstone, which suggests that the aquifer began to become unconfined in 1992. The influence of unconfined storage may explain why drawdown in the Selby wellfield appears to have stabilized at this time (Fig. 2). Drawdown will have occurred much more slowly once the aquifer became unconfined. The modal value for the confined storage coefficient measured in pumping tests for the Sherwood Sandstone in NE England is around 10^{-3}, whereas the un-confined storage coefficient is typically 0.15–0.2

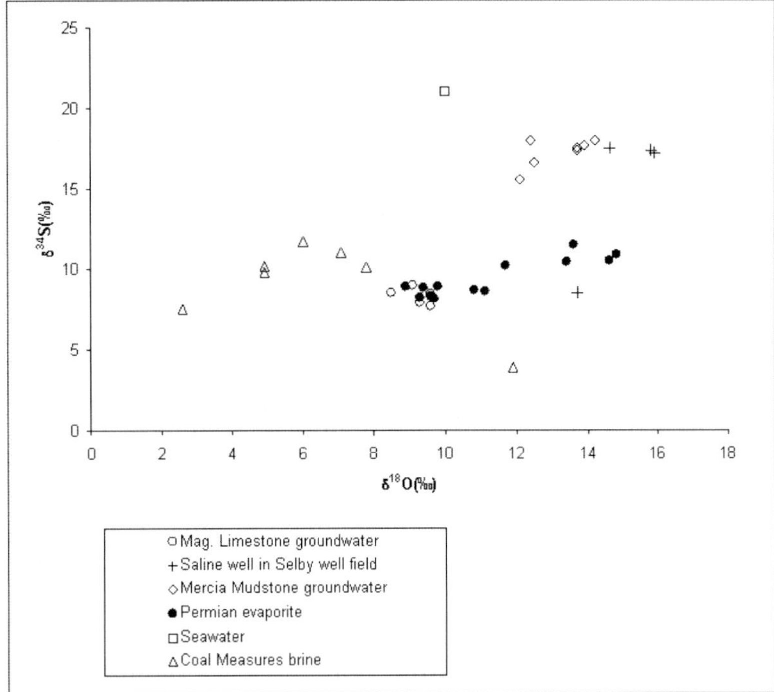

Fig. 9. Sulphate isotopic compositions of Selby wellfield groundwaters affected by salinity increase compared with sulphate isotopic compositions of the various possible salinity sources.

(Allen *et al.* 1997). Hence, the rate of increase in drawdown should reduce by two orders of magnitude once the top of the aquifer is reached.

Well capture zones. Several abstraction wells in the Selby wellfield have been contaminated by saline groundwater, with chloride levels reaching about 900 mg l^{-1} in the late 1990s. Capture zones for wells showing saline contamination are shown in Figure 10a; those for wells that had not shown saline contamination by 1996 are shown in Figure 10b. It is striking that three of the four wells showing contamination by saline water capture at least some water from the NE quadrant (the fourth contaminated well captures water mainly from the north). In contrast, uncontaminated wells do not capture any of their water from the NE quadrant. This distribution of capture zones strongly suggests that the source of the saline contamination lies to the north and east of the Selby wellfield.

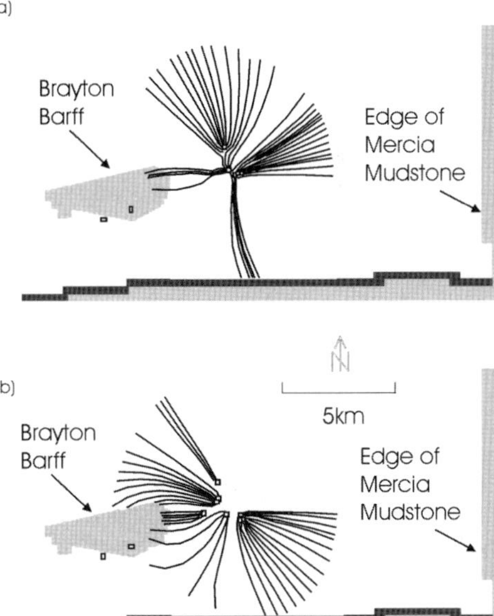

Fig. 10. Hundred-year capture zone calculated for wells in the Selby wellfield. (**a**) Four wells affected by salinity increase; and (**b**) wells remaining uncontaminated.

Discussion

Origin of salinity in the Selby wellfield

Waters in the Sherwood Sandstone aquifer and Selby wellfield are oxidizing, with no evidence of sulphate reduction (Stagg 1995; Yoshida 2000); thus there is no risk of modification of sulphate isotopic compositions within the aquifer. The sulphate compositions can therefore be treated as conservative tracers of the origin of sulphate and, hence, the wellfield salinity. Figure 9 shows the $\delta^{34}S$–SO_4 and $\delta^{18}O$–SO_4 compositions of potential saline-water sources and the saline waters in the Selby wellfield. Three of the affected Selby wells show a tight grouping, but one has lighter $\delta^{34}S$ as a result of mixing of drift water at the well (see Sherwood Sandstone, above, and Table 3). We thus take the isotopic compositions of the group as representative of sulphate associated with the source of salinity.

Sulphate in modern seawater has a $\delta^{34}S$ of +20‰ and $\delta^{18}O$ of +10‰ (e.g. Claypool *et al.* 1980). This is significantly different to the Selby wellfield saline-water composition (Fig. 9) indicating that the source of the saline water is not seawater. Hence, the origin is not modern or Quaternary seawater intrusion (which had closely similar values, according to Claypool *et al.* 1980).

The sulphur and oxygen isotope compositions of local Permian evaporite sulphate are +8–+11‰ and +9–+15‰, respectively. Groundwater samples taken from the Permian Magnesian Limestone aquifer have sulphate with $\delta^{34}S$ between +7.7 and +8.5‰, and $\delta^{18}O$ between +8.5 and +9.6‰. These are all significantly different to the saline waters (Fig. 9), implying that the origin of saline water is not in this case from the underlying Permian strata, as has happened in the Sherwood Sandstone further to the north in Cleveland (e.g. Bottrell *et al.* 1996).

Samples of Coal Measures brines contain sulphate with $\delta^{34}S$ in the range +3.9–+11.7‰ and $\delta^{18}O$ in the range +2.6–+11.9‰. These differ considerably from the saline water sulphate values (Fig. 9), and hence the source of saline water is not Coal Measures brine.

Triassic marine evaporite sulphate minerals have similar $\delta^{34}S$ and $\delta^{18}O$ values to the saline waters in the Selby wellfield (Fig. 9). The isotope signatures of one groundwater sample taken from a borehole in the Mercia Mudstone reflect the local expression of this source and are also close to those of the Selby wellfield saline waters. Therefore, it is likely that the source of saline groundwater in the Selby area is inflow of

water influenced by the dissolution of evaporites present in the Mercia Mudstone. The Mercia Mudstone overlies the Sherwood Sandstone and crops out to the east of the Selby wellfield (Fig. 1). Groundwaters contaminated by leakage from the Mercia Mudstone could enter the catchment of the Selby wellfield as a result of the creation by abstraction of a westwards-directed hydraulic gradient. Alternatively, a Triassic evaporite sulphate isotopic 'fingerprint' could be present in evolved saline groundwaters in the deep confined aquifer (gained by long-term aquifer groundwater interaction with evaporite minerals in the sandstone sequence). Again, creation of a westwards head gradient would draw such waters into the Selby abstractions.

Flow pathway of saline water to affected wells

The 100-year particle traces for wells showing saline contamination, based on the 1992–1996 steady-state condition (Fig. 10), are approximately 5 km long extending to the east and north of Selby. The direction of these traces suggests that the source of the saline contamination lies to the east and north of Selby, which is consistent with the identified Mercia Mudstone isotopic signature for sulphate. However, it is interesting to note that, despite the 100 year travel times, the flowlines still extend less than half the distance to the Mercia Mudstone feather edge. However, preferential flowpaths, such as E–W trending fault zones, are known to be present in the area but were not incorporated into the modelling. Hence, it is possible that saline waters from the confined zone of the Sherwood Sandstone could have been drawn into the Selby wellfield as a result of groundwater abstraction. Alternatively, the particle-tracking analysis may indicate that saline waters were present within the aquifer between Selby and the Mercia Mudstone feather-edge before pumping began.

Conclusions

The isotopic composition of sulphate associated with salinity in affected wells casts doubt on the previously presumed source of saline contamination in the Selby wellfield. Rather than matching the Permian evaporite source, the sulphate isotopes point to a Triassic source for the salinity, probably from the Mercia Mudstones overlying the aquifer.

Groundwater modelling of the Selby area has

demonstrated that water sources of the Triassic Sherwood Sandstone aquifer are recharge, both via the unconfined aquifer at Brayton Barff and via the glacial moraine sediments in the north of the area. The areas covered by glacio-lacustrine sediments contribute relatively little. A small component of leakage from the River Ouse was incorporated in the model representation based on geochemical evidence. Calibration was achieved for steady-state conditions based on average piezometric levels between 1992 and 1996. It is probable that the reason that the deepening of the cone of drawdown slowed suddenly in 1996 was because the aquifer started to become unconfined in the wellfield area. The model also shows that the development of the cone produced an E–W hydraulic gradient in the Selby aquifer to the east of the wellfield. Well capture-zone analysis using MODPATH demonstrates that Selby wellfield boreholes that show saline contamination collect the groundwaters from the NE, whereas uncontaminated wells do not collect water from this quadrant. This is consistent with the Mercia Mudstone evaporite source indicated by the sulphur and oxygen isotope data.

This study was only possible with the kind co-operation of RJB Mining (UK) Ltd (later UK Coal) who allowed access for sampling of drill cuttings and mine waters, and gave practical assistance with sampling and permission to publish these results. We also thank those companies with abstraction wells who permitted sampling, and the Environment Agency for access to observation wells. Isotope facilities at Leeds were provided by NERC grant GR3/8134 to S. H. Bottrell.

References

ALDRICK, R.J. 1976. *The Hydrogeology of the Triassic Sandstone Within the South-eastern Division of the Yorkshire Water Authority.* Yorkshire Water Authority Internal Report.

ALLEN, D.L., BREWERTON, L.J. *ET AL.* 1997. *The Physical Properties of Major Aquifers in England and Wales.* British Geological Survey Technical Report, **WD/97/34**.

BARKER, A.P. 1996. *Isotopic studies of groundwater.* PhD thesis, University of Leeds.

BISHOP, P.K. 1990. Precipitation of dissolved carbonate species from natural waters for $\delta^{13}C$ analysis – a critical appraisal. *Chemical Geology*, **80**, 251–259.

BOTTRELL, S.H., LEOSSON, M.A. & NEWTON, R.J. 1996. Origin of brine inflows at Boulby potash mine, Cleveland, England. *Transactions of the Institution of Mining and Metallurgy*, **105**, B151–B164.

CLAYPOOL, G.E., HOLSER, W.T., KAPLAN, I.R., SAKAI, H. & ZAK, I. 1980. The age curves of sulphur and oxygen in marine sulfate and their mutual interpretation. *Chemical Geology*, **28**, 199–260.

COOPER, A.H. & GIBSON, A. 2003. *Geology of the Leeds District.* British Geological Survey.

EDMUNDS, W.M. 1975. Geochemistry of brines in the Coal Measures of northeast England. *Transactions of the Institution of Mining and Metallurgy*, **84**, B39–B52.

EDWARDS, W. 1950. *Geology of the District North and East of Leeds.* British Geological Survey Sheet Memoir, **70**.

HAITJEMA, H., KELSON, V. & LANGE, W. 2001. Selecting MODFLOW cell sizes for accurate flow field. *Groundwater*, **39**, 931–938.

HALAS, S., SHAKUR, A. & KROUSE, H.R. 1982. A modified method for SO_2 extraction from sulphates for isotopic analysis using $NaPO_3$. *Isotopenpraxis*, **18**, 11–13.

HARVEY, M.J. & STEWART, S.A. 1998. Influence of salt on the structural evolution of the Channel Basin. *In*: UNDERHILL, J.R. (ed.) *Development, Evolution and Petroleum Geology of the Wessex Basin.* Geological Society, London, Special Publications, **133**, 241–266.

HUGHES, A.J. 1998. *Identification of natural and anthropogenic sources of sulphate in two UK aquifers using S and O isotopes.* PhD thesis, Birmingham University.

HUGHES, A.J., TELLAM, J.H., LLOYD, J.W., STAGG, K.A., BOTTRELL, S.H., BARKER, A.P. & BARRETT, M.H. 1999. *Sulphate Isotope Signatures in Borehole Waters From Three Urban Triassic Sandstone Aquifers, UK.* IAHS Publication, **259**, 143–149.

INSTITUTE OF GEOLOGICAL SCIENCES. 1973. *Geological Map of Selby (Solid and Drift)*, Sheet 71. Institute of Geological Sciences, London, UK.

JACKSON, D. & LLOYD, J.W. 1983. Groundwater chemistry of the Birmingham Triassic sandstone aquifer and its relation to structure. *Quarterly Journal of Engineering Geology*, **16**, 135–142.

KIMBLIN, R.T. 1995. The chemistry and origin of groundwater in Triassic sandstone and Quaternary deposits, northwest England and some UK comparisons. *Journal of Hydrology*, **172**, 293–311.

LLOYD, R.M. 1968. Oxygen isotope behavior in the sulfate-water system. *Journal of Geophysical Research*, **73**, 6099–6110.

LOVELOCK, P.E.R. 1977. *Aquifer Properties of the Permo-Triassic Sandstones of the United Kingdom.* Bulletin of the Geological Survey of Great Britain, **56**.

McCARTHY, M.D.B., NEWTON, R.J. & BOTTRELL, S.H. 1998. Oxygen isotopic composition of sulphate from coals: implications for primary sulphate sources and secondary weathering processes. *Fuel*, **76**, 677–682.

McDONALD, M.G. & HARBAUGH, A.W. 1988. *A Modular Three-dimensional Finite-difference Groundwater Flow Model.* US Geological Survey Techniques of Water-resources Investigations, Book 6, Chapter A1.

McCREA, J.M. 1950. On the isotope chemistry of carbonates and a paleotemperature scale. *Journal of Chemical Physics*, **18**, 849–857.

OWEN, M., HEADWORTH, H.G. & MORGAN-JONES, M. 1991. Groundwater in basin management. *In*:

DOWNING, R.A. & WILKINSON, W.B. (eds) *Applied Groundwater Hydrology*. Clarendon Press, Oxford, 17–34.

PLUMMER, L.N., PRESTEMON, E.C. & PARKHURST, D.L. 1991. *An Interactive Code (NETPATH) for Modeling Net Geochemical Reactions Along a Flow Path*. US Geological Survey Water Resources Investigations Report, **91-4078**.

POLLOCK, D.W. 1989. *Documentation of Computer Programs to Complete and Display Pathlines Using Results From the US Geological Survey Modular Three-dimensional Finite-difference Ground-water Model*. US Geological Survey Open File Report, **89-381**.

REEVES, M.J., SKINNER, A.C. & WILKINSON, W.B. 1975. The relevance of aquifer flow mechanisms to exploration and development of groundwater resources. *Journal of Hydrology*, **25**, 1–21.

ROBINSON, B.W. & BOTTRELL, S.H. 1997. Discrimination of sulfur sources in pristine and polluted New Zealand river catchments using stable isotopes. *Applied Geochemistry*, **12**, 305–319.

SHEPPARD, S.M.F. & LANGLEY, K.M. 1984. Origin of saline formation waters in northeast England: application of stable isotopes. *Transactions of the Institution of Mining and Metallurgy*, **93**, B195–B201.

STAGG, K.A. 1995. *Mechanisms for the intrusion of contaminates into the Sherwood Sandstone*. MSc thesis, University of Leeds.

SULLIVAN, L.A. & KOPPI, A.J. 1995. Micromorphology of authigenic celestobarite in a duripan from central Australia. *Geoderma*, **64**, 357–361.

TAYLOR, S.R. 1983. A stable isotope study of the Mercia Mudstones (Keuper Marl) and associated sulphate horizons in the English Midlands. *Sedimentology*, **30**, 11–31.

TELLAM, J.H. 1995. Hydrochemistry of the saline groundwaters of the lower Mersey Basin Permo-Triassic sandstone aquifer, UK. *Journal of Hydrology*, **165**, 45–84.

TELLAM, J.H. & LLOYD, J.W. 1986. Problems in the recognition of seawater intrusion by chemical means: an example of apparent chemical equivalence. *Quarterly Journal of Engineering Geology*, **19**, 389–398.

TODD, D.K. 1980. *Groundwater Hydrology*, 2nd edn. Wiley, Chichester.

TRUSS, S.W. 2004. *Characterisation of sedimentary structure and hydraulic behaviour within the unsaturated zone of the Triassic Sherwood Sandstone aquifer in North East England*. PhD thesis, University of Leeds.

YOSHIDA, K. 2000. *Groundwater vulnerability of the Triassic sandstone aquifer in the Selby area, North Yorkshire*. PhD thesis, University of Leeds.

Index

Page numbers in *italics* denote figures. Page numbers in **bold** denote tables.

From: Barker, R. D. & Tellam, J. H. (eds) 2006. *Fluid Flow and Solute Movement in Sandstones: The Onshore UK Permo-Triassic Red Bed Sequence*. Geological Society, London, Special Publications, **263**, 339–346.
0305–8719/06/$15 © The Geological Society of London 2006.